NUMERICAL METHODS
IN CHEMISTRY

UNDERGRADUATE CHEMISTRY

A Series of Textbooks

edited by
J. J. Lagowski
Department of Chemistry
The University of Texas at Austin

NUMERICAL METHODS IN CHEMISTRY

K. JEFFREY JOHNSON

Department of Chemistry
University of Pittsburgh
Pittsburgh, Pennsylvania

MARCEL DEKKER, INC.   New York and Basel

Library of Congress Cataloging in Publication Data

Johnson, K. Jeffrey   [Date]
    Numerical methods in chemistry.

    (Undergraduate chemistry ; v. 7)
    Includes bibliographical references and index.
    1. Chemistry--Data processing. 2. Chemistry--
Computer programs. I. Title.
QD39.3.E46J63   542'.8   79-25444
ISBN 0-8247-6818-3

MARCEL DEKKER, INC.
270 Madison Avenue, New York, New York 10016

Current printing (last digit):
10 9 8 7 6 5 4 3 2 1

PRINTED IN THE UNITED STATES OF AMERICA

To My Student Collaborators

PREFACE

The digital computer has become one of the most important and versatile tools
available to chemists. Crystallographers and quantum chemists were among the first
to use computers on a large scale. Today chemists are using computers not only for
such classical "number-crunching" applications as crystal structure determination
and ab initio calculations, but also for routine data reduction, on-line data
acquisition and control of experiments, computer-assisted instruction, information
retrieval, and even synthesis of organic compounds. An operational knowledge of
computers and computer programming is rapidly becoming a requirement for chemists.

   This book is intended to serve primarily as a textbook for a numerical methods
and computer applications course for chemistry students. The emphasis is on software (computer programming) rather than hardware. Also, the results of numerical
analysis and linear algebra are presented and applied to the solution of chemistry
problems. The mathematics have not been derived. There are more than 50 computer
programs in this book, 43 of which have been fully documented.

   The book can be logically divided into three parts. Chapters 1 and 2 provide
a review of the Fortran programming language and a collection of programs which have
closed-form solutions. Chapters 3 through 8 introduce and apply various numerical
methods to the solution of chemistry problems. Chapter 9 contains an overview of
some additional applications of computers in chemistry, and includes a relatively
extensive bibliography.

   The computer programs are all written in Fortran. The particular dialect of
Fortran is Fortran-10, provided by the Digital Equipment Corporation. The programs
have been implemented on a DECSystem-1099 computer. Slight modifications may be
required to implement some of these programs on other computers. The execution
times cited do not include compilation time.

   This book has evolved with the senior-level elective course, "Numerical Methods
in Chemistry," which has been offered several times at the University of Pittsburgh.
I am deeply indebted to the students who have participated in the development of

this course. I particularly wish to acknowledge my graduate and undergraduate student
collaborators who have helped me develop the software documented in these pages. I
am happy to acknowledge the Pitt Computer Center staff for their cooperation and
assistance, and the University for the computer time. I am grateful to my colleagues
for providing me the opportunity to write this book. Mrs. Barbara Hunt deserves
special commendation for typing this book and its earlier draft. Finally, I am
particularly grateful to my wife, Sharyn, for her patience and encouragement.

K. Jeffrey Johnson

CONTENTS

CHAPTER 1

COMPUTER PROGRAMMING AND FORTRAN

A block diagram of a relatively large, general purpose computer is given in Figure 1.1. The "brain" of the computer is the <u>central processing unit</u> (CPU). The "muscle" is the <u>arithmetic</u> and <u>logical</u> unit. The CPU is the control center for all computer the <u>arithmetic</u> and <u>logical</u> units. The CPU is the control center for all computer operations. These operations are executed using the binary number system (0,1). The rate of execution of these binary operations is on the order of a million to a billion operations per second.

The memory of the computer can be divided into two types, core memory and auxiliary memory. <u>Core memory</u> is a rapid-access device for storage and retrieval of information under control of the CPU. <u>Auxiliary memory</u> is slower (and less expensive) and consists of tapes, disks, drums, and other devices. Core memory consists of an array of computer words, each word consisting, in turn, of a fixed number of bistable memory elements, or bits (binary digits). These unique magnetic or electronic states are correlated with the 0 and 1 of the binary number system. The DECSystem-10 computer contains 36-bit words. The Digital Equipment Corporation sells core memory units for the DECSystem-10 computer in 64k blocks ($k = 2^{10} = 1024$). Each of these 64k words is addressable by the CPU.

All large computer centers have extensive auxiliary storage facilities. These include magnetic tapes, disks, drums, etc. These devices have slower access time than core memory because the information is stored sequentially on the device. For example, magnetic tape must be moved to the appropriate location on the tape to access information. The time required to retrieve information from core memory using the DECSystem-1099 computer is less than $1\,\mu$sec. Access times for auxiliary devices range from microseconds to seconds. The auxiliary devices are used for long-term storage of computer programs and data.

The computer outlined in Figure 1.1 is configured for three types of operations: (a) batch processing, (b) remote job entry (RJE), and (c) time sharing. The input and output (I/O) devices for a batch-processing computer usually consist of card readers, magnetic tape drives, and line printers. The jobs (computer programs) are queued and executed sequentially according to some scheduling algorithm (computer-based procedure). Except for large compute-bound (as opposed to I/O-

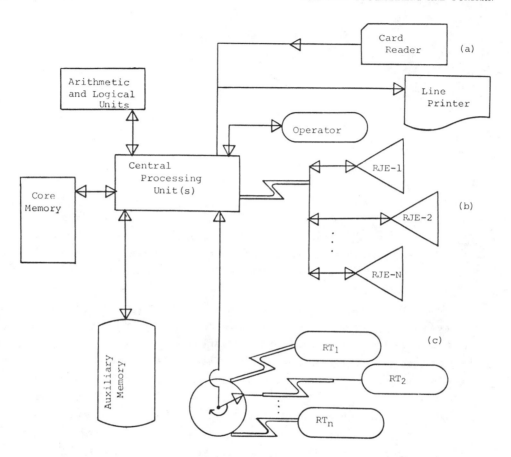

Figure 1.1 Block diagram of a computer showing (a) batch processing, (b) remote job entry, and (c) time sharing.

bound) jobs, the CPU is idle much of the time in batch processing because the CPU is several orders of magnitude faster than the mechanical I/O devices.

RJE is an extension of batch processing designed to increase the utilization of the CPU and to facilitate access to the system by the users. Each RJE station contains I/O devices. In an RJE configuration the CPU has a series of I/O queues to process. The jobs are read, executed sequentially according to the priority assigned by the scheduling algorithm, and the results directed to the appropriate RJE station for printing. However, a powerful CPU will still be idle for a significant fraction of the time, and therefore can support a number of remote terminals, providing a time-sharing environment. The arrow in Figure 1.1 revolves around the ring making contact with each remote user for a small fraction of a second.

During that time information can be read from the terminal, processing of information can be completed, or results printed at the terminal. If the average response
time for the time-sharing system is less than 1 sec, it appears to the user at the
terminal that he is the sole user of the system. He may in reality be sharing the
computer with 50 other time-sharing users, 10 RJE stations, and a background batch
queue.

1.1 PROGRAMMING LANGUAGE

Users communicate with computers using one of four classes of computer languages.
The most primitive of these is <u>machine language</u>, in which the instructions (store,
add, branch, etc.) are coded in binary or some other numeric code. The second
class of computer languages is <u>assembly language</u>, where the instructions are coded
using mnemonics to represent the numeric codes, for example, STR for store. Each
computer has its own set of assembly instructions. Programs written in the
assembly language are compiled, that is, translated into machine language by a
program called the <u>assembler</u>. The resulting machine language program is then
executed. The third class is the <u>compiler languages</u>, for example, Fortran, COBAL,
and PL/1. These higher level languages contain key words, for example, READ, GO
TO, and DO. Programs written in Fortran (source programs) are translated by a
Fortran compiler into machine code (the object program) which is then executed.
The fourth class is the set of <u>interpretive languages</u>, for example, BASIC and APL.
These languages operate in time-sharing mode, are relatively easy to learn and use,
and allow the user a maximum of control over the writing, debugging, and execution
of the program.

Fortran is the programming language most widely used by scientists and
engineers in the United States. The word Fortran derives from <u>formula translation</u>,
and indeed Fortran statements closely resemble algebraic formulas. For example,
the formulas $P = nRT/V$ and $A = A_o e^{-kt}$ are coded in Fortran as follows:

P=N*R*T/V   and   A=AØ*EXP(-K*T)

Here "*" is the multiplication symbol, "Ø" is used to differentiate the number
zero and the letter "O," and "EXP" refers to the exponential function. Each computer has its own dialect of Fortran. The programs in this book have been compiled
using the Fortran-10 compiler on a DECSystem-1099 computer. Some modifications may
be required for successful compilation on other computers.

1.2 ELEMENTS OF THE FORTRAN LANGUAGE

The Fortran compiler reads Fortran statements and translates them into machine-
readable code. The set of Fortran statements will be called the <u>source deck</u> and
the output of the Fortran compiler will be called the <u>object deck</u>. Source decks

are frequently punched on 80-column cards. The following column convention is
used:

Column 1	A "C" in column 1 indicates a comment card; comment cards are ignored by the compiler.
Columns 1-5	Statement number; statement numbers may range from 1 to 99999.
Column 6	Continuation; any character (except a blank) in column 6 identifies the card as a continuation of the preceding card.
Columns 7-12	The Fortran source statement.
Columns 73-80	Available for identification, sequence numbers, remarks, etc.; these eight columns are ignored by the Fortran compiler.

Consider the following Fortran statements:

```
C
C      CALCULATE THE DETERMINANT OF A 3X3 MATRIX
C
C       10        20        30        40        50        60
C23456789012345678901234567890123456789012345678901234567890
C
  100 DETOFA=A(1,1)*A(2,2)*A(3,3) + A(1,2)*A(2,3)*A(3,1)
     1 + A(1,3)*A(2,1)*A(3,2) - A(1,2)*A(2,1)*A(3,3)
     2 - A(1,1)*A(2,3)*A(3,2) - A(1,3)*A(2,2)*A(3,1)
```

The first six cards are comment cards. The first character of the Fortran
statement starts in column 7. The "1" in column six of the second statement
indicates that it is a continuation of the preceding statement. The Fortran-10
compiler will accept up to 19 consecutive continuation cards. The statement
"DETOFA=..." has been arbitrarily assigned the statement number 100.

1.2.1 Constants

Six types of constants are considered here: integer, real (single precision),
double precision, complex, logical, and literal. Integer constants are signed
numbers without a decimal point. The DECSystem-10 computer has a 36-bit word (36
binary digits per word). The sign of the integer is stored in the first bit (0 for
positive, 1 for negative). The binary equivalent of the integer value is stored in
the remaining 35 bits. The available range is

$$-2^{35} \leq \text{ integer value } \leq 2^{35}-1$$

or

$$-34,359,738 \leq \text{ integer value } \leq 34,359,738,367$$

The following are examples of valid and invalid integer constants,

Valid	Invalid	Reason
101	135.	Decimal point
-67	-6,830	Comma
34359738367	34359738368	Too large

Real constants are signed numbers written either with a decimal point or in exponential notation. Exponential notation is illustrated by the following examples:

$$0.157 = 1.57E-1 \qquad 9756 = 9.756E3$$
$$= 15.8E-2 \qquad\qquad = 97.56E2$$
$$= 157.E-3 \qquad\qquad = 975.6E1$$

The 36-bit word is divided into two parts to represent real constants. The first 9 bits contain the binary equivalent of the signed exponential part, and the remaining 27 digits contain the binary equivalent of the signed fractional part of the constant. The precision of 27 binary bits is approximately 8 decimal digits. The range (in absolute value) of real constants is

$$1.4 \times 10^{-39} \quad < \quad \text{real constant} \quad < \quad 1.7 \times 10^{38}$$

The following are examples of valid and invalid real constants.

Valid	Invalid	Reason
1.7E-5	-3.4E41	Exponent too large
1.76.937	96,457.02	Comma
1.2345678	107	No decimal point

Double-precision constants use two 36-bit words. Only 9 bits are assigned to the exponential part, so the range of magnitude of double- and single-precision values is the same. However, the prcision is increased to approximately 16 digits. Double-precision constants contain the letter "D" rather than "E." Examples of valid and invalid double-precision constants follow.

Valid	Invalid	Reason
1.00D-3	6.74	D missing
0.0D0	0.0E0	D missing
6.63D-27	0.64D74	Exponent too large

Complex constants are represented by a pair of integer or real constants separated by a comma and enclosed in parentheses. The first number is the real part, and the second number is the imaginary part of the complex constant. For example,

Valid	Invalid	Reason
(0.0,0.0)	6.3,2.4	Parentheses missing
(-5.9E-4,6.3E-3)	(6.9E-4)	Imaginary part missing
(2,-5)	(8.8E72,9.6E-67)	Exponents too large in magnitude

The two logical constants are .TRUE. and .FALSE. The logical constants must be delimited by periods.

Literal constants are denoted either by a pair of apostrophes or with the specification "nH," where n denotes the number of characters in the literal constant and H stands for "Hollerith." Literal constants can contain one or more of the 26 alphabetic letters, the nine numerals, and/or the set of special characters (",#,$,%,&,etc.). For example,

Valid	Invalid	Reason
'DATA'	'Y(OBS)	Closing apostrophe missing
10H 62 PART A	6HY(CALC)	Should be 7H...

1.2.2 Variables and Specification Statements

Variable names in Fortran consist of between one and six alphameric (alphabetic and numeric) characters, the first of which must be alphabetic. Five types of variables will be considered here: integer, real, double-precision, complex and logical. By convention, variable names which begin with one of the letters I, J, K, L, M, or N are considered integer variables. Variable names which begin with any other letter in the alphabet are real (single-precision) variable names. Valid integer variable names include KJJ, MASS, IGO, J100, and LETITB. Valid real variable names include X, DATA, ENTROP, Q74, and Z18XY.

Other types of variables must be explicitly declared using a specification statement. The general form is

$$\text{type } v_1, v_2, v_3, \ldots$$

where TYPE is INTEGER, REAL, DOUBLE PRECISION, COMPLEX, or LOGICAL, and v_1, v_2, etc., are variable names. For example,

```
INTEGER SUM, Z10, R20
REAL J,K10,IJUMP
DOUBLE PRECISION A,B,C
COMPLEX C1,C2,C3,Z
LOGICAL L, LOG, FLAG
```

The IMPLICIT specification statement can be used to override the standard Fortran convention. This statement allows the user to declare all variable names starting with a given letter to be of an arbitrary type. For example,

```
IMPLICIT INTEGER (A-Z)
```

The effect of this statement is to declare all variable names in the program to be of integer type. All variables that are single precision (real) by default can be declared double precision by the statement

```
IMPLICIT DOUBLE PRECISION (A-H, O-Z)
```

Variables may be subscripted by the DIMENSION statement

DIMENSION INDEX(25),Y(100),A(10,10)

The effect of this statement is to allocate 25 integer values to the subscripted
name INDEX, 100 real values to be stored in the vector Y, and 100 real values to
be stored (columnwise) in the array A.

1.2.3 Operators and Expressions

The five arithmetic operators and a subset of the functions available in Fortran
are given in Table 1.1.

Arithmetic expressions consist of arithmetic operators, constants, variables,
and function references. Some examples of assignment statements containing
expressions are

1. DISCR=B*B-4.*A*C
2. ROOT1=(-B + SQRT(DISCR))/(2*A)
3. J=J+1
4. P=(A*B)/(C*D)
5. A=A0*EXP(-CONST*T)
6. Z(I)=(X(I)-Y(I))*ABS(SIN(J*PI*Z))

Statement 1 contains only real variable names and a real constant (4.). Statement
2 is a mixed-mode expression because it contains both real variable names and the
integer constant 2. When this statement is compiled, Fortran-10 converts the
integer constant to a real constant. Statement 3 is a mathematical absurdity,
but in Fortran it simply means the replacement of the value of J by the value of
J + 1. Statement 4 contains parentheses to avoid the following ambiguity:

Fortran expression	Equivalent algebraic expression
A*B/C*D	ABD/C
A*B/C/D	AB/CD
(A*B)/(C*D)	AB/CD

Table 1.1 Arithmetic Operators and Some Functions

Operator or function	Definition	Example	Comment
+	Addition	Y=A+B	
-	Subtraction	Y=A-B	
*	Multiplication	Y=A*B	
/	Division	Y=A/B	B not 0
**	Exponentiation	Y=A**B	
ABS(arg)	Absolute value	Y=ABS(A)	
SQRT(arg)	Square root	Y=SQRT(A)	arg > 0
EXP(arg)	Exponential, e	Y=EXP(A)	
SIN(arg)	Sine	Y=SIN(A)	arg in radians
ALOG(arg)	ln	Y=ALOG(A)	arg > 0
ALOG10(arg)	log	Y=ALOG10(A)	arg > 0

Statements 5 and 6 show the two versions of the "-" operator. In statement 5, the "-" indicates the unary operation of negation, and statement 6 contains the binary operation of subtraction.

Care should be taken when mixing modes with the exponentiation operator. If the exponent is real, the base must be positive to avoid imaginary numbers. For example, $-4**0.5$ is 2i, where i is the square root of -1. Also, A**B and A**I are evaluated differently. The former uses the identity,

$$A^B = \exp(B\ln A)$$

and the latter will use repetitive multiplication of I < 10.

Logical expressions consist of relational and logical operators, constants, variables and arithmetic expressions. Six relational operators and five logical operators are given in Table 1.2. The truth table for the five logical operators is given in Table 1.3. Examples of logical expressions assuming I and J are integers, X and Y are single precision, and P and Q are logical variables are

        I.LT.J
        (X.LT.Y) .OR. (X.EQ.0.)
        P .AND. (I.EQ.J)

If a Fortran statement contains several operators, the execution is determined by the following hierarchy:

        functions > ** > (*,/) > (+,-) > (<,≤,>,≥,=,≠)
                  > .NOT. > .AND. > .OR. > (.EQV.,.XOR.)

In the case of equal operator precedence, computation proceeds from left to right. For example, the statement

        Y=A+B/SQRT(C)*D**E+COS(F)

Table 1.2 Relational and Logical Operators

Relation or operation	Operator	Example (X and Y are single precision; P and Q are logical)
>	.GT.	X.GT.Y
≥	.GE.	X.GE.Y
<	.LT.	X.LT.Y
≤	.LE.	X.LE.Y
=	.EQ.	X.EQ.Y
≠	.NE.	X.NE.Y
AND	.AND.	P.AND.Q
Inclusive OR	.OR.	P.OR.Q
Exclusive OR	.XOR.	P.XOR.Q
Equivalence	.EQV.	P.EQV.Q
Complementation	.NOT.	P.NOT.Q

Table 1.3 Truth Table (T = true, F = false)

P	Q	P.AND.Q	P.OR.Q	P.XOR.Q	P.EQV.Q
T	T	T	T	F	T
T	F	F	T	T	F
F	T	F	T	T	F
F	F	F	F	F	T

is executed as follows. Let G=SQRT(C) and H=COS(F),

then	Y=A+B/G*D**E+H
Now let	P=D**E,
then	Y=A+B/G*P+H
Next, let	Q=B/G,
then	Y=A+P*Q+H
Finally, let	R=P*Q,
then	Y=A+R+H

1.3 CONTROL STATEMENTS

The following Fortran statements are considered in this section: GO TO (branching),
IF (conditional branching), DO (looping), CONTINUE, PAUSE, STOP, and END. The
execution of a Fortran program proceeds sequentially from the first statement to
the last (END) unless one of these control statements is encountered.

1.3.1 The GO TO Statement

Branching is accomplished in Fortran programs with one of three variants of
the GO TO statement:

```
1.  Unconditional    GO TO n1
2.  Computed          GO TO (n1, n2,n3,...,nk),i
3.  Assigned          ASSIGN 25 to i
                      .............
                      GO TO i
                      or
                      GO TO i, (12,25,24)
```

Here n1, n2, ..., nk are statement numbers and i is an integer variable name. The
unconditional GO TO statement causes control to branch to the specified statement
number. The computed GO TO transfers control to n1 if i = 1, to n2 if i = 2, etc.
If i is not in the range $1 \leq i \leq k$, then the next statement is executed. The
assigned GO TO transfers control to the statement number corresponding to the
value of i. If i is not assigned one of the values in the list (12,25,24), then
the next statement is executed. Some examples of GO statements follow.

```
GO TO 100                     ASSIGN 85 to JJ
..........                    .............
IGO=4                         GO TO JJ
.....                         ........
GO TO (10,20,15,75,88),IGO    GO TO JJ, (15,35,85,75)
```

In these examples control is branched to statement number 75 by the computed GO TO, and to statement number 85 by the assigned GO TO statement.

1.3.2 The IF Statement

There are two IF statements in Fortran for conditional branching, the arithmetic IF and the logical IF. The general form of the arithmetic IF is

IF (arithmetic expression) n1,n2,n3

Control is branched as follows:

If the arithmetic expression is <0, branch to statement n1,
If the arithmetic expression is =0, branch to statement n2,
If the arithmetic expression is >0, branch to statement n3.

Statement numbers n1 and n2, or n2 and n3 may be equal, providing the conditions ≤ and ≥, respectively. For example,

IF(FOFX)20,25,25
IF((ABS(X-XOLD)/X)-TOL)95,95,91

The general form of the logical IF is

IF(logical expression) statement

Here if the logical expression is true, then the statement is executed. Otherwise, control passes to the next statement. Consider the following examples:

1. IF(N.LE.0)STOP
2. IF(IPLT.EQ.1)CALL PLOT(X,Y,N,M)
3. X=P+Q
    IF(X.GE.Y)X=P-Q

In the third example, the value assigned to X is P + Q if X is less than Y, P - Q otherwise.

1.3.3 The DO and CONTINUE Statements

Loops (iterative procedures) are coded in Fortran using the DO statement,

DO n i=j1,j2,j3

Here n is the statement number designating the terminal statement of the loop, i is an integer variable name, referred to as the index variable, and j1, j2, and j3 are integer constants or variables denoting, respectively, the initial value, final value, and increment of the index variable. The increment is 1 by default. For example,

```
    SUM=0
    DO 10 I=1,N
    SUM=SUM+X(I)
10 CONTINUE
20 AVG=SUM/N
```

The execution of this code proceeds as follows:

1. Set the value of SUM to zero.

2. Set the value of I to 1.

3. Add the values of SUM and X(1), and store the result in SUM

4. Increment the value of I by 1.

5. If the value of I exceeds N, branch to statement 20.

6. Otherwise, add the values of SUM and X(I), and store the result in SUM.

7. Repeat steps 4 through 6 until the value of I exceeds N.

The CONTINUE statement is commonly used to indicate the end of the range of a DO loop. Its use is not required.

DO loops may be nested as long as the range of the loops do not overlap. For example,

```
Valid nesting          Invalid nesting
    DO 50 ...               DO 50 ...
    DO 30 ...               DO 30 ...
    .........               .........
30 CONTINUE                 DO 40 ...
    DO 40 ...               .........
    .........          30 CONTINUE
40 CONTINUE            40 CONTINUE
50 CONTINUE            50 CONTINUE
```

Branches may be made within and out of nested DO loops, but a transfer cannot be made into the range of a DO loop from outside its range. Valid and invalid transfers are indicated in Figure 1.2.

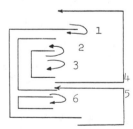

Figure 1.2 Diagram of nested DO loops: branches 2, 3, and 4 are valid; 1, 5, and 6 are invalid.

1.3.4 Other Control Statements

The PAUSE statement causes a suspension of the execution of a program (in the time-sharing mode). The statement causes the following to be printed at the terminal:

    TYPE G TO CONTINUE, X TO EXIT, T TO TRACE.

The alternate form,

    PAUSE 'literal string'

is useful for both debugging and user control over the execution of the program. For example, the statement

    PAUSE 'INDEX IS NOW 5'

causes the following to be printed at the terminal:

    INDEX IS NOW 5
    TYPE G TO CONTINUE, X TO EXIT, T TO TRACE.

The interested reader is referred to the DECSystem-10 Fortran-10 manual for a description of the TRACE feature.

The STOP statement terminates execution of the program. For example,

    IF(DISCR.LT.O.)STOP

The END statement is physically the last statement in the program.

1.4 I/O

I/O statements allow the programmer to transfer information from one component of the computing system to another. Input information is usually transferred using card readers, terminals, or magnetic tape. Output information is usually transferred to the line printer or the terminal. Disk I/O is also useful and is described in Section 1.6.

The programs in the book use only three I/O statements: READ, WRITE, and FORMAT. The general forms of these statements are

    READ(u,n)list
    WRITE(v,n)list
    FORMAT(S1,S2,S3,...,Sn)

Here u is an integer designating a logical unit number assigned for input by the computer center, n is a FORMAT statement number, v is a logical unit number assigned for output by the computer center, list is a set of variable names constituting the I/O record, and S1, S2, etc., represent FORMAT specification codes. Logical units 5 and 6 are frequently assigned to the card reader (input) and the line printer (output). In a time-sharing environment, logical units 5 and 6 often refer to the

terminal. The most frequently used specification codes for numeric work are the
following:

Specification Code	Use
iIw	Integer
$iFw.d$	Floating point (single or double precision)
$iEw.d$	Single precision with E exponent
$iDw.d$	Double precision with D exponent

Here i, w, and d are positive integers representing an optional repetition factor,
the width of the field (number of columns), and the number of significant digits,
respectively. For example, consider the following program:

```
      DOUBLE PRECISION C
      J=132
      A=123.456
      B=7.89E-16
      C=1.234567890123D25
      WRITE(6,11)J,A,B,C
   11 FORMAT(I5,F10.3,E15.5,D20.11)
      END
```

The output of this program would appear as follows:

```
                  10        20        30        40        50
Column:   12345678901234567890123456789012345678901234567890
            132   123.456   0.78900E-15   0.12345678901D+26
```

These specification codes are right justified, i.e., if the width exceeds that
required (for example, I5 exceeds the specification required for 123), blanks are
imbedded in the left-most columns of the output field.

The format I5 will suffice for all integer values in the range

$$-9999 \leq \text{integer value} \leq 99999$$

The format F5.1 will be adequate for all floating-point values in the range

$$-99.9 \leq \text{floating-point value} \leq 999.9$$

The width w should be at least d + 2 for F specification codes because one column
is required for the decimal point and a second for the sign if the number is
negative. The width should be at least d + 7 for E and D specification codes,
since seven additional columns are required,

```
±0.ddd...ddE±ee
123        4567
```

The following example will assign these values to the indicated variable
names:

ITMAX ← 10, TOL ← 0.0001, and AVAG ← 6.02E23

```
      READ(5,22)ITMAX,TOL,AVAG
   22 FORMAT(I3,F10.0,E15.0)
```

The values would be punched on a card or entered at the terminal as follows:

```
              10        20        30
Column:  12345678901234567890123456789O
           10    .0001          6.02E23
```

The 10 (ITMAX) must appear in columns 2 and 3. If the 1 is in column 1, then
ITMAX will be assigned the value 100. If the 1 is punched in column 3, then ITMAX
will be 1. The specification F10.0 means that the value for TOL will appear
between columns 4 and 13 inclusive. The decimal point overrides the format
specification code, so the number with its decimal point may be punched anywhere
in these 10 columns. The specification E15.0 means that columns 14 through 28
inclusive are reserved for the variable AVAG. The value 6.02E23 can be punched
anywhere in these 15 columns.

The single quote specification is used to print character strings. For
example,

```
      WRITE(6,33)DISCR,ROOT1,ROOT2
   33 FORMAT(//'  DISCR, ROOT1 AND ROOT2 ARE:',3E15.3)
```

The slash (/) is used to skip a line in the output. In this example two lines
will be skipped, then starting in column 3, the text "DISCR, ROOT1, AND ROOT2 ARE:"
will be printed, and then 45 columns will be used to print the three numeric values
using the E15.3 format.

The skip specification code X provides additional formatting control. For
example, the following statement prints a table heading.

```
      WRITE(6,55)
   55 FORMAT(//5X,'I',10X,'X(I)',10X,'Y(I)'/)
```

Here two lines will be skipped, the letter "I" will be printed in column 6, "X(I)"
will appear in columns 17-20, "Y(I)" will appear in columns 31-34, and another line
will be skipped.

The DECSystem's Fortran-10 allows free-format I/O. For example,

```
      READ(5,22)ITMAX,TOL,AVAG
   22 FORMAT(I,2E)
```

The data could be entered as follows:

```
      10,  0.0001,  6.02E23
```

The comma is used in this example to delimit the input values. Any other character
including the blank can be used. Most of the programs in this book use free format
for input. Also, the format G12.3 is frequently used for output, For example,

```
      A=0.5678
      B=6.789E-15
      C=65.43
      D=1234.5678
      WRITE(6,11)A,B,C,D
   11 FORMAT(1X,5('1234567890'),/4G12.3)
      END
```

The sample execution of this program is

```
12345678901234567890123456789012345678901234567890
  0.568        0.679E-14     65.4        0.123E+04
```

The following conversions are made if free formatting is used for output:

I becomes I15

F becomes F15.7

E becomes E15.7

D becomes D25.16

1.5 SUBPROGRAMS

It is often convenient in writing Fortran programs to use subprograms for certain
sections of code. Four varieties of subprograms are available in Fortran: the
statement function and the FUNCTION, SUBROUTINE, AND BLOCK DATA subprograms. The
first three are considered in this section, the fourth (BLOCK DATA) is discussed
in the following section.

1.5.1 Statement Functions

The <u>statement function</u> is used to define an algorithm in a single line. The
general form of the statement function is

        name(a1,a2,...,an)=expression

Here "name" is the statement function name; a1, a2, ..., an   are nonsubscripted
arguments; and "expression" represents either an arithmetic or logical expression.
For example,

```
      FUNK(A,B,C)=SQRT(A*A+B*B+C*C)
      ...............................
      Q=FUNK(X,Y,Z)
      ...........
      R=Q-EXP(Z+FUNK(D,17.6,Z))
```

Here Q is assigned the value $(X^2 + Y^2 + Z^2)^{1/2}$ and R is assigned the value $Q - e^{Z+T}$,
where $T = (D^2 + 17.6^2 + Z^2)^{1/2}$. The type(integer, single precision, double precision,
or logical) can be specified. For example,

```
      DOUBLE PRECISION LORNTZ,A,B,C,X
      LORNTZ(A,B,C,X)=A*B/(B*B+(X-C)**@)
```

and

```
      LOGICAL TEST
      TEST(FN)=(FN.GT.DVGTOL .OR. FN.LT.CVGTOL)
```

In the second example, FN, DVGTOL, and CVGTOL are single-precision values. The
currently assigned values of DVGTOL and CVGTOL are used when TEST(FN) is referenced.
The arguments of the statement function must agree in type, number, and order with
the arguments in the expression referencing the statement function. The statement
function cannot be recursive, that is, it cannot reference itself.

1.5.2 FUNCTION Subprograms

Two of the limitations of statement functions are removed with the FUNCTION
subprogram. These restrictions are that only one line of code may be used and that
the arguments cannot be subscripted. FUNCTION subprograms have the following
general form:

```
      type FUNCTION name(a1,a2,...,an)
      ................................
      name=expression
      ....................
      RETURN
      ......
      END
```

Here type is optional and may be INTEGER, REAL, DOUBLE PRECISION, LOGICAL, or
COMPLEX: "name" is a symbolic name that must be assigned a value at least once
in the subprogram; and a1, a2, ..., an are arguments. Control is returned to the
main program or calling subprogram with the RETURN statement. For example,

Calling program	FUNCTION subprogram
.... XX=X(I) FRCTN=GAUSS(WIDTH,PKPOSN,XX)	FUNCTION GAUSS(SIGMA,XO,X) ARG=-0.5*((X-XO)/SIGMA)**2 GAUSS=EXP(ARG)/(2.506627*SIGMA) RETURN

The values assigned to WIDTH, PKPOSN, and XX are transferred to SIGMA, XO, and X,
respectively. After the statement

```
      FRCTN=GAUSS(WIDTH,PKPOSN,XX)
```

is executed, FRCTN will have the value

$$\frac{1}{s\sqrt{2\pi}} \exp \left[-\frac{1}{2} \left(\frac{x-m}{s}\right)^{\overline{2}} \right]$$

where s, WIDTH, and SIGMA all refer to the standard deviation and m, PKPOSN, and
XO refer to the average value. The FUNCTION subprogram is used to code an
algorithm that is frequently referenced in the main or calling program. If the
values of the arguments are reassigned in the subprogram, the new values are
returned to the calling program. The arguments may be subscripted. The FUNCTION

subprogram is not recursive, that is, it can neither call itself, nor call another subprogram in which it is directly or indirectly referenced.

1.5.3 SUBROUTINE Subprograms

The general form of the SUBROUTINE subprogram is

```
SUBROUTINE name(a1,a2,a3,...,an)
..............................
RETURN
......
END
```

The subroutine is referenced by the CALL statement,

```
CALL name(a1,a2,a3,...,an)
```

For example,

Calling program	Subprogram
	SUBROUTINE QUAD(A,B,C,X1,X2)
	....
CALL QUAD(X,Y,Z,R1,R2)	DISCR=B*B-4.*A*C
	....
....	D=SQRT(DISCR)
	....
	X1=(-B+D)/(2.*A)
	X2=(-B-D)/(2.*A)
	....
	RETURN
	......
	END

The arguments A, B, and C in the subroutine QUAD are assigned the values of X, Y, and Z in the calling program. The values assigned to X1 and X2 in QUAD are referenced as R1 and R2 in the calling program.

Subroutines are commonly used in Fortran programming. It is convenient when writing a large program to segment the code into several logical components, to write and debug these segments as subroutines, and then to merge them into one program, if appropriate, when all the parts have been thoroughly tested. Computer centers usually have libraries of subroutines available to perform such standard numeric operations as matrix inversion and diagonalization. Subroutines, like the other two subprograms, are not recursive.

1.6 ADDITIONAL FORTRAN FEATURES

The Fortran that has been reviewed to this point is sufficient to read most of the programs in this book. This section draws attention to a few additional features of Fortran that are used in some of the more complex programs.

1.6.1 COMMON and EQUIVALENCE

COMMON is a memory allocation feature that allows the user to reserve an area of memory that can be shared by the main program and one or more subprograms. The general form of COMMON is

COMMON/$name_1$/v_1,v_2,...,v_n/$name_2$/u_1,u_2,...,u_n/...

Here $name_1$ and $name_2$ are optional labels and the v_1 and u_1 are variable names which may be subscripted. If the labels are used, the storage area is said to be labeled, otherwise, it is blank or unlabeled. For example,

```
COMMON X(100),Y(100),A(10,10),ITMAX,TOL     (unlabeled)
COMMON/STOR/B(10,10),C(100),TEMP            (labeled)
```

In the first example 302 words are reserved in memory; in the second example 201 words are reserved and assigned the label STOR. By using COMMON, the programmer can dispense with variable names in CALL and SUBROUTINE statements. For example,

Without COMMON	With COMMON
DIMENSION X(100),Y(100),Z(100)	DIMENSION Z(100)
....	COMMON X(100),Y(100), ITMAX,N,TOL
	......
CALL SUBQ(X,Y,ITMAX,N,TOL)	CALL SUBQ
....	....
SUBROUTINE SUBQ(A,B,JJ,M,CVG)	SUBROUTINE SUBQ
DIMENSION A(100),B(100)	COMMON A(100),B(100),JJ,M,CVG
....	....
END	END

In both versions of this program 203 memory locations are reserved for the five variables that are referenced in both the main program and in the subroutine. When the statement

```
CALL SUBQ(X,Y,ITMAX,N,TOL)
```

is executed, only the addresses of the variable names are transferred to the subroutine. B(1) in SUBQ and Y(1) in the main program refer to the same location in memory. The effect of COMMON is to allocate at compilation time a common block of memory as follows:

```
X(100)    or   A(100)
Y(100)    or   B(100)
ITMAX     or   JJ
N         or   M
TOL       or   CVG
```

COMMON relieves the programmer of the burden of keeping the arguments in the CALL and SUBROUTINE in order and saves the CPU time required to transfer addresses to the subroutine(s).

1.6.2 EQUIVALENCE

   The EQUIVALENCE statement allows the programmer to refer to the same values
in memory by two or more names. The general form is

   EQUIVALENCE $(v_1, v_2, \ldots), (u_1, u_2, \ldots)$

where the v_i and u_i are variable names which may be subscripted. For example,

   EQUIVALENCE (A(100),STOR(100)),(TEMP1,HOLD,PLACE)

Here the references A(I) and STOR(I), where I = 1, 2, 3, ..., 100, will address
the same memory location. TEMP1, HOLD, and PLACE will all have the same value.
The statements

   DIMENSION X(5),Y(5)
   EQUIVALENCE (X(1),Y(3))

will cause the following storage allocation:

   Y(1)
   Y(2)
   Y(3)   or   X(1)
   Y(4)   or   X(2)
   Y(5)   or   X(3)
   X(4)
   X(5)

The statements

   COMMON X(3),Q(2),ZAP
   DIMENSION Y(5)
   EQUIVALENCE (X(3),Y(1))

will extend the length of COMMON as follows:

   X(1)
   X(2)
   X(3)   or   Y(1)
   Q(1)   or   Y(2)
   Q(2)   or   Y(3)
   ZAP   or   Y(4)
   Y(5)

EQUIVALENCE statements which force an upward extension of a common block are not
allowed. For example,

   COMMON,IGO,NCNT,ICGS
   DIMENSION A(4)
   EQUIVALENCE (IGO,A(3))

This would try to extned the common block to include A(1) and A(2). This is not
allowed.

1.6.3 DATA and BLOCK DATA

The DATA statement is a convenient initialization statement. The general form
is

DATA $list_1$/$data_1$/,$list_2$/$data_2$/,...

Here $list_1$, $list_2$, etc., are lists of variable names which may be subscripted, and
$data_1$, $data_2$, etc., are the associated values to be assigned to the variables. For
example,

DATA A(100),CVGTOL,DVGTOL,ITMAX/100*0.,1.E-6,1.E6,25/

This statement is equivalent to

```
      DO 10 I=1,100
   10 A(I)=0.
      CVGTOL=1.E-6
      DVGTOL=1.E6
      ITMAX=25
```

The BLOCK DATA feature is a subprogram that is required to initialize variables in COMMON using the DATA statement. For example,

```
BLOCK DATA
COMMON/ARRAYS/A(100),B(50)/SCALRS/P,Q,R
COMMON TOL,ITMAX
DATA A/100*0./,B/20*1.,20*2.,10*1.0E-5/
DATA P,R,TOL,ITMAX/4.75,9.,1.E-6,25/
END
```

Here the 150 values in the labeled COMMON, ARRAYS, are initialized as indicated.
Four of the five scalars, 3 in labeled COMMON and two in unlabeled COMMON, are
also initialized. The BLOCK DATA subprogram may contain only the following statements: BLOCK DATA, DATA, DIMENSION, EQUIVALENCE, COMMON, END, and the specification statements INTEGER, DOUBLE PRECISION, etc.

1.6.4 Strings (Literals)

Strings are conveniently initialized using the DATA statement. For example,

```
DOUBLE PRECISION ORDNT,ABSCA
DATA SYMBOL(4)/'*','+','-',' '/
DATA ORDNT,ABSCA/'Y-AXIS','X-AXIS'/
```

The apostrophe is the delimiting character used to define a string value using the
DATA statement. Alternatively, the Hollerith specification nH may be used:

DATA ORDNT/6HY-AXIS/

In Fortran up to five characters can be stored (left-justified) in a single-
precision or integer variable and up to 10 characters in double precision. Strings
are printed using the A format specification code. For example,

```
        WRITE(6,11)(SYMBOL(I),I=1,4),ORDNT,ABSCA
     11 FORMAT(4A5,2A10)
```

The output would be

```
                    10        20        30        40
     Column:  12345678901234567890123456789012345678 90
              *    +    -         Y-AXIS    X-AXIS
```

1.6.5 Disk I/O

It is frequently convenient to store data in disk files. These disk files
may contain either input or output information. For example, a program may read
data from a disk file and write results to another disk file. Fortran-10 allows
the following unit designations for disk I/O:

```
     READ(u,n)list
     WRITE(v,n)list
```

Here u and v are integers and designate the name of the disk file to be read or
written. For example,

```
     ....................
     READ(22,110)X(I),Y(I)
     ..................
     WRITE(24,220)X(I),Y(I),YCALC,DIFF
     ............................
```

Here the unit designation 22 indicates that the input file is the disk file
FOR22.DAT, and the unit designation 24 indicates that the output file is
FOR24.DAT.

1.6.6 Other Fortran Features

The serious Fortran programmer will acquire the Fortran manual provided by
the vendor (Digital Equipment Corporation, International Business Machines,
Control Data Corporation, etc.). The Fortran-10 manual for the DECSystem-10 lists
the following features that are not considered in this chapter:

ACCEPT	FOROTS
Adjustable dimensions	Fortran-10 compiler switches
BACKFILE	Global optimizer
BACKSPACE	Multi-statement line
CLOSE	Multiple returns
Control characters	NAMELIST
for the printer	OPEN
DECODE	PRINT
ENCODE	PUNCH
ENDFILE	REREAD
ENTRY	REWIND
Extended range of DO	SKIPFILE
EXTERNAL	Tab
FIND	TRACE
Floating point DO loops	TYPE
FORDDTS	UNLOAD

1.7 PROGRAMMING CONSIDERATIONS

The computer should be used to solve only a subset of the numerical problems in chemistry. One should seek diligently for an analytical solution to a given problem before resorting to numerical methods. If the computer has to be used, the solution should proceed in the following sequence:

1. Define the problem completely (I/O, boundary conditions, special cases, degree of precision, etc.).

2. Do the required numerical analysis by hand (if possible), and consider the translation of the algorithm to Fortran.

3. Prepare a flow chart (if necessary), and write the Fortran code with pencil and paper.

4. Prepare the code in machine-readable form, debug (if necessary), and optimize the program.

6. Test the program with sample data.

7. Flog the program as thoroughly as time allows.

8. Execute the program and interpret the results.

Novice programmers frequently underestimate the importance of steps 1 through 4 and, consequently, spend much of their time debugging. Experienced programmers usually maximize their efficiency by carefully planning, writing, and hand-testing the code prior to interacting with the computer.

1.7.1 Round-off and Truncation Errors

One should never impute oracular powers of infallibility to computer output. It is more prudent to take the defensive position that every computer program has at least one bug. Even if there are no logical errors, one should always be wary of round-off and truncation errors in computing.

Round-off errors arise due to the finite length of the computer word. For example, the DECSystem-10 has a 36-bit word, which limits the precision of a single-precision value to approximately eight decimal digits. The following program illustrates one of the problems due to finite word size. This example shows two solutions to the round-off problem associated with the calculation of the expression $y = 1 - \cos x$. As x approaches zero, cos x approaches 1, and round-off errors become significant. One solution is to use double precision. The other is to use the trigonometric identity

$$1 - \cos x = 2 \sin^2 \frac{x}{2}$$

```
C     PROGRAM TO ILLUSTRATE ROUND-OFF ERROR
C
C     Y = 1-COS(X)  = 2*SIN(X/2)**2
C
      DOUBLE PRECISION DX,DY
      WRITE(6,11)
      DO 10 I=1,10
      PWR=-1.-0.3*I
      X=10.**(PWR)
      DX=10.0D0**(PWR)
      Y=1.0-COS(X)
      DY=1.0D0-DCOS(DX)
      SY=2.0*SIN(X/2.)**2
   10 WRITE(6,22)I,X,Y,DY,SY
   11 FORMAT(//'  I',6X,'X',13X,'Y',17X,'DY',15X,'SY'/)
   22 FORMAT(I3,E10.2,3G18.8)
      END
```

The sample execution of this program follows.

I	X	Y	DY	SY
1	0.50E-01	0.12556911E-02	0.12556804D-02	0.12556803E-02
2	0.25E-01	0.31546503E-03	0.31546207D-03	0.31546206E-03
3	0.13E-01	0.79251826E-04	0.79243611D-04	0.79243612E-04
4	0.63E-02	0.19915402E-04	0.19905291D-04	0.19905291E-04
5	0.32E-02	0.50067902E-05	0.49999958D-05	0.49999997E-05
6	0.16E-02	0.12591481E-05	0.12559429D-05	0.12559432E-05
7	0.79E-03	0.31292439E-06	0.31547866D-06	0.31547868E-06
8	0.40E-03	0.81956387E-07	0.79244656D-07	0.79244653E-07
9	0.20E-03	0.14901161E-07	0.19905357D-07	0.19905358E-07
10	0.10E-03	0.74505806E-08	0.50000000D-08	0.49999999E-08

Truncation errors result from evaluating numeric functions, e.g., logarithms, exponentials, and trigonometric functions, using a finite series approximation. For example, the Maclaurin series might be used to approximate the sine function for a certain range of the angle (in radians),

$$\sin x = x - \frac{x^3}{3!} + \frac{x^5}{5!} - \cdots$$

The following program shows the intermediate values in the approximation to the sine using the Maclaurin series.

```
C
C        PROGRAM TO ILLUSTRATE TRUNCATION ERROR
C
C        THE MACLAURIN SERIES APPROXIMATION TO THE SINE FUNCTION IS
C
C        SIN(X) = X - X**3/3! + X**5/5! - ...
C
         DOUBLE PRECISION DX,DSINX,DDSIN,DDSINX
         X=1.0
         DX=1.0D0
         SINX=SIN(X)
         DSINX=DSIN(DX)
         DDSINX=DDSIN(DX,1.0D-10)
         WRITE(6,11)X,SINX,DSINX,DDSINX
      11 FORMAT(//'  X = ',G12.3,21X,'SIN(X) = ',E20.12,//
      1 '  DSIN(X) = ', D20.12,5X,'DDSIN(X) = ', D20.12//)
         END
C
C        MACLAURIN SERIES APPROXIMATION TO SIN(X)
C
         DOUBLE PRECISION FUNCTION DDSIN(X,ERR)
         DOUBLE PRECISION X,ERR,T
         DDSIN=X
         T=X
         WRITE(6,11)
         DO 10 I=3,100,2
         T=-T*X**2/FLOAT(I*(I-1))
         DDSIN=DDSIN+T
         WRITE(6,22)I,T,DDSIN
      10 IF(DABS(T).LT.ERR)RETURN
         WRITE(6,33)
         RETURN
      11 FORMAT(//'  I',20X,'T',27X,'DDSIN'/)
      22 FORMAT(I3,2D30.12)
      33 FORMAT(//'  CONVERGENCE FAILURE'//)
         END
```

The following sample execution shows the effort that must be expended to reduce the truncation error to 1.E-10.

I	T	DDSIN
3	-0.166666666667D+00	0.833333333333D+00
5	0.833333333333D-02	0.841666666667D+00
7	-0.198412698413D-03	0.841468253968D+00
9	0.275573192240D-05	0.841471009700D+00
11	-0.250521083854D-07	0.841470984648D+00
13	0.160590438368D-09	0.841470984809D+00
15	-0.764716373182D-12	0.841470984808D+00

X =   1.00   SIN(X) =   0.841470970000E+00

DSIN(X) =   0.841470984808D+00   DDSIN(X) =   0.841470984808D+00

Computer programmers should always be prepared for the consequences of accumulated round-off and truncation errors. It is wise to monitor these errors so that the user of a program will be alerted if they exceed acceptable limits. For example, a matrix, A, can be inverted to give the inverse, A^{-1}. The product AA^{-1}, should be the identity matrix, I (Section 4.3.2). The diagonal elements of the identity matrix should all be exactly 1, and the off-diagonal elements should be exactly zero. One can test a matrix inversion routine by simply summing all the absolute values of the elements in the matrix and comparing the sum with N, the order of the matrix. This procedure is used in the program LINSYS (Section 4.3.3).

1.7.2 Debugging Techniques

One cannot overemphasize the importance of careful planning and adequate "paper-and-pencil work" prior to coding a program. When the program is coded, it should be segmented into logical parts and coded one part at a time. The initial version should be copiously sprinkled with debug statements. For example,

```
      ........
   25 CONTINUE
      ISTMT=25
      WRITE(6,666)ISTMT,A,B,C,D,E
  666 FORMAT(/' ISTMT,A,B,C,D,E: ',I3,5G12.3)
      PAUSE
      .....
```

These statements allow the user to monitor the execution of his code. The debug statements are easily removed after the program has been thoroughly tested.

Also, it is a good practice to echo input values for verification purposes. For example,

```
      ..............................
      READ(5,11)ITMAX,XO,CVGTOL,DVGTOL
      WRITE(6,22)ITMAX,XO,CVGTOL,DVGTOL
      ..............................
```

Once a program compiles successfully it should be subjected to a series of tests to prove that the code executes correctly for all conditions that can be anticipated. For example, if a matrix inversion routine tests for singularity, then a singular matrix should be provided to verify that the program executes as designed.

A time-sharing environment is very convenient for correcting obvious syntax and logical errors. However, valuable computer and terminal time can be wasted by novice programmers "banging their heads against a logical wall" trying to spot and correct subtle bugs at a terminal. Usually, it is more efficient to obtain a current listing of the program, a sample execution corresponding to that listing, log off the system, and search for the bug(s) off-line. After finding a bug, the programmer should resist the urge to run back to the terminal (the "Eureka complex") to see if the program works with this bug fixed. Instead, he or she should make sure that the correction does not have adverse ramifications on the remainder of the code, and then look for additional errors before returning to the computer.

1.7.3 Pitfalls of Fortran Programming

There are a number of mistakes commonly made by beginning Fortran programmers. The letters "O" and "l" often are interchanged with the numbers "0" and "1" until the fundamental difference between them becomes part of the programmer's conscious-ness. The default specification in Fortran is

```
IMPLICIT INTEGER(I-N), REAL(A-H,O-Z)
```

A frequent source of error occurs with mixed mode expressions, for example,

```
A=L/Q
```

Many Fortran compilers will accept mixed mode expressions. The programmer must be aware that the results will not necessarily be the same as the statement

```
A=FLOAT(L)/Q
```

For example, if $Q > 0$, and $0 \le L < Q$, the mixed mode expression will assign a zero value to A.

A programmer should never forget to initialize all variables before he uses them. Many compilers do not check for uninitialized variables, and the system may automatically assign a value of zero. For example,

```
          ...........
          DO 10 I=1,N
       10 SUM=SUM+X(I)
          ...........
```

Here the statement SUM=0. should be inserted prior to the DO statement.

The programmer should be careful to avoid subscripting errors in arrays. For example,

Calling Program	Subprogram
DIMENSION A(10,10)	SUBROUTINE MAT(P,R,M)
....	DIMENSION P(3,3)
CALL MAT(A,B,N)	....
....	

The compiler and run-time system may or may not check for improperly subscripted arrays in subprograms and/or for indices exceeding the bounds of arrays. The program may execute without error, and yet return meaningless results because of improper subscripting.

These and a multitude of other errors can, and more than likely will, occur
unless the programmer makes a concerted effort to think about the code and pain-
stakingly checks the code for bugs before attempting a compilation.

1.7.4 Optimization and Documentation

If a computer program is going to be used on a production rather than a "one-
shot" basis, then it deserves extra programming effort to optimize the code. For
example, consider the evaluation of the polynomial

$$y = x^4 + 2x^3 + 3x^2 + 4x + 5$$

Here are two programs which correctly calculate this polynomial over the range
$1 \leq x \leq 10$:

Program 1	Program 2
`............`	`............`
`DO 10 I=1,10`	`DO 10 I=1,10`
`X=I`	`X=I`
`Y=X**4+2.*X**3+3.*X**2+4.*X+5`	`Y=(((X+2.)*X+3.)*X+4.)*X+5.`
`.......................`	`.........................`

Assuming the computer interprets X**4 as X*X*X*X rather than as exp(4*lnX),
program 1 requires nine multiplications for each evaluation of Y. Program 2
achieves the same results with only three multiplications. An effort has been
made in this book to optimize the code as much as possible. The optimization
features are discussed as part of the documentation of the program. For some
examples, see PLOT (Section 2.2), EDTA (Section 3.1.1), POLREG (Section 5.6),
and TRAP (Section 6.4).

Conscientious programmers document their programs. A set of comments should
be inserted at the beginning of the program to describe its function, the I/O
specifications, and a glossary of the more important variable names. Comments
should be sprinkled liberally throughout the program to facilitate reading the
code. Effective documentation significantly increases the value of the program
to others, and often to the author several months after he has written it. The
documentation format for the 44 programs included in this book is the following:

    Introduction
    Method
    Sample execution
    Listing
    Discussion

The introduction is a paragraph describing the objective of the program. The method
section varies in length depending on the complexity of the chemistry, the numerical
analysis, and the Fortran code required to solve it. All the programs in this book
have been executed in a time-sharing environment. The user is prompted by the program for the input values. The discussion section draws attention to optimization
features and comments on the significance of the sample execution where appropriate.

1.8 FORTRAN FUNCTIONS

R=Real   I=Integer   D=Double Precision   C=Complex

Function	Name	Type Argument	Type Function	Comment
Absolute value				
R	ABS	R	R	
I	IABS	I	I	ARG
D	DABS	D	D	
C	CABS	C	C	$(x^2 + y^2)^{1/2}$
Arc Functions				
Sine	ASIN	R	R	
Cosine	ACOS	R	R	
Tangent	ATAN	R	R	
	DTAN	D	D	
	ATAN2	R(2)	R	arctan(ARG1/ARG2)
	DATAN2	D(2)	D	
Complex Conjugate				
	CONJG	C	C	
Conversion				
I → R	FLOAT	I	R	
R → I	IFIX	R	I	Result is largest
D → R	SNGL	D	R	integer ⩽ ARG
R → D	DBLE	R	D	
C → R	REAL	C	R	Converts real part
C → R	AIMAG	C	R	Converts imag. part
R → C	CMPLX	R(2)	C	C=ARG1+i*ARG2
Cosine				
R	COS	R	R	ARG in radians
R	COSD	R	R	ARG in degrees
D	DCOS	D	D	
C	CCOS	C	C	
Exponential				
R	EXP	R	R	
D	DEXP	D	D	
C	CEXP	C	C	
Hyperbolic				
Sine	SINH	R	R	
Cosine	COSH	R	R	
Tangent	TANH	R	R	
Logarithm				
R	ALOG	R	R	ln(ARG)
R	ALOG10	R	R	log(ARG)
D	DLOG	D	D	
D	DLOG10	D	D	
C	CLOG	C	C	

| Function | Name | Type | | Comment |
		Argument	Function	
Maximum Value				
R	AMAX0	I($\geq$2)	R	
I	MAX0	I "	I	
R	AMAX1	R "	R	Max [ARG1,ARG2,...]
I	MAX1	R "	I	
D	DMAX1	D "	D	
Minimum Value				
R	AMIN0	I($\geq$2)	R	
R	AMIN1	R "	R	
I	MIN0	I "	I	Min [ARG1,ARG2,...]
I	MIN1	R "	I	
D	DMIN1	D "	D	
Positive Difference				
R	DIM	R	R	ARG1-Min[ARG1,ARG2]
I	IDIM	I	I	
Ramdon Number				
	RAN	R	R	Random number in the range: $0 < \# < 1$
Remaindering				
R	AMOD	R(2)	R	Quotient of
I	MOD	I(2)	I	ARG1/ARG2
D	DMOD	D(2)	D	
Sine				
R	SIN	R	R	ARG in radians
R	SIND	R	R	ARG in degrees
D	DSIN	D	D	
C	CSIN	C	C	
Square Root				
R	SQRT	R	R	
D	DSQRT	D	D	
C	CSQRT	C	C	
Transfer of Sign				
R	SIGN	R(2)	R	
I	ISIGN	I(2)	I	Sign(ARG2)*$\lvert$ARG1$\rvert$
D	DSIGN	D(2)	D	Sign(ARG2)*$\lvert$ARG1$\rvert$
Truncation				
I $\rightarrow$ R	AINT	I	R	Sign of ARG*largest
R $\rightarrow$ I	INT	R	I	integer $\leq \lvert$ARG$\rvert$
D $\rightarrow$ I	IDINT	D	I	

CHAPTER 2

PROGRAMS WITH CLOSED-FORM ALGORITHMS

This chapter documents nine Fortran programs, none of which uses special numerical techniques. Seven of them simply calculate a dependent variable, y, as a function of an independent variable, x. The functionality can be represented as follows:

$$y = f(x; \alpha_1, \alpha_2, \ldots \alpha_p)$$

Here α_1, α_2, ..., α_p represent adjustable parameters. The simulation programs allow the user to vary the parameters (input data) and observe the response of the dependent variable. The first program is a plotting subprogram which is used subsequently by some of the simulation programs. The final program is a data reduction program. In this case the input consists of the observed variables x and y and from these values two parameters are calculated. A brief description of these programs is given in Table 2.1.

Table 2.1 Programs with Closed-Form Algorithms

Program	Description
PLOT	Teletype or line-printer plot routine
ABCKIN	Simulates the kinetic system $A \rightleftharpoons B \rightleftharpoons C$
SOL	Calculates the solubility of the salt M A as a function of pH. The anion A^{z-} hydrolyzes in solution for form z acids, $HA^{(z-1)-}$, $H_2A^{(z-2)-}$, ..., H_zA. Parameters: ionic charges, solubility product, and the acid dissociation constants.
NERNST	Calculates the potential of the $M\|M^{z+}$ reversible electrode in an aqueous ammonia solution as a function of pH. Parameters: the charge and coordination number of the metal, the standard reduction potential, and the stepwise metal-ammine formation constants.
RADIAL	Calculates and plots the radial distribution function for hydrogen-like atoms. Parameters: the orbital, the effective nuclear charge, and the maximum radial extension.
CONTOUR	Calculates and displays using "teletype graphics" electron density contour plots for certain atomic, hybrid, and molecular orbitals. Parameters: the orbital, the effective nuclear charge, and (for molecular orbitals) the internuclear distance.
EQUIL	Calculates the equilibrium constant of a gas-phase equilibrium system by finding the minimum value of the total free energy as a function of extent of reaction. Parameters: temperature, stoichiometry, and chemical potentials.

Table 2.1 Continued

Program	Description
NMR	Simulates AB, AB_2, ABX, A_2X_2, and A_2B_3 proton nuclear magnetic resonance spectra. Parameters: chemical shifts and coupling constants.
TITR	Calculates the two thermodynamic dissociation constants of an amino acid. TITR reads student pH and volume of titrant data and uses the limiting Debye-Huckel law to estimate the activity coefficients.

This chapter also contains a number of computer-drawn figures using an incremental plotter. The program that prepared these plots is CPLOT, which is documented in Section 9.1.1.

2.1 PLOT

Introduction

PLOT is a SUBROUTINE subprogram that prints a teletype or line-printer plot of up to three dependent variables of the same independent variable,

$$v_1 = f_1(x)$$
$$v_2 = f_2(x)$$
$$v_3 = f_3(x)$$

The values of the dependent and independent variables are assigned in the main program. They are subscripted variables defined in the following statement:

DIMENSION X(50),Y(50,3)

Three assumptions are made:

1. The three dependent variables are functions of the same independent variable,

2. X is linear, i.e., $X(I + 1)/X(I)$ is a constant,

3. X is sorted, i.e., $X(I) < X(I + 1)$ or $X(I) > X(I + 1)$.

Method

The calling sequence is

CALL PLOT(X,Y,M,N)

Here X and Y are the subscripted independent and dependent variables; M is 1, 2, or 3; and N is the number of values in X and Y. For the purposes of this documentation, the calling program calculates the fraction of the diprotic acid H_2A and the associated conjugate bases HA^- and A^{2-} as a function of pH [1].

$$H_2A \rightleftharpoons H^+ + HA^- \qquad K_1$$
$$HA^- \rightleftharpoons H^+ + A^{2-} \qquad K_2$$

$$f_{H_2A} = \frac{[H^+]^2}{D}$$

$$f_{HA^-} = \frac{K_1[H^+]}{D}$$

$$f_{A^{2-}} = \frac{K_1K_2}{D}$$

where

$$D = [H^+]^2 + K_1[H^+] + K_1K_2$$

The calling program reads pK_1 and pK_2, calculates the three fractions as a function of pH, prints a table, and calls the plotting subroutine.

The dimensions of the plot are arbitrarily set to 50 x 50. PLOT first finds the maximum and minimum values in Y. These values are stored in YMAX and YMIN and are used to calculate the scaling factor YSCALE. For example, if YMIN = -20, and YMAX = 180, then the scaling factor for a plot of width 50 is

YSCALE = 50/(180 + 20) = 1/4

The position of the plotting symbol in the output line is computed as follows:

NN(J) = (Y(I,J)-YMIN) * YSCALE + 1.5

For example, if YMIN = -20, YMAX = 180, and YSCALE = 1/4,

Y(I,J)	NN(J)
-20	1
20	11
100	31
180	51

The value 1.5 is added for two reasons:

1. The value 1.0 is added so that NN(0) is never referenced,

2. The value 0.5 is added so that the floating point value will be rounded, for example,

NN(J) = 45.6 + 1.5
      = 47.1
      = 47

Sample Execution [Malonic Acid, $CH_2(COOH)_2$]

ENTER PK1 AND PK2
>2.83 5.69

PK1 =   2.830   AND   PK2 =   5.690

PH	F(H2A)	F(HA(-))	F(A(2-))
0.0	0.999	0.148E-02	0.302E-08
1.0	0.985	0.146E-01	0.298E-06
2.0	0.871	0.129	0.263E-04
3.0	0.403	0.596	0.122E-02
4.0	0.621E-01	0.919	0.188E-01
5.0	0.558E-02	0.826	0.169
6.0	0.222E-03	0.329	0.671
7.0	0.316E-05	0.467E-01	0.953
8.0	0.330E-07	0.487E-02	0.995
9.0	0.331E-09	0.490E-03	1.00
10.0	0.331E-11	0.490E-04	1.00
11.0	0.331E-13	0.490E-05	1.00
12.0	0.331E-15	0.490E-06	1.00
13.0	0.331E-17	0.490E-07	1.00
14.0	0.331E-19	0.490E-08	1.00

```
   X(I)                                    Y(I)

            0.331E-19              0.500                        1.00
                 :                  :                            :
                 :+----+----+----+----+----+----+----+----+----+----+
 0.000          :*                                               +
 0.500          :*                                               +
 1.00           :*-                                             +
 1.50           :* -                                           +
 2.00           :*      -                                    +
 2.50           :*           -            +
 3.00           :*               +     -
 3.50           :*         +                              -
 4.00           : * +                                       -
 4.50           : + *                                       -
 5.00           :+       *                                -
 5.50           :+           *       -
 6.00           :+       -       *
 6.50           :+     -                            *
 7.00           :+ -                                      *
 7.50           :+-                                       *
 8.00           :-                                        *
 8.50           :-                                        *
 9.00           :-                                        *
 9.50           :-                                        *
 10.0           :-                                        *
 10.5           :-                                        *
 11.0           :-                                        *
 11.5           :-                                        *
 12.0           :-                                        *
 12.5           :-                                        *
 13.0           :-                                        *
 13.5           :-                                        *
 14.0           :-                                        *
```

Listing

```
C
C                     PLOT
C
C   THE EXAMPLE USED TO DOCUMENT THE SUBPROGRAM
C   PLOT IS THE DIPROTIC ACID SYSTEM.
C
C       H2A    <===>  H(+)  +  HA(-)       K1
C       HA(-)  <===>  H(+)  +  A(2-)       K2
C
C   THE THREE FRACTIONS, F(H2A),F(HA(-)), AND
C   F(A(2-)) ARE PLOTTED AS A FUNCTION OF PH
C
C          0 .LE. PH .LE. 14      BY 0.5
C
C   GLOSSARY (MAIN PROGRAM):
C
C   NAME                  DESCRIPTION
C
C   K1,PK1                K1,PK1
C   K2,PK2                K2,PK2
C   H                     [H(+)]
C   X(I)                  PH
C   Y(I,1)                F(H2A)
C   Y(I,2)                F(HA(-))
C   Y(I,3)                F(A(2-))
C
C
C
C       AUTHOR:  K. J. JOHNSON
C
      DIMENSION X(50),Y(50,3)
      REAL K1,K2
C
C   READ PK1 AND PK2
C
      WRITE(6,11)
      READ(5,22)PK1,PK2
      WRITE(6,33)PK1,PK2
      K1=10.**(-PK1)
      K2=10.**(-PK2)
      T3=K1*K2
C
C   CALCULATE THE THREE FRACTIONS
C
      DO 100 I=1,29
      PH=(I-1.)/2.
      H=10.**(-PH)
      T1=H*H
      T2=K1*H
      D=T1+T2+T3
      X(I)=PH
      Y(I,1)=T1/D
      Y(I,2)=T2/D
      Y(I,3)=T3/D
  100 CONTINUE
```

```
C
C    PRINT 15 OF THE 29 VALUES
C
      DO 200 I=1,29,2
  200 WRITE(6,44)X(I),(Y(I,J),J=1,3)
C
C   CALL PLOT
C
      CALL PLOT(X,Y,29,3)
      STOP
   11 FORMAT(//'  ENTER PK1 AND PK2'/)
   22 FORMAT(2F)
   33 FORMAT(//'  PK1 = ',F6.3,'   AND  PK2 = ',F6.3,
     1///7X,'PH',7X,'F(H2A)',9X,'F(HA(-))',7X,'F(A(2-))'/)
   44 FORMAT(F10.1,3G15.3)
      END
C
C
C                    SUBROUTINE PLOT
C
C
C    THIS SUBROUTINE PRINTS A TELETYPE OR LINE-PRINTER
C    PLOT OF UP TO THREE DEPENDENT VARIABLES OF THE SAME
C    INDEPENDENT VARIABLE
C
C
C    USAGE:   CALL PLOT(X,Y,N,M)
C
C   GLOSSARY:
C
C    NAME          DESCRIPTION
C
C   Y(50,3)   DEPENDENT VARIABLE(S)
C   X(50)     INDEPENDENT VARIABLE
C   N         LENGTH OF X AND Y
C   M         NUMBER OF DEPENDENT VARIABLES (1,2 OR 3)
C   CHAR(4)   PLOTTING SYMBOLS AND BLANK ("+","-","*" AND " ")
C   LINE(51)  VECTOR CONTAINING BLANKS AND PLOTTING CHARACTERS
C   NN(3)     POSITION IN LINE FOR THE PLOTTING CHARACTERS
C   LENGTH    ARBITRARY LENGTH OF THE PLOT (50 LINES)
C   WIDTH     ARBITRARY WIDTH OF THE PLOT (50 COLUMNS)
C   YMIN      MINIMUM OF Y(I,J),J=1,M; I=1,N
C   YMAX      MAXIMUM OF Y(I,J),J=1,M; I=1,N
C   YSCALE    SCALING FACTOR OF WIDTH
C   ISKIP     SCALING FACTOR FOR LENGTH
C
C
C    IT IS ASSUMED THAT X IS LINEAR, I.E., THAT X(I+1)/X(I)
C    IS CONSTANT; AND THAT X IS SORTED, I.E., THAT
C    X(I) .LT. X(I+1) OR THAT X(I) .GT. X(I+1)
C
C    THE LENGTH AND WIDTH OF THE PLOT ARE ARBITRARILY SET
C    BY LENGTH = 50   AND WIDTH = 50
C
C    AUTHOR:  K. J. JOHNSON
C
      SUBROUTINE PLOT(X,Y,N,M)
      INTEGER CHAR(4)
```

```
      DIMENSION X(50),Y(50,3),LINE(51),NN(3)
      DATA LINE/51*' '/,CHAR/'+','-','*',' '/
      DATA LENGTH,WIDTH/50,50./
C
C     FIND MINIMUM AND MAXIMUM Y AND SET SCALING FACTORS
C
      YMIN=Y(1,1)
      YMAX=Y(1,1)
      DO 10 I=1,N
      DO 10 J=1,M
      IF(Y(I,J).GT.YMAX)YMAX=Y(I,J)
   10 IF(Y(I,J).LT.YMIN)YMIN=Y(I,J)
      YSCALE=WIDTH/(YMAX-YMIN)
      YMID=(YMAX+YMIN)/2.
      ISKIP=N/LENGTH +1
      IF(MOD(N,LENGTH).EQ.0)ISKIP=ISKIP-1
      WRITE(6,11)YMIN,YMID,YMAX
C
C     PLOT ONE LINE AT A TIME
C
      DO 40 I=1,N,ISKIP
      DO 30 J=1,M
      NN(J)=(Y(I,J)-YMIN)*YSCALE + 1.5
   30 LINE(NN(J))=CHAR(J)
      NMAX=AMAX0(NN(1),NN(2),NN(3))
      WRITE(6,22) X(I), (LINE(JJ),JJ=1,NMAX)
      DO 25 J=1,M
   25 LINE(NN(J))=CHAR(4)
   40 CONTINUE
      RETURN
   11 FORMAT(///'     X(I)',30X,'Y(I)',//,9X,G12.3,12X,G12.3,
     1   12X,G12.3,/14X':',24X,':',24X,':',/,
     2 13X,':+----+----+----+----+----+----+----+----+----+----+')
   22 FORMAT(G12.3,' :',51A1)
      END
```

Discussion

The main program and the subprogram require 388 and 273 words of memory, respectively. The execution time for this sample execution was approximately 0.3 sec. PLOT contains an optimization feature worthy of note. The use of the statements

```
NMAX=AMAX0(NN(1),NN(2),NN(3))
WRITE(6,22) X(I), (LINE(JJ),JJ=1,NMAX)
```

makes sure that the trailing blanks will not be plotted.

A discussion of computer graphics, including several literature references, is given in Section 9.1.

2.2 ABCKIN

Introduction

ABCKIN prints the closed-form solution to the following first-order reversible kinetic system:

$$A \xrightleftharpoons[k_{21}]{k_{12}} B \xrightleftharpoons[k_{32}]{k_{23}} C$$

The program reads the four rate constants and tabulates the concentrations of A, B, and C as a function of time.

Method

The time dependence of the three concentrations is given by the solution to the following system of differential equations:

$$\frac{dA}{dt} = k_{21}B - k_{12}A$$

$$\frac{dB}{dt} = k_{12}A + k_{32}C - (k_{21} + k_{23})B$$

$$\frac{dC}{dt} = k_{23}B - k_{32}C$$

These differential equations can be solved in closed-form [2] if the following initial values obtain.

$$t = 0 \qquad A(0) = A_0 \qquad B(0) = C(0) = 0.0$$

The closed-form solution to these equations is

$$\frac{A(t)}{A(0)} = T_1 + T_2 e^{-\lambda_2 t} + T_3 e^{-\lambda_3 t}$$

$$\frac{B(t)}{A(0)} = T_4 + T_5 e^{-\lambda_2 t} + T_6 e^{-\lambda_3 t}$$

$$\frac{C(t)}{A(0)} = A(0) - [A(t) + B(t)]$$

where

$$T_1 = \frac{k_{21}k_{32}}{\lambda_2 \lambda_3}$$

$$T_2 = \frac{k_{12}(\lambda_2 - k_{23} - k_{32})}{\lambda_2(\lambda_2 - \lambda_3)}$$

$$T_3 = \frac{k_{12}(k_{23} + k_{32} - \lambda_3)}{\lambda_3(\lambda_2 - \lambda_3)}$$

$$T_4 = \frac{k_{12}k_{32}}{\lambda_2 \lambda_3}$$

$$T_5 = \frac{k_{12}(k_{32} - \lambda_2)}{\lambda_2(\lambda_2 - \lambda_3)}$$

$$T_6 = \frac{k_{12}(\lambda_3 - k_{32})}{\lambda_3(\lambda_2 - \lambda_3)}$$

$$\lambda_2 = \frac{P + Q}{2}$$

$$\lambda_3 = \frac{P - Q}{2}$$

$$P = k_{12} + k_{21} + k_{23} + k_{32}$$

$$Q = [P^2 - 4(k_{12}k_{23} + k_{21}k_{32} + k_{12}k_{32})]^{1/2}$$

ABCKIN assumes $A(0) = 1.0$. The input includes the four rate constants and the time interval,

$$0 < t < nh$$

where h is the time increment and n is the number of.time increments desired in the output.

Sample Execution

```
              K12     K23           A(0)=1, B(0)=C(0)=0
         A <==> B <==> C
              K21     K32              ( 0 < T < N*H )


      ENTER N,H,K12,K21,K23,AND K32
     >30 .25 1 .5 .75 .25

        N =      30              H =    0.250
       K12 =     1.00           K21 =   0.500
       K23 =     0.750          K32 =   0.250


         T               A              B              C

      0.000            1.00           0.000          0.000
      0.250            0.791          0.190          0.191E-01
      0.500            0.643          0.294          0.630E-01
      0.750            0.537          0.345          0.118
      1.00             0.458          0.366          0.176
      1.25             0.397          0.370          0.232
      1.50             0.350          0.365          0.285
      1.75             0.312          0.354          0.333
      2.00             0.282          0.342          0.376
      2.25             0.257          0.329          0.414
      2.50             0.235          0.317          0.448
      2.75             0.218          0.305          0.477
      3.00             0.203          0.294          0.503
      3.25             0.190          0.285          0.525
      3.50             0.179          0.277          0.544
      3.75             0.170          0.270          0.561
      4.00             0.161          0.263          0.575
```

```
      4.25         0.155         0.258         0.588
      4.50         0.149         0.253         0.599
      4.75         0.143         0.249         0.608
      5.00         0.139         0.245         0.616
      5.25         0.135         0.242         0.623
      5.50         0.132         0.239         0.629
      5.75         0.129         0.237         0.634
      6.00         0.127         0.235         0.638
      6.25         0.124         0.233         0.642
      6.50         0.123         0.232         0.646
      6.75         0.121         0.230         0.649
      7.00         0.120         0.229         0.651
      7.25         0.119         0.228         0.653
      7.50         0.118         0.227         0.655
```

ENTER 1 IF A PLOT IS DESIRED, 0 OTHERWISE
>1

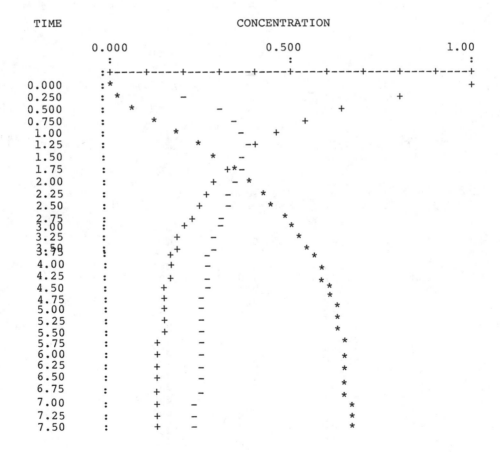

```
C
C     INITIAL CONDITIONS:   A(0)=1, B(0)=C(0)=0
C
C       SEE REF. 2, P. 175
C
```

Listing
```
C
C            ABCKIN SOLVES THE KINETIC SYSTEM
C
C                 K12     K23
C              A <==> B <==> C        ( 0<T<N*H )
C                 K21     K32
C
C
C     INITIAL CONDITIONS:  A(0)=1, B(0)=C(0)=0
C
C
C     SEE REF. 2, P. 175
C
C     INPUT SPECIFICATIONS:
C
C           N -   OF TIME STEPS  ( N < 51 )
C           H - STEP SIZE
C           K12,K21,K23,K32 - SPECIFIC RATE CONSTANTS
C
C     THE SUBROUTINE PLOT IS CALLED ON OPTION TO
C     PLOT A(T),B(T), AND C(T)
C
C     AUTHOR:  K. J. JOHNSON
C
      IMPLICIT REAL (K,L)
      DIMENSION X(51),Y(51,3)
      DATA A,Y(1,1),T,B,C,Y(1,2),Y(1,3),X(1)/2*1.,6*0./
C
C     INPUT
C
      WRITE(6,11)
      READ(5,22)N,H,K12,K21,K23,K32
      IF(N.LE.0 .OR. N.GT.50)STOP
      WRITE(6,55)N,H,K12,K21,K23,K32
C
C     SET CONSTANTS
C
      NP1=N+1
      T0=K12+K21+K23+K32
      T1=T0*T0
      T2=4.*(K12*K23+K21*K32+K12*K32)
      T3=T1/T2
      IF(T3.LT.1.000001)GO TO 95
      T2=SQRT(T1-T2)
      L2=(T0+T2)/2.
      L3=(T0-T2)/2.
      IF(L2.EQ.0.OR.L3.EQ.0)GO TO 95
      T1=K21*K32/(L2*L3)
      T2=K12*(L2-K23-K32)/(L2*(L2-L3))
      T3=K12*(K23+K32-L3)/(L3*(L2-L3))
      T4=K12*K32/(L2*L3)
      T5=K12*(K32-L2)/(L2*(L2-L3))
      T6=K12*(L3-K32)/(L3*(L2-L3))
      WRITE(6,66)
      WRITE(6,77)T,A,B,C
```

```
C
C      MAIN LOOP  -   STEP T FROM H TO N*H
C
       DO 50 I=1,N
       IP1=I+1
       T=T+H
       E2=EXP(-L2*T)
       E3=EXP(-L3*T)
       A=T1+T2*E2+T3*E3
       B=T4+T5*E2+T6*E3
       C=1.-A-B
       X(IP1)=T
       Y(IP1,1)=A
       Y(IP1,2)=B
       Y(IP1,3)=C
    50 WRITE(6,77)T,A,B,C
C
C    IS A PLOT DESIRED?
C
       WRITE(6,99)
       READ(5,22)IPLOT
       IF(IPLOT.EQ.1)CALL PLOT(X,Y,NP1,3)
       STOP
    95 WRITE(6,88)T0,T1,T2,T3
       STOP
    11 FORMAT(//10X,'K12',4X,'K23',9X,'A(0)=1, B(0)=',
      1'C(0)=0'/7X,'A <==> B <==> C'/10X,'K21',4X,'K32',
      211X,'( 0 < T < N*H )'//'  ENTER N,H,K12,K21,K23,AND K32'/)
    22 FORMAT(I,5F)
    55 FORMAT(/'    N =',I6,11X,'  H =',G12.3,
      1/'  K12 =',G12.3,5X,'K21 =',G12.3,
      2/'  K23 =',G12.3,5X,'K32 =',G12.3/)
    66 FORMAT(/'       T'12X,'A',10X,'B',11X,'C'/)
    77 FORMAT(4G12.3)
    88 FORMAT(//'  ERROR CONDITION! T0,T1,T2 AND T3 ARE:'/
      14G12.3/)
    99 FORMAT(//'  ENTER 1 IF A PLOT IS DESIRED, 0 OTHERWISE'/)
       END
```

Discussion

The main program and PLOT1 require 603 and 264 words of memory, respectively. This sample execution took approximately 0.3 sec of CPU time. The solution to the following related kinetic systems can be obtained by entering zero for the appropriate rate constant(s):

$$A \rightleftharpoons B \longrightarrow C$$
$$A \longrightarrow B \rightleftharpoons C$$
$$A \longrightarrow B \longrightarrow C$$

Problem 2.1   Plot B(t) versus t for several sets of rate constants to demonstrate the sensitivity of the intermediate species concentration [B] to the four rate constants.

Solution

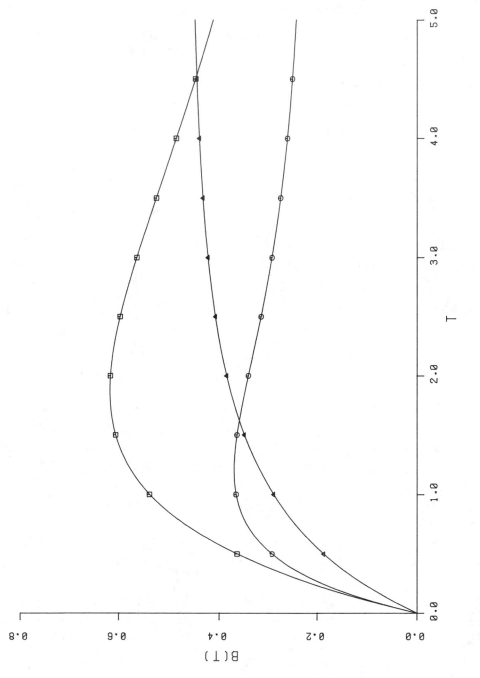

Figure 2.1 B(t) versus t (see text for the rate constants)

The following values for the four rate constants were used:

Case	k_{12}	k_{21}	k_{23}	k_{32}
I. ($\square$)	1.0	0.05	0.25	0.05
II. ($\bigcirc$)	1.0	0.5	0.75	0.25
III. ($\triangle$)	0.5	0.25	0.5	0.75

Problem 2.2 Show that a maximum in $B(t)$ will appear if $k_{32} < k_{12}$. Show that a maximum in $B(t)$ will not appear if $k_{32} > k_{12}$.

Problem 2.3 Under what conditions will the steady-state approximation

$$dB(t)/dt = 0$$

be valid for this chemical system?

2.3 SOL

Introduction

SOL calculates the solubility of the sparingly soluble salt M_nA_p as a function of pH.

$$M_nA_p \rightleftharpoons nM^{y+} + pA^{z-}$$
$$K_{sp} = [M^{y+}]^n[A^{z-}]^p$$

The solubility is a function of pH because the anion A^{z-} hydrolyzes in solution to form z conjugate acids. For example, for $Cd_3(PO_4)_2$

$$Cd_3(PO_4)_2 \rightleftharpoons 3Cd^{2+} + 2PO_4^{3-} \qquad K_{sp}$$
$$PO_4^{3-} + H_2O \rightleftharpoons HPO_4^{2-} + OH^- \qquad K_{h1}$$
$$HPO_4^{2-} + H_2O \rightleftharpoons H_2PO_4^- + OH^- \qquad K_{h2}$$
$$H_2PO_4^- + H_2O \rightleftharpoons H_3PO_4 + OH^- \qquad K_{h3}$$

Here $K_{h1} = K_w/K_3$, $K_{h2} = K_w/K_2$, $K_{h3} = K_w/K_1$, and K_w is the water dissociation constant.

The solubility of $Cd_3(PO_4)_2$ is strongly pH dependent. Indeed, most sparingly soluble salts contain anions that are conjugate bases of weak acids.

Method

Ideal solution behavior is assumed. It is further assumed that the metal does not hydrolyze and that all other equilibrium reactions, for example, complexation reactions, can be safely ignored.

Consider the case for z = 3.

$$M_nA_p \rightleftharpoons nM^{y+} + pA^{3-}$$

Let S be the solubility of M_nA_p in moles per liter. Then

$$[M^{y+}] = nS$$

and

$$A_{tot} = pS = [A^{3-}] + [HA^{2-}] + [H_2A^-] + [H_3A] \qquad (2.1)$$

An expression relating only $[A^{3-}]$ to S is derived as follows:

$$K_3 = \frac{[H^+][A^{3-}]}{[HA^{2-}]}$$

so

$$[HA^{2-}] = \frac{[H^+][A^{3-}]}{K_3}$$

$$K_2 = \frac{[H^+][HA^{2-}]}{[H_2A^-]}$$

so

$$[H_2A^-] = \frac{[H^+][HA^{2-}]}{K_2} = \frac{[H^+]^2[A^{3-}]}{K_2K_3}$$

$$K_1 = \frac{[H^+][H_2A^-]}{H_3A}$$

so

$$[H_3A] = \frac{[H^+][H_2A^-]}{K_1} = \frac{[H^+]^3[A^{3-}]}{K_1K_2K_3}$$

With the appropriate substitutions, Eq. (2.1) becomes

$$pS = [A^{3-}]Q$$

where

$$Q = 1 + \frac{[H^+]}{K_3} + \frac{[H^+]^2}{K_2K_3} + \frac{[H^+]^3}{K_1K_2K_3}$$

Then

$$[A^{3-}] = \frac{pS}{Q} \qquad [M^{y+}] = nS$$

$$K_{sp} = [M^{y+}]^n[A^{3-}]^p$$

$$= (nS)^n(\frac{pS}{Q})^p$$

$$= \frac{n^n p^p S^{n+p}}{Q^p}$$

Finally,

$$S = \left[\frac{K_{sp}Q^p}{n^n p^p}\right]^{1/(n+p)}$$

The conjugate bases with $z = 2$ and $z = 1$ are special cases of the above, so one function suffices for the three systems. The input to the program includes the ionic charges y and z, the negative logarithm of the solubility product, pK_{sp}, and the negative logarithms of the stepwise dissociation constants, pK_i, $i = 1$ to z.

SOL determines the stoichiometric coefficients n and p from the ionic charges
y and z. For example,

$$CaC_2O_4 \rightleftharpoons Ca^{2+} + C_2O_4^{2-} \qquad (y = z = 2;\ n = p = 1)$$

$$Ag_3AsO_4 \rightleftharpoons 3Ag^+ + AsO_4^{3-} \qquad (y = 1,\ z = 3;\ n = 3,\ p = 1)$$

This is done by finding the largest common divisor of y and z. If L is the largest
common divisor of y and z, then n = z/L and n = y/L.

SOL contains a subroutine, PLOT1, which is a modification of PLOT (Section 2.1).
PLOT1 plots a single dependent variable, y = f(x).

Sample Execution $[Cd_3(PO_4)_2]$

```
                       SOLUBILITY OF M A
                                     N P



                  M A    <===> NM(Y+)   +   PA(Z-)
                   N P


                 AS A FUNCTION OF PH


   ENTER Y,Z,PKSP,(PK(I),I=1,Z)   (Y=0 ENDS PROGRAM)
   >2 3 32.66 2.22 7.11 11.52



                M(3)A(2)     <===>    3M(2+)    +    2A(3-)


        PK(1)  =     2.22        K(1)  =    6.03E-03
        PK(2)  =     7.11        K(2)  =    7.76E-08
        PK(3)  =    11.52        K(3)  =    3.02E-12
        PKSP   =    32.66        KSP   =    2.19E-33


        PH                    S                   LOG(S)

        0.0               2.53E+01                 1.40
        1.0               1.63E+00                 0.211
        2.0               1.21E-01                -0.917
        3.0               1.38E-02                -1.86
        4.0               2.07E-03                -2.68
        5.0               3.27E-04                -3.49
        6.0               5.33E-05                -4.27
        7.0               1.03E-05                -4.99
        8.0               3.09E-06                -5.51
        9.0               1.18E-06                -5.93
       10.0               4.73E-07                -6.33
       11.0               2.07E-07                -6.68
       12.0               1.29E-07                -6.89
       13.0               1.17E-07                -6.93
       14.0               1.15E-07                -6.94
```

```
    ENTER 1 FOR A PLOT OF LOG(S) VS. PH, 0 OTHERWISE
    >1


        PH                                  LOG(S)


              -6.94                 -2.77                    1.40
                  :                    :                       :
              :+----+----+----+----+----+----+----+----+----+----+
    0.000     :                                                  *
    1.00      :                                             *
    2.00      :                                        *
    3.00      :                                    *
    4.00      :                                *
    5.00      :                            *
    6.00      :                        *
    7.00      :                    *
    8.00      :                *
    9.00      :            *
    10.0      :         *
    11.0      :      *
    12.0      :*
    13.0      :*
    14.0      :*
```

Listing

```
C
C                        SOL
C
C
C
C     THIS PROGRAM CALCULATES THE SOLUBILITY OF THE SALT M A
C                                                        N P
C
C         M A    <====>    NM(Y+) + PA(Z-)
C          N P
C
C     AS A FUNCTION OF PH.   THE ANION, A(Z-), HYDROLYZES TO FORM
C     Z CONJUGATE ACIDS, HA(1+Z-),H2A(2+Z-)....HZA
C
C     THE EXPRESSION FOR THE SOLUBILITY, S, IS:
C
C         S=(KSP/(N**N*P**P))**(1./(N+P))*
C
C     (1.+(H(+))/K(Z) + ...+ (H(+))**Z/(K(1)*K(2)...*K(Z)))**(P/(P+N))
C
C               WHERE N AND P ARE RESPECTIVELY THE NUMBER OF
C               ATOMS OF M AND A PRESENT IN THE MOLECULE M A
C                                                          N P
C
C
C               AND Y AND Z ARE RESPECTIVELY THE CHARGE OF THE
C               CATION AND THE ANION
C
```

```
C
C
C
C      INPUT:              Y--CHARGE OF THE CATION
C                          Z--CHARGE OF THE ANION   (Z.LE.6)
C                   PKSP--NEGATIVE LOG OF THE SOLUBILITY PRODUCT
C          PK(I)I=1,2...Z--NEGATIVE LOGS OF THE STEPWISE ACID
C                          DISSOCIATION CONSTANTS OF HZA
C
C
C      OUTPUT:  PH,S,LOG(S)
C
C
C      A PLOT OF -LOG(S) VS. PH IS AVAILABLE ON OPTION
C      THE PLOTTING SUBROUTINE, PLOT1, IS A MODIFICATION OF
C      THE SUBRPROGRAM PLOT.  PLOT1 IS DESIGNED
C      TO PLOT ONLY ONE DEPENDENT VARIABLE.
C
C        AUTHORS:  S. B. LEVITT AND K. J. JOHNSON
C
       DIMENSION K(6),PK(6),X(15),YY(15)
       REAL K,KSP,KPDT
       INTEGER  Z,Y,P
       WRITE (6,11)
       READ(5,13)Y,Z,PKSP,(PK(I),I=1,Z)
       IF(Y.EQ.0)STOP
       IF(Z.LT.0) Z=-Z
       IF(Z .GT. 6) GO TO 90
C
C      FIND THE STOICHIOMETRIC COEFFICIENTS, N AND P,
C      OF THE MOLECULAR FORMULA GIVEN THE
C      CHARGES, Y AND Z.   (L IS THE LARGEST
C      COMMON DIVISOR OF Y AND Z)
C
C
       L=1
       BIG=AMAX0(Y,Z)
       SMALL=AMIN0(Y,Z)
       IF(AMOD(BIG,SMALL).NE.0)GO TO 88
       L=BIG+0.0001
       IS=SMALL+0.0001
    8  IF(IS.EQ.0)GO TO 88
       IR=MOD(L,IS)
       L=IS
       IS=IR
       GO TO 8
   88  P=Y/L
       N=Z/L
       WRITE(6,15)N,P,N,Y,P,Z
C
C      DEFINE CONSTANTS
C
       DO 30 J=1,Z
       K(J)=10.**(-PK(J))
   30  WRITE(6,19)J,PK(J),J,K(J)
       PWR=1./FLOAT(P+N)
       IF(PKSP.GT.36)GO TO 52
       KSP=10.**(-PKSP)
       WRITE(6,21)PKSP,KSP
```

```
            FCTR=(KSP/FLOAT(P**P*N**N))**PWR
            GO TO 55
      52 FLOG=PWR*(-PKSP-ALOG10(FLOAT(P**P*N**N)))
            FCTR=10.**(FLOG)
            WRITE(6,23)PKSP
      55 PWR=FLOAT(P)*PWR
            WRITE(6,22)
C
C       MAIN LOOP (10) STEPS PH
C
            DO 10 I=0,14
            PH=I
            H=10.**(-PH)
            KPDT=1.0
            FACTR=1.0
            L=Z
C
C       INNER LOOP (20) CALCULATES THE FRACTION OF
C           TOTAL A PRESENT AS A(Z-)
C
            DO 20 II=1,Z
            FACTR=FACTR*H/K(L)
            KPDT=KPDT+FACTR
            L=L-1
      20 CONTINUE
C
C       CALCULATION OF SOLUBILITY (S) AND LOGARITHM OF S (SLOG)
C
            S=FCTR*KPDT**PWR
            SLOG=ALOG10(S)
            WRITE (6,18) PH,S,SLOG
            X(I+1)=I
            YY(I+1)=SLOG
      10 CONTINUE
C
C       IS A PLOT DESIRED?
C
            WRITE(6,66)
            READ(5,13)IPLOT
            IF(IPLOT.EQ.1)CALL PLOT1(X,YY,15)
            STOP
      90 WRITE(6,91)
            STOP
      11 FORMAT (/,22X,'SOLUBILITY OF M A ',/
         1 37X,'N',1X,'P',///,22X,'M A ',
         2 ' <===> NM(Y+)  +  PA(Z-)',/,23X,'N',1X,'P',//,22X,
         3 'AS A FUNCTION OF PH',
         4//' ENTER Y,Z,PKSP,(PK(I),I=1,Z)  (Y=0 ENDS PROGRAM)'/)
      13 FORMAT(2I,7E)
      15 FORMAT(//20X,'M(',I1,')A(',I1,
         1')  <===>  ',I1,'M(',I1,'+)  +  ',
         2I1,'A(',I1,'-)'//)
      18 FORMAT(F10.1,12X,1PE14.2,11X,G10.3)
      19 FORMAT(8X,'PK(',I1,') = ',F6.2,10X,'K(',I1,') = ',1PE10.2)
      21 FORMAT(8X,'PKSP  = ',F6.2,10X,'KSP  = ',1PE10.2,//)
      22 FORMAT (7X,'PH',22X,'S',16X,'LOG(S)'/)
      23 FORMAT(8X,'PKSP  = ',F6.2//)
      66 FORMAT(//' ENTER 1 FOR A PLOT OF LOG(S) VS. PH',
         1', 0 OTHERWISE'/)
      91 FORMAT(' SORRY, Z CANNOT EXCEED 6'/)
            END
C
```

```
C
C                       SUBROUTINE PLOT1
C
C
C       THIS SUBROUTINE PRINTS A TELETYPE OR LINE-PRINTER PLOT
C
C
C       USAGE:   CALL PLOT1(X,Y,N)
C
C       WHERE     X(N)      = INDEPENDENT VARIABLE
C                 Y(N)      = DEPENDENT VARIABLE
C                 N         = LENGTH OF X, Y
C
C
C       IT IS ASSUMED THAT X IS LINEAR, I.E., THAT
C       X(I+1)/X(I) IS CONSTANT,
C       AND THAT X IS SORTED, I.E., THAT X(I) .LT. X(I+1) OR THAT
C       X(I+1) .GT.X(I)
C
C       THE LENGTH AND WIDTH OF THE PLOT ARE ARBITRARILY SET
C          BY LENGTH = 50   AND WIDTH = 50
C
C        AUTHOR:   K. J. JOHNSON
C
        SUBROUTINE PLOT1(X,Y,N)
        DIMENSION X(15),Y(15),LINE(51)
        INTEGER BLANK,CHAR
        DATA LINE,BLANK,CHAR/52*' ',' ','*'/
        LENGTH=50
        WIDTH=50.
C
C       FIND MINIMUM AND MAXIMUM Y AND SET SCALING FACTORS
C
        YMIN=Y(1)
        YMAX=Y(1)
        DO 10 I=1,N
        IF(Y(I).GT.YMAX)YMAX=Y(I)
     10 IF(Y(I).LT.YMIN)YMIN=Y(I)
        YSCALE=WIDTH/(YMAX-YMIN)
        YMID=(YMAX+YMIN)/2.
        WRITE(6,11)YMIN,YMID,YMAX
C
C       PLOT ONE LINE AT A TIME
C
        DO 40 I=1,N
        NN=(Y(I)-YMIN)*YSCALE + 1.5
        LINE(NN)=CHAR
        WRITE(6,22) X(I),  (LINE(JJ),JJ=1,NN)
        LINE(NN)=BLANK
     40 CONTINUE
        RETURN
     11 FORMAT(//'     PH',30X,'LOG(S)',//,9X,G12.3,12X,G12.3,
        1  12X,G12.3,/14X':',24X,':',24X,':',/,
        2 13X,':+----+----+----+----+----+----+----+----+----+----+')
     22 FORMAT(G12.3,' :',51A1)
        END
```

Discussion

SOL and PLOT1 require 548 and 213 words of memory, respectively. This sample execution took approximately 0.2 sec of CPU time.

SOL contains an optimization feature worthy of note. The 20 loop, nested inside the 10 loop, calculates the following expression:

$$1 + \frac{[H^+]}{K_3} + \frac{[H^+]^2}{K_2 K_3} + \frac{[H^+]^3}{K_1 K_2 K_3}$$

The loop is written so that the factors $[H^+]/K_i$ are computed only once for a given pH.

Figure 2.2 was drawn using the output from SOL.

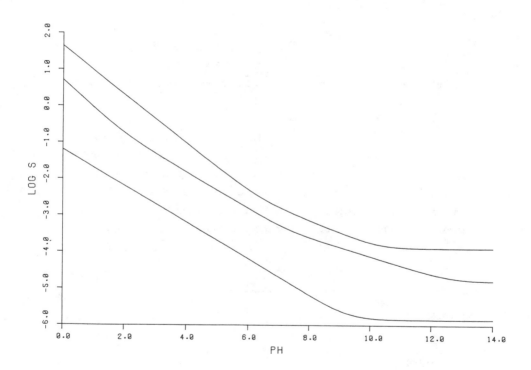

Figure 2.2 Solubility of Ag_2CO_3 (top: pK_{sp} = 11.09, pK_1 = 6.37, pK_2 = 10.25), Ag_3PO_4 (middle: pK_{sp} = 17.75, pK_1 = 2.12, pK_2 = 7.21, pK_3 = 12.67), and AgCN (bottom: pK_{sp} = 11.70, pK_a = 9.31).

It is clear that hydrolysis must be considered if the pH is less than 10 for these salts.

Problem 2.4   Use SOL to generate the data and construct a series of plots similar to Figure 2.2. Interpret the results.

Problem 2.5   Modify SOL so that the hydrolysis of the metal ion is also considered.

Problem 2.6   Write a program to calculate the solubility of a salt as a function of free NH_3 concentration. Here the metal ion forms a series of ammine complexes in solution.

Problem 2.7   Write a program to calculate the solubility of a salt as a function of free anion concentration, $[A^{z-}]$, demonstrating both the common-ion effect and the effect of complexation.

Problem 2.8   Write a program to calculate the solubility of a salt as a function of the ionic strength of the solution.

2.4   NERNST

Introduction

NERNST computes the potential of the following electrode as a function of pH:

$$M \mid M^{n+} \ (0.01 \ \underline{M}), \ NH_3 \ (1.0 \ \underline{M}), \ H^+ \ (x \ \underline{M})$$

The metal ion forms a series of complexes with ammonia. The pH is varied from 0 to 14. At each pH the concentration of free (uncomplexed) NH_3 is calculated, then the concentration of free M^{n+}, and finally the electrode potential using the Nernst equation. The input to the program includes the number of electrons transferred (n), the standard reduction potential, the coordination number of the metal ion, and the logarithms of the stepwise formation constants of the metal-ammine complexes.

Method

The Nernst equation for the electrode reaction

$$M^{n+} + ne^- \rightleftharpoons M$$

at 25°C is

$$E = E^o + \frac{0.0592}{n} \log [M^{n+}] \tag{2.2}$$

Let the coordination number of the metal ion be x. Then the following stepwise equilibria will be established.

$$M^{n+} + NH_3 \rightleftharpoons M(NH_3)^{n+} \qquad K_1$$

$$M(NH_3)^{n+} + NH_3 \rightleftharpoons M(NH_3)_2^{n+} \qquad K_2$$

$$\cdots\cdots\cdots\cdots\cdots\cdots\cdots$$

$$M(NH_3)_{x-1}^{n+} + NH_3 \rightleftharpoons M(NH_3)_x^{n+} \qquad K_x$$

The concentration of the free metal ion is a function of the free ammonia concentration,

$$C_m = [M^{n+}] + [M(NH_3)^{n+}] + [M(NH_3)_2^{n+}] + \ldots + [M(NH_3)_x^{n+}]$$
$$= [M^{n+}]Q$$

where

$$Q = 1 + \beta_1[NH_3] + \beta_2[NH_3]^2 + \ldots + \beta_x[NH_3]^x$$

and

$$\beta_1 = K_1, \ \beta_2 = K_1K_2, \text{ and } \beta_x = K_1K_2\ldots K_x.$$

So

$$[M^{n+}] = \frac{C_m}{Q} \tag{2.3}$$

Here C_m, the total M^{n+} concentration, is assigned the value 0.01 $\underline{M}$. The concentration of free ammonia is calculated as follows:

$$Ca = [NH_3] + [NH_4^+] + [M(NH_3)^{n+}]$$
$$+ 2[M(NH_3)_2^{n+}] + \ldots + x[M(NH_3)_x^{n+}]$$

If $C_a \gg C_m$, then to a first approximation,

$$C_a = [NH_3] + [NH_4^+]$$

Here $C_a = 1.0 \ \underline{M}$, $C_m = 0.01 \ \underline{M}$, and this approximation is made. The pH dependence of the $[NH_3]$ is due to the following equilibrium:

$$NH_4^+ \rightleftharpoons H^+ + NH_3 \qquad K_a = \frac{[H^+][NH_3]}{[NH_4^+]}$$

$$Ca = [NH_3] + [NH_4^+] = [NH_3](1 + \frac{[H^+]}{K_a})$$

Therefore,

$$[NH_3] = \frac{K_a C_a}{K_a + [H^+]} \tag{2.4}$$

NERNST varies the pH from 0 to 14, calculates the $[H^+]$ from the pH, calculates the $[NH_3]$ from Eq. (2.4), calculates the $[M^{n+}]$ using Eq. (2.3), and then calculates the electrode potential using Eq. (2.2).

Sample Execution [Hg^{2+}]

                              NERNST

ENTER THE NUMBER OF ELECTRONS TRANSFERRED, THE STANDARD
REDUCTION POTENTIAL, AND THE COORDINATION NUMBER OF HIGHEST COMPLEX
   (NUMBER OF ELECTRONS TRANSFERRED = 0 ENDS PROGRAM)
> 2 .851 4

 ENTER THE LOGS OF THE STEPWISE FORMATION CONSTANTS
 (LOG(Kl), LOG(K2), ... , LOG(KM)
> 8.8 8.7 1 .78

THE NUMBER OF ELECTRONS TRANSFERRED IS 2
THE STANDARD REDUCTION POTENTIAL IS 0.851 V.
THE COORDINATION NUMBER OF THE HIGHEST COMPLEX IS 4

LOG K(1) =   8.80   K(1) =   6.31E+08
LOG K(2) =   8.70   K(2) =   5.01E+08
LOG K(3) =   1.00   K(3) =   1.00E+01
LOG K(4) =   0.78   K(4) =   6.03E+00

PH	[NH3]	[M(2+)]	EMF
0.0	5.62E-10	6.88E-03	0.787
0.5	1.78E-09	3.21E-03	0.777
1.0	5.62E-09	6.88E-04	0.757
1.5	1.78E-08	8.92E-05	0.731
2.0	5.62E-08	9.66E-06	0.703
2.5	1.78E-07	9.90E-07	0.673
3.0	5.62E-07	9.98E-08	0.644
3.5	1.78E-06	1.00E-08	0.614
4.0	5.62E-06	1.00E-09	0.585
4.5	1.78E-05	1.00E-10	0.555
5.0	5.62E-05	1.00E-11	0.525
5.5	1.78E-04	1.00E-12	0.496
6.0	5.62E-04	9.97E-14	0.466
6.5	1.77E-03	9.87E-15	0.436
7.0	5.59E-03	9.57E-16	0.406
7.5	1.75E-02	8.69E-17	0.376
8.0	5.32E-02	6.56E-18	0.342
8.5	1.51E-01	3.58E-19	0.305
9.0	3.60E-01	1.97E-20	0.268
9.5	6.40E-01	2.41E-21	0.241
10.0	8.49E-01	8.29E-22	0.227
10.5	9.47E-01	5.47E-22	0.222
11.0	9.83E-01	4.75E-22	0.220
11.5	9.94E-01	4.53E-22	0.219
12.0	9.98E-01	4.47E-22	0.219
12.5	9.99E-01	4.45E-22	0.219
13.0	1.00E+00	4.44E-22	0.219
13.5	1.00E+00	4.44E-22	0.219
14.0	1.00E+00	4.44E-22	0.219

ENTER 1 FOR A PLOT OF EMF VS. PH, 0 OTHERWISE
>1

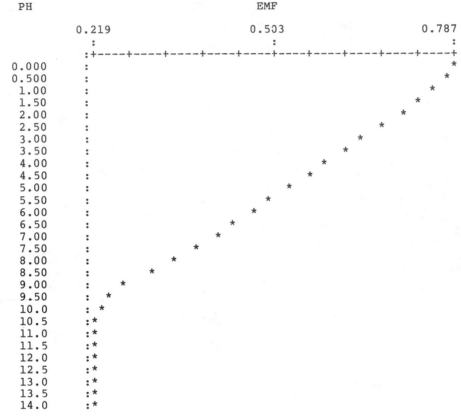

```
    PH                                    EMF

          0.219                0.503                0.787
            :                    :                    :
            :+----+----+----+----+----+----+----+----+----+----+
  0.000     :                                                  *
  0.500     :                                                *
  1.00      :                                              *
  1.50      :                                            *
  2.00      :                                          *
  2.50      :                                       *
  3.00      :                                    *
  3.50      :                                  *
  4.00      :                               *
  4.50      :                             *
  5.00      :                          *
  5.50      :                        *
  6.00      :                      *
  6.50      :                   *
  7.00      :                 *
  7.50      :              *
  8.00      :            *
  8.50      :         *
  9.00      :       *
  9.50      :     *
 10.0       :   *
 10.5       :*
 11.0       :*
 11.5       :*
 12.0       :*
 12.5       :*
 13.0       :*
 13.5       :*
 14.0       :*
```

Listing

```
C                               NERNST
C
C       PROGRAM TO SIMULATE THE POTENTIAL OF A M/M(N+) ELECTRODE
C       IN AN AQUEOUS AMMONIA SOLUTION AS A FUNCTION OF PH.
C
C
C       ASSUMPTIONS
C
C           (1)   THE TOTAL CONCENTRATION OF M(N+) = 0.01 M.
C           (2)   THE TOTAL CONCENTRATION OF NH(3) = 1.00 M.
C           (3)   THE CONCENTRATION OF COMPLEXED NH(3) CAN BE NEGLECTED
C           (4)   THE M/M(N+) ELECTRODE IS REVERSIBLE
C           (5)   IDEAL SOLUTION BEHAVIOR
C           (6)    NH4(+)   <====>  NH3  +   H(+)      K = 5.62E-10
C           (7)   ALL COMPETING EQUILIBRIA CAN BE IGNORED
C
```

```
C       GLOSSARY
C
C          N       THE NUMBER OF ELECTRONS TRANSFERRED
C          E       THE STANDARD REDUCTION POTENTIAL OF M(N+)
C          M       THE COORDINATION NUMBER OF THE HIGHEST COMPLEX
C          NH3     THE FREE AMMONIA CONCENTRATION IN THE SOLUTION
C          METAL   THE METAL ION CONCENTRATION (UNCOMPLEXED)
C          EMF     THE EMF OF THE CELL AT THE SPECIFIED PH
C
C       A PLOT IS AVAILABLE ON OPTION
C
C     AUTHORS:  D. L. DOERFLER AND K. J. JOHNSON
C
        REAL NH3, NNH3,METAL,LOGK(9)
        DIMENSION X(29),Y(29)
C
C        INPUT
C
    10 WRITE(6,11)
        READ(5,24) N,E,M
        IF(N.LE.0)STOP
        IF(M.LE.0 .OR. M.GE.9)GO TO 200
        WRITE(6,22)
        READ(5,25) (LOGK(I),I=1,M)
C
C     WRITE OUT INPUT
C
        WRITE(6,30) N,E,M
C
C     CHECK TO SEE IF ANY LOG K'S HAVE BEEN LEFT OUT.
C     IF SO, WRITE WARNING MESSAGE
C
        DO 40 I=1,M
        IF(LOGK(I) .EQ. 0.0) WRITE(6,38)I
C
C     TAKE ANTILOG OF LOG K TO GET K
C
        SUM=LOGK(I)
        LOGK(I)=10.0**SUM
    40 WRITE(6,43) I,SUM,I,LOGK(I)
        WRITE(6,45)
C
C     LOOP THROUGH 29 VALUES OF PH        0<PH<14
C
        CONST=0.0592/N
        DO 72 I=1,29
        PH=(I-1)/2.0
C
C     FIND CONC OF NH3
C
C     ASSUME TOTAL CONC OF NH3 = CONC OF FREE NH3 + NH4(+)
C
        NH3=5.62E-10/(5.62E-10+10.0**(-PH))
        PKN=1.0
        SUM=1.
        DO 70 J=1,M
        PKN=LOGK(J)*NH3*PKN
    70 SUM=SUM+PKN
C
```

```
C      FIND THE CONC OF FREE M(N+)
C
       METAL=.01/SUM
C
C      USE NERNST EQUATION TO FIND HALF CELL POTENTIAL
C
       EMF=E+CONST*ALOG10(METAL)
       X(I)=PH
       Y(I)=EMF
   72 WRITE(6,75) PH,NH3,METAL,EMF
C
C   PLOT EMF VS PH ON OPTION
C
       WRITE(6,66)
       READ(5,24)IPLOT
       IF(IPLOT.EQ.1)CALL PLOT1(X,Y,29)
       STOP
C
C      ERROR MESAGES
C
  200 WRITE(6,201) M
       GO TO 10
       STOP
   11 FORMAT(//25X,'NERNST',//'  ENTER THE NUMBER OF ELECTRONS',
      .' TRANSFERRED, THE STANDARD',/,'  REDUCTION POTENTIAL, ',
      .'AND THE COORDINATION NUMBER OF HIGHEST COMPLEX  '
      .,/,5X,'( NUMBER OF ELECTRONS TRANSFERRED = 0 ENDS PROGRAM)'/)
   22 FORMAT(/'  ENTER THE LOGS OF THE STEPWISE FORMATION CONSTANTS ',
      1/'  ( LOG(K1), LOG(K2), ... , LOG(KM)'/)
   24 FORMAT(I,F,I)
   25 FORMAT(9F)
   30 FORMAT(//,' THE NUMBER OF ELECTRONS TRANSFERRED IS',I2,/,
      .' THE STANDARD REDUCTION POTENTIAL IS', F6.3,' V.',/,
      .' THE COORDINATION NUMBER OF THE HIGHEST COMPLEX IS',I2,/)
   38 FORMAT(//,' WARNING - LOG K(',I1,') IS ZERO')
   43 FORMAT(' LOG K(',I1,') = ',0PF5.2,12X,'K(',I1,') = ',1PE10.2)
   45 FORMAT(//,2X,'PH',9X,'<NH3>',9X,'<M(',I2,'+)>',9X,'EMF',/)
   66 FORMAT(//'  ENTER 1 FOR A PLOT OF EMF VS. PH, 0 OTHERWISE')
   75 FORMAT(0PF5.1,4X,2(1PE11.2,4X),G12.3)
  201 FORMAT(' A COMPLEX COORDINATION NUMBER OF ',I3,' IS NOT
      . APPROPRIATE FOR THIS PROGRAM',/,' STARTING OVER')
       END
```

Discussion

   NERNST and PLOT1 require 513 and 212 words of core, respectively. This sample
execution took approximately 0.3 sec of CPU time.

   The 70 loop, nested inside the 72 loop, calculates the following expression
without redundant multiplications.

$$1 + K_1[NH_3] + K_1K_2[NH_3]^2 + \cdots + K_1K_2\cdots K_x[NH_3]^x$$

The potential-pH plots for Zn^{2+}, Cd^{2+}, and Hg^{2+} are shown in Figure 2.3.

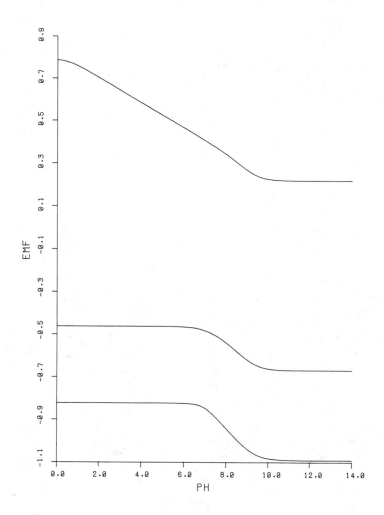

Figure 2.3 Potential-pH plots for Hg^{2+} (top), Cd^{2+}, and Zn^{2+} (bottom) in aqueous ammonia solution.

Problem 2.9 Use NERNST to prepare a series of plots similar to Figure 2.3, and interpret the results.

Problem 2.10 Modify NERNST so that C_m and C_a are input variables. Estimate the error associated with the approximation $C_a \gg C_m$.

Problem 2.11 Modify NERNST so that some of the competing equilibria are considered.

2.5 RADIAL

Introduction

RADIAL calculates and plots normalized radial distribution functions for the
1s, 2s, 2p, 3s, 3p, and 3d hydrogen-like atomic orbitals. The user enters an
integer indicating which one of these six orbitals is desired, the effective nuclear
charge, and the maximum value of the radial distance r in angstroms ($\overset{o}{A}$). RADIAL
prints the input data for verification purposes, a table containing r and the
normalized radial distribution function $r^2 R(r)^2$, and a teletype plot of $r^2 R(r)^2$
versus r.

Method

The following radial wavefunctions are used [3]:

Orbital	R(r)	Orbital	R(r)
1s	$ce^{-\rho}$	2p	$c\rho e^{-\rho/2}$
2s	$c(2 - \rho)e^{-\rho/2}$	3p	$c\rho(6 - \rho)e^{-\rho/3}$
3s	$c(27 - 18\rho + 2\rho^2)e^{-\rho/3}$	3d	$c\rho^2 e^{-\rho/3}$

Here c is a normalizing constant, $\rho = Zr/a_o$, Z is the effective nuclear charge,
and a_o = 0.529 $\overset{o}{A}$. The normalizing constants are chosen so that the maximum value
of $r^2 R(r)^2$ is 1.0.

The effective nuclear charge can be computed using Slater's rules [4,5].
Table 2.2 gives the effective nuclear charges for several elements in the first
three periods calculated by Slater's rules.

Table 2.2 Effective Nuclear Charges

Atom	1s	2s,2p	Atom	2s,2p	3s,3p	Atom	3s,3p	3d
H	1.0							
He	1.7							
Li	2.7	1.3	Na	6.85	2.2	K	7.75	
Be	3.7	1.95	Mg	7.85	2.85	Ca	8.75	
B	4.7	2.6	Al	8.85	3.5	Sc	9.75	3.0
C	5.7	3.25	Si	9.85	4.15	Ti	10.75	3.65
N	7.7	4.55	S	11.85	5.45	Cr	12.75	5.6
F	8.7	5.2	Cl	12.85	6.1	Fe	14.75	6.25
Ne	9.7	5.85	Ar	13.85	6.75	Br	23.75	13.85

<u>Sample Execution</u> (Cl, 3s)

                              RADIAL

ENTER THE NUMBER CORRESPONDING TO THE DESIRED ORBITAL, THE EFFECTIVE
NUCLEAR CHARGE, AND THE MAXIMUM VALUE OF R IN ANGSTROMS

NUMBER	ORBITAL	NUMBER	ORBITAL
1	1S	4	3S
2	2S	5	3P
3	2P	6	3D

                    (0   TERMINATES EXECUTION)

>4   6.1   2.5

          EFF. NUCLEAR CHARGE = 6.10   MAX. R =   2.50 ANGSTROMS

RADIAL DISTRIBUTION FUNCTION   (R**2)*(R(R)**2) FOR THE 3S ORBITAL:


```
      R          RDF      0.0                 0.5                 1.0
      -          ---      !+----+----+----+----+----+----+----+----+
   0.000       0.000      !*
   0.100       0.978E-01  !    *
   0.200       0.344E-01  !  *
   0.300       0.307      !              *
   0.400       0.357      !                *
   0.500       0.151      !         *
   0.600       0.308E-02  !*
   0.700       0.867E-01  !    *
   0.800       0.353      !                *
   0.900       0.659      !                          *
   1.00        0.890      !                                 *
   1.10        0.993      !                                      *
   1.20        0.977      !                                      *
   1.30        0.877      !                                 *
   1.40        0.733      !                            *
   1.50        0.580      !                       *
   1.60        0.438      !                  *
   1.70        0.318      !              *
   1.80        0.224      !          *
   1.90        0.153      !        *
   2.00        0.102      !       *
   2.10        0.668E-01  !    *
   2.20        0.428E-01  !  *
   2.30        0.270E-01  !  *
   2.40        0.167E-01  !*
   2.50        0.103E-01  !*
```

Listing

```
C                              RADIAL
C
C
C              PROGRAM TO CALCULATE AND PLOT RADIAL DISTRIBUTION
C              FUNCTIONS FOR THE 1S, 2S, 2P, 3S, 3P, AND 3D
C              HYDROGEN-LIKE ATOMIC ORBITALS.
C
C          GLOSSARY
C
C
C      R(25)      - DISTANCE FROM THE NUCLEUS IN ANGSTROMS.
C      RDF(25)    - THE RADIAL DISTRIBUTION FUNCTION IS NORMALIZED
C                   BY THE CONSTANT C SO THAT THE MAX. VALUE IS 1.0
C      ORBITL(6) - NAME OF THE ATOMIC ORBITAL USED.
C      LINE(41)   - PLOTTING ARRAY
C
C
C       INPUT:
C
C
C         NUMBER     -   DESIGNATES WHICH RADIAL DISTRIBUTION
C                        FUNCTION IS TO BE CALCULATED AND PLOTTED:
C
C                  NUMBER :   1   2   3   4   5   6   0
C                  --------:----------------------------------
C                  ORBITAL:   1S  2S  2P  3S  3P  3D  (ENDS PROG.)
C
C         Z         -    EFFECTIVE NUCLEAR CHARGE
C
C         RANGE     -    MAXIMUM VALUE OF R IN ANGSTROMS.
C
C    AUTHORS:  M. FELICE, D. E. HAWKINS,JR., AND K. J. JOHNSON
C
C
      DIMENSION RDF(25),R(25),ORBITL(6),LINE(41)
      INTEGER STAR,BLANK
      DATA ORBITL/'1S','2S','2P','3S','3P','3D'/,STAR/'*'/,BLANK/' '/,
     1LINE/41*' '/,ZERO/0.0/
      WRITE(6,200)
      READ(5,1) NUMBER,Z,RANGE
      IF(NUMBER.EQ.0 .OR. NUMBER .GE.7)STOP
      WRITE(6,400) Z,RANGE
      C1=Z/0.529
      C2=2*C1
      C3=-C2/3
      RANGE=RANGE/25.0
      WRITE(6,300) ORBITL(NUMBER)
      GO TO(21,22,24,23,25,26),NUMBER
C
C    *********    1S    *********
C
C    C=Z**2 * EXP(2) / .529**2
C
   21 C=Z**2*26.40448
      DO 11 I=1,25
      R(I)=FLOAT(I)*RANGE
   11 RDF(I)=C*R(I)**2*EXP(-C2*R(I))
      GO TO 100
C
```

```
C        *********    2S     *********
C
C   C=Z**2 * EXP(3+SQRT(5))  /(.529 * (3+SQRT(5)) * (-1-SQRT(5)) )**2
C
   22 C=Z**2*2.339043
      DO 12 I = 1,25
      R(I)=FLOAT(I)*RANGE
   12 RDF(I)=C*(R(I)*(2-R(I)*Cl))**2*EXP(-R(I)*Cl)
      GO TO 100
C
C        *********    3S     *********
C
C        D = 13.07403
C
C        C = Z**2 * EXP(2*D/3) / ((27-18*D + 2*D**2) * D*.529)**2
C
   23 C=Z*Z*7.15228E-3
      DO 13 I=1,25
      R(I)=FLOAT(I)*RANGE
   13 RDF(I)=C*(((C2*R(I)-18)*R(I)*Cl+27)*R(I))**2*EXP(R(I)*C3)
      GO TO 100
C
C        *********    2P     *********
C
C        C=Z**4*EXP(4.)/256./.529**4
C
   24 C=Z**4*2.723424
      DO 14 I=1,25
      R(I)=FLOAT(I)*RANGE
   14 RDF(I)=C*R(I)**4*EXP(-R(I)*Cl)
      GO TO 100
C
C        *********    3P     *********
C
C        C = Z**2 * EXP(8) / (72**2 * 144 * .529**2)
C
   25 C=Z**2*.01426977
      DO 15 I=1,25
      R(I)=FLOAT(I)*RANGE
   15 RDF(I)=C*((-R(I)*Cl+6)*R(I)**2*Cl)**2*EXP(C3*R(I))
      GO TO 100
C
C        *********    3D     *********
C
C        C=Z**6*EXP(6.)/9.**6/.529**6
C
   26 C=Z**6*.03464
      DO 16 I=1,25
      R(I)=FLOAT(I)*RANGE
   16 RDF(I)=C*R(I)**6*EXP(C3*R(I))
C
C        PLOT RDF VERSUS R
C
  100 WRITE(6,84) ZERO,ZERO
      DO 44 I=1,25
      N1=RDF(I)*40+1
      LINE(N1)=STAR
      WRITE(6,85) R(I),RDF(I),(LINE(K),K=1,N1)
   44 LINE(N1)=BLANK
```

```
      STOP
    1 FORMAT(I,2F)
   84 FORMAT(//5X,'R',11X,'RDF',5X,'0.0',17X,'0.5',17X,
     1'1.0',/5X,'-',11X,'---',5X,'!+----+----',
     2'+----+----+----+----+----+----+',/1X,2G11.3,2X,'!*')
   85 FORMAT(1X,2G11.3,2X,'!',41A1)
  200 FORMAT(//25X,'RADIAL'//,'  ENTER THE NUMBER CORRESPONDING TO
     1 THE DESIRED ORBITAL, THE EFFECTIVE'/'  NUCLEAR CHARGE, AND
     2 THE MAXIMUM VALUE OF R IN ANGSTROMS '//
     3 1X,T10,'NUMBER',T25,'ORBITAL',
     4T40,'NUMBER',T55,'ORBITAL'//1X,T12,'1',T27,'1S',T42,'4',T57,
     5'3S'/1X,T12,'2',T27,'2S',T42,'5',T57,'3P'/1X,T12,'3',T27,'2P',
     6T42,'6',T57,'3D'/1X,T25,'(0  TERMINATES EXECUTION)'//)
  300 FORMAT(/1X,' RADIAL DISTRIBUTION FUNCTION  (R**2)*(R(R)**2)
     1 FOR THE ',A2,' ORBITAL:')
  400 FORMAT(10X,'EFF. NUCLEAR CHARGE =',F5.2,5X
     1,'MAX. R =',F6.2,' ANGSTROMS'/)
      END
```

Discussion

RADIAL requires 516 words of memory and executes in approximately 0.2 sec of CPU time. Figure 2.4 was drawn using data provided by RADIAL.

Problem 2.12 Extend RADIAL so that the orbitals with principal quantum numbers 4, 5, and 6 are included. Prepare plots similar to Figure 2.4 to illustrate the features of these radial functions.

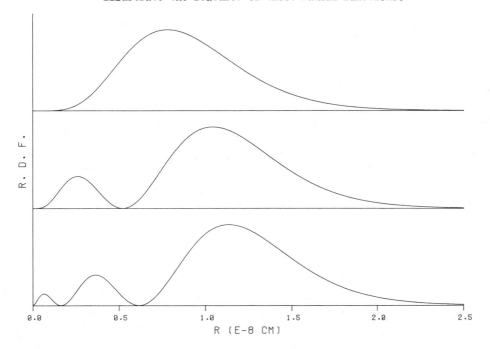

Figure 2.4 The 3s (bottom), 3p, and 3d (top) radial distribution functions for the chlorine atom.

2.6 CONTOUR

Introduction

CONTOUR produces teletype or lineprinter plots of electron density contours for a set of atomic, hybrid, and molecular orbitals. The following table indicates which orbitals have been included in CONTOUR.

Key	Atomic	Hybrid	Molecular
1	1s	sp	1s $\pm$ 1s
2	2s	sp^2	2s $\pm$ 2s
3	3s	sp^3	2px $\pm$ 2px (π)
4	4s	dsp^2	2py $\pm$ 2py (σ)
5	$2p_x$	d^2sp^3	2s $\pm$ 2py (σ)
6	$3p_x$		
7	$4p_x$		
8	$3d_{z^2}$		
9	$3d_{x^2-y^2}$		

The input includes the following: a key which designates the type of orbital (atomic, hybrid, or molecular); the effective nuclear charge(s) calculated using Slater's rules (Section 2.5); and, for molecular orbitals, the internuclear distance in angstroms.

The program calculates the relative electron density, ψ^2/ψ^2_{max}, as a function of position in the x-y plane (the x-z plane for the $3d_{z^2}$ orbital), and plots the resulting contours using a set of typewriter symbols to represent ranges in relative electron density. The y axis is the molecular axis.

Method

CONTOUR uses one-electron, hydrogen-like wavefunctions [3]. Slater's rules are used to calculate the effective nuclear charges [4,5]. The procedures for generating the atomic, hybrid, and molecular orbital contour plots are considered separately.

1. _Atomic orbitals._ The atomic wavefunctions have the form

$$\psi(n,l,m) = f(Z,r,\theta,\phi)$$

Here n, l, and m are the quantum numbers designating the atomic orbital; Z is the effective nuclear charge; and r, θ, and ϕ are the spherical polar coordinates. The atomic functions used by CONTOUR are given in Table 2.3. Here $\rho = Zr/a_o$, $a_o = 0.529$ Å, and c is a normalizing constant.

All atomic orbitals except $3d_{z^2}$ are plotted in the x-y plane, for which $\theta = 90°$. The $3d_{z^2}$ orbital is plotted in the x-z plane.

Table 2.3 Atomic Wavefunctions

Atomic Orbital	n	ℓ	m	Atomic Wavefunction
1s	1	0	0	$ce^{-\rho}$
2s	2	0	0	$ce^{-\rho/2}(2 - \rho)$
3s	3	0	0	$ce^{-\rho/3}(27 - 18\rho + 2\rho^2)$
4s	4	0	0	$ce^{-\rho/4}(192 - 144\rho + 24\rho^2 - \rho^3)$
$2p_x$	2	1	1	$ce^{-\rho/2} \sin \theta \cos \phi$
$2p_y$	2	1	-1	$ce^{-\rho/2} \sin \theta \sin \phi$
$2p_z$	2	1	0	$ce^{-\rho/2}\rho \cos \theta$
$3p_x$	3	1	1	$ce^{-\rho/3} (6 - \rho) \sin \theta \cos \phi$
$4p_x$	4	1	1	$ce^{-\rho/4}(80 - 20\rho + \rho^2) \sin \theta \cos \phi$
$3d_{z^2}$	3	2	0	$ce^{-\rho/3}\rho^2(3 \cos^2 \theta - 1)$
$3d_{x^2-y^2}$	3	2	-2	$ce^{-\rho/3}\rho^2 \sin^2 \theta \cos 2\phi$

The coordinates in the x-y plane vary over the following ranges:

$-2.4 \leqslant x \leqslant 2.4$ (step size = 0.08)

$-2.67 \leqslant y \leqslant 2.67$ (step size = 0.133)

All distances are in angstroms (Å). The increased spacing in the y direction is necessary to compensate for the nonuniform vertical and horizontal spacing of the teletype or line printer. The program varies x and y systematically over these ranges, and applies the following transformation equations (except for the $3d_{z^2}$ orbital):

$$r = (x^2 + y^2)^{1/2} \qquad\qquad \sin \theta = 1$$

$$\cos \phi = \frac{x}{r} \qquad\qquad\qquad \cos \theta = 0$$

$$\sin \phi = \frac{y}{r} \qquad\qquad\qquad \cos 2\phi = 2 \cos^2\phi - 1$$

The values for $\psi(n,\ell,m)$ and $\psi^2(n,\ell,m)$ are calculated as a function of x and y. The program utilizes the fourfold symmetry of the atomic orbitals.

The subroutine ATORB calculates ψ over the ranges

$-2.4 \leqslant x \leqslant 0$ and $-2.67 \leqslant y \leqslant 0$ (lower left quadrant)

These values are then replaced by ψ^2/ψ^2_{max}, where ψ_{max} is the maximum value calculated for $\psi(x,y)$. The corresponding values in the other three quadrants are assigned by symmetry,

$$PP(-x,-y) = PP(-x,y) = PP(x,-y) = PP(x,y)$$

where PP represents ψ^2/ψ^2_{max}.

The subroutine SYMBOL prints the following symbols to represent ranges in relative electron density (PP):

Range of PP	Symbol
$PP < 0.01$	(blank)
$0.01 \leq PP < 0.02$	: (colon)
$0.02 \leq PP < 0.10$	/ (slash)
$0.10 \leq PP < 0.25$	0 (zero)
$0.25 \leq PP < 0.50$	& (ampersand)
$0.50 \leq PP < 1.0$	# (pound)

2. <u>Hybrid orbitals</u>. The contour plots for the five hybrid orbitals are calculated using the following hybridization schemes [6, 7]:

Hybrid	Hybridization Scheme
sp	$(2s + 2p_x)/\sqrt{2}$
sp^2	$[2s + \sqrt{2}(2p_x)]/\sqrt{3}$
sp^3	$[2s + \sqrt{3}(2p_x)]/\sqrt{2}$
dsp^2	$0.5(3d_{x^2-y^2}) + 0.5(4s) + (4p_x + 4p_y)/\sqrt{2}$
d^2sp^3	$0.5(3d_{x^2-y^2}) + 3d_{z^2}/\sqrt{12} + 4s/\sqrt{6} + (4p_x + 4p_y)/\sqrt{2}$

The procedure for calculating ψ^2/ψ^2_{max} for these hybrid orbitals is the same as that described above for the atomic orbitals. The hybrid orbitals are plotted in the x-y plane, with x the major axis. The twofold symmetry about the line y = 0 is utilized.

3. <u>Molecular orbitals</u>. CONTOUR calculates 10 molecular orbitals, 5 bonding and 5 antibonding, for the diatomic molecule A-B. The following linear combinations of atomic orbitals are used:

Bonding	Antibonding
$\sigma_{1s} = 1s_a + 1s_b$	$\sigma^*_{1s} = 1s_a - 1s_b$
$\sigma_{2s} = 2s_a + 2s_b$	$\sigma^*_{2s} = 2s_a - 2s_b$
$\sigma_{2p_y} = 2p_{ya} - 2p_{yb}$	$\sigma^*_{2p_y} = 2p_{ya} + 2p_{yb}$
$\pi_{2p_x} = 2p_{xa} + 2p_{xb}$	$\pi^*_{2p_x} = 2p_{xa} - 2p_{xb}$
$\sigma_{sp} = 2s_a + 2p_{yb}$	$\sigma^*_{sp} = 2s_a - 2p_{yb}$

The positions of the nuclei (a and b) are along the line x = 0 and at the positions $-R_{ab}/2$ and $R_{ab}/2$, respectively. The molecular orbitals have fourfold symmetry if $Z_a = Z_b$, otherwise they have twofold symmetry about the line x = 0.

CONTOUR consists of the following routines:

1. Main program: read input and control the execution,
2. FPSI: evaluate the atomic wavefunctions,
3. SYMBOL: convert numeric values to symbols and print one line at a time,
4. ATORB, HYORB, MOLORB: calculate ψ^2 for atomic, hybrid, and molecular orbitals.

Sample Executions

1. Atomic orbital.

                              CONTOUR

ENTER AN INTEGER BETWEEN 1 AND 4:

    INTEGER: 1   2   3   4

    ORBITAL: ATOMIC   HYBRID   MOLECULAR   STOP
>1

 ENTER AN INTEGER BETWEEN 1 AND 9:

INTEGER: 1   2   3   4   5   6   7   8   9

ATOMIC
ORBITAL: 1S   2S   3S   4S   2PX 3PX 4PX 3DZ2   3DX2-Y2
>6

 ENTER THE EFFECTIVE NUCLEAR CHARGE
>6.1

THE 3PX   ATOMIC ORBITAL   Z =   6.10

                                              X ----->

        -2.40   -1.60   -0.80   0.00   0.80   1.60   2.40
          +---------+---------+---------+---------+---------+---------+
 2.67   !
 2.53   !
 2.40   !
 2.27   !
 2.13   !
 2.00   !
 1.87   !
 1.73   !
 1.60   !
 1.47   !

```
 1.33  !
 1.20  !
 1.07  !                    ::::::::              ::::::::
 0.93  !                 :::////////::         ::://////:::
 0.80  !                ::://////////:        :////////////::
 0.67  !               ::://///////////:      :////////////::
 0.53  !              :://///////////        ////////////::
 0.40  !             ::///////////         ////////////::
 0.27  !             :://////000///  //0/ /0//  //000/////::
 0.13  !             :://////000//: /0&#& &#&0/ ://000/////::
 0.00  !             ://////0000//   0&### ###&0  //0000/////:
-0.13  !             :://////000//: /0&#& &#&0/ ://000//////::
-0.27  !             :://////000///  //0/ /0//  //000/////::
-0.40  !             ::///////////         ////////////::
-0.53  !              :://///////////        ////////////::
-0.67  !               ::://///////:       :////////////::
-0.80  !                :::////////:        :////////////:::
-0.93  !                 :::///////::         ::://////:::
-1.07  !                    ::::::::              ::::::::
-1.20  !
-1.33  !
-1.47  !
-1.60  !
-1.73  !
-1.87  !
-2.00  !
-2.13  !
-2.27  !
-2.40  !
-2.53  !
-2.67  !
        +---------+---------+---------+---------+---------+---------+
```

2. Hybrid orbital.

                          CONTOUR

ENTER AN INTEGER BETWEEN 1 AND 4:

    INTEGER: 1   2   3   4

    ORBITAL: ATOMIC   HYBRID MOLECULAR STOP
>2

ENTER AN INTEGER BETWEEN 1 AND 5:

    INTEGER: 1   2   3   4   5
    HYBRID
   ORBITAL: SP   SP2 SP3 DSP2 D2SP3
>5

ENTER THE EFFECTIVE NUCLEAR CHARGE
>6.25

THE D2SP3 HYBRID ORBITAL   Z = 6.25


```
                                                    X ----->

            -2.40       -1.60       -0.80        0.00        0.80        1.60        2.40
             +---------+---------+---------+---------+---------+---------+---------+
    2.67  !
    2.53  !
    2.40  !
    2.27  !
    2.13  !
    2.00  !
    1.87  !                                                    ::::
    1.73  !                                                 ::::::::::::
    1.60  !                                               :::::::::::::::::
    1.47  !                                               :::::::::::::::::
    1.33  !                                               ::::::::::::::::
    1.20  !                                                :::::::::::::
    1.07  !                                                 :::::::::::::
    0.93  !            ::::             ::::               ::::://:::::::
    0.80  !         ::::::::::        ://///:      ::/:::    :::://///::::
    0.67  !        :::::::::::::     //////:     ://///////:   :::///::::::
    0.53  !       :::::://////////::   ://////  //00000//:    :::///::::
    0.40  !      :::::://////////////:   ://00/: /00&&000//:   :::///::::
    0.27  !      :::://////////////////: /000/ /0&&&&&00//:   :::///::::
    0.13  !      :::://///////////////0000&&00 00/ /&&##&&00/:   :::///::::
    0.00  !      :::://///////////////0000&&####/0/ /&&###&&00//   :::///::::
   -0.13  !      :::://///////////////0000&&00 00/ /&&##&&00/:   :::///::::
   -0.27  !      ::::://////////////: /000/ /0&&&&&00//:   :::///::::
   -0.40  !      :::::://////////////:  ://00/: /00&&000//:   :::///::::
   -0.53  !       :::::://////////::   ://////  //00000//:    :::///::::
   -0.67  !        :::::::::::::     //////:     ://///////:   :::///::::::
   -0.80  !         ::::::::::        ://///:      ::/:::    :::://///::::
   -0.93  !            ::::             ::::               ::::://:::::::
   -1.07  !                                                 ::::::::::::::
   -1.20  !                                                :::::::::::::
   -1.33  !                                                ::::::::::::::::
   -1.47  !                                               :::::::::::::::::
   -1.60  !                                               :::::::::::::::::
   -1.73  !                                                 ::::::::::::
   -1.87  !                                                    ::::
   -2.00  !
   -2.13  !
   -2.27  !
   -2.40  !
   -2.53  !
   -2.67  !
             +---------+---------+---------+---------+---------+---------+---------+
```

3. Molecular orbital.

CONTOUR

ENTER AN INTEGER BETWEEN 1 AND 4:

    INTEGER: 1   2   3   4

     ORBITAL: ATOMIC   HYBRID   MOLECULAR   STOP
>3

ENTER AN INTEGER BETWEEN 1 AND 5:

    INTEGER: 1   2   3   4   5

    MOLECULAR 1S   2S   2PX   2PY   2S
    ORBITAL: +/-   +/-   +/-   +/-   +/-
               2S   2S   2PX   2PY   2PY

         NOTE: Y IS THE MOLECULAR AXIS
>4

ENTER THE INTERNUCLEAR SEPARATION IN ANGSTROMS,
  AND THE TWO EFFECTIVE NUCLEAR CHARGES
>1.35 3.25 3.25

ENTER +1 FOR THE BONDING ORBITAL OR
      -1 FOR THE ANTIBONDING ORBITAL
>1

THE INTERNUCLEAR SEPARATION IS   1.35
ZA = 3.25   AND ZB = 3.25

2PYA - 2PYB

```
                                      X ----->

          -2.40      -1.60      -0.80       0.00       0.80       1.60       2.40
            +---------+---------+---------+---------+---------+---------+
   2.67  !
   2.53  !
   2.40  !
   2.27  !
   2.13  !
   2.00  !                             :::::::::::::
   1.87  !                           :::://///////////:::
   1.73  !                          :::////////////////::
   1.60  !                         :::://////////////////:::
   1.47  !                         :://///000000000/////::
   1.33  !                         ://///00000000000/////:
   1.20  !                         ://///0000&&&&0000/////:
   1.07  !                         ::////00&&&&&&&00////:
   0.93  !                          :::///00&&&&&00///:::
   0.80  !                            ://///000///:
   0.67  !                            ::::://///::::
   0.53  !                        ::://///00&&&#&&&00/////:::
   0.40  !                     :::////000&&#########&&000////::
   0.27  !                     :////000&&##########&&000////:
   0.13  !                   ::://///00&&&###########&&&00////::
   0.00  !                   ://///000&&&############&&000////:
  -0.13  !                   :::////00&&&##########&&&00////:::
  -0.27  !                     :////000&&&##########&&000////:
  -0.40  !                     :::////000&&########&&000////::
  -0.53  !                        ::://////00&&&#&&&00/////::
  -0.67  !                            ::::://///::::
  -0.80  !                            ://///000///:
  -0.93  !                          :::///00&&&&&00///::
  -1.07  !                         ://///00&&&&&&&00////:
  -1.20  !                         ://///0000&&&&0000////:
  -1.33  !                         :://///00000000000/////:
  -1.47  !                         ::://///000000000/////::
  -1.60  !                         :::://////////////////:::
  -1.73  !                          :::////////////////::
  -1.87  !                           :::://///////////:::
  -2.00  !                             :::::::::::::
  -2.13  !
  -2.27  !
  -2.40  !
  -2.53  !
  -2.67  !
            +---------+---------+---------+---------+---------+---------+
```

Listing

```
C
C                                     CONTOUR
C
C       FUNCTION OF PROGRAM: TO CALCULATE AND PLOT ELECTRON DENSITY
C                            DISTRIBUTIONS FOR ATOMIC, HYBRID, AND
C                            MOLECULAR ORBITALS
C
C          I/O: THE INPUT INCLUDES THE FOLLOWING KEYS TO THE DESIRED ORBITALS,
C
```

```
C         ORBITAL TYPE :  ATOMIC      HYBRID      MOLECULAR
C                  KEY :    1           2             3
C
C     KEY DESIGNATING SPECIFIC ORBITAL
C
C     KEY       ATOMIC       HYBRID       MOLECULAR
C
C      1        1S           SP          1S +/- 1S
C      2        2S           SP2         2S +/1 2S
C      3        3S           SP3         2PX +/- SPX  (PI)
C      4        4S           DSP2        2PY +/- 2PY  (SIGMA)
C      5        2PX          D2SP3       2S  +/- 2PY  (SIGMA)
C      6        3PX
C      7        4PX
C      8        3DZ2
C      9        3DX2-Y2
C
C         ALSO, THE EFF. NUCLEAR CHARGE(S), AND FOR MOLECULAR
C         ORBITALS,THE INTERNUCLEAR DISTANCE AND THE BOND TYPE
C         (BONDING OR ANTIBONDING)
C              DISTANCE FOR THE MOLECULAR ORBITALS.
C
C     PROGRAM STRUCTURE
C
C     CONTOUR     :  THE MAIN PROGRAM
C
C     BLOCK DATA : DEFINES THE CONSTANTS NEEDED FOR THE SLATER ORBITAL
C                  WAVEFUNCTIONS AND THE SYMBOLS WHICH WILL DESIGNATE
C                  THE DIFFERENT MAGNITUDES OF THE ELECTRON DENSITY
C
C
C     ATORB       : DETERMINES THE ELECTRON DENSITY MAP FOR THE
C                   DESIGNATED ATOMIC ORBITAL AND OUTPUTS THE SYMBOL MAP
C
C     HYBRID      : PERFORMES THE SAME FUNCTION AS ATORB BUT FOR THE HYBRID
C                   ORBITALS
C
C     MOLORB      : PERFORMES THE SAME FUNCTION AS ATORB BUT FOR THE
C                   MOLECULAR ORBITALS
C
C     FPSI        : DEFINES THE HYDROGEN ATOM ORBITALS   (SLATER WAVE FNS.)
C
C     SYMBOL      : STORES FOR EACH X AND Y COORDINATE THE OUTPUT SYMBOL
C                   REPRESENTING ITS ELECTRON DENSITY
C
C     GLOSSARY
C
C     PSI(41,61) COMMON ARRAY TO STORE THE ELECTRON DENSITY AT Y(I),X(I)
C     X(61)       COMMON ARRAY TO STORE VALUES OF X(I)
C     Y(41)       COMMON ARRAY TO STORE VALUES OF Y(I)
C      RAD        SPHERICAL COORDINATE R
C     BOHRAD      SPHERICAL COORDINATE R DIVIDED BY BOHR RADIANS (.529)
C     COSPHI      COSINE OF THE SPHERICAL COORDINATE PHI
C     COSTH       COSINE OF THE SPHERICAL COORDINATE THETA
C     COS2PI      COSINE OF TWICE THE SPHERICAL COORDINATE PHI
C     SINPHI      SINE OF THE SPHERICAL COORDINATE PHI
C     ZA          EFFECTIVE NUCLEAR CHARGE ON ATOM A
C     ZB          EFFECTIVE NUCLEAR CHARGE ON ATOM B
C     ICNTRL      DETERMINES WHICH SUBROUTINE IS CALLED
C     IORB        DETERMINES WHICH ORBITAL IS DESIRED
```

```
C        ISYM         WHICH OF 3 POSSIBLE SYMMETRIES   OBTAINS
C        PSIMAX       MAXIMUM ELECTRON  DENSITY
C        RAB          INTERNUCLEAR DISTANCE
C
C        AUTHORS:  S. B. LEVITT  AND  K. J. JOHNSON
C
      COMMON /ARRAYS/PSI(41,61),X(61),Y(41)
      COMMON /CONST/CONST(14)
      COMMON/SYMB/BLANK,COLON,SLASH,ZERO,ANDSGN,POUND
      COMMON/FPARAM/BOHRAD,COSPHI,SINPHI,COS2PI,COSTH,IMDZ2
C
C        INPUT
C
  10 WRITE(6,11)
      READ(5,22) ICNTRL
      IF (ICNTRL .LT.1 .OR. ICNTRL .GE.4)  GO TO 120
      GO TO (20,30,40),ICNTRL
C
C        ATOMIC ORBITAL
C
  20 WRITE(6,33)
      READ(5,22)IORB
      CALL ATORB(IORB)
      GO TO 10
C
C        HYBRID ORBITAL
C
  30 WRITE(6,44)
      READ(5,22)IORB
      CALL HYBORB(IORB)
      GO TO 10
C
C        MOLECULAR ORBITAL
C
  40 WRITE(6,55)
      READ(5,22)IORB
      CALL MOLORB(IORB)
      GO TO 10
C
C        FORMATS
C
  11 FORMAT(//25X,'CONTOUR'//,'  ENTER AN INTEGER BETWEEN 1 AND 4:'
     1,// '     INTEGER:   1       2       3        4'//,
     2'    ORBITAL: ATOMIC   HYBRID  MOLECULAR  STOP'/)
  22 FORMAT(I)
  33 FORMAT(//'  ENTER AN INTEGER BETWEEN 1 AND 9:'//,
     1' INTEGER:  1  2  3  4  5  6  7  8    9'//,
     2' ATOMIC'/,
     3' ORBITAL:  1S  2S  3S  4S  2PX 3PX 4PX 3DZ2  3DX2-Y2'/)
  44 FORMAT(//'  ENTER AN INTEGER BETWEEN 1 AND 5:'//,
     1'    INTEGER:   1   2   3   4     5'/,
     2'    HYBRID'/,
     3'   ORBITAL:  SP SP2  SP3  DSP2  D2SP3'/)
  55 FORMAT(//'  ENTER AN INTEGER BETWEEN 1 AND 5:'//,
     1'    INTEGER:   1      2      3      4      5'//,
     2'    MOLECULAR  1S   2S     2PX    2PY    2S'/,
     3'    ORBITAL:   +/-   +/-    +/-    +/-    +/-'/,
     4'               2S   2S     2PX    2PY    2PY'//,
     55X,'   NOTE:  Y IS THE MOLECULAR AXIS'/)
 120 STOP
      END
```

```
C
C
C
C      -----------------------------------------------------------
C
C
C          BLOCK DATA
C
C          INITIALIZATION OF VALUES IN CONST(14)
C
       BLOCK DATA
       COMMON/CONST/CONST(14)
       DATA CONST(1)/3.1415926/
       DATA CONST(2)/0.5641896/
       DATA CONST(3)/0.099735572/
       DATA CONST(4)/0.0040214198/
       DATA CONST(5)/0.0098504268/
       DATA CONST(6)/0.0049252134/
       DATA CONST(7)/0.0028435732/
       DATA CONST(8)/0.00036731094/
       DATA CONST(9)/0.00049279932/
       DATA CONST(10)/0.57735026/
       DATA CONST(11)/0.81649658/
       DATA CONST(12)/0.4082483/
       DATA CONST(13)/0.70710678/
       DATA CONST(14)/0.28867514/
       END
C
C
C
C      -----------------------------------------------------------
C
C
C                         ATORB
C
C        ATORB CALCULATES ATOMIC ORBITALS
C
       SUBROUTINE ATORB(IORB)
       COMMON /FPARAM/BOHRAD,COSPHI,SINPHI,COS2PI,COSTH,IMDZ2
       COMMON/ARRAYS/PSI(41,61),X(61),Y(41)
       DIMENSION ORBNAM(10)
       DATA ORBNAM/'1S','2S','3S','4S','2PX','3PX','4PX','3DZ2',
      1'   3D','X2-Y2'/
C
C        INPUT THE EFFECTIVE NUCLEAR CHARGE
C
   10  WRITE(6,11)
       READ(5,22)ZA
       IF(IORB.EQ.9)GO TO 20
       WRITE(6,33)ORBNAM(IORB),ZA
       GO TO 25
   20 WRITE(6,34)ORBNAM(9),ORBNAM(10),ZA
C
C        INITIALIZE
C
   25 PSIMAX=0.0
       ZAA=(ZA/.529)**1.5
       NY=21
       NX=31
       X(1)=-2.40
       Y(1)=2.667
```

```
      DO 30 IY=2,NY
   30 Y(IY)=Y(IY-1)-0.1333
      DO 35 IX=2,NX
   35 X(IX)=X(IX-1)+0.080
      IMDZ2=0
      IF(IORB.EQ.8)IMDZ2=1
      IGO=3
      IF(IORB.LE.8)IGO=2
      IF(IORB.LE.4)IGO=1
C
C        THE MAIN CALCULATION LOOP (UPPER LEFT QUADRANT)
C
      DO 110 I=1,NY
      DO 100 J=1,NX
C
      RAD=SQRT(Y(I)**2+X(J)**2)
      IF(ABS(RAD) .LT. 1.0E-20)  GOTO 80
      BOHRAD=RAD/.529
C
C        BRANCH TO 50 IF 1S,2S,3S OR 4S; 60 IF 2PX,3PX,4PX OR
C      3DZ2; 70 IF 3DX2-Y2 ORBITAL DESIRED.
C
      GO TO (50,60,70),IGO
      RETURN
C
C        THE TRIGONOMETRIC FUNCTIONS ARE CALCULATED AND FPSI IS CALLED
C      TO EVALUATE THE WAVEFUNCTION.
C
   50 PSI(I,J)=(FPSI(IORB,ZA,ZAA))**2
      GO TO 90
   60 COSPHI=-X(J)/RAD
      COSTH=COSPHI
      PSI(I,J)=(FPSI(IORB,ZA,ZAA))**2
      GO TO 90
   70 COSPHI=-X(J)/RAD
      COS2PI=2.*COSPHI*COSPHI-1.
      PSI(I,J)=FPSI(IORB,ZA,ZAA)**2
      GO TO 90
   80 PSI(I,J)=0.
   90 IF(PSI(I,J).GT.PSIMAX)PSIMAX=PSI(I,J)
  100 CONTINUE
  110 CONTINUE
C
C        THE SUBROUTINE SYMBOL IS CALLED FOR NORMALIZATION OF THE
C      ELECTRON DENSITY VALUES AND FOR OUTPUT
C
      CALL SYMBOL(NY,NX,PSIMAX,1)
C
C        FORMATS
C
   11 FORMAT(/'   ENTER THE EFFECTIVE NUCLEAR CHARGE '/)
   22 FORMAT(F)
   33 FORMAT(//'   THE ',A5,' ATOMIC ORBITAL    Z = ',F5.2)
   34 FORMAT(//'   THE ',2A5,' ATOMIC ORBITAL    Z = ',F5.2)
      RETURN
      END
C
C
C    ------------------------------------------------------------
C
```

```fortran
C
C
C                               HYBORB
C
C           HYBORB CALCULATES HYBRID ORBITALS
C
        SUBROUTINE HYBORB(IORB)
        COMMON /FPARAM/BOHRAD,COSPHI,SINPHI,COS2PI,COSTH,IMDZ2
        COMMON/CONST/CONST(14)
        COMMON/ARRAYS/PSI(41,61),X(61),Y(41)
        DIMENSION ORBNAM(5)
        DATA ORBNAM/'SP','SP2','SP3','DSP2','D2SP3'/
C
C
C           INPUT THE EFFECTIVE NUCLEAR CHARGE
C
        WRITE(6,11)
        READ(5,22)ZA
        WRITE(6,33)ORBNAM(IORB),ZA
C
C
C            INITIALIZATION
C
        PSIMAX=0.
        NY=21
        NX=61
        ZAA=(ZA/.529)**1.5
        Y(1)=2.667
        X(1)=-2.40
        DO 10 IY=2,NY
   10   Y(IY)=Y(IY-1)-0.1333
        DO 20 IX=2,NX
   20   X(IX)=X(IX-1)+0.080
C
C
C           THE MAIN CALCULATION LOOP
C
        DO 110 I=1,NY
        DO 100 J=1,NX
C
C
C           THE RADIAL SPHERICAL COORDINATE,RAD, AND COSPHI ARE CALCULATED
C
        RAD=SQRT(Y(I)**2+X(J)**2)
        BOHRAD=RAD/.529
        IF(ABS(RAD) .LE. 1.0E-20)   GOTO 80
        COSPHI=-X(J)/RAD
C
C
C           PROGRAM WILL BRANCH TO  30 IF SP; 40 IF SP2; 50 IF SP3;
C                              60 IF DSP2; 70 IF D2SP3 ORBITAL DESIRED
C
        GO TO (30,40,50,60,70),IORB
C
C
C           THE TRIGONOMETRIC FUNCTIONS ARE EVALUATED AND FPSI IS CALLED TO
C           EVALUATE THE WAVEFUNCTION
C
C
C
C           *********************  SP  **********************************
C
C             PSI=( 2S +2PX )/ 1.4142136
C
   30   PSI(I,J)=((FPSI(2,ZA,ZAA)+FPSI(5,ZA,ZAA))/CONST(13))**2
        GO TO 90
C
C           *********************  SP2  *********************************
```

```
C
C          PSI=( 2S + 2PX * (2.)**.5)/(3.)**.5
C
C
   40  PSI(I,J)=(FPSI(2,ZA,ZAA)*CONST(10)+FPSI(5,ZA,ZAA)*CONST(11))**2
       GO TO 90
C
C       *************** SP3 IN PLANE DEFINED BY THETA=54* 44**   *********
C
C        PSI=( 2S - 2PX + 2PY + 2PZ)/.5
C
   50  COSPHI=COSPHI/.81647
       SINPHI=Y(I)/(RAD*.81647)
       PSI(I,J)=((FPSI(2,ZA,ZAA)+(-FPSI(5,ZA,ZAA)+FPSI(10,ZA,ZAA))
      1 *.81467+FPSI(11,ZA,ZAA)*.57738)*.5)**2
       GO TO 90
C
C       ******************** DSP2  *************************************
C
C          PSI=(4S*.5 + 3DX2-Y2*.5 + 4PX*(2)**.5)
C
   60  SINPHI=Y(I)/RAD
       COS2PI=2.*COSPHI*COSPHI-1.
       PSI(I,J)=(FPSI(4,ZA,ZAA)*.5+FPSI(7,ZA,ZAA)*CONST(13)
      1+FPSI(9,ZA,ZAA)*.5)**2
       GO TO 90
C
C       ******************* D2SP3  *************************************
C
C          PSI=( 4S*(6)**.5 + 4PX*(2)**.5 + 3DX2-Y2*.5 - 3DZ2*(12)**.5)
C
   70  SINPHI=Y(I)/RAD
       COS2PI=2.*COSPHI*COSPHI-1.
       COSTH=0.
       PSI(I,J)=(FPSI(4,ZA,ZAA)*CONST(12)+FPSI(7,ZA,ZAA)*CONST(13)
      1-FPSI(8,ZA,ZAA)*CONST(14)+.5*FPSI(9,ZA,ZAA))**2
       GO TO 90
   80  PSI(I,J)=0.
   90  IF(PSI(I,J) .GT. PSIMAX)PSIMAX=PSI(I,J)
  100  CONTINUE
  110  CONTINUE
C
C      THE SUBROUTINE SYMBOL IS CALLED FOR NORMALIZATION OF THE
C      ELECTRON DENSITY VALUES AND FOR OUTPUT
C
       CALL SYMBOL(NY,NX,PSIMAX,0)
C
C       FORMATS
C
   11  FORMAT(/'  ENTER THE EFFECTIVE NUCLEAR CHARGE '/)
   22  FORMAT(F)
   33  FORMAT(/'  THE ',A5,' HYBRID ORBITAL    Z = ',F5.2)
       RETURN
       END
C
C
C      ------------------------------------------------------------
C
C
C                      MOLORB
```

```
C
C          MOLORB CALCULATES MOLECULAR ORBITALS
C
           SUBROUTINE MOLORB(IORB)
           COMMON/FPARAM/BOHRAD,COSPHI,SINPHI,COS2PI,COSTH,IMDZ2
           COMMON/ARRAYS/PSI(41,61),X(61),Y(41)
           COMMON/CONST/CONST(14)
           DIMENSION ORBNAM(10)
           DATA ORBNAM/'1SA','1SB','2SA','2SB',
          1'2PXA','2PXB','2PYA','2PYB','  2SA','2PYB'/
           DATA SIGNN,SIGNP/' - ',' + '/
C
C          INPUT THE INTERNUCLEAR DISTANCE AND THE EFFECTIVE
C          NUCLEAR CHARGES, ZA AND AB
C
           WRITE(6,11)
           READ(5,22)RAB,ZA,ZB
C
C          ABONDING OR AN ANTIBONDING ORBITAL IS CALCULATED DEPENDING ON THE
C          VALUE OF BNDTYP, +1 FOR A BONDING MO, -1 FOR ANTIBONDING
C
           WRITE(6,33)
           READ(5,22)BNDTYP
           WRITE(6,44) RAB,ZA,ZB
C
C          INITIALIZATION
C
           ISYM=1
           IF(ZA.NE.ZB .OR. IORB .EQ.5 ) ISYM=-1
           IF(IORB-3)8,8,5
     5     BNDTYP=-BNDTYP
           SIGN=SIGNN
           IF(BNDTYP.EQ.1)SIGN=SIGNP
           GO TO 9
     8     SIGN=SIGNP
           IF(BNDTYP.LT.1)SIGN=SIGNN
     9     PSIMAX=0.
           Z1A=(ZA/.529)**1.5
           Z2B=(ZB/.529)**1.5
           X(1)=-2.4
           Y(1)=2.667
           NX=31
           NY=21
           IF(ISYM.EQ.-1 .OR. IORB .EQ.5 ) NY=41
           DO 10 IY=2,NY
    10     Y(IY)=Y(IY-1)-0.1333
           DO 15 IX=2,NX
    15     X(IX)=X(IX-1)+.08
C
C          THE KEYS TO THE FUNCTION SUBROUTINE CALL ARE SET
C
           IF(IORB-4)20,30,40
    20     IF1=5
           IF2=5
           GO TO 50
    30     IF1=10
           IF2=10
           GO TO 50
    40     IF1=2
           IF2=10
```

```
   50    IMOL2 = IORB*2
         IMOL21 = IMOL2 - 1
         WRITE(6,55) ORBNAM(IMOL21),SIGN,ORBNAM(IMOL2)
C
C        THE MAIN CALCULATION LOOP
C
C
C        THE TRIGONOMETRIC FUNCTIONS ARE CALCULATED AND FPSI IS CALLED
C        TO EVALUATE THE WAVEFUNCTION.
C         THE(X,Y) COORDINATES OF A AND B ARE (0,-RAB/2),
C         AND(0,+RAB/2) RESPECTIVELY
C
         DO 140 I=1,NY
         YA=Y(I)-RAB/2
         YB=YA+RAB
         YA2=YA*YA
         YB2=YB*YB
         DO 130 J=1,NX
         XJ2=X(J)*X(J)
         IF(IORB .GT. 2)   GO TO 60
C
C        MO'S WITH 1S AND 2S FUNCTIONS
C
         BOHRAD=(SQRT(YA2+XJ2))/.529
         TEMP1=FPSI(IORB,ZA,Z1A)
         BOHRAD=(SQRT(YB2+XJ2))/.529
         PSI(I,J)=(TEMP1+BNDTYP*FPSI(IORB,ZB,Z2B))**2
         GO TO 120
C
C        MO'S WITH 2P FUNCTIONS
C
   60    RAD=SQRT(YA2+XJ2)
         BOHRAD=RAD/.529
         IF(ABS(RAD) .LE. 1.0E-20)   GOTO 70
         SINPHI=YA/RAD
         COSPHI=-X(J)/RAD
         GO TO 80
   70    SINPHI=0.
         COSPHI=0.
   80    TEMP1=FPSI(IF1,ZA,Z1A)
         RAD=SQRT(YB2+XJ2)
         BOHRAD=RAD/.529
         IF(ABS(RAD) .LE. 1.0E-20)   GOTO 90
         SINPHI=YB/RAD
         COSPHI=-X(J)/RAD
         GO TO 100
   90    TEMP2=0.
         GO TO 110
  100    TEMP2=FPSI(IF2,ZB,Z2B)
  110    PSI(I,J)=(TEMP1+BNDTYP*TEMP2)**2
  120    IF(PSI(I,J).GT. PSIMAX)   PSIMAX=PSI(I,J)
  130    CONTINUE
  140    CONTINUE
C
C        THE SUBROUTINE SYMBOL IS CALLED FOR NORMALIZATION OF
C        ELECTRON DENSITY VALUES AND FOR OUTPUT
C
         CALL SYMBOL(NY,NX,PSIMAX,ISYM)
         RETURN
C
```

```
C        FORMATS
C
   11 FORMAT(/'   ENTER THE INTERNUCLEAR SEPARATION IN ANGSTROMS,'/
      1'    AND THE TWO EFFECTIVE NUCLEAR CHARGES   '/)
   22 FORMAT(3F)
   44 FORMAT(/'   THE INTERNUCLEAR SEPARATION IS ',G12.3,/
      1'  ZA = ',F5.2,'   AND ZB = ',F5.2)
   33 FORMAT(/'   ENTER +1 FOR THE BONDING ORBITAL OR',/
      1 '          -1 FOR THE ANTIBONDING ORBITAL'/)
   55 FORMAT(//20X,A5,A3,A5//)
      END
C
C
C
C        --------------------------------------------------------------
C
C
C
C                           FPSI
C
C        SLATER-TYPE HYDROGEN-LIKE WAVEFUNCTIONS ARE DEFINED
C
C        PARAMETERS:
C                IFUNC- KEY TO THE DESIRED WAVEFUNCTION
C                Z     -THE EFFECTIVE NUCLEAR CHARGE
C                ZZ    -(Z/.529)**1.5
C
      FUNCTION FPSI(IFUNC,Z,ZZ)
      COMMON/FPARAM/BOHRAD,COSPHI,SINPHI,COS2PI,COSTH,IMDZ2
      COMMON/CONST/CONST(14)
      GO TO (10,20,30,40,50,60,70,80,90,100,110),IFUNC
C
C     ************************  1S  *****************************
C
C        PARAMETERS: BOHRAD,Z,ZZ
C
   10 FPSI=ZZ*EXP(-Z*BOHRAD)*CONST(2)
      RETURN
C
C     ************************  2S  *****************************
C
C        PARAMETERS: BOHRAD,Z,ZZ
C
   20 FPSI=ZZ*EXP(-Z*BOHRAD/2.)*(2.-Z*BOHRAD)*CONST(3)
      RETURN
C
C     ************************  3S  *****************************
C
C        PARAMETERS: BOHRAD,Z,ZZ
C
   30 FPSI=ZZ*(27.-18.*Z*BOHRAD+(2.*(BOHRAD*Z)**2))*
      1EXP(-Z*BOHRAD/3.)*CONST(4)
      RETURN
C
C     ************************  4S  *****************************
C
C        PARAMETERS: BOHRAD,Z,ZZ
C
   40 FPSI=ZZ*CONST(8)*(192.-144.*BOHRAD*Z+24.*((BOHRAD*Z)**2)
      1-(BOHRAD*Z)**3)*EXP(-BOHRAD*Z/4.)
```

```
      RETURN
C
C        ************************ 2PX **********************************
C
C        PARAMETERS: BOHRAD,Z,ZZ,COSPHI
C
   50 FPSI=ZZ*BOHRAD*Z*EXP(-Z*BOHRAD/2.)*CONST(3)*COSPHI
      RETURN
C
C        ************************ 3PX **********************************
C
C        PARAMETERS: BOHRAD,Z,ZZ,COSPHI
C
   60 FPSI=ZZ*Z*BOHRAD*(6.-BOHRAD*Z)*EXP(-BOHRAD*Z/3.)*
     1COSPHI*CONST(5)
      RETURN
C
C        ************************ 4PX **********************************
C
C        PARAMETERS: BOHRAD,Z,ZZ,COSPHI
C
   70 FPSI=ZZ*CONST(9)*(80.-20.*Z*BOHRAD+(Z*BOHRAD)**2)*Z*
     1BOHRAD*EXP(-Z*BOHRAD/4.)*COSPHI
      RETURN
C
C        ************************ 3DZ2 *********************************
C
C        PARAMETERS: BOHRAD,Z,ZZ,COSPHI
C
   80 FPSI=ZZ*(BOHRAD*Z)**2*EXP(-BOHRAD*Z/3.)*CONST(7)
     1*(3.*(COSTH**2)-1.)
      RETURN
C
C        ************************ 3DX2-Y2 ******************************
C
C        PARAMETER=BOHRAD,Z,ZZ,COS2PI
C
   90 FPSI=ZZ*(BOHRAD*Z)**2*EXP(-BOHRAD*Z/3.)*COS2PI*CONST(6)
      RETURN
C
C        ************************ 2PY **********************************
C
C        PARAMETERS: BOHRAD,Z,ZZ,SINPHI
C
  100 FPSI=ZZ*Z*BOHRAD*EXP(-BOHRAD*Z/2.)*SINPHI*CONST(3)
      RETURN
C
C        ************************ 2PZ **********************************
C
C        USED ONLY BY HYBORB  (SP3)
C
C        PARAMETERS: BOHRAD,Z,ZZ
C
  110 FPSI=ZZ*Z*BOHRAD*EXP(-Z*BOHRAD/2.)*CONST(3)
      RETURN
      END
C
C
```

```
C
C       -----------------------------------------------------------------
C
C                  SYMBOL
C          PARAMETERS:
C
C              PSI-THE ELECTRON DENSITY FOR EACH POINT; AFTER
C                  EXECUTION IT HOLDS THE SYMBOLS THAT REPRESENT THE
C                  MAGNITUDE OF THE ELECTRON DENSITY
C              PSIMAX-THE MAXIMUM ELECTRON DENSITY FOUND FOR ANY POINT
C              IC - OUTPUT CONTROL VECTOR
C
C          THEELECTRON DENSITIES ARE NORMALIZED BY DIVIDING BY PSIMAX
C          (0 .GE. PSI .LE. 1).   QQ HOLDS THE NORMALIZED ELECTRON DENSITY.
C          THEFOLLOWING SYMBOLS REPRESENT THE VARYING RANGES OF THE
C          ELECTRON DENSITY=
C
C          ELECTRON DENSITY                  SYMBOL          OUTPUT FORM
C
C              QQ .LT. .01              BLANK
C          .01 .GE. QQ .LT. .02           COLON                  :
C          .02 .GE. QQ .LT. .10           SLASH                  /
C          .10 .GE. QQ .LT. .25           ZERO                   0
C          .25 .GE. QQ .LT. .50           AND                    &
C          .50 .GE. QQ                    POUND                  #
C
C          ISYM      SYMMETRY
C          1         4-FOLD  (ATORB AND MOLORB WITH ZA=ZB)
C           0        2-FOLD  (HYBORB)
C          -1        2-FOLD  (MOLORB WITH ZA .NE. ZB)
C
C
C          IF ISYM=1, THEN ONLY WORK ON UPPER LEFT QUADRANT
C
C
          SUBROUTINE SYMBOL(NY,NX,PSIMAX,ISYM)
          DATA BLANK,COLON,SLASH,ZERO,ANDSGN,POUND/
         1' ',':','/','0','&','#'/
          COMMON/FPARAM/BOHRAD,COSPHI,SINPHI,COS2PI,COSTH,IMDZ2
          COMMON/ARRAYS/PSI(41,61),X(61),Y(41)
          DIMENSION IC(41)
          X(41)=-X(21)
          X(51)=-X(11)
          X(61)=-X(1)
          IF(IMDZ2.EQ.0)WRITE(6,11)
          IF(IMDZ2.EQ.1)WRITE(6,13)
          WRITE(6,14)(X(JJ),JJ=1,61,10)
          WRITE(6,12)
          DO 70 I=1,NY
          KI=1
          IFLAG=1
          DO 60 J=1,NX
          QQ=PSI(I,J)/PSIMAX
          IF(QQ .LT. .01)   GO TO 50
          IF (QQ .GE. .01 .AND. QQ .LT. .02)   GO TO 40
          IF(QQ .GE. .02 .AND. QQ .LT. .1)   GO TO 30
          IF(QQ .GE. .1 .AND. QQ .LT. .25)   GO TO 20
          IF(QQ .GE. .25 .AND. QQ .LT. .5)   GO TO 10
          PSI(I,J)=POUND
          GO TO 59
```

```
10   PSI(I,J)=ANDSGN
     GO TO 59
20   PSI(I,J)=ZERO
     GO TO 59
30   PSI(I,J)=SLASH
     GO TO 59
40   PSI(I,J)=COLON
     GO TO 59
50   PSI(I,J)=BLANK
     IF(IFLAG.EQ.0)GO TO 60
     KI=J
     GO TO 60
59   IFLAG=0
60   CONTINUE
     IF(ISYM.EQ.0)GO TO 68
     IC(I)=61-KI
     IF(KI.EQ.31)IC(I)=1
C
C        IF ISYM = +/- 1, CALCULATE SYMMETRIC RIGHT HALF
C
     DO 65 IX=1,31-KI
65   PSI(I,IX+NX)=PSI(I,NX-IX)
     WRITE(6,66)Y(I),(PSI(I,JJ),JJ=1,IC(I))
     GO TO 70
C
C        HYBRID ORBITAL -  PRINT 1 OR 61 COLS.
C
68   IC(I)=1
     IF(KI.NE.61)IC(I)=61
     WRITE(6,66)Y(I),(PSI(I,JJ),JJ=1,IC(I))
70   CONTINUE
     IF(ISYM.LT.0)GO TO 95
C
C        PRINT SYMMETRIC BOTTOM HALF
C
     DO 90 I=1,NY-1
     N=NY-I
     YN=-Y(N)
90   WRITE(6,66)YN,(PSI(N,J),J=1,IC(N))
     WRITE(6,12)
     RETURN
95   WRITE(6,12)
     RETURN
11   FORMAT(/50X,'X ----->'/)
12   FORMAT(9X,'+',6('---------+'))
13   FORMAT(/50X,'Z ----->'/)
14   FORMAT(1X,7F10.2)
66   FORMAT(1X,F5.2,2X,'!',61A1)
     END
```

Discussion

CONTOUR requires approximately 5000 words. The execution times are approximately 0.3 sec for the atomic orbitals, 1 sec for the hybrid orbitals, and 0.5 sec for the molecular orbitals. Two optimization features are worthy of note. The three sub-routines, ATORB, HYBORB, and MOLORB utilize the two- or fourfold symmetry of the corresponding orbitals. The printing subroutine SYMBOL utilizes the work vector IC(21) to suppress the printing of blank characters.

Problem 2.13 Extend CONTOUR to include more atomic, hybrid, and molecular orbitals. For example, plot the molecular orbitals for CO_2 and related linear triatomics in both the σ and the π planes.

The interested reader is referred to a number of sources for additional information on orbital plotting techniques [8-15].

2.7 EQUIL

Introduction

EQUIL calculates the total free energy as a function of extent of reaction [16] for the gas-phase reaction,

$$aA + bB \rightleftharpoons cC + dD$$

The input includes the stoichiometric coefficients (a, b, c, and d), the absolute temperature (T), and the standard chemical potentials at the specified temperature (μ^o_A, μ^o_B, μ^o_C, and μ^o_D). The program tabulates the total free energy, approximates the minimum value of the total free energy, and using the extent of reaction corresponding to this minimum, approximates the equilibrium constant.

Method

EQUIL assumes ideal gas behavior, a total pressure of 1.0 atm, and initially,

$$n_A = a, \qquad n_B = b, \qquad \text{and} \qquad n_C = n_D = 0.0$$

Here n is the number of moles. The program first calculates the standard free energy change, ΔG^o, and the equilibrium constant using the expressions,

$$\Delta G^o = c\mu^o_C + d\mu^o_D - (a\mu^o_A + b\mu^o_B)$$

and

$$K_p = \exp\frac{-\Delta G^o}{RT}$$

Here K_p is the equilibrium constant expressed in partial pressures, and R is the gas constant. The total free energy is

$$G_{tot} = n_A\mu_A + n_B\mu_B + n_C\mu_C + n_D\mu_D$$

where

$$n_i = \text{number of moles of species i}$$
$$\mu_i = \mu^o_i + RT \ln a_i$$
$$= \mu^o_i + RT \ln p_i$$
$$= \mu^o_i + RT \ln X_i$$

Here a_i, p_i, and X_i are the activity, partial pressure, and mole fraction of species i, respectively. For ideal gases the partial pressure is equal to the activity ($p_i = a_i$). Here the partial pressure is also equal to the mole fraction because the total pressure is 1.0 atm. When the extent of reaction is x,

$$n_A = a(1 - x) \qquad \mu_A = \mu_A^o + RT \ln \frac{a(1 - x)}{W}$$

$$n_B = b(1 - x) \qquad \mu_B = \mu_B^o + RT \ln \frac{b(1 - x)}{W}$$

$$n_C = cx \qquad \mu_C = \mu_C^o + RT \ln \frac{cx}{W}$$

$$n_D = dx \qquad \mu_D = \mu_D^o + RT \ln \frac{dx}{W}$$

where

$$W = \text{total moles} = (1 - x)(a + b) + x(c + d)$$

EQUIL evaluates the function

$$G_{tot}(x) = FA + FB + FC + FD$$

where

$$FA = a(1 - x) \, (\mu_A^o + RT \ln \frac{a(1 - x)}{W})$$

$$FB = b(1 - x) \, (\mu_B^o + RT \ln \frac{b(1 - x)}{W})$$

$$FC = cx(\mu_C^o + RT \ln \frac{cx}{W})$$

$$FD = dx(\mu_D^o + RT \ln \frac{dx}{W})$$

$G_{tot}(x)$ is calculated over the range $0 \le x \le 1$ in steps of 0.005. The minimum value of $G_{tot}(x)$ provides an initial approximation to x_{eq}, the equilibrium extent of reaction. EQUIL then calculates a better approximation to x_{eq} by varying x in the region of the minimum in steps of 0.0002. The equilibrium constant is then approximated using the equation

$$K_p = \frac{p_C^c p_D^d}{p_A^a p_B^b}$$

$$= \frac{x_C^c x_D^d}{x_A^a x_B^b}$$

$$= \frac{(n_C/W)^c (n_D/W)^d}{(n_A/W)^a (n_B/W)^b}$$

$$= \frac{c^c d^d x_{eq}^{c+d} W_{eq}^{a+b-c-d}}{a^a b^b (1 - x_{eq})^{a+b}}$$

where

$$W_{eq} = (1 - x_{eq})(a + b) + x_{eq}(c + d)$$

Finally, the standard free energy change is approximated using

$$\Delta G^{\circ} = -RT \ln K_p$$

The values for K_p and ΔG° should agree (approximately) with the input values.

Sample Execution   ($2HI \rightleftharpoons H_2 + I_2$)

FOR THE REACTION   A   +   B   <=====>   C   +   D

ENTER THE STOICHIOMETRIC COEFFICIENTS OF A, B, C, D
(A=0 ENDS PROGRAM)
>2 0 1 1

ENTER THE STANDARD CHEMICAL POTENTIALS OF A, B, C, D IN CALORIES.
> -2713

ENTER THE ABSOLUTE TEMPERATURE
> 666.8

COMPOUND	A	B	C	D
COEFFICIENT	2.0	0.0	1.0	1.0
STD. CHEMICAL POTENTIAL	-2713.0	0.0	0.0	0.0

TEMPERATURE =   666.8 K

THE CALCULATED FREE ENERGY CHANGE IS   5426.0 CALORIES,

THE CALCULATED EQUILIBRIUM CONSTANT IS   1.666D-02

DEGREE OF REACTION	GTOT (CALORIES)
0.000	-5426.0
0.050	-5772.7
0.100	-5928.6
0.150	-6007.9
0.200	-6034.4
0.250	-6019.1
0.300	-5968.2
0.350	-5885.7
0.400	-5.74.0

```
        0.450         -5634.7
        0.500         -5468.5
        0.550         -5275.8
        0.600         -5056.2
        0.650         -4809.0
        0.700         -4532.6
        0.750         -4224.5
        0.800         -3881.0
        0.850         -3495.6
        0.900         -3057.4
        0.950         -2542.5
        1.000         -1837.0
```

THE APPROXIMATE MINIMUM VALUE OF G IS   -6034.6 CALORIES

THE CORRESPONDING DEGREE OF REACTION IS 0.2052

THE APPROX. EQUILIBRIUM CONSTANT IS   1.666D-02

AND THE APPROXIMATE STANDARD DELTA G IS   5425.6 CALORIES

Listing

```
C
C                                EQUIL
C
C
C THIS PROGRAM CALCULATES THE TOTAL FREE ENERGY FOR VARIOUS
C DEGREES OF REACTION FOR THE FOLLOWING GAS-PHASE REACTION:
C
C          A  +  B   <==>   C  +  D
C
C    IT THEN APPROXIMATES THE MINIMUM VALUE OF THE TOTAL FREE ENERGY
C    AND THE CORRESPONDING DEGREE OF REACTION.   FROM THIS
C    MINIMUM AN EQUILIBRIUM CONSTANT AND A STANDARD DELTA G ARE
C    CALCULATED.
C
C    GLOSSARY:
C
C          OLDG,GTOTL        = VALUES OF GTOT FOR TABLE
C          AL                = DEGREE OF REACTION    0 ==> 1
C          LA                = 1 - AL
C          AA,BB,CC,DD       = STOICHIOMETRIC COEFF. OF A,B,C,D
C          UA,UB,UC,UD       = STD. CHEM. POTENTIAL OF A,B,C,D
C          T                 = ABSOLUTE TEMPERATURE
C          RT                = R * T      1.98726 * T
C          W                 = TOTAL NUMBER OF MOLES PRESENT
C          GTOT              = FUNCT. TO CALC TOTAL FREE ENERGY
C
C    ASSUMPTIONS:
C
C          (1)     IDEAL GASES - ACTIVITY = PARTIAL PRESSURE
C          (2)     TOTAL PRESSURE IS CONSTANT AT 1.00 ATM.
C          (3)     INITIALLY, NUMBER OF MOLES OF A+B = AA+BB
C          (4)     INITIALLY,  NUMBER OF MOLES OF C = D = 0
```

```
C
C      INPUT:
C                    AA,BB,CC,DD
C                    UA,UB,UC,UD
C                    T
C
C      AUTHORS:  D. L. DOERFLER AND K. J. JOHNSON
C
C
C
C  THE FUNCTION FOR THE GENERAL RELATIONSHIP BETWEEN FREE ENERGY (G)
C  AND THE DEGREE OF REACTION (ALPHA) IS:
C
C G(ALPHA)=LA AA (UA + RT LN(LA AA/W)) + LA BB (UB + RT LN(LA BB/W))
C        + AL CC (UC + RT LN(AL CC/W)) + AL DD (UD + RT LN(AL DD/W))
C
       IMPLICIT DOUBLE PRECISION (A-H,O-Z)
       DOUBLE PRECISION LA,KEQ
       COMMON AA,BB,CC,DD,AB,CD,UA,UB,UC,UD,RT
C
C
C    INPUT
C
       WRITE(6,11)
       READ(5,22)  AA,BB,CC,DD
       IF(AA.EQ.0.)STOP
       WRITE(6,33)
       READ(5,22)  UA,UB,UC,UD
   20  WRITE(6,44)
       READ(5,22)  T
       IF(T .LE. 0.0D+0) GO TO 90
C
C      SET CONSTANTS
C
       RT = 1.98726D+0 * T
       AB=AA+BB
       CD=CC+DD
       IFLAG=0
C
C  CALCULATE EQUILIBRIUM CONSTANT AND STANDARD FREE
C  ENERGY CHANGE FROM INPUT DATA.  PRINT INPUT DATA
C  FOR VERIFICATION PURPOSES
C
       DGCALC=CC*UC+DD*UD- (AA*UA+BB*UB)
       KEQ=DEXP(-DGCALC/RT)
       WRITE(6,55) AA,BB,CC,DD,UA,UB,UC,UD,T,DGCALC,KEQ
C
C      PART I.     PRINT TABLE OF TOTAL FREE ENERGY AS A FUNCTION
C                  OF THE EXTENT OR REACTION.  GET A FIRST
C                  APPROXIMATION TO THE MINIMUM IN GTOT(ALPHA)
C                  STEP SIZE: 0.05
C
       OLDG=1.0D38
       DO 30 I=1,201
       IM1=I-1
       AL=IM1*5.0D-3
       GTOTL=GTOT(AL)
       IF(GTOTL.GE.OLDG)GO TO 30
```

```
      OLDG=GTOTL
      ALMIN=AL
   30 IF(MOD(IM1,10).EQ.0)WRITE(6,88)AL,GTOTL
C
C     PART II.      FIND THE APPROXIMATE MINIMUM FREE ENERGY
C                   STEP SIZE: 0.0002
C
      OLDG=1.0D38
      ALLO=ALMIN-5.0D-3
      IF(ALLO.LT.0.0D0)ALLO=0.0D0
      IF(ALLO.EQ..995D0)ALLO=.990D0
      DO 40 I=1,51
      AL=ALLO+(I-1)*2.0D-4
      GTOTL=GTOT(AL)
      IF(GTOTL.GT.OLDG)GO TO 50
      OLDG=GTOTL
      ALMIN=AL
   40 CONTINUE
C
C   SPECIAL CASE:  0.9998 < AL < 1.0
C   CALCULATE MINIMUM KEQ
C
      IFLAG=1
      AL=0.9998
      GO TO 60
   50 AL=AL-2.0D-4
      IF(AL.NE.0.0D0)GO TO 60
C
C  SPECIAL CASE:   0.0 < AL < 0.0002
C  CALCULATE MAXIMUM KEQ
C
      IFLAG=-1
      AL=0.0002D0
   60 LA=1.0-AL
      W = LA*AB + AL*CD
      GMIN=GTOT(AL)
C
C     CHECK FOR THE POSSIBILITY OF 0**0
C     IF SO, SET COEFFICIENT TO 1
C
      IF(BB.EQ.0.)BB=1.
      IF(CC.EQ.0.)CC=1.
      IF(DD.EQ.0.)DD=1.
C
C   CALCULATE APPROXIMATE KEQ,DGCALC
C
      KEQ=((CC**CC*DD**DD*AL**CD)/(AA**AA*BB**BB*LA**AB))*(W**(AB-CD))
      DGCALC = -RT*DLOG(KEQ)
C
C   SPECIAL CASES
C
      IF(IFLAG.EQ.-1)WRITE(6,101)KEQ,DGCALC
      IF(IFLAG.EQ.0)WRITE(6,111)GMIN,AL,KEQ,DGCALC
      IF(IFLAG.EQ.1)WRITE(6,122)KEQ,DGCALC
      STOP
   90 WRITE(6,133)
      GO TO 20
   11 FORMAT(//,' FOR THE REACTION    A  +  B   <=====>   C  +  D'
     1//,' ENTER THE STOICHIOMETRIC COEFFICIENTS OF A, B, C, D '
     2/,' (A=0 ENDS PROGRAM)'/)
```

```
   22 FORMAT( 4D )
   33 FORMAT(/,' ENTER THE STD. CHEMICAL POTENTIALS OF A, B,',
      1' C AND D IN CALORIES.'/)
   44 FORMAT(/,' ENTER THE ABSOLUTE TEMPERATURE'/)
   55 FORMAT(//,' COMPOUND',11X,'A',10X,'B',10X,'C',10X,'D',//,
      1' COEFFICIENT',6X,F4.1,3(7X,F4.1),//,' STD. CHEMICAL',/,
      .' POTENTIAL',
      2 F14.1,3(1X,F10.1),//,' TEMPERATURE =',F9.1,' K',
      3//' THE CALCULATED FREE ENERGY CHANGE IS',F9.1,
      4 ' CALORIES,'//' THE CALCULATED EQUILIBRIUM'
      5 ' CONSTANT IS ',1PD12.3,
      6//,' DEGREE OF',7X,'GTOT',/,' REACTION      (CALORIES)'
      1/)
   88 FORMAT( F9.3, F14.1 )
  101 FORMAT(//' THE MINIMUM VALUE OF GTOT WAS NOT FOUND.',
      1 ' THE EQUILIBRIUM EXTENT'/'  OF REACTION  IS BETWEEN',
      2 ' 0.0 AND 0.0002.    USING ALPHA = 0.0002,'//
      3 '  THE APPROXIMATE EQUILIBRIUM CONSTANT IS',1PD12.3,
      4 //'  THE APPROXIMATE STANDARD DELTA G IS',0PF9.1,
      5 ' CALORIES.'/)
  111 FORMAT(//' THE APPROXIMATE MINIMUM VALUE OF G IS', 0PF9.1,
      1' CALORIES'//' THE CORRESPONDING DEGREE OF REACTION IS',
      2 F7.4,//,' THE APPROX. EQUILIBRIUM CONSTANT IS',1PD12.3,
      3//' AND THE APPROXIMATE STANDARD DELTA G IS',0PF9.1,' CALORIES')
  122 FORMAT(//' THE MINIMUM VALUE OF GTOT WAS NOT FOUND.'
      1 /' THE EQUILIBRIUM EXTENT OF REACTION IS BETWEEN ',
      2 ' 0.9998 AND 1.0.  USING ALPHA = 0.9998,'/, '  THE APPROX. EQ.'
      3 ' CONSTANT IS',1PD12.3,/'  AND THE APPROX. STD. DELTA G IS',
      4 0PF9.1,'  CALORIES'/)
  133 FORMAT(/,' THE TEMPERATURE CANNOT BE LESS THAN OR EQUAL',
      1 ' TO 0.0'//)
      END
C
C
C                          GTOT
C
C

      DOUBLE PRECISION FUNCTION GTOT(ALPHA)
      IMPLICIT DOUBLE PRECISION (A-Z)
      COMMON AA,BB,CC,DD,AB,CD,UA,UB,UC,UD,RT
C
C     FUNCTION TO CALCULATE THE TOTAL FREE ENERGY, G
C
      LA=1.0D+0-ALPHA
      W=LA*AB+ALPHA*CD
C
      ARG1=LA*AA/W
      ARG2=LA*BB/W
      ARG3=ALPHA*CC/W
      ARG4=ALPHA*DD/W
C
C     CHECK FOR DLOG(0.0)
C
      IF( ARG1 .GT. 0.0D+0 ) ARG1=DLOG(ARG1)
      IF( ARG2 .GT. 0.0D+0 ) ARG2=DLOG(ARG2)
      IF( ARG3 .GT. 0.0D+0 ) ARG3=DLOG(ARG3)
      IF( ARG4 .GT. 0.0D+0 ) ARG4=DLOG(ARG4)
C
      GTOT=LA*(AA*(UA+RT*ARG1)+BB*(UB+RT*ARG2))+ALPHA*(CC*(UC+RT*ARG3)
      .+DD*(UD+RT*ARG4))
C
      RETURN
      END
```

Discussion

EQUIL and GTOT require 709 and 120 words of core, respectively. This sample execution took approximately 0.3 sec of CPU time.

Two special cases had to be considered in this program:

$$0 \leq x_{eq} \leq 0.0002 \qquad \text{and} \qquad 0.9998 \leq x_{eq} \leq 1.0$$

Since 0.0002 is the increment in the search for the minumum value of $G_{tot}(x)$, the minimum will not be found for these cases. EQUIL prints an appropriate message and calculates either an upper limit for K_{eq} (using $x_{eq} = 0.0002$) or a lower limit for K_{eq} (using $x_{eq} = 0.9998$).

Problem 2.14 Plot $G_{tot}(x)$ for a given reaction at two or more temperatures on the same graph.

Solution. See Figure 2.5.

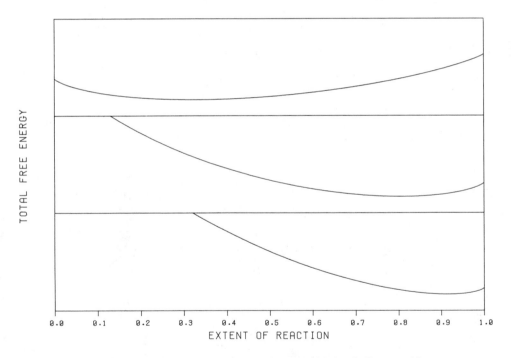

Figure 2.5 Total free energy as a function of extent of the reaction $2NH_3 \rightleftharpoons N_2 + 3H_2$: T = 400K (top), 500K, and 550K (bottom). It is assumed that the enthalpy and entropy changes are constant.

2.8 NMR

Introduction

NMR simulates the following proton nuclear magnetic resonance spin systems
[17-19]: AB, AB_2, AB_3, A_2B_3, A_2X_2, and ABX. The input includes a number indicating
which one of the six spin systems is to be simulated, and the corresponding chemical
shifts and coupling constants. NMR prints the input data for verification purposes,
a table containing the transition frequencies and intensities of all allowed transitions, and, on option, a teletype plot of the spectrum.

Method

Closed-form expressions for the relative intensities and transition frequencies
as a function of chemical shifts and coupling constants have been derived [9, 10].
For example, the AB spectrum has four allowed transitions:

Line	Transition frequency	Relative intensity
1	$-J/2 - C$	$1 - Q$
2	$J/2 - C$	$1 + Q$
3	$-J/2 + C$	$1 + Q$
4	$J/2 + C$	$1 - Q$

Here J is J_{AB}, the spin-spin coupling constant, and

$$C = \frac{1}{2}[D^2 + J^2]$$

where D is the difference in the chemical shifts of the A and B protons, and

$$Q = \frac{J}{2C}$$

NMR reads J and D, calculates and prints the four transition frequencies in
order of increasing frequency and, optionally, prints a teletype plot of the
spectrum.

The input requirements and the number of transitions for each spectrum are
given in the following table:

Spectrum	Key	N	Input
AB	1	4	D, J_{AB}
AB_2	2	8	$\delta_A, \delta_B, J_{AB}$
AB_3	3	14	$\delta_A, \delta_B, J_{AB}$
A_2B_3	4	20	$\delta_A, \delta_B, J_{AB}$
A_2X_2	5	12	$\delta_A, \delta_X, J_{AA}, J_{AX}, J_{AX'}, J_{XX}$
ABX	6	14	$\delta_A, \delta_B, \delta_X, J_{AB}, J_{AX}, J_{BX}$

Here δ_A, δ_B, and δ_X are the chemical shifts of protons A, B, and X, and J_{ij} is the
coupling constant.

The teletype plot of the spectrum is calculated by assuming overlapping Lorentzian bands of constant half-width at half-maximum intensity,

$$I_{tot}(\nu) = \sum_{j=1}^{N} \frac{I_j w}{w^2 + (\nu - \nu_{oj})^2}$$

Here $I_{tot}(\nu)$ is the total intensity as a function of frequence ν, N is the number of allowed transitions, I_j and ν_{oj} are the calculated transition intensity and frequency of the jth transition, and w is the constant half-width and half-maximum intensity. The value 0.2 has been arbitrarily assigned to w.

The dimensions of the plot are determined by the values assigned to NLINES and NWIDE, as indicated below,

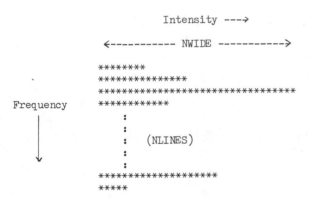

NWIDE has been set to 58 which is optimal for 72-column output, NLINES varies with the spin system:

Spin system	AB	AB_2	AB_3	A_2B_3	A_2X_2	ABX
NLINES	61	91	106	91	101	56 (AB) 51 (X)

Low-intensity transitions are excluded to improve the resolution of the plot.

The program is modular in structure. The modules are delimited by statement numbers as follows: AB spectrum, 40-52; AB_2 spectrum, 60-70; A_2B_3 spectrum, 70-80; A_2X_2 spectrum, 80-90; ABX spectrum, 90-100; code to sort frequencies, 100; code to calculate overlapping Lorentzian bands, 175; and code to plot the spectrum, 195.

The AB_2 and AB_3 sections of the program use trigonometric identies to express $\sin\theta$ and $\cos\theta$ in terms of sine 2θ and cos 2θ, where t stands for the angle theta. First,

$$\sin\theta = \pm\left[\frac{1 - \cos 2\theta}{2}\right]^{1/2}$$

Since θ is by definition between 0 and π radians, the positive square root is always correct. Second,

$$\tan 2\theta = \frac{\sin 2\theta}{\cos 2\theta} = \frac{2 \tan \theta}{1 - \tan^2\theta} \qquad (2.5)$$

and

$$\tan^2\theta = \frac{1 - \cos 2\theta}{1 + \cos 2\theta} \qquad (2.6)$$

Substituting Eq. (2.6) into (2.5), and further substituting $(\sin \theta)/(\cos \theta)$ for $\tan \theta$,

$$\cos \theta = \frac{2 \sin \theta \cos 2\theta}{\sin 2\theta \, [1 - (1 - \cos 2\theta)/(1 + \cos 2\theta)]}$$

Sample Execution (AB$_2$ case: 2,6-lutidine [20])

NMR SPECTRUM SIMULATION PROGRAM

    THIS PROGRAM SIMULATES THE FOLLOWING NMR SPECTRA:

1 - AB   2 - AB$_2$ 3 - AB$_3$ 4 - A$_2$B$_3$ 5 - A$_2$X$_2$ 6 - ABX

ENTER THE NUMBER CORRESPONDING TO THE SPECTRUM YOU WISH TO SIMULATE.
   (ZERO TERMINATES THE PROGRAM)
>2

THE AB$_2$ CASE

ENTER THE CHEMICAL SHIFTS OF THE A AND B PROTONS (in Hz.)
>100   121.9

ENTER THE COUPLING CONSTANT Jab.
>8.20

               AB$_2$ NMR SPECTRUM

THE CHEMICAL SHIFTS ARE: A = 100.00   B = 121.90

THE COUPLING CONSTANTS ARE: Jab =   8.20   Jbb = IMMATERIAL

```
        FREQUENCY      INTENSITY

           90.57        0.467        A
           97.04        0.682        A
          100.00        1.000        A
          106.48        1.853        A
          118.29        2.850        B
          119.03        2.533        B
          125.51        1.465        B
          127.72        1.147        B
          146.76        0.003        *
```

 * INDICATES A COMBINATION (FORBIDDEN) TRANSITION.

DO YOU WANT A PLOT OF THE CALCULATED SPECTRUM?
  ENTER 1 FOR YES, 0 FOR NO.
>1

TO IMPROVE THE RESOLUTION OF THE GRAPHICAL OUTPUT,
WEAK COMBINATION TRANSITIONS ARE IGNORED.


```
     88.24     *
               *
               *
               **
               ***
     90.57     **************
               ***
               **
               *
               *
     92.89     *
               *
               *
               *
               *
     95.21     *
               *
               **
               ****
               ******************
     97.53     ****
               **
               **
               **
               ****
     99.85     ******************
               *********
               ***
               **
               **
    102.18     *
```

```
               *
               *
               *
               *
               *
104.50         **
               **
               ***
               ******
               ************************************
106.82         *************
               ****
               **
               **
               *
109.14         *
               *
               *
               *
               *
111.47         *
               *
               *
               *
               *
113.79         *
               *
               *
               **
               **
116.11         **
               ***
               ****
               ******
               **********************
118.43         ***********************************************************
               *********************************************************
               *********************
               ******
               ****
120.76         ***
               **
               **
               **
               **
123.08         **
               **
               **
               ***
               ******
125.40         ****************************
               **********
               ****
               ***
               ******
127.72         *****************************
               ******
               ***
               **
               **
130.04         *
```

NMR SPECTRUM SIMULATION PROGRAM

    THIS PROGRAM SIMULATES THE FOLLOWING NMR SPECTRA:

1 - AB   2 - AB   3 - AB   4 - A B   5 - A X   6 - ABX
              2   3   2 3   2 2

ENTER THE NUMBER CORRESPONDING TO THE SPECTRUM YOU WISH TO SIMULATE.
    (ZERO TERMINATES THE PROGRAM)
  >0

```
C
C
C                              NMR
C
C        NMR SIMULATES AB,AB2,AB3,A2B3,A2X2 AND ABX NMR SPECTRA
C
C        THE PROGRAM ACCEPTS AS INPUT THE CHEMICAL SHIFTS AND
C        COUPLING CONSTANTS AND PRINTS A TABLE OF THE
C        CALCULATED FREQUENCIES AND INTENSITIES, AND THE
C        LINE ASSIGNMENTS.  A PLOT OF THE CALCULATED SPECTRUM IS
C        AVAILABLE AS AN OPTION.  THE EXACT INPUT AND ANY
C        ASSUMPTIONS MADE BY THE PROGRAM ARE GIVEN IN THE COMMENTS AT
C        THE BEGINNING OF EACH CASE.  THE PLOTTING ROUTINE
C        ASSUMES LORENTZIAN LINE SHAPES AND 0.2 Hz PEAK WIDTH
C        AT HALF-HEIGHT.  THIS PEAK WIDTH IS ARBITRARY AND MAY BE
C        MODIFIED BY ASSIGNING A DIFFERENT VALUE TO PW.
C
C         REFERENCES TO PSB ARE PAGE NUMBERS IN REF. 17.
C         THE REFERENCE TO EMS IS REF. 18.
C
C        GLOSSARY:
C
C        A, B, X  --  CHEMICAL SHIFTS OF THE RESPECTIVE PROTONS OR ATOMS
C        FREQ (20)  --   THE CALCULATED TRANSITION FREQUENCIES
C        INTEN (20)  --   THE CORRESPONDING INTENSITIES
C        CHAR (20)  --   THE CORRESPONDING TRANSITION ASSIGNMENTS   (A2)
C        NFREQ  --   THE NUMBER OF TRANSITIONS OR LINES FOR A GIVEN CASE
C
C        VARIABLES ASSOCIATED WITH THE OPTIONAL PLOT INCLUDE
C
C        NLINES  --   THE NUMBER OF LINES USED FOR A PARTICULAR PLOT
C        LINE (121)  --   THE SUMMED INTENSITIES OF ALL TRANSITIONS FOR A
C                         GIVEN LINE OF THE PLOT
C        MAXI  --   THE INTENSITY OF THE STRONGEST LINE IN A GIVEN PLOT
C        LENGTH  --   THE NUMBER OF PLOTTING CHARACTERS FOR A LINE OF THE
C                     PLOT
C        NFLOW  --   THE NUMBER OF THE LOWEST NON-FORBIDDEN TRANSITION
C        NFBDNH  --   THE NUMBER OF FORBIDDEN TRANSITIONS OCCURRING AT
C                     HIGHEST FREQUENCIES
C        INCRE  --   THE FREQUENCY DIFFERENCE BETWEEN SUCCESSIVE LINES OF
C                     THE PLOT
C        LOW  --   ORIGINALLY THE LOWEST FREQUENCY LINE OF THE PLOT;  LOW
C                  IS INCREMENTED IN STEPS OF 5*INCRE AND USED TO LABEL
C                  EVERY FIFTH LINE OF THE PLOT
C
C  AUTHORS:  W.F. SLIWINSKI, D.L. DOERFLER, AND K.J. JOHNSON
C
```

```
C
        DIMENSION LINE(121), FREQ(20),INTEN(20),CHAR(20)
        REAL INCRE,INTEN,J,JAA,JAB,JAX,JAXX,JBX,JXX,LINE,LOW,MAXI
        REAL K,L,M,N,J2,J2D,JD2,JD
        DATA ALET,AP,BLANK,BLET,BP,STAR/'A ','A+','  ','B ','B+','* '/
        DATA XLET,XSTAR,YES /'X ','X*','1 '/
C
C          SWITCH IS USED IN THE ABX CASE TO CAUSE THE PLOT TO BE
C          PRINTED IN TWO PARTS -- THE AB PART AND THE X PART.
C          IT IS SET TO .FALSE. EVERYWHERE ELSE.
C
        LOGICAL SWITCH
        SWITCH = .FALSE.
C
C          WIDE CONTROLS THE WIDTH OF THE  PLOT.  IT IS THE
C          NUMBER OF PLOTTING CHARACTERS USED FOR THE STRONGEST PEAK
C          IN THE CALCULATED SPECTRUM.  FOR PROPER OUTPUT, WIDE MUST BE
C          SET TO AT LEAST TWELVE CHARACTERS LESS THAN THE LENGTH OF AN
C          OUTPUT LINE AND MUST NOT EXCEED 118.
C
        WIDE = 58
C
C          PW IS THE PEAK WIDTH AT HALF-HEIGHT AND IS USED
C          IN THE LORENTZIAN BAND FUNCTION
C
        PW=0.2
        PWSQ=PW**2
C
C          WRITE OUT OPTIONS AVAILABLE, AND ASK FOR DESIRED SPECTRUM.
C          KEY CONTROLS THE PROGRAM EXECUTION AS FOLLOWS:
C
C             KEY = 1    AB SPECTRUM
C             KEY = 2    AB2 SPECTRUM
C             KEY = 3    AB3 SPECTRUM
C             KEY = 4    A2B3 SPECTRUM
C             KEY = 5    A2X2 SPECTRUM
C             KEY = 6    ABX SPECTRUM
C             KEY = ANY OTHER VALUE TERMINATES EXECUTION
C
     10 WRITE (6,20)
        READ (5,25) KEY
        IF( KEY.EQ.0 .OR. KEY .GT.6)STOP
        NFLOW = 1
        NFBDNH = 0
        GO TO (40,50,60,70,80,90),KEY
        STOP
C
C
C          THE AB SPECTRUM
C
C          INPUT:  THE CHEMICAL SHIFT DIFFERENCE BETWEEN A AND B  (SHIFT)
C                  THE COUPLING CONSTANT Jab  (J)
C          REFERENCE: PSB, P 119 ff
C
C
C          WRITE TITLE;  READ IN AND WRITE OUT INPUT DATA
C
     40 WRITE (6,42)
        READ (5,43) SHIFT
```

```
        WRITE (6,53)
        READ (5,43) J
        WRITE (6,45) J,SHIFT
C
C       CALCULATE THE FREQUENCIES OF THE TRANSITIONS
C
        IF (SHIFT .EQ. 0.0 .OR. J .EQ. 0.0)  GO TO 200
        CONST = SQRT(SHIFT*SHIFT+J*J)/2.0
        J2 = ABS(J/2.0)
        FREQ(1) = -J2 - CONST
        FREQ(2) = -J2 + CONST
        FREQ(3) = -FREQ(2)
        FREQ(4) = -FREQ(1)
        NFREQ = 4
C
C       CALCULATE THE INTENSITIES
C
        INTEN(1) = 1.0 - J2/CONST
        INTEN(2) = 1.0 + J2/CONST
        INTEN(3) = INTEN(2)
        INTEN(4) = INTEN(1)
C
C       BLANK OUT ASSIGNMENTS
C
        DO 47 I=1,4
   47 CHAR(I) = BLANK
        NLINES = 56
        GO TO 100
C
C
C       THE AB  SPECTRUM
C             2
C
C       ASSUMPTION: BOTH COUPLING CONSTANTS Jab ARE THE SAME.
C       INPUT:   THE CHEMICAL SHIFTS OF A AND B  (A,B)
C                THE COUPLING CONSTANT Jab  (JAB)
C       REFERENCE:  PSB, P 123 ff
C
C
C       WRITE TITLE;   READ IN AND WRITE OUT INPUT DATA
C
   50 WRITE (6,51)
        WRITE (6,52)
        READ (5,43)  A,B
        WRITE (6,53)
        READ (5,43) JAB
        WRITE (6,54)  A,B,JAB
C
C       CALCULATE USEFUL CONSTANTS
C       X CORRESPONDS TO THETA-SUB-PLUS AND Y TO THETA-SUB-MINUS IN PSB
C       NOTATION.
C
        IF (JAB .EQ. 0.0 .OR. A-B .EQ. 0.0)  GO TO 200
        CPLUS =SQRT((A-B)**2 + (A-B)*JAB +JAB*JAB*9/4)/2
        CMINUS =SQRT((A-B)**2 - (A-B)*JAB + JAB*JAB*9/4)/2
        SS = (2.0**0.5)
        ARG = JAB/SS
        SIN2X = ARG/CPLUS
        COS2X = (A-B+JAB/2)/CPLUS*0.5
```

```
      SINX = SQRT((1-COS2X)/2)
      TANSQX = (1-COS2X)/(1+COS2X)
      COSX = 2*SINX*COS2X/SIN2X/(1-TANSQX)
      SIN2Y = ARG/CMINUS
      COS2Y = (A-B-JAB/2)/CMINUS*0.5
      SINY = SQRT((1-COS2Y)/2)
      TANSQY = (1-COS2Y)/(1+COS2Y)
      COSY = 2*SINY*COS2Y/SIN2Y/(1-TANSQY)
      QQ = SINX*COSY - COSX*SINY
      RR = COSX*COSY + SINX*SINY
C
C         CALCULATE THE FREQUENCIES OF THE TRANSITIONS
C
      FREQ(1) = (A+B)/2.0 + JAB*3/4 + CPLUS
      FREQ(2) = B + CPLUS + CMINUS
      FREQ(3) = A
      FREQ(4) = (A+B)/2 + CMINUS - JAB*3/4
      FREQ(5) = B + CPLUS - CMINUS
      FREQ(6) = (A+B)/2 + JAB*3/4 - CPLUS
      FREQ(7) = B - CPLUS + CMINUS
      FREQ(8) = (A+B)/2 - JAB*3/4 - CMINUS
      FREQ(9) = B -CPLUS - CMINUS
      NFREQ = 9
C
C         CALCULATE THE INTENSITIES
C
      INTEN(1) = (SS*SINX - COSX)**2
      INTEN(2) = (SS*QQ + COSX*COSY)**2
      INTEN(3) = 1.0
      INTEN(4) = (SS*SINY + COSY)**2
      INTEN(5) = (SS*RR + COSX*SINY)**2
      INTEN(6) = (SS*COSX + SINX)**2
      INTEN(7) = (SS*RR - SINX*COSY)**2
      INTEN(8) = (SS*COSY - SINY)**2
      INTEN(9) = (SS*QQ + SINX*SINY)**2
C
C         ASSIGN TRANSITIONS;  PREPARE FOR POSSIBLE PLOT.
C
      NLINES = 91
      IF (A .LT. B) GO TO 59
      DO 58 I=1,4
      CHAR(I) = ALET
   58 CHAR(I+4) = BLET
      CHAR(9) = STAR
      NFLOW = 2
      GO TO 100
   59 CHAR(1) = BLET
      CHAR(4) = BLET
      CHAR(5) = BLET
      CHAR(7) = BLET
      CHAR(2) = STAR
      CHAR(3) = ALET
      CHAR(6) = ALET
      CHAR(8) = ALET
      CHAR(9) = ALET
      NFBDNH = 1
      GO TO 100
C
```

```
C
C                 THE AB  SPECTRUM
C                      3
C
C                 ASSUMPTION: ALL COUPLING CONSTANTS Jab ARE EQUAL.
C                 INPUT:   THE CHEMICAL SHIFTS OF A AND B  (A,B)
C                          THE COUPLING CONSTANT Jab  (JAB)
C                 REFERENCES:  PSB, P 128 ff  AND  EFS, P 329 ff
C
C
C                 WRITE TITLE;  READ IN AND WRITE OUT INPUT DATA
C
   60 WRITE (6,61)
      WRITE (6,52)
      READ (5,43)  A,B
      WRITE (6,53)
      READ (5,43) JAB
      WRITE (6,62) A,B,JAB
C
C                 CALCULATE USEFUL CONSTANTS.  P,Z,M AND ZP CORRESPOND TO PLUS,
C                 ZERO, MINUS AND ZERO-PRIME IN STANDARD NOTATION. THUS DP STANDS
C                 FOR D-SUB-PLUS, COS2P STANDS FOR COS(2*THETA-SUB-PLUS), ETC.
C
      DIFF = A - B
      IF (DIFF .EQ. 0.0 .OR. JAB .EQ. 0.0)  GO TO 200
      SUM2 = (A+B)/2
      DP = SQRT (DIFF*DIFF + 2.0*JAB*DIFF + 4.0*JAB*JAB)/2
      DZ = SQRT (DIFF*DIFF + 4.0*JAB*JAB)/2
      DM = SQRT (DIFF*DIFF - 2.0*JAB*DIFF + 4.0*JAB*JAB)/2
      DZP = SQRT (DIFF*DIFF + JAB*JAB)/2
      SS = 3.0**0.5
      COS2P = (DIFF+JAB)/(2.0*DP)
      SIN2P = SS*JAB/(2.0*DP)
      COS2M = (DIFF-JAB)/(2.0*DM)
      SIN2M = SS*JAB/(2.0*DM)
      COS2ZP = DIFF/(2.0*DZP)
      SIN2ZP = JAB/(2.0*DZP)
      COS2Z =DIFF/(2.0*DZ)
      SIN2Z = JAB/DZ
      SINP = SQRT((1.0-COS2P)/2)
      SINZ = SQRT((1.0-COS2Z)/2)
      SINM = SQRT((1-COS2M)/2)
      SINZP = SQRT((1-COS2ZP)/2)
      TANSQP = (1-COS2P)/(1+COS2P)
      TANSQM = (1-COS2M)/(1+COS2M)
      TANSQZ = (1-COS2Z)/(1+COS2Z)
      TANSZP = (1-COS2ZP)/(1+COS2ZP)
      COSP = 2*SINP*COS2P/SIN2P/(1-TANSQP)
      COSM = 2*SINM*COS2M/SIN2M/(1-TANSQM)
      COSZ = 2*SINZ*COS2Z/SIN2Z/(1-TANSQZ)
      COSZP = 2*SINZP*COS2ZP/SIN2ZP/(1-TANSZP)
C
C                 CALCULATE THE FREQUENCIES OF THE TRANSITIONS
C
      FREQ(1) = JAB + DP + SUM2
      FREQ(2) = -DIFF/2 + DP + DZ + SUM2
      FREQ(3) = JAB/2 + DZP + SUM2
      FREQ(4) = -DIFF/2 + DZ + DM + SUM2
      FREQ(5) = -JAB/2 + DZP + SUM2
```

```
      FREQ(6) = -JAB + DM + SUM2
      FREQ(7) = -DIFF/2 + DZ - DM + SUM2
      FREQ(8) = JAB/2 - DZP + SUM2
      FREQ(9) = -DIFF/2 +DP - DZ + SUM2
      FREQ(10) = JAB - DP + SUM2
      FREQ(11) = -DIFF/2 + DZ - DP + SUM2
      FREQ(12) = -DIFF/2 + DM - DZ + SUM2
      FREQ(13) = -JAB/2 - DZP + SUM2
      FREQ(14) = -JAB - DM + SUM2
      FREQ(15) = -DIFF/2 - DZ - DM + SUM2
      FREQ(16) = -DIFF/2 - DP - DZ + SUM2
      NFREQ = 16
C
C         CALCULATE THE INTENSITIES
C
      INTEN(1) = (SS*SINP - COSP)**2
      INTEN(2) = (-2*SINZ*COSP + COSZ*COSP + SS*COSZ*SINP)**2
      INTEN(3) = 2 - 4*COSZP*SINZP
      INTEN(4) = (-2*SINZ*COSM - COSZ*COSM + SS*COSZ*SINM)**2
      INTEN(5) = 2 + 4*SINZP*COSZP
      INTEN(6) = (COSM + SS*SINM)**2
      INTEN(7) = (SS*COSZ*COSM + SINM*COSZ + 2*SINM*SINZ)**2
      INTEN(8) = INTEN(5)
      INTEN(9) = (2*COSZ*COSP +SINZ*COSP + SS*SINZ*SINP)**2
      INTEN(10) = (SS*COSP + SINP)**2
      INTEN(11) = (SS*COSZ*COSP + 2*SINZ*SINP - COSZ*SINP)**2
      INTEN(12) = (2*COSZ*COSM - SINZ*COSM + SS*SINZ*SINM)**2
      INTEN(13) = INTEN(3)
      INTEN(14) = (SS*COSM - SINM)**2
      INTEN(15) = (-SS*SINZ*COSM - SINZ*SINM + 2*COSZ*SINM)**2
      INTEN(16) = (-2*COSZ*SINP - SINZ*SINP + SS*SINZ*COSP)**2
C
C         ASSIGN TRANSITIONS;  PREPARE FOR POSSIBLE PLOT.
C
      NLINES = 106
      DO 65 I=1,16
   65 CHAR(I) = ALET
      IF (A .LT. B) GO TO 68
      DO 66 I=7,14
   66 CHAR(I) = BLET
      CHAR(15) = STAR
      CHAR(16) = STAR
      NFLOW = 3
      GO TO 100
   68 DO 69 I=1,11,2
   69 CHAR(I) = BLET
      CHAR(6) = BLET
      CHAR(12) = BLET
      CHAR(2) = STAR
      CHAR(4) = STAR
      NFBDNH = 2
      GO TO 100
C
C
C         THE A B  SPECTRUM
C            2 3
C
C         ASSUMPTION:  ALL COUPLING CONSTANTS Jab ARE EQUAL.
C         INPUT:  THE CHEMICAL SHIFTS OF A AND B  (A,B)
```

```
C                    THE COUPLING CONSTANT Jab   (JAB)
C          REFERENCE:  PSB, P 154 ff
C
C
C          WRITE TITLE;  READ IN AND WRITE OUT INPUT DATA
C
   70 WRITE (6,71)
      WRITE (6,52)
      READ (5,43)  A,B
      WRITE (6,53)
      READ (5,43) JAB
      WRITE (6,72) A,B,JAB
C
C          CALCULATE USEFUL CONSTANTS
C
      DIFF = A - B
      IF (DIFF .EQ. 0.0 .OR. JAB .EQ. 0.0)  GO TO 200
      J2 = JAB*JAB
      J2D = J2/DIFF
      JD2 = J2D/2
      JD = JAB/DIFF
      HJ = JAB/2
C
C          CALCULATE THE FREQUENCIES OF THE TRANSITIONS
C
      FREQ(1)  = 3*HJ + 3*JD2 + A
      FREQ(2)  = 3*HJ + A
      FREQ(3)  = HJ + 2*J2D + A
      FREQ(4)  = HJ + 3*JD2 + A
      FREQ(5)  = HJ + JD2 + A
      FREQ(6)  = HJ + A
      FREQ(7)  = -HJ + 2*J2D + A
      FREQ(8)  = -HJ + 3*JD2 + A
      FREQ(9)  = -HJ +JD2 + A
      FREQ(10) = -HJ + A
      FREQ(11) = -3*HJ + 3*JD2 + A
      FREQ(12) = -3*HJ + A
      FREQ(13) = JAB + JD2 + B
      FREQ(14) = JAB - JD2 + B
      FREQ(15) = JAB - 3*JD2 + B
      FREQ(16) = -JD2 + B
      FREQ(17) = B
      FREQ(18) = -JAB + JD2 + B
      FREQ(19) = -JAB -JD2 + B
      FREQ(20) = -JAB - 3*JD2 + B
      NFREQ = 20
C
C          CALCULATE THE INTENSITIES
C
      INTEN(1)  = 2 - 6*JD
      INTEN(2)  = INTEN(1)
      INTEN(3)  = 2 - 2*JD
      INTEN(4)  = INTEN(3)
      INTEN(5)  = 4 - 4*JD
      INTEN(6)  = INTEN(5)
      INTEN(7)  = 2 + 2*JD
      INTEN(8)  = INTEN(7)
      INTEN(9)  = 4 + 4*JD
      INTEN(10) = INTEN(9)
      INTEN(11) = 2 + 6*JD
```

```
      INTEN(12) = INTEN(11)
      INTEN(13) = 3 + 6*JD
      INTEN(14) = 6 + 12*JD
      INTEN(15) = INTEN(13)
      INTEN(16) = 12
      INTEN(17) = 12
      INTEN(18) = 3 - 6*JD
      INTEN(19) = 6 - 12*JD
      INTEN(20) = INTEN(18)
      DO 74 I=1,20
      IF (INTEN(I) .LT. 0.0)  INTEN(I) = 0.0
   74 CONTINUE
C
C         ASSIGN TRANSITIONS;  PREPARE FOR POSSIBLE PLOT.
C
      DO 75 I=1,12
   75 CHAR(I) = ALET
      CHAR(5) = AP
      CHAR(6) = AP
      CHAR(9) = AP
      CHAR(10) = AP
      CHAR(13) = BLET
      CHAR(14) = BP
      CHAR(15) = BLET
      CHAR(16) = BP
      CHAR(17) = BP
      CHAR(18) = BLET
      CHAR(19) = BP
      CHAR(20) = BLET
      NLINES = 81
      GO TO 100
C
C
C         THE A X   SPECTRUM
C             2 2
C
C         ASSUMPTION:  THE CHEMICAL SHIFT DIFFERENCE IS LARGE
C                      (>10) RELATIVE TO THE COUPLING CONSTANTS BETWEEN
C                      THE A AND X PROTONS.
C         INPUT:  THE CHEMICAL SHIFTS OF A AND X   (A,X)
C                 THE FOUR COUPLING CONSTANTS Jaa, Jax, Jax' AND Jxx
C                 (JAA, JAX, JAXX, JXX)
C         REFERENCE:  PSB, P 140 ff
C
C
C         WRITE TITLE;   READ IN AND WRITE OUT INPUT DATA
C
   80 WRITE (6,81)
      READ (5,43) A,X
      WRITE (6,82)
      READ (5,43) JAA,JAX,JAXX,JXX
      WRITE (6,84) A,X,JAA,JAX,JAXX,JXX
C
C         CALCULATE USEFUL CONSTANTS.  K, L, M AND N AS DEFINED HERE ARE
C         HALF THE VALUES OF PSB, FOR COMPUTATIONAL EFFICIENCY.
C
      IF (A-X .EQ. 0.0)  GO TO 200
      IF (JAX .EQ. 0.0 .AND. JAXX .EQ. 0.0)  GO TO 200
      K = (JAA + JXX)/2
      L = (JAX - JAXX)/2
```

```
        M = (JAA - JXX)/2
        N = (JAX + JAXX)/2
        B = SQRT(K*K + L*L)
        C = SQRT(M*M + L*L)
C
C          CALCULATE THE FREQUENCIES OF THE TRANSITIONS
C
        FREQ(1) = N
        FREQ(2) = N
        FREQ(3) = -N
        FREQ(4) = -N
        FREQ(5) = B + K
        FREQ(6) = B - K
        FREQ(7) = K - B
        FREQ(8) = -K - B
        FREQ(9) = C + M
        FREQ(10) = C - M
        FREQ(11) = M - C
        FREQ(12) = -M - C
        NFREQ = 12
C
C          CALCULATE THE INTENSITIES
C
        INTEN(1) = 1
        INTEN(2) = 1
        INTEN(3) = 1
         INTEN(4) = 1
        INTEN(5) = (-K/B + 1)/2
        INTEN(6) = (K/B + 1)/2
        INTEN(7) = INTEN(6)
        INTEN(8) = INTEN(5)
        INTEN(9) = (-M/C + 1)/2
        INTEN(10) = (M/C + 1)/2
        INTEN(11) = INTEN(10)
        INTEN(12) = INTEN(9)
C
C          BLANK OUT PROTON ASSIGNMENTS;  PREPARE FOR POSSIBLE PLOT.
C
        DO 85 I=1,12
     85 CHAR(I) = BLANK
        NLINES = 101
        GO TO 100
C
C
C          THE ABX SPECTRUM
C
C          ASSUMPTION:  THE CHEMICAL SHIFT DIFFERNCES BETWEEN A - X AND
C                       B - X ARE LARGE (>10) RELATIVE TO THE RESPECTIVE
C                       COUPLING CONSTANTS Jax AND Jbx
C          INPUT:  THE CHEMICAL SHIFTS OF A,B AND X   (A,B,X)
C                  THE COUPLING CONSTANTS Jab, Jax AND Jbx  (JAB,JAX,JBX)
C          REFERENCE:  PSB, P 132 ff
C
C
C          WRITE TITLE;  READ IN AND WRITE OUT INPUT DATA
C
     90 WRITE (6,91)
        READ (5,43) A,B,X
        WRITE (6,93)
        READ (5,43) JAB,JAX,JBX
```

```
      WRITE (6,94) A,B,X,JAB,JAX,JBX
C
C         CALCULATE USEFUL CONSTANTS.
C         P AND M CORRESPOND TO PHI-PLUS AND PHI-MINUS IN PSB NOTATION.
C
      IF (A .EQ. X .OR. B .EQ. X .OR. A .EQ. B)  GO TO 200
      IF (JAX .EQ. 0.0 .AND. JBX .EQ. 0.0)  GO TO 200
      AB = A-B
      C = (JAX - JBX)/2
      D = JAX + JBX
      E = 2*JAB
      SUM2 = (A + B)/2
      DPLUS = SQRT((AB + C)**2 + JAB*JAB)/2
      DMINUS = SQRT((AB - C)**2 + JAB*JAB)/2
      SIN2P = JAB/2/DPLUS
      SIN2M = JAB/2/DMINUS
      COS2P = (AB + C)/2/DPLUS
      COS2M = (AB - C)/2/DMINUS
C
C         CALCULATE THE FREQUENCIES OF THE TRANSITIONS
C
      FREQ(1) = SUM2 - (D + E)/4 - DMINUS
      FREQ(2) = SUM2 + (D - E)/4 - DPLUS
      FREQ(3) = SUM2 + (E - D)/4 - DMINUS
      FREQ(4) = SUM2 + (E + D)/4 - DPLUS
      FREQ(5) = SUM2 - (E + D)/4 + DMINUS
      FREQ(6) = SUM2 + (D - E)/4 + DPLUS
      FREQ(7) = SUM2 + (E - D)/4 + DMINUS
      FREQ(8) = SUM2 + (D + E)/4 + DPLUS
      FREQ(9) = X - D/2
      FREQ(10) = X + DPLUS - DMINUS
      FREQ(11) = X - DPLUS + DMINUS
      FREQ(12) = X + D/2
      FREQ(13) = X - DPLUS - DMINUS
      FREQ(14) = X + DPLUS + DMINUS
      FREQ(15) = 2*SUM2 - X
      NFREQ = 15
C
C         CALCULATE THE INTENSITIES
C
      INTEN(1) = 1 - SIN2M
      INTEN(2) = 1 - SIN2P
      INTEN(3) = 1 + SIN2M
      INTEN(4) = 1 + SIN2P
      INTEN(5) = INTEN(3)
      INTEN(6) = INTEN(4)
      INTEN(7) = INTEN(1)
      INTEN(8) = INTEN(2)
      INTEN(9) = 1
      INTEN(10) = (1 + COS2P*COS2M + SIN2P*SIN2M)/2
      INTEN(11) = INTEN(10)
      INTEN(12) = 1
      INTEN(13) = (1 - COS2P*COS2M - SIN2P*SIN2M)/2
      INTEN(14) = INTEN(13)
      INTEN(15) = 0
C
C         ASSIGN TRANSITIONS;  PREPARE FOR POSSIBLE PLOT.
C
      IF (A .GT. B)  GO TO 96
      CHAR1 = ALET
```

```
         CHAR2 = BLET
         GO TO 97
   96 CHAR1 = BLET
         CHAR2 = ALET
   97 DO 98 I=1,4
         CHAR(I) = CHAR1
         CHAR(I+4) = CHAR2
   98 CHAR(I+8) = XLET
         CHAR(13) = XSTAR
         CHAR(14) = XSTAR
         CHAR(15) = STAR
         NLINES = 56
C
C        SORT THE FREQUENCIES IN INCREASING ORDER
C
  100 II = NFREQ - 1
         DO 140 I=1,II
         KK = I
         JJ = KK + 1
  110 IF (FREQ(KK) - FREQ(JJ)) 140,140,120
  120 TEMP = FREQ(JJ)
         FREQ(JJ) = FREQ(KK)
         FREQ(KK) = TEMP
         TEMP = INTEN(JJ)
         INTEN(JJ) = INTEN(KK)
         INTEN(KK) = TEMP
         TEMP = CHAR(JJ)
         CHAR(JJ) = CHAR(KK)
         CHAR(KK) = TEMP
         KK = KK - 1
         IF (KK) 140,140,130
  130 JJ = KK + 1
         GO TO 110
  140 CONTINUE
C
C        WRITE OUT THE TRANSITIONS AND THEIR INTENSITIES
C
         WRITE (6,150)
         WRITE (6,151) (FREQ(I),INTEN(I),CHAR(I),I=1,NFREQ)
         GO TO (160,152,152,154,156,158), KEY
  152 WRITE (6,153)
         GO TO 160
  154 WRITE (6,155)
         GO TO 160
  156 WRITE (6,157) A,X
         GO TO 160
  158 WRITE (6,159)
         SWITCH = .TRUE.
C
C        CHECK PLOTTING OPTION
C
  160 WRITE (6,35)
         READ (5,36) ANSWER
         IF (ANSWER .NE. YES) GO TO 10
         WRITE (6,1620)
         IF (KEY .EQ. 2 .OR. KEY .EQ. 3)  WRITE (6,1605)
C
C
C        CONSTRUCT AND OUTPUT GRAPH
C
```

```
C
  161 DO 162 I=1,NLINES
  162 LINE(I) = 0
C
C          IF NOT INITIALIZING THE PLOTTING OF AN ABX SPECTRUM   GO TO 170
C
      IF (.NOT. SWITCH)   GO TO 170
C
C          SEPARATE ABX CASE INTO TWO PLOTS
C
C          THE ABX SPECTRUM IS PLOTTED IN TWO PARTS.  FIRST THE AB PART,
C          THEN THE X PART.  THE PLOT GRAPHS THE TRANSITIONS FROM NFLOW TO
C          NFREQ, SO THIS SECTION SETS THESE LIMITS PROPERLY FOR THE AB
C          PART AND STORES THE CORRESPONDING LIMITS FOR THE X PART OF THE
C          SPECTRUM IN NFLSAV AND NFHSAV.
C
      WRITE(6,1595)
      IF (A .LT. X)   GO TO 166
      NFLOW = 7
      NFREQ = 14
      NFHSAV = 6
      NFLSAV = 1
      GO TO 171
  166 NFLOW = 2
      NFREQ = 9
      NFHSAV = 15
      NFLSAV = 10
C
C          CONSTRUCT GRAPH
C
  170 NFREQ = NFREQ - NFBDNH
  171 INCRE = (FREQ(NFREQ) - FREQ(NFLOW))/(NLINES-11)
      LOW = FREQ(NFLOW) - 5.0*INCRE
      DO 180 I=NFLOW,NFREQ
      POINT = LOW
      DO 175 II=1,NLINES
      LINE(II) = LINE(II) + (INTEN(I)*PW)/((POINT-FREQ(I))**2+PWSQ)
  175 POINT = POINT + INCRE
  180 CONTINUE
C
C          SCALE GRAPH
C
      MAXI = LINE(6)
      II = NLINES - 5
      DO 185 I=7,II
  185 IF (LINE(I) .GT. MAXI) MAXI = LINE(I)
      SCALE = MAXI/(WIDE-1.0)
C
C          OUTPUT GRAPH
C
      DO 195 I=1,NLINES
      LENGTH = (LINE(I)+SCALE*.5)/SCALE + 1
      IF (MOD(I-1,5) .EQ. 0) GO TO 190
      WRITE (6,188) (STAR,II=1,LENGTH)
      GO TO 195
  190 IF (I .EQ. 51 .AND. KEY .EQ. 5) GO TO 193
      WRITE (6,192) LOW,(STAR,II=1,LENGTH)
      GO TO 1945
  193 WRITE (6,194) (STAR,II=1,LENGTH)
```

```
 1945 LOW = LOW + 5*INCRE
  195 CONTINUE
C
C         IF FINISHED PLOTTING, ASK FOR NEXT SPECTRUM
C
      IF (.NOT. SWITCH) GO TO 10
C
C         PLOT THE X PART OF THE SPECTRUM FOR THE ABX CASE
C
      NLINES = 51
      NFLOW = NFLSAV
      NFREQ = NFHSAV
      WRITE (6,199)
      SWITCH = .FALSE.
      GO TO 161
C
C         RIDICULOUS INPUT FOUND.  CALCULATION OF SPECTRUM ABORTED.
C
  200 WRITE (6,1610)
      GO TO 10
C
C         FORMAT STATEMENTS
C
   20 FORMAT(///' NMR SPECTRUM SIMULATION PROGRAM'//5X,'THIS PROGRAM ','
     .SIMULATES THE FOLLOWING NMR SPECTRA:'//' 1 - AB    2 - AB
     .3 - AB    4 - A B    5 - A X    6 - ABX '/16X,'2',9X,'3',8X,'2
     .3',8X,'2 2'// ' ENTER THE NUMBER CORRESPONDING TO THE SPECTRUM YO
     .U WISH TO SIMULATE.'/'    (ZERO TERMINATES THE PROGRAM)'/)
   25 FORMAT (I)
   35 FORMAT (/' DO YOU WANT A PLOT OF THE CALCULATED SPECTRUM?'/
     .'    ENTER 1 FOR YES, 0 FOR NO.'/)
   36 FORMAT (A1)
   42 FORMAT ( ///,' THE AB CASE',//,' ENTER THE CHEMICAL SHIFT DIFFEREN
     .CE BETWEEN A AND B (in Hz.)  '/)
   43 FORMAT (4F)
   45 FORMAT(///,15X,'AB NMR SPECTRUM'//7X,'Jab =',F6.2,5X,'SHIFT DIFFER
     .ENCE =',F7.2//)
   51 FORMAT ( ///' THE AB  CASE'/'         2'//)
   52 FORMAT (' ENTER THE CHEMICAL SHIFTS OF THE A AND B PROTONS (in Hz.
     .)  '/)
   53 FORMAT (/' ENTER THE COUPLING CONSTANT Jab. '/)
   54 FORMAT ( ///15X,'AB  NMR SPECTRUM'/17X,'2'//' THE CHEMICAL SHIFTS
     .ARE:',9X,'A =',F7.2,7X,'  B =',F7.2//' THE COUPLING CONSTANTS ARE:
     .Jab =',F7.2,7X,'Jbb = IMMATERIAL'//)
   61 FORMAT ( ///' THE AB  CASE'/'         3'//)
   62 FORMAT ( ///15X,'AB  NMR SPECTRUM'/17X,'3'//' THE CHEMICAL SHIFTS
     .ARE:',9X,'A =',F7.2,7X,'  B =',F7.2//' THE COUPLING CONSTANTS ARE:
     .Jab =',F7.2,7X,'Jbb = IMMATERIAL'//)
   71 FORMAT ( ///' THE A B  CASE'/'        2 3'//)
   72 FORMAT ( ///15X,'A B  NMR SPECTRUM'/16X,'2 3'//' THE CHEMICAL SHIF
     .TS ARE:',9X,'A =',F7.2,9X,'B =',F7.2//' THE COUPLING CONSTANT IS:
     .',5X,'Jab =',F7.2//)
   81 FORMAT ( ///' THE A X  CASE'/'        2 2'//' ENTER THE CHEMICAL SHI
     .FTS OF THE A AND X PROTONS (IN HZ.) '/)
   82 FORMAT(/'  ENTER THE COUPLING CONSTANTS Jaa,Jax,Jax'' AND Jxx.'/)
   84 FORMAT ( ///15X,'A X  NMR SPECTRUM'/16X,'2 2'//' THE CHEMICAL SHIF
     .TS ARE:',8X,'A =',F7.2,9X,'X =',F7.2//' THE COUPLING CONSTANTS ARE
     .:   Jaa =',F7.2,7X,'Jax =',F7.2//30X,6HJax' =,F7.2,7X,'Jxx =',F7.2
     .//)
```

```
 91 FORMAT ( ///' THE ABX CASE'//' ENTER THE CHEMICAL SHIFTS OF THE A,
   .B AND X PROTONS (in Hz.)   '/)
 93 FORMAT (' ENTER THE COUPLING CONSTANTS Jab, Jax AND Jbx. '/)
 94 FORMAT ( ///15X,'ABX NMR SPECTRUM'///' THE CHEMICAL SHIFTS ARE:',
   .7X,'A =',F7.2,5X,'B =',F7.2,5X,'X ='F7.2//' THE COUPLING CONSTANTS
   .ARE:  Jab =',F7.2,3X,'Jax =',F7.2,3X,'Jbx =',F7.2//)
150 FORMAT (/14X,'FREQUENCY    INTENSITY'/)
151 FORMAT(9X,F12.2,F13.3,6X,A2)
153 FORMAT (/'  * INDICATES A COMBINATION (FORBIDDEN) TRANSITION.'/)
155 FORMAT (/'  + INDICATES THE CALCULATED INTENSITY IS THE SUM OF TWO
   .OR MORE'/' COINCIDENT TRANSITIONS.  THE SECOND ORDER PERTURBATION
   .APPROXIMATION '/' USED TO CALCULATE THIS SPECTRUM ALSO GIVES NEGA
   .TIVE RATHER THAN ZERO'/' INTENSITY FOR CERTAIN TRANSITIONS;  ALL N
   .EGATIVE INTENSITIES HAVE BEEN'/' SET TO ZERO.  FORBIDDEN TRANSITION
   .S HAVE BEEN IGNORED.  THE EXACT '/' SOLUTION INVOLVES AN ADDITIONA
   .L 9 COMBINATION (FORBIDDEN) TRANSITIONS.'/)
157 FORMAT(/' THIS IS THE A2 HALF OF THE CALCULATED SPECTRUM.'/' THE C
   .ALCULATED FREQUENCIES ARE RELATIVE TO A ='F7.2,4H Hz./' THE X2 HAL
   .F OF THE SPECTRUM IS IDENTICAL TO THE A2 HALF'/' BUT IS CENTERED A
   .BOUT X =',F8.2,' HZ.'/)
159 FORMAT ( /' THE TRANSITIONS DESIGNATED * AND X* ARE COMBINATION (F
   .ORBIDDEN)'/' TRANSITIONS.'/)
188 FORMAT (13X,118A1)
192 FORMAT (1X,F9.2,3X,118A1)
194 FORMAT ('    CENTER   ',118A1)
199 FORMAT (///' THE X PART OF THE CALCULATED SPECTRUM'//)
1595 FORMAT (' THE AB PART OF THE CALCULATED SPECTRUM'//)
1605 FORMAT ( ' TO IMPROVE THE RESOLUTION OF THE GRAPHICAL OUTPUT,'
   ./' WEAK COMBINATION TRANSITIONS ARE IGNORED.'//)
1610 FORMAT (//' RIDICULOUS INPUT FOUND.'/' EITHER THE CHEMICAL SHIFT D
   .IFFERENCE(S) OR THE COUPLING CONSTANT(S)'/' OR BOTH ARE ZERO.  CAL
   .CULATION OF THIS SPECTRUM ABORTED.'//)
1620 FORMAT (//3X)
    END
```

Discussion

NMR requires 3107 words of core memory. The approximate execution times, including plots, for the six spin systems are

Spin system	AB	AB_2	AB_3	A_2B_3	A_2X_2	ABX
Execution time (sec)	0.2	0.3	0.3	0.3	0.3	0.3

Problem 2.16 Use NMR to find the set of chemical shifts and coupling constants that give a good fit to an observed nmr spectrum.

Solution. Figure 2.6 was drawn from data provided by NMR.

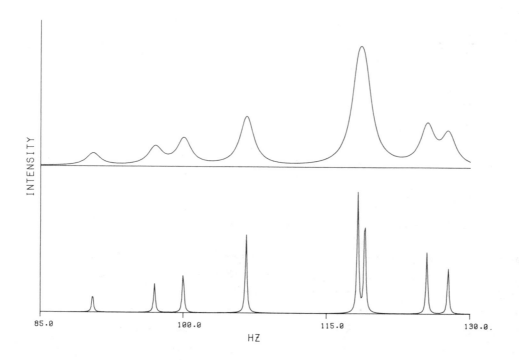

<u>Figure 2.6</u> NMR spectrum for the AB$_2$ case with chemical shifts A = 100.0 and
B = 121.9 and coupling constant J$_{AB}$ = 8.2. The spectrum on top was calculated by
modifying the line width parameter until a reasonable fit to the experimental
spectrum of 2,6-lutidine [20] was obtained.

<u>Problem 2.17</u> Write a computer program to simulate first-order electron spin
resonance spectra [21].

2.9 TITR

<u>Introduction</u>

TITR is a data reduction program that approximates the two thermodynamic
dissociation constants of an amino acid from experimental titration data. The two
dissociation reactions are

$$H_3NCHRCOOH^+ \rightleftharpoons H^+ + H_3NCHRCOO \qquad K_1$$

$$H_3NCHRCOO \rightleftharpoons H^+ + H_2NCHRCOO^- \qquad K_2$$

Here H$_3$NCHRCOO represents an amino acid in the neutral or Zwitterion form. The
equilibrium constants are

$$K_1 = \frac{(a_{H^+})(a_{H_3NCHRCOO})}{a_{H_3NCHRCOOH^+}}$$

and

$$K_2 = \frac{(a_{H^+})(a_{H_2NCHRCOO^-})}{a_{H_3NCHRCOO}}$$

Here a stands for activity, not concentration [22].

Two titrations are performed:

1. $H_3NCHRCOO + H^+ \longrightarrow H_3NCHRCOOH^+$

2. $H_3NCHRCOO + OH^- \longrightarrow H_2O + H_2NCHRCOO^-$

A standard HCl solution is used as the titrant in part 1, and a standard NaOH solution is the titrant in part 2. The input data includes the initial concentration and volume of the amino acid solution, the concentration of the titrant (HCl or NaOH), and the titration data, i.e., the volume of titrant added and the corresponding observed pH. TITR echoes the input data for verification purposes and prints a table containing observed volume, observed pH, the ionic strength of the solution, the univalent activity coefficient (see below), and the calculated thermodynamic dissociation constants. TITR also calculates the average value and standard deviation of the dissociation constant.

Method

To simplify notation, let

$$HA^+ = H_3NCHRCOOH^+ \qquad A = H_3NCHRCOO$$

and

$$A^- = H_2NCHRCOO^-$$

Then,

$$HA^+ \rightleftharpoons H^+ + A \qquad K_1 = \frac{(a_{H^+})(a_A)}{a_{HA^+}}$$

and

$$A \rightleftharpoons H^+ + A^- \qquad K_2 = \frac{(a_{H^+})(a_{A^-})}{a_A}$$

The hydrogen ion activity a_{H^+} is measured directed using the glass electrode. The activities a_{HA^+} and a_{A^-} are approximated using the Debye-Huckel limiting law [22],

$$a_{HA^+} = [HA^+]f \qquad \text{and} \qquad a_{A^-} = [A^-]f$$

where f is the univalent activity coefficient, given by

$$\log f = -0.51 \sqrt{u}$$

where μ is the ionic strength of the solution,

$$\mu = 0.5 \sum_{i=1}^{n} c_i z_i^2$$

c_i and z_i are the molar concentration and ionic charge of species i, respectively. The activity of the neutral species HA is assumed to be equal to its molar concentration.

1. <u>HCl titration</u>. Let C_a and V_o be the initial concentration and volume of A, and let C_{HCl} be the concentration of the titrant HCl solution. The charge balance equation in part 1 is

$$[H^+] + [HA^+] = [Cl^-] + [OH^-]$$

So

$$[HA^+] = [Cl^-] + [OH^-] - [H^+]$$

The mass balance equation is

$$A_{tot} = [HA^+] + [A] + [A^-]$$

Assuming in part 1, $[A^-] \ll [HA^+] + [A]$,

then

$$[A] = A_{tot} - [HA^+]$$

or

$$[A] = A_{tot} - [Cl^-] - [OH^-] + [H^+]$$

Finally

$$K_1 = \frac{(a_{H^+})(a_A)}{a_{HA^+}}$$

$$= \frac{(a_{H^+})[A]}{[HA^+]f}$$

$$= \frac{a_{H^+}}{f} \left\{ \frac{A_{tot} - [Cl^-] - [OH^-] + [H^+]}{[Cl^-] + [OH^-] - [H^+]} \right\} \qquad (2.6)$$

Here

$$a_{H^+} = 10^{-pH} \qquad\qquad [Cl^-] = \frac{C_{HCl}V_{HCl}}{V_{tot}}$$

$$[H^+] = \frac{a_{H^+}}{f} \qquad\qquad A_{tot} = \frac{C_a V_o}{V_{tot}}$$

$$[OH^-] = \frac{K_w}{[H^+]} \qquad V_{tot} = V_o + V_{HCl}$$

TITR first reads C_a, V_o, and C_{HCl}. Then a series of data pairs (V_{HCl}, pH) are read. For each data pair the program calculates K_1 in the following sequence of operations,

1. $V_{tot} = V_o + V_{HCl}$.

2. $A_{tot} = C_a V_o / V_{tot}$.

3. $\mu = 0.5 \sum_{i=1}^{n} c_i z_i^{\;2}$

4. $\log f = -0.51 \sqrt{\mu}$.

5. $a_{H^+} = 10^{-pH}$.

6. $[H^+] = a_{H^+}/f$

7. $[OH^-] = a_{OH^-}/f = 1.0 \times 10^{-14}/(a_{H^+})f$.

8. Equation (2.6) is used to calculate K_1.

For example, $V_o = 100$ ml, $C_a = 0.05$ $\underline{M}$, and $C_{HCl} = 0.010$ $\underline{M}$. When $V_{HCl} = 20.0$ ml and pH = 3.55, then

1. $V_{tot} = 100 + 20 = 120$ ml

2. $A_{tot} = 0.05 \times 100/120 = 0.0417$ $\underline{M}$

3. $[Cl^-] = 0.1 \times 20/120 = 0.0167$ $\underline{M}$

   $\mu = 0.5([HA^+] + [H^+] + [Cl^-] + [OH^-])$

Assume

$[HA^+] = [Cl^-] \qquad [H^+] \ll [Cl^-]$

and

$[OH^-] \ll [H^+]$

then $\mu = [Cl^-] = 0.0167$ $\underline{M}$

4. $\log f = -0.51 \sqrt{0.0167} = -0.0658$

   $f = 10^{-0.0658} = 0.859$

5. $a_{H^+} = 10^{-pH} = 10^{-3.55} = 0.000282$

6. $[H^+] = a_{H^+}/f = 0.000328$

7. $[OH^-] = K_w/(a_{H^+})f = 1.0 \times 10^{-14}/(.000282 \times 0.859)$

   $= 4.13 \times 10^{-11}$

8. $K_1 = \dfrac{2.82 \times 10^{-4}}{0.859} \left\{ \dfrac{0.0417 - 0.0167 - 4.13 \times 10^{-11} + 3.28 \times 10^{-4}}{0.0167 + 4.13 \times 10^{-11} - 3.28 \times 10^{-4}} \right\}$

$\quad = 5.06 \times 10^{-5}$

2. <u>NaOH Titration</u>. Let C_a and V_o be the initial concentration and volume of A, and let C_{NaOH} be the concentration of the NaOH solution. The charge balance equation in part 2 is

$$[Na^+] + [H^+] = [A^-] + [OH^-]$$

So

$$[A^-] = [Na^+] + [H^+] - [OH^-]$$

The mass balance equation is

$$A_{tot} = [HA^+] + [A] + [A^-]$$

Assuming in part 2

$$[HA^+] \ll [A] + [A^-]$$

then

$$[A] = A_{tot} - [A^-]$$

or

$$[A] = A_{tot} - [Na^+] - [H^+] + [OH]$$

therefore,

$K_2 = \dfrac{(a_{H^+})(a_{A^-})}{a_A}$

$\quad = \dfrac{(a_{H^+})[A^-]f}{[A]}$

$\quad = (a_{H^+})f \left\{ \dfrac{[Na^+] + [H^+] - [OH^-]}{A_{tot} - [Na^+] - [H^+] + [OH^-]} \right\}$

The program executes the same sequence of operations discussed above for part 1. For example, V_o = 100 ml, C_a = 0.050 $\underline{M}$, and C_{NaOH} = 0.10 $\underline{M}$. When V_{NaOH} = 20.0 ml and pH = 8.50, then

1. V_{tot} = 100 + 20 = 120 ml

2. $A_{tot} = \dfrac{0.05 \times 100}{120}$ = 0.0417 $\underline{M}$

3. $[Na^+] = \dfrac{0.1 \times 20}{120}$ = 0.0167 $\underline{M}$

$\quad \mu = 0.5([Na^+] + [H^+] + [OH^-] + [A^-])$

Assume

$$[A^-] = [Na^+] \qquad [OH^-] \ll [Na^+]$$

and

$$[OH^-] \ll [Na^+]$$

Then $\mu = [Na^+] = 0.0167 \underline{M}$

4. $\log f = -0.51\sqrt{\mu} = -0.51\sqrt{0.0167} = -0.0658$

   $f = 10^{-0.0658} = 0.859$

5. $a_{H^+} = 10^{-pH} = 10^{-8.50} = 3.15 \times 10^{-9}$

6. $[H^+] = \dfrac{a_{H^+}}{f} = 3.68 \times 10^{-9}$

7. $[OH^-] = \dfrac{K_w}{(a_{H^+})f}$

   $= \dfrac{1.0 \times 10^{-14}}{3.16 \times 10^{-9} \times 0.859} = 3.68 \times 10^{-6}$

8. $K_2 = (3.16 \times 10^{-9} \times 0.859)\left\{\dfrac{0.0167 + 3.68 \times 10^{-9} - 3.68 \times 10^{-6}}{0.0417 - 0.0167 - 3.68 \times 10^{-9} + 3.68 \times 10^{-6}}\right\}$

   $= 1.81 \times 10^{-9}$

At the end of each part, TITR calculates an average value for K and the standard deviation of this average using the following equations:

$$K_{avg} = \frac{\sum\limits_{i=1}^{N} K_i}{N}$$

and

$$K_{avg} = \left[\frac{\sum\limits_{i=1}^{N} (K_i - K_{avg})^2}{N - 1}\right]^{1/2}$$

Here K_i is one of the N values for K_1 calculated in part 1 and for K_2 calculated in part 2, and K_{avg} is the standard deviation of the average.

Sample Execution

HCL TITRATION

ENTER THE INITIAL CONCENTRATION OF B-ALANINE
>.04

ENTER THE INITIAL VOLUME OF THE B-ALANINE SOLUTION
>100

ENTER THE CONCENTRATION OF THE TITRANT
>.1

TITRATION OF 100.00 ML OF 0.400E-01 MOLAR B-ALANINE
WITH 0.100   MOLAR TITRANT

ENTER THE VOLUME OF TITRANT AND THE CORRESPONDING PH
(VOLUME AND PH = 0   INDICATES END OF DATA)

		VOLUME	PH	IONIC STRENGTH	ACTIVITY COEFFICIENT	K
>7.66	3.6					
		7.66	3.60	7.11E-03	0.906	1.23E-03
>11.49	3.42					
		11.49	3.42	1.03E-02	0.888	1.13E-03
>15.32	3.28					
		15.32	3.28	1.33E-02	0.873	1.04E-03
>19.15	3.11					
		19.15	3.11	1.61E-02	0.862	1.09E-03
>22.98	2.98					
		22.98	2.98	1.87E-02	0.852	1.06E-03
>26.81	2.82					
		26.81	2.82	2.11E-02	0.843	1.13E-03
>30.64	2.67					
		30.64	2.67	2.35E-02	0.835	1.19E-03
>0	0					

AVERAGE K =   1.13E-03   +/-   6.7E-05

                         NAOH TITRATION

ENTER THE INITIAL CONCENTRATION OF B-ALANINE
>.04

ENTER THE INITIAL VOLUME OF THE B-ALANINE SOLUTION
>100

ENTER THE CONCENTRATION OF THE TITRANT
>.1

TITRATION OF 100.00 ML OF 0.400E-01 MOLAR B-ALANINE
WITH 0.100   MOLAR TITRANT

ENTER THE VOLUME OF TITRANT AND THE CORRESPONDING PH
(VOLUME AND PH = 0   INDICATES END OF DATA)

| | | IONIC | ACTIVITY | |
VOLUME	PH	STRENGTH	COEFFICIENT	K
>8.02 7.38				
8.02	7.38	7.42E-03	0.904	9.45E-09
>12.03 7.62				
12.03	7.62	1.07E-02	0.885	9.13E-09
>16.04 7.8				
16.04	7.80	1.38E-02	0.871	9.24E-09
>20.05 7.98				
20.05	7.98	1.67E-02	0.859	9.04E-09
>24.06 8.15				
24.06	8.15	1.94E-02	0.849	9.07E-09
>28.07 8.34				
28.07	8.34	2.19E-02	0.840	9.03E-09
>32.08 8.52				
32.08	8.52	2.43E-02	0.833	1.02E-08
>0 0				

AVERAGE K =   9.31E-09   +/-   4.1E-10

Listing

```
C
C           TITR
C
C    TITR READS PH-VOLUME DATA FOR THE TITRATION OF BETA-
C    ALANINE WITH BOTH STANDARD HCL AND STANDARD NAOH
C
C    TITR CALCULATES THE IONIC STRENGTH OF THE SOLUTION,
C    APPROXIMATES THE ACTIVITY COEFFICIENT USING
C    THE DEBYE-HUCKEL LIMITING LAW, CALCULATES THE APPROPRIATE
C    DISSOCIATION CONSTANT FOR EACH POINT ON THE TITRATION
C    CURVE, THE AVERAGE VALUE AND THE STANDARD DEVIATION
C    OF THE DISSOCIATION CONSTANT.
C
C  GLOSSARY:
C
C  CONBAL   -  MOLARITY OF BETA-ALANINE
C  CONTIT   -  MOLARITY OF THE TITRANT (HCL OR NAOH)
C   VOL     -  INITIAL VOLUME OF BETA-ALANINE SOLUTION
C   BAL     -  NUMBER OF MOLES OF BETA-ALANINE INITIALLY
C    V      -  VOLUME OF TITRANT
C    PH     -  OBSERVED PH
C    CA     -  CONCENTRATION OF ACIDIC SPECIES CORRECTED FOR DILUTION
C    CB     -  CONCENTRATION OF BASIC SPECIES CORRECTED FOR DILUTION
C    Fl     -  UNIVALENT ACTIVITY COEFFICIENT
C    H      -  ACTIVITY OF H(+)
C    K      -  THERMODYNAMIC DISSOCIATION CONSTANT
C  SIGMA    -  STANDARD DEVIATION OF K
C
C     AUTHORS:  S. B. LEVITT AND K. J. JOHNSON
C
```

```
C
      REAL K
C
C    INITIALIZATION
C
      IWHICH=0
   10 SUM=0
      SUMSQ=0.
      I=0
C
C    INPUT
C
      IF(IWHICH.EQ.0)WRITE(6,11)
      IF(IWHICH.EQ.1)WRITE(6,22)
      READ(5,33)CONBAL
      WRITE(6,44)
      READ(5,33)VOL
      WRITE(6,55)
      READ(5,33)CONTIT
      WRITE(6,66)VOL,CONBAL,CONTIT
      BAL=CONBAL*VOL
C
C    READ VOLUME AND CORRESPONDING PH
C
   20 READ(5,33)V,PH
C
C    IF VOLUME = 0, NO MORE DATA EXPECTED
C
      IF(V.EQ.0. .AND. PH.EQ.0. )GO TO 90
C
C    CALCULATE THE CONCENTRATION OF THE  ACIDIC AND BASIC SPECIES, THE
C    ACTIVITY COEFFICIENT, AND THE HYDROGEN ION CONCENTRATION
C
      VTOT=VOL+V
      CA=(V*CONTIT)/VTOT
      CB=(BAL-V*CONTIT)/VTOT
      F1=10.**(-.51*SQRT(CA))
      H=10.**(-PH)
C
C    IF IWHICH .GT. 1, TITRATION WITH NAOH, OTHERWISE WITH HCL
C
      IF(IWHICH .EQ. 1) GO TO 50
      FUDGE=H/F1
      C1=CB
      C2=CA
      GO TO 51
   50 FUDGE=H*F1
      C1=CA
      C2=CB
C
C    CALCULATE THE THERMODYNAMIC DISSOCATION CONSTANT
C
   51 K=FUDGE*(C1+(H/F1)-((1.0E-14)/(F1*H)))/
     1 (C2-(H/F1)+((1.0E-14)/(F1*H)))
      WRITE(6,77)V,PH,CA,F1,K
C
C    AVERAGE K AND ITS STANDARD DEVIATION FOUND
C
```

```
          I=I+1
          SUM=SUM+K
          SUMSQ=SUMSQ+K*K
          GO TO 20
   90     XB=SUM/I
          SQSUM=SUM**2
          SIGMA=(SQRT(ABS(SUMSQ-SQSUM/I)/(I-1)))
          WRITE(6,88)XB,SIGMA
C
C    IF IWHICH=1 STOP
C
          IF(IWHICH .EQ. 1 ) STOP
          IWHICH=1
          GO TO 10
C
C    FORMATS
C
   11 FORMAT(//25X,'HCL TITRATION'///'  ENTER THE INITIAL',
      1 ' CONCENTRATION OF B-ALANINE'/)
   22 FORMAT(//25X,'NAOH TITRATION'///'  ENTER THE INITIAL',
      1 ' CONCENTRATION OF B-ALANINE'/)
   33 FORMAT(2F)
   44 FORMAT(//'  ENTER THE INITIAL VOLUME OF THE B-ALANINE',
      1 ' SOLUTION'/)
   55 FORMAT(//'  ENTER THE CONCENTRATION OF THE TITRANT'/)
   66 FORMAT(//'  TITRATION OF ',F6.2,' ML OF ',G10.3,
      1 ' MOLAR B-ALANINE'/,'  WITH',G10.3,' MOLAR TITRANT'//,
      2 '  ENTER THE VOLUME OF TITRANT AND THE CORRESPONDING PH',
      3 /'  (VOLUME AND PH = 0  INDICATES END OF DATA)'//,
      4 37X,'IONIC      ACTIVITY',
      5 /15X,'  VOLUME      PH      STRENGTH  COEFFICIENT',8X,'K'//)
   77    FORMAT(15X,F7.2,3X,F7.2,3X,1PE10.2,0PF7.3,7X,1PE10.2,/)
   88 FORMAT(//'  AVERAGE K = ',1PE12.2,3X,'+/-',E12.1)
          STOP
          END
```

Discussion

TITR requires 521 words of core memory. The execution time for parts 1 and 2, each with 10 volume-pH pairs, was approximately 0.3 sec.

TITR contains two optimization features worth mentioning. First, an earlier version of this program contained two very nearly identical sections of code, one for part 1 and the other for part 2. The code was reduced by nearly a factor of 2 by using the common variable name CION and introducing the flag IPART. Second, the program calculates the standard deviation of the average (σ) as follows (the summations extend from i = 1 to i = N):

$$X_{avg} = \frac{\sum\limits_{i=1}^{N} X_i}{N}$$

$$\sigma^2 = \frac{\sum\limits_{i=1}^{N} (X_i - X_{avg})^2}{N - 1}$$

$$\sigma^2(N - 1) = \sum_{i=1}^{N} (X_i^2 + 2X_i X_{avg} + X_{avg}^2)$$

$$= \sum_{i=1}^{N} X_i^2 - \frac{2\left(\sum\limits_{i=1}^{N} X_i\right)^2}{N} + \frac{\left(\sum\limits_{i=1}^{N} X_i\right)^2}{N}$$

$$= \sum_{i=1}^{N} X_i^2 - \frac{\left(\sum\limits_{i=1}^{N} X_i\right)^2}{N}$$

So

$$\sigma = \left[\frac{\sum\limits_{i=1}^{N} X_i^2 - \left(\sum\limits_{i=1}^{N} X_i\right)^2 /N}{N - 1}\right]^{1/2}$$

It is therefore not necessary to use subscripted variable names. The program accumulates $\sum\limits_{i=1}^{N} X_i$ in SUM and $\sum\limits_{i=1}^{N} X_i^2$ in SUMSQ.

2.10 ADDITIONAL PROBLEMS

1. Write programs to generate the data for distribution and formation curves for polyprotic acids and complexes [23].

2. Calculate the possible elemental compositions corresponding to a measured ion mass in mass spectrometry [24, 25].

3. Write a computer program to simulate a set of reaction mechanisms for which closed-form expressions are available [2, 26, 27].

4. Calculate all the bond angles in molecules of the form AX_nY_m given some of the angles [28].

5. Determine all possible molecular formulas consistent with a given set of analytical and molecular weight data [29-31]. For example, a compound contains the following elemental composition (% by weight):

$$5.80 \leq C \leq 58.8$$
$$5.3 \leq H \leq 6.1$$
$$1.8 \leq N \leq 2.2$$
$$4.0 \leq S \leq 4.9$$
$$28.0 \leq O \leq 30.9$$

The molecular weight range is between 680 and 750. One possible compound is $C_{34}H_{37}NSO_{13}$, which has molecular weight 699.706.

6. Perform a numerical experiment with a large set of random numbers to show various statistical properties of the numbers, for example, the average, standard deviation, confidence intervals, etc. [32].

7. See Refs. 32-37 for examples of problems that can be solved using closed-form algorithms.

REFERENCES

1. H. A. Laitinen, "Chemical Analysis," McGraw-Hill, New York, 1960, p. 35ff.

2. A. A. Frost and R. G. Pearson, "Kinetics and Mechanism," Wiley, New York, 1961, pp. 175-177.

3. J. C. Davis, Jr., "Advanced Physical Chemistry," Ronald, New York, 1965, pp. 161-165.

4. J. C. Slater, Phys. Rev., 36, 57 (1930).

5. J. A. Pople and D. L. Beveridge, "Approximate Molecular Orbital Theory," McGraw-Hill, New York, 1970, p. 27ff.

6. L. Pauling, "The Nature of the Chemical Bond," 3d, ed., Cornell, Ithica, N.Y., 1960.

7. C. A. Coulson, "Valence," Oxford University Press, Oxford, England, 1961.

8. A Streitwieser, Jr., and P. H. Owens, "Orbital and Electron Density Diagrams," Macmillan, New York, 1973.

9. J. W. Moore and W. G. Davies, Illustration of some consequences of the indistinguishability of electrons -- use of computer-generated dot density diagrams, J. Chem. Educ., 53, 426 (1976).

10. S. L. Holgren and J. S. Evans, Accurate contours for sp(alpha) hybrid orbitals, J. Chem. Educ., 51, 189 (1974).

11. M. J. S. Dewar and J. Kelemen, LCAO MO theory illustrated by application to H_2, J. Chem. Educ., 48, 494 (1971).

12. K. E. Banyard, Electron correlation in atoms and molecules, J. Chem. Educ., 47, 669 (1970).

13. W. T. Bordass and J. W. Linnett, A new way of presenting atomic orbitals, J. Chem. Educ., 47, 672 (1970).

14. I. Cohen and J. B. Bene, Hybrid orbitals in molecular orbital theory, J. Chem. Educ., 46, 487 (1969).

15. E. A. Ogryzlo and G. B. Porter, Contour surfaces for atomic and molecular orbitals, J. Chem. Educ., 40, 256 (1963).

16. C. J. Nyman and R. E. Hamm, "Chemical Equilibrium," Raytheon Education Co., 1968, pp. 28-39.

17. J. A. Pople, W. G. Schneider, and H. J. Bernstein, "High Resolution Nuclear Magnetic Resonance," McGraw-Hill, New York, 1959, Chapter 6.

18. J. W. Emsley, J. Feeney, and L. H. Sutcliffe, "High Resolution Nuclear Magnetic Resonance Spectroscopy," Vol. 1, Pergamon Press, London, 1965, Chapter 8.

19. K. B. Wiberg and B. J. Nist, "The Interpretation of NMR Spectra," Benjamin, Menlo Park, Calif., 1962.

20. J. A. Pople et al., cp. cit., pp. 123-128.

21. A. C. Ling, A computer program for simulating first-order electron spin resonance spectra, J. Chem. Educ., 51, 174 (1974).

22. H. A. Laitinen, cp. cit., pp. 7-13.

23. J. N. Butler, "Ionic Equilibrium," Addison-Wesley, Reading, Mass., 1964.

24. H. M. Bell, Computer analysis of isotope clusters in mass spectrometry, J. Chem. Educ., 51, 548 (1974).

25. B. Mattson and E. Carberry, A new program for the calculation of mass spectrum isotope peaks, J. Chem. Educ., 50, 511 (1973).

26. C. H. Bamford and C. Tippur, "Comprehensive Chemical Kinetics," vol. 2, Elsevier, New York, 1969.

27. W. C. Child, Jr., A computer simulation of a kinetics experiment, J. Chem. Educ., 50, 290 (1973).

28. W. Li and T. Mak, Bond angle relationships in some AX_nY_m molecules, J. Chem. Educ., 51, 571 (1974).

29. G. Reade and D. J. Sne, A computer programme to aid in the determination of molecular formulae, J. Chem. Soc., C, 906 (1966).

30. D. A. Usher, J. Z. Gougoutas, and R. B. Woodward, Digital computer program for calculation of molecular formulae, Anal. Chem., 37, 330 (1965).

31. J. Lederberg, "Computation of Molecular Formulas for Mass Spectrometry,"
     Holden-Day, San Francisco, 1964.

32. G. Beech, "Fortran IV in Chemistry," Wiley, New York, 1975.

33. T. R. Dickson, "The Computer and Chemistry," Freeman, San Francisco, 1968.

34. T. L. Isenhour and P. C. Jurs, "Introduction to Computer Programming for
     Chemists," Allyn and Bacon, Boston, 1972.

35. R. R. Roskos, "Problem Solving in Physical Chemistry," West, New York, 1975.

36. L. Soltzberg, A. Shah, J. Saber, and E. Carty, "BASIC and Chemistry," Houghton
     Mifflin, Boston, 1975.

37. P. A. Cauchon, "Chemistry with a Computer," Educomp., Hartford, Conn., 1976.

CHAPTER 3

ROOTS OF EQUATIONS

Finding the roots of an equation or a system of simultaneous equations is one of the most important applications of numerical analysis for scientists and engineers. A function of x, f(x), is said to have a root, x_r, if $f(x_r) = 0$. For example, the radial wavefunction R(r) for the 3s electron in an excited hydrogen atom is [1]

$$R(r) = c(27 - \frac{18r}{a_o} + \frac{2r^2}{a_o^2}) \ e^{-r/3a_o}$$

Here r is the radial distance of the electron from the nucleus (the proton) in angstroms, c is a normalizing constant, and $a_o = 0.529$ Å. Let c = 1 and $x = r/a_o$, then the radial function becomes

$$f(x) = (27 - 18x + 2x^2)e^{-x/3}$$

Figure 3.1 is a plot of this function over the range

$$0 \le x \le 15$$

There are two roots in this region,

$$f(1.90) = f(7.10) = 0.$$

Therefore, there are two nodes in the 3s radial function for the hydrogen atom, at 1.01 (0.529 x 1.90) and 3.76 (0.529 x 7.10) Å.

Consider the polynomial,

$$f(x) = a_o + a_1x + a_2x^2 + \cdots + a_mx^m$$

There are closed-form expressions for m = 2 (quadratic), m = 3 (cubic), and m = 4 (quartic). Except in special cases, iterative techniques must be used to find the roots of polynomials of order greater than 4. Iterative methods must also be used to solve such transcendental equations as the 3s radial function.

Eight computer programs that find roots of polynomial and transcendental equations are documented in this chapter. Brief descriptions of these programs are given in Table 3.1.

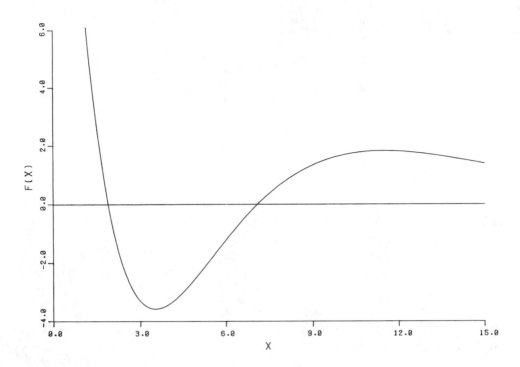

Figure 3.1   The 3s radial function.

Table 3.1   Programs That Find Roots of Equations

Program	Description
EDTA	Solves the M^{2+}-EDTA titration system; a quadratic equation is solved.
VDWGAS	Calculates the compressibility of a van der Waals gas; a subroutine, CUBIC, finds the real and complex roots of a cubic equation.
BINBIS	Program to find the root of an equation by the binary bisection method.
BOX	Uses the binary bisection method to calculate the eigenvalues and eigenvectors of a particle in a finite box; the associated eigenvectors are also calculated.
NEWTON	Program to find a root of a function using the Newton-Raphson method.
HNA	Calculates the equilibrium concentrations of all species in the chemical systems involving the weak acids HA-NaA, H_2A-NaHA-Na_2A, and H_3A-NaH$_2$A-Na_2HA-Na_3A.
HNATRN	Simulates the three weak acid-strong base titration systems HA-NaOH, H_2A-NaOH, and H_3A-NaOH.
GASEQ	Solves gas-phase equilibrium problems by calculating the equilibrium extent of reaction.

3.1 CLOSED-FORM SOLUTIONS

Two equations for which closed-form expressions are available will be considered
here:

    1. Quadratic equations, $f(x) = ax^2 + bx + c = 0$

    2. Cubic equations, $f(x) = ax^3 + bx^2 + cx + d = 0$

The two roots of the quadratic equation are

$$x\underline{+} = \frac{-b \pm x\sqrt{b^2 - 4ac}}{2a} \tag{3.1}$$

This solution is disarmingly simple. A computer program to solve quadratic
equations should be prepared to handle a number of contingencies. For example,

    1. If $a = 0$, Eq. (3.1) will fail, $x = -c/b$.

    2. If $b = 0$, Eq. (3.1) will fail if $4ac > 0$.

    3. If $c = 0$, one of the roots will be 0.

    4. If $b^2 - 4ac < 0$, the roots are imaginary and the program must avoid
        passing a negative argument to the SQRT function.

    5. If $b^2 \gg 4ac$, then both truncation and round-off errors may accumulate
        in the positive root,

$$x_+ = \frac{-b + (b + \epsilon)}{2a} = \frac{g}{2a}$$

        Here $b + \epsilon = \sqrt{b^2 - 4ac}$ and g is a relatively small number which possibly
        contains significant error due to both truncation (SQRT) and round-off
        $(-b + b)$ errors. To minimize this error the program should use double
        precision.

The following program contains code to solve a quadratic equation.

3.1.1 EDTA

Introduction

    The program EDTA simulates M^{2+}-EDTA titrations [2]. The shape of the titration
curve depends on three parameters: the pH at which the titration system is buffered,
the formation constant of the M^{2+}-EDTA complex, and the initial concentration of the
metal ion. The input includes values for these three parameters, the initial volume
of the metal ion solution, and the concentration of the titrant EDTA solution. The
program prints the input data for verification purposes and a table containing
percent of titration, the negative logarithm of the free metal ion concentration
(pM), the volume of EDTA added, and the concentrations of MY^{2-} and Y^{4-} at each
point $(H_4Y = EDTA)$.

Method

Let C_m and V_m be the initial concentration and volume of the metal ion solution, and C_y and V be the concentration and volume of the titrant EDTA solution. Then the mass action and mass balance equations are

$$H_4Y \rightleftharpoons H^+ + H_3Y^- \qquad K_1$$

$$H_3Y^- \rightleftharpoons H^+ + H_2Y^{2-} \qquad K_2$$

$$H_2Y^{2-} \rightleftharpoons H^+ + HY^{3-} \qquad K_3$$

$$HY^{3-} \rightleftharpoons H^+ + Y^{4-} \qquad K_4$$

$$M^{2+} + Y^{4-} \rightleftharpoons MY^{2-} \qquad K_f$$

$$M_{tot} = \frac{C_m V_m}{V_m + V} = [M^{2+}] + [MY^{2-}] \tag{3.2}$$

$$Y_{tot} = \frac{C_y V}{V_m + V} = [MY^{2-}] + U \tag{3.3}$$

Here U is the total concentration of uncomplexed EDTA,

$$U = [H_4Y] + [H_3Y^-] + [H_2Y^{2-}] + [HY^{3-}] + [Y^{4-}]$$

Equations (3.2) and (3.3) are the mass balance equations for total M^{2+} and EDTA, respectively. It is assumed that MY^{2-} is the only complex formed between M^{2+} and EDTA. It is also assumed that volumes are additive, the pH is constant, that activity effects can be ignored, and that the stepwise dissociation constants for EDTA are $pK_1 = 2.18$, $pK_2 = 2.73$, $pK_3 = 6.0$, and $pK_4 = 10.0$.

The output of this program is the tabulated function

$$pM = f(V) \qquad 0 \leq V \leq V_{200}$$

Here $pM = -\log [M^{2+}]$ and V_{200} is the volume of the titrant (EDTA) that corresponds to 200% of titration. There are two approaches to this problem. The first, which corresponds more closely to the experimental procedure, is to vary the volume of titrant EDTA over the specified range by an appropriate increment and calculate the concentration of M^{2+} remaining in the solution. The alternative approach is to derive the inverse function in closed form, that is,

$$V = f(pM) \qquad pM_0 \leq pM \leq pM_{200}$$

Here pM_0 and pM_{200} are the negative logarithms of the initial and final M^{2+} concentrations, respectively. The latter approach is used here for reasons discussed below.

First, an expression relating U to $[Y^{4-}]$ is derived.

$$U = [H_4Y] + [H_3Y^-] + [H_2Y^{2-}] + [HY^{3-}] + [Y^{4-}]$$

$$= [Y^{4-}]\left\{ 1 + \frac{[H^+]}{K_1} + \frac{[H^+]^2}{K_1K_2} + \frac{[H^+]^3}{K_1K_2K_3} + \frac{[H^+]^4}{K_1K_2K_3K_4} \right\}$$

Let f_4 be the fraction of uncomplexed EDTA present as Y^{4-}. Then,

$$f_4 = \frac{[Y^{4-}]}{U}$$

$$= \frac{K_1K_2K_3K_4}{[H^+]^4 + K_1[H^+]^3 + K_1K_2[H^+]^2 + K_1K_2K_3[H^+] + K_1K_2K_3K_4} \tag{3.4}$$

Equation (3.3) becomes

$$\frac{C_yV}{V_m + V} = [MY^{2-}] + \frac{[Y^{4-}]}{f_4}$$

$$= [MY^{2-}] + \frac{[MY^{2-}]}{f_4K_f[M^{2+}]}$$

Solving for $[MY^{2-}]$,

$$[MY^{2-}] = \frac{f_4K_f[M^{2+}]C_yV}{(1 + f_4K_f[M^{2+}])(V_m + V)}$$

Equation (3.2) becomes

$$\frac{C_mV_m}{V_m + V} = [M^{2+}] + \frac{f_4K_f[M^{2+}]C_yV}{(1 + f_4K_f[M^{2+}])(V_m + V)}$$

Finally, solving for V,

$$V = \frac{-V_m([M^{2+}]K_f + 1/f_4 - C_m/f_4[M^{2+}]) - C_mK_f}{C_yK_f + [M^{2+}]K_f + 1/f_4} \tag{3.5}$$

Here $[M^{2+}]$ is considered the independent variable, and is varied over the range corresponding to 200% of titration. At 0% of titration, $V = 0$ and $[M^{2+}] = C_m$. At 200% of titration, 2 equiv of EDTA have been added. The relevant equilibrium is

$$MY^{2-} \rightleftharpoons M^{2+} + Y^{4-}$$

Let x by the equilibrium concentration of M^{2+} at 200% of titration. Then

$$[MY^{2-}] = \frac{C_mV_m}{V_m + V_{200}} - x$$

and

$$[Y^{4-}] = f_4 \left\{ \frac{C_m V_m}{V_m + V_{200}} + x \right\}$$

where

$$V_{200} = \frac{2C_m V_m}{C_y} \qquad (2 \text{ equiv of EDTA})$$

Let

$$Q = \frac{C_m V_m}{V_m + V_{200}}$$

then

$$K_f = \frac{[MY^{2-}]}{[M^{2+}][Y^{4-}]} = \frac{Q - x}{f_4 x(Q + x)}$$

Usually, to a good approximation, $x \ll Q$, so

$$x = [M^{2+}] \simeq \frac{1}{f_4 K_f}$$

However, if this approximate value for x exceeds $0.01Q$ the quadratic equation is solved,

$$x^2 + \frac{Q + 1}{f_4 K_f} x - \frac{Q}{f_4 K_f} = 0$$

Only the positive root is physically significant, so

$$x = [M^{2+}] = \frac{-b + \sqrt{b^2 - 4ac}}{2a}$$

where

$$a = 1 \qquad b = Q + \frac{1}{f_4 K_f} \qquad c = \frac{-Q}{f_4 K_f}$$

To summarize, the execution of EDTA proceeds in the following steps:

1. Read pH, C_m, V_m, and C_y.
2. Calculate f_4 using Eq. (3.4), pM_o, and pM_{200} using the quadratic formula if necessary.
3. Increment pM in 20 steps from pM_o to pM_{200}; for each value of pM use Eq. (3.5) to calculate the volume of titrant EDTA that has to be added to obtain that pM.
4. Print a table containing percent of titration, pM, V, $[Y^{4-}]$, and $[MY^{2-}]$.

Sample Execution (Ca^{2+})

ENTER THE CONCENTRATION AND VOLUME OF THE M(2+) SOLUTION
> .1 50

ENTER THE CONCENTRATION OF THE EDTA SOLUTION
> .1

ENTER THE PH AND THE COMMON LOGARITHM OF THE
FORMATION CONSTANT OF THE M(2+)-EDTA COMPLEX
> 7 10.7

PH =   7.00   LOG(KF) =   10.70

M(2+) CONC. =   1.00E-01   VOL. OF M(2+) =   50.00

AND CONC. OF EDTA =   1.00E-01

% TITR	PM	V	Y(4-)	MY(2-)
0.000	1.000	0.000	0.00E+00	0.00E+00
36.443	1.332	18.222	1.14E-11	2.67E-02
64.341	1.664	32.171	3.60E-11	3.92E-02
81.641	1.995	40.821	8.87E-11	4.49E-02
91.008	2.327	45.504	2.02E-10	4.76E-02
95.709	2.659	47.854	4.45E-10	4.89E-02
97.980	2.991	48.990	9.67E-10	4.95E-02
99.058	3.323	49.529	2.09E-09	4.98E-02
99.568	3.654	49.784	4.49E-09	4.99E-02
99.816	3.986	49.908	9.65E-09	4.99E-02
99.952	4.318	49.976	2.07E-08	5.00E-02
100.058	4.650	50.029	4.45E-08	5.00E-02
100.201	4.981	50.100	9.55E-08	4.99E-02
100.466	5.313	50.233	2.05E-07	4.99E-02
101.017	5.645	50.508	4.38E-07	4.97E-02
102.191	5.977	51.095	9.36E-07	4.95E-02
104.707	6.309	52.354	1.98E-06	4.88E-02
110.107	6.640	55.053	4.15E-06	4.76E-02
121.698	6.972	60.849	8.44E-06	4.51E-02
146.581	7.304	73.291	1.63E-05	4.06E-02
200.000	7.636	100.000	2.88E-05	3.33E-02

Listing

```
C
C
C                    EDTA
C
C         SIMULATES THE M(2+) - EDTA TITRATION
C
C     ASSUMPTIONS:
C                    1.   CONSTANT PH
```

```
C                    2.   FOR EDTA, PK(I),I=1,4 ARE:
C                              2.18,2.73,6.20,10.0
C                    3.   VOLUMES ARE ADDITIVE
C                    4.   MY(2- IS THE ONLY COMPLEX FORMED
C                    5.   ACTIVITY EFFECTS CAN BE IGNORED
C
C     GLOSSARY:
C
C       PH          PH
C       KF          FORMATION CONSTANT FOR MY(2-)
C       CM          INITIAL CONC. OF M(2+)
C       VM          INITIAL VOLUME OF M(2+)
C       CY          CONCENTRATION OF TITRANT EDTA
C       HPLUS       <H(+)>
C       M           <M(2+)>
C       MY          <MY(2-)>
C       Y4          <Y(4-)>
C       PRCT        % OF TITRATION
C       V           VOLUME OF EDTA ADDED
C       ALPHA4      FRACTION OF TOTAL UNCOMPLEXED EDTA AS Y(4-)
C
C     INPUT SPECIFICATIONS:  PH,KF,CM,VM,CY    (5F)
C
C       AUTHORS:   D. L. DOERFLER AND K. J. JOHNSON
C
      IMPLICIT DOUBLE PRECISION (A-H,O-Z)
      DOUBLE PRECISION M,MY,KF
C
C     INPUT
C
      WRITE(6,11)
      READ(5,22)CM,VM
      WRITE(6,12)
      READ(5,22)CY
      WRITE(6,13)
      READ(5,22)PH,KF
      IF(CM.EQ.0)STOP
      WRITE(6,33)PH,KF,CM,VM,CY
C
C     SET CONSTANTS
C
      HPLUS=10.**(-PH)
      KF=10.**KF
      ALPHA4=1./(((((1.289D21*HPLUS+8.511D18)*HPLUS
     1 + 1.584D16)*HPLUS + 1.0D10)*HPLUS + 1.0)
C
C     INITIAL PM
C
      PMO=-DLOG10(CM)
C
C     FINAL PM
C
      V=2*CM*VM/CY
      Q=CM*VM/(VM+V)
      M=1./KF/ALPHA4
      IF(M/Q.LT.0.001)GO TO 50
      B=Q+1/(KF*ALPHA4)
      C=Q/KF/ALPHA4
      M=(-B+DSQRT(B*B+4.0*C))/2
```

```
 50  PM=-DLOG10(M)
C
C   SET CONSTANTS
C
     DLTAPM=(PM-PMO)/20
     PM=PMO
     V=0
     CONST1=1./ALPHA4-CM*KF
     CONST2=CY*KF+1/ALPHA4
     CONST3=100.*CY/CM/VM
     CONST4=CM*VM
     PRCT=0.
     Y4=0.
     MY=0.
     WRITE(6,66)
     WRITE(6,55)PRCT,PMO,V,Y4,MY
C
C   INTERMEDIATE VALUES
C
     DO 100 I=1,20
     PM=PM+DLTAPM
     M=10**(-PM)
     V=-VM*(M*KF-CM/ALPHA4/M+CONST1)/(M*KF+CONST2)
     PRCT=V*CONST3
     VTOT=VM+V
     MY=CONST4/VTOT-M
     Y4=MY/(M*KF)
100  WRITE(6,55)PRCT,PM,V,Y4,MY
     STOP
C
 11  FORMAT(//'  ENTER THE CONCENTRATION AND VOLUME OF THE',
    1 ' M(2+) SOLUTION')
 12  FORMAT(//'  ENTER THE CONCENTRATION OF THE EDTA SOLUTION')
 13  FORMAT(//'  ENTER THE PH AND THE COMMON LOGARITHM OF THE'
    1/'  FORMATION CONSTANT OF THE M(2+)-EDTA COMPLEX')
 22  FORMAT(2F)
 33  FORMAT(/'   PH = ',F6.2,15X,'LOG(KF) = ',F6.2//'  M(2+) CONC.
    1 = ',1PE10.2,10X,'VOL. OF M(2+) = ',0PF6.2//'  AND CONC. OF
    2 EDTA = ',1PE10.2/)
 55  FORMAT(0P3F10.3,1P2E14.2)
 66  FORMAT(//4X,'% TITR',7X,'PM',8X,'V',9X,'Y(4-)',9X,'MY(2-)'/)
     END
```

Discussion

EDTA requires 451 words of memory and executes in approximately 0.2 sec. The calculation of f_4 [Eq. (3.4)] presents interesting programming problems. If the pH is 10 or more, then $[H^+]^4 \leq 10^{-40}$, a value that will cause an exponential underflow on a DECSystem-10 computer. EDTA uses the substitutions,

$$b_1 = \frac{1}{K_4} \qquad b_2 = \frac{1}{K_3 K_4}$$

$$b_3 = \frac{1}{K_2 K_3 K_4} \qquad b_4 = \frac{1}{K_1 K_2 K_3 K_4}$$

Then

$$f_4 = \frac{1}{b_4[H^+]^4 + b_3[H^+]^3 + b_2[H^+]^2 + b[H^+] + 1}$$

$$= \frac{1}{(((b_4[H^+] + b_3)[H^+] + b_2)[H^+] + b_1)[H^+] + 1}$$

Only four multiplications are required and the underflow problem is eliminated for all reasonable values of pH.

There are two advantages to using the inverse function method. The first is efficiency of code. The alternative approach would involve solving the quadratic equation associated with the equilibrium reaction

$$MY^{2-} \rightleftharpoons M^{2+} + Y^{4-}$$

The quadratic equation would have to be solved for each point on the titration curve except the trivial case of 0% of titration. The second advantage is that the most interesting region of the titration curve is the region

$$90 < \% \text{ titration} < 110$$

The use of the inverse function approach guarantees that this region of the titration receives the attention it deserves.

Problem 3.1. Use EDTA to demonstrate the effect of the three parameters, pH, K_f, and the initial metal ion concentration on the shape of the titration curve.

Solution. See Figure 3.2.

Problem 3.2. Modify EDTA so that the titration system includes a ligand or chelating agent as a masking agent [3].

Problem 3.3. Calculate the minimum pH for an effective EDTA titration as a function of K_f and C_m [4].

The second closed-form solution considered here is for the cubic equation [5].

$$y^3 + py^2 + qy + r = 0 \qquad (3.6)$$

Equation (3.6) can be simplified to

$$x^3 + ax + b = 0$$

by the substitutions

$$x = y + \frac{p}{3} \qquad a = \frac{3q - p^2}{3}$$

and

$$b = \frac{2p^3 - 9pq + 27r}{27}$$

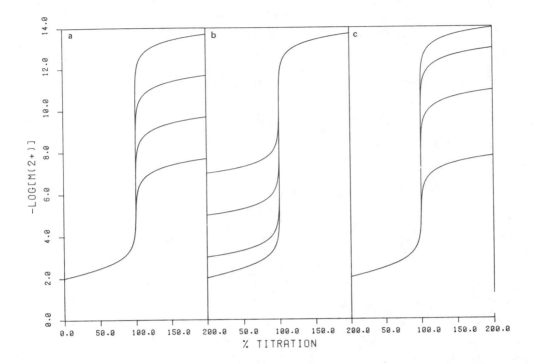

Figure 3.2 (a) EDTA titration curves as a function of K_f: $[M^{2+}]_o$ = 0.01 M; pH = 10; log K_f = 8 (bottom), 10, 12, and 14 (top).

The three roots of Eq. (3.6) are

$$x_1 = A + B$$
$$x_2 = 0.5[-(A + B) + (A - B)\sqrt{-3}]$$
$$x_3 = 0.5[-(A + B) - (A - B)\sqrt{-3}]$$

where

$$A = (-\frac{b}{2} + \sqrt{c})^{1/3}$$
$$B = (-\frac{b}{2} - \sqrt{c})^{1/3}$$
$$c = \frac{b^2}{4} + \frac{a^3}{27}$$

The roots of the original cubic equation (3.6) are

$$y_i = x_i - \frac{p}{3} \qquad i = 1, 2, 3$$

There are three possibilities for cubic equations:

1. If $c > 0$, there is one real root and two conjugate imaginary roots.
2. If $c = 0$, there are three real roots, two of which are equal.
3. If $c < 0$, there are three real and unequal roots.

A trigonometric solution can be used in the third case:

Let

$$\cos \theta = \frac{-b}{2 \sqrt{-a^3/27}} \quad \text{and} \quad d = \sqrt{\frac{-a}{3}}$$

Then the three roots are

$$x_1 = 2d \cos \frac{\theta}{3}$$

$$x_2 = 2d \cos \frac{\theta + 2\pi}{3}$$

$$x_3 = 2d \cos \frac{\theta + 4\pi}{3}$$

This method requires an arc cosine function,

$$\theta = \cos^{-1}\left\{ \frac{-b}{2\sqrt{-a^3/27}} \right\}$$

<u>Problem 3.4.</u> Find the roots of the following cubic equations:

$$
\begin{aligned}
&\text{a.} \quad y^3 - y^2 + 9y - 9 = 0 \qquad (c > 0) \\
&\text{b.} \quad y^3 + y^2 - y - 1 = 0 \qquad (c = 0) \\
&\text{c.} \quad y^3 - 2y^2 - y + 2 = 0 \qquad (c < 0)
\end{aligned}
$$

<u>Solution.</u>

a. $y = 1, -3i, 3i$ $\qquad (i = \sqrt{-1})$

b. $y = -1, 1, 1$

c. $y = -1, 1, 2$

The following program contains the subroutine CUBIC which finds the real and imaginary roots of cubic equations.

3.1.2 VDWGAS

Introduction

VDWGAS uses the closed-form solution to the cubic equation to calculate the molar volume of a real gas using the van der Waals equation of state,

$$\left(P + \frac{a}{V^2}\right)(V - b) = RT$$

Here a and b are the van der Waals parameters for the gas, and P, V, R, and T are the pressure in atmospheres, the molar volume in liters, the ideal gas constant,

0.08205 liter·atm/mol·K, and the absolute temperature, respectively. VDWGAS reads
values for a, b, T, P_i, and P_f, where P_i and P_f define the range of P,

$$P_i \leq P \leq P_f$$

The program varies P over this range and prints a table containing the pressure
and the three roots $(V_1, V_2,$ and $V_3)$ of the cubic equation.

Method

The van der Waals equation of state is cubic in V,

$$V^3 + pV^2 + qV + r = 0 \tag{3.7}$$

where

$$p = -b - \frac{RT}{P} \qquad q = \frac{a}{P} \qquad r = \frac{-ab}{P}$$

VDWGAS computes the critical temperature T_c, the temperature above which the gas
cannot be liquified,

$$T_c = \frac{8a}{27Rb}$$

If $T \geq T_c$, there will be only one real root for Eq. (3.7). Otherwise, there will
be either one or three real roots, depending on the values of P and T. If there
are three real roots, $V_1 < V_2 < V_3$, then V_1 and V_3 are the liquid and vapor molar
volumes, respectively, and V_2 has no physical significance [6, 7]. The program
prints a warning message if $T < T_c$.

VDWGAS uses the subprogram CUBIC to find the three roots of Eq. (3.7). The
calling sequence is

        CALL CUBIC(P,Q,R,X1,X2,X3,CX2,CX3,ICPLX)

Here P, Q, and R are the coefficients p, q, and r of Eq. (3.7); X1, X2, and X3 are
the real roots of the cubic; CX2 and CX3 are the conjugate imaginary roots; and
ICPLX is assigned the value 0 if the three roots are real, and 1 if they are complex.

Sample Execution (H_2O at 700 and 298 K)

```
ENTER A,B,T,PI,PF
   (UNITS:  ATM, LITERS, KELVIN;  A=0 TERMINATES THE PROGRAM)
>5.464   0.03049   700   .1   10


A =     5.46          B =     0.305E-01      T =      700.

THE CRITICAL TEMPERATURE IS     647.
```

P	Vl	V2 (REAL & IMAGINARY)		V3 (REAL & IMAGINARY)	
0.100	574.	0.476E-01	0.302E-01	0.476E-01	-0.302E-01
1.20	47.8	0.476E-01	0.253E-01	0.476E-01	-0.253E-01
2.30	24.9	0.476E-01	0.253E-01	0.476E-01	-0.253E-01
3.40	16.8	0.477E-01	0.253E-01	0.477E-01	-0.253E-01
4.50	12.7	0.477E-01	0.253E-01	0.477E-01	-0.253E-01
5.60	10.2	0.477E-01	0.253E-01	0.477E-01	-0.253E-01
6.70	8.51	0.478E-01	0.253E-01	0.478E-01	-0.253E-01
7.80	7.30	0.478E-01	0.253E-01	0.478E-01	-0.253E-01
8.90	6.39	0.478E-01	0.254E-01	0.478E-01	-0.254E-01
10.0	5.68	0.479E-01	0.254E-01	0.479E-01	-0.254E-01

```
ENTER A,B,T,PI,PF
     (UNITS:  ATM, LITERS, KELVIN;  A=0 TERMINATES THE PROGRAM)
> 5.464  0.03049  298  .1 10
```

```
A =     5.46          B =     0.305E-01       T =     298.
```

THE CRITICAL TEMPERATURE IS   647.

WARNING: T IS LESS THAN THE CRITICAL TEMPERATURE

P	Vl	V2 (REAL & IMAGINARY)		V3 (REAL & IMAGINARY)	
0.100	244.	0.364E-01	0.000	0.187	0.000
1.20	20.2	0.364E-01	0.000	0.189	0.000
2.30	10.4	0.364E-01	0.000	0.191	0.000
3.40	6.99	0.364E-01	0.000	0.192	0.000
4.50	5.23	0.364E-01	0.000	0.194	0.000
5.60	4.16	0.364E-01	0.000	0.196	0.000
6.70	3.45	0.364E-01	0.000	0.198	0.000
7.80	2.93	0.364E-01	0.000	0.200	0.000
8.90	2.54	0.364E-01	0.000	0.202	0.000
10.0	2.23	0.364E-01	0.000	0.205	0.000

Listing

```
C
C
C                        VDWGAS
C
C      VDWGAS CALCULATES THE MOLAR VOLUME OF A VAN DER WAALS
C      GAS AS A FUNCTION OF THE ABSOLUTE TEMPERATURE (T),
C      PRESSURE (P) AND THE VAN DER WAALS PARAMETERS A AND B.
C
C      THE VAN DER WAALS EQUATION OF STATE IS
C
C          (P+A/V**2)*(V-N*B)  = RT
C
C    INPUT:  A,B  -  VAN DER WAALS PARAMETERS
C             T   -  ABSOLUTE TEMPERATURE
C             PI  -  INITIAL PRESSURE (ATM)
C             PF  -  FINAL PRESSURE (ATM)
C
```

```
C    THE SUBROUTINE CUBIC IS CALLED TO CALCULATE THE
C    REAL AND IMAGINARY ROOTS OF THE CUBIC EQUATION.
C
C       CALL CUBIC(P,Q,R,X1,XW,X3,CX2,CX3,ICPLX)
C
C    HERE P,Q,AND R ARE THE COEFFICIENTS OF THE CUBIC EQUATION,
C    X1,X2, AND X3 ARE THE REAL ROOTS AND
C    CX2 AND CX3 ARE IMAGINARY ROOTS.  ICPLX IS SET TO 0
C    IF THE ROOTS ARE ALL REAL, 1 OTHERWISE.
C
C     AUTHOR:   K. J. JOHNSON
C
        COMPLEX CV2,CV3
C
C      INPUT
C
   10 WRITE(6,11)
      READ(5,22)A,B,T,PI,PF
      IF(A.EQ.0.0)STOP
C
C   SET CONSTANTS
C
      R=0.08205
      ZERO=0.0
      TC=8.*A/(27.*R*B)
      WRITE(6,33)A,B,T,TC
      IF(T.LT.TC)WRITE(6,35)
      PINC=(PF-PI)/9.
      WRITE(6,36)
C
C    MAIN LOOP
C
      DO 20 I=1,10
      P=PI+PINC*(I-1)
      CP=-(B+R*T/P)
      CQ=A/P
      CR=-A*B/P
      CALL CUBIC(CP,CQ,CR,V1,V2,V3,CV2,CV3,ICPLX)
      IF(ICPLX.EQ.1)GO TO 30
C
C    HAVE REAL ROOTS
C
      WRITE(6,55)P,V1,V2,ZERO,V3,ZERO
      GO TO 20
C
C    HAVE 1 REAL AND 2 CONJUGATE IMAGINARY ROOTS
C
   30 WRITE(6,55)P,V1,CV2,CV3
   20 CONTINUE
      GO TO 10
   11 FORMAT(//'  ENTER A,B,T,PI,PF',/3X,'  (UNITS: ',
     1'ATM, LITERS, KELVIN;  A=0 TERMINATES THE PROGRAM)'/)
   22 FORMAT(5F,I)
   33 FORMAT(/'  A = ',G12.3,5X,'  B = ',G12.3,5X,'  T = ',
     1G12.3,//'  THE CRITICAL TEMPERATURE IS ',G12.3//)
   36 FORMAT(//5X,'P',10X,'V1',9X,'V2 (REAL & IMAGINARY)',3X,
     1 'V3 (REAL & IMAGINARY)'/)
   35 FORMAT(/'  WARNING:  T IS LESS THAN THE CRITICAL TEMPERATURE'//)
   55 FORMAT(6G12.3)
      END
```

```
C
C
C                    CUBIC
C
C     CUBIC CALCULATES THE   ROOTS OF THE CUBIC EQUATION
C
C          X**3 + P*X**2 + Q*X + R = 0
C
C
C     THE ARC COSINE FUNCTION IS REQUIRED
C
      SUBROUTINE CUBIC(P,Q,R,X1,X2,X3,CX2,CX3,ICPLX)
      COMPLEX CX2,CX3
      ICPLX=0
      A=(3.*Q-P*P)/3.
      B=(2.*P*P*P-9.*P*Q+27.*R)/27.
      P3=P/3.
      A3=A*A*A/27.
      DISCR=B*B/4. + A3
      IF(DISCR)20,30,40
   20 CONTINUE
C
C     HAVE THREE REAL AND UNEQUAL ROOTS
C
      COSPHI=-B/(2.*SQRT(-A3))
      PHI=ACOS(COSPHI)
      CONST=2*SQRT(-A/3.)
      PHI3=PHI/3.
      X1=CONST*COS(PHI3)-P3
      X2=CONST*COS(PHI3+2.094395)-P3
      X3=CONST*COS(PHI3+4.188790)-P3
      GO TO 80
   30 CONTINUE
C
C     HAVE 3 REAL ROOTS   2 OF WHICH ARE EQUAL
C
      A=(-B/2.)**(1./3.)
      X1=2.*A-P3
      X2=-A-P3
      X3=X2
      GO TO 80
   40 CONTINUE
C
C     HAVE 1 REAL ROOT AND 2 CONJUGATE IMAGINARY ROOTS
C
      ICPLX=1
      ARG1=-B/2.
      ARG2=SQRT(DISCR)
      A=(ARG1+ARG2)**(1./3.)
      B=(ARG1-ARG2)**(1./3.)
      CPART=(A-B)*0.8660254
      RPART=-(A+B)/2. - P3
      X1=A+B-P3
      CX2=CMPLX(RPART,CPART)
      CX3=CMPLX(RPART,-CPART)
   80 RETURN
      END
```

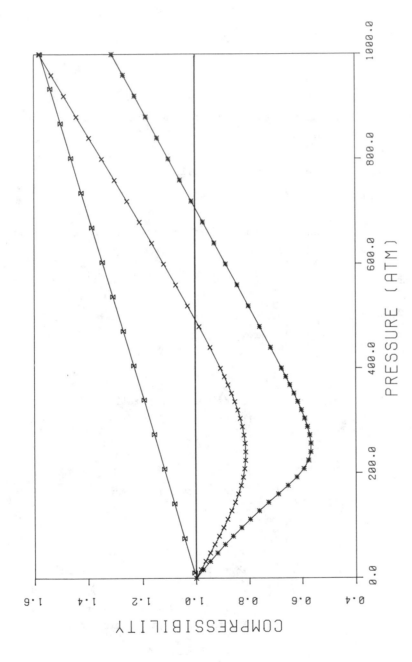

<u>Figure</u> 3.3   Compressibility curves for He (top), CO_2 and NH_3 (bottom) at 500 K.

Discussion

VDWGAS and CUBIC require 257 and 221 words of core, respectively. The execution time is approximately 0.2 sec. Note that the van der Waals equation of state, while cubic in volume, is linear in pressure,

$$P = \frac{RT}{V - b} - \frac{a}{V^2}$$

One could readily write an algorithm to generate a table of pressure as a function of volume. Such an algorithm would be more efficient than VDWGAS, but the output would not be linear in pressure.

Problem 3.5. Use VDWGAS to calculate the compressibility,

$$z = \frac{PV}{RT}$$

for a series of gases as a function of pressure. Plot the compressibility curves on the same graph.

Solution. See Figure 3.3.

Problem 3.6. Use VDWGAS to show how a real gas approaches ideal behavior at low pressure and high temperature.

3.2 THE BINARY BISECTION METHOD

This section and the following section are concerned with iterative rather than closed-form solutions to equations. The binary bisection or half-interval method [8] requires two values of x, x_1 and x_2, that bracket the root x_r,

$$x_1 \leq x_r \leq x_2$$

This interval is bisected,

$$x_3 = \frac{x_1 + x_2}{2}$$

and the function is evaluated at these three points. The procedure is illustrated in Figure 3.4. Here $f(x_1) < 0$ and $f(x_3) > 0$, so the root lies between x_1 and x_3. Now this interval is bisected,

$$x_4 = \frac{x_1 + x_3}{2}$$

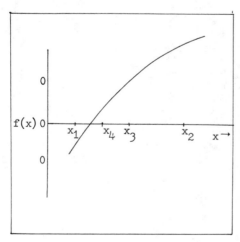

Figure 3.4 The binary bisection method.

Since $f(x_1) < 0$ and $f(x_4) > 0$, the root is now bracketed by x_1 and x_4. This procedure is continued until the interval

$$\frac{x_i + x_j}{x_i} \leq \varepsilon$$

where ε is an arbitrary convergence criterion, or until

$$\left| f(x_i) \right| \leq \varepsilon'$$

where ε' is a value arbitrarily close to 0.

The binary bisection method will converge if the initial values bracket at least one of the roots of the function. This iterative procedure is coded in the following program.

3.2.1 BINBIS

Introduction

BINBIS uses the binary bisection method to find a root of the function $f(x)$. The function is coded as an arithmetic statement function. The input to the program includes the following

ITMAX, the maximum number of iterations allowed

TOL, the relative convergence criterion

XLOW and XHI, the values of x which bracket the root

The program prints a table showing the progress of the binary bisection algorithm.

Method

Let

$$P_1 = f(x_{lo})f(x_{mid}) \qquad\qquad P_2 = f(x_{mid})f(x_{hi})$$

where

$$x_{mid} = \frac{x_{lo} + x_{hi}}{2}$$

If $P_1 < 0$, then the function changes sign on the interval

$$x_{lo} < x < x_{mid}$$

and the root is between x_{lo} and x_{mid}. In that case, the program reassigns x_{hi} and x_{mid},

$$x_{hi} = x_{mid} \qquad\qquad x_{mid} = \frac{x_{lo} + x_{hi}}{2}$$

and continues. On the other hand, if $P_1 > 0$, then the root is in the interval

$$x_{mid} < x < x_{hi}$$

and the program reassigns these as follows:

$$x_{lo} = x_{mid} \qquad x_{mid} = \frac{x_{lo} + x_{hi}}{2}$$

If either P_1 or P_2 is zero, then the root has been found.

   BINBIS iteratively evaluates P_1 and P_2 and modifies the limits x_{lo} and x_{ni} until the method converges, that is, until

$$\left| \frac{x_{ni} - x_{mid}}{x_{mid}} \right| \leq TOL$$

The following sample execution finds one of the two roots of the equation

$$f(x) = (27 - 18x + 2x^2)^{-x/3}$$

Sample Execution

ENTER ITMAX, TOL, XLO, AND XHI   (XLO AND XHI MUST BRACKET THE ROOT)
 >20 .0001 0 5

   ITMAX =   20   TOL =   0.100E-03

I	XLO	YLO	XMID	YMID	XHI	YHI
0	0.000	27.0	2.50	-2.39	5.00	-2.46
1	0.000	27.0	1.25	5.03	2.50	-2.39
2	1.25	5.03	1.88	0.151	2.50	-2.39
3	1.88	0.151	2.19	-1.35	2.50	-2.39
4	1.88	0.151	2.03	-0.666	2.19	-1.35
5	1.88	0.151	1.95	-0.275	2.03	-0.666
6	1.88	0.151	1.91	-0.665E-01	1.95	-0.275
7	1.88	0.151	1.89	0.409E-01	1.91	-0.665E-01
8	1.89	0.409E-01	1.90	-0.131E-01	1.91	-0.665E-01
9	1.89	0.409E-01	1.90	0.139E-01	1.90	-0.131E-01
10	1.90	0.139E-01	1.90	0.377E-03	1.90	-0.131E-01
11	1.90	0.377E-03	1.90	-0.635E-02	1.90	-0.131E-01
12	1.90	0.377E-03	1.90	-0.299E-02	1.90	-0.635E-02
13	1.90	0.377E-03	1.90	-0.131E-02	1.90	-0.299E-02

          THE ROOT IS   1.902

          AND THE FUNCTION EVALUATED AT THE ROOT IS   -0.465E-03

Listing
```
C
C
C                      BINBIS
C
C      PROGRAM TO FIND A ROOT OF AN EQUATION
C      BY THE BINARY BISECTION METHOD
C
C      THE EQUATION IS CODED IN THE STATEMENT FUNCTION, F(XX)
C
C      INPUT:  ITMAX  -  MAX.   OF ITERATIONS
C              TOL    -  RELATIVE CONVERGENCE TOLERANCE
```

```
C               XLO    -  LOWER LIMIT OF ROOT
C               XHI    -  UPPER LIMIT OF ROOT
C
C     AUTHOR:  K. J. JOHNSON
C
      DIMENSION X(3),Y(3)
      F(XX)=(27.-18.*XX+2.*XX**2)*EXP(-XX/3.)
C
C     INPUT
C
      WRITE(6,11)
      READ(5,22) ITMAX,TOL,X(1),X(3)
      WRITE(6,33)ITMAX,TOL
C
C     INITIALIZE VALUES
C
      X(2)=(X(1)+X(3))/2.
      DO 10 I=1,3
      XX=X(I)
   10 Y(I)=F(XX)
      I=0
      WRITE(6,44)I,(X(J),Y(J),J=1,3)
C
C     ITERATION LOOP
C
      DO 100 I=1,ITMAX
      P1=Y(1)*Y(2)
      P2=Y(2)*Y(3)
      IF(P1.LE.0.0)GO TO 20
C
C     THEN P2 SHOULD BE NEGATIVE.  CHECK FOR AN ERROR
C
      IF(P2.GT.0.0)GO TO 200
C
C   OK.  THEN  X(2) .LE. ROOT .LE. X(3)
C
      X(1)=X(2)
      Y(1)=Y(2)
      GO TO 90
C
C   THEN  X(1) .LE. ROOT .LE. X(2)
C
   20 X(3)=X(2)
      Y(3)=Y(2)
   90 X(2)=(X(1)+X(3))/2.
      XX=X(2)
      Y(2)=F(XX)
      IF(ABS((X(3)-X(2))/X(2)).LE.TOL)GO TO 150
  100 WRITE(6,44)I,(X(J),Y(J),J=1,3)
      WRITE(6,55)
      STOP
  150 WRITE(6,66)X(2),Y(2)
      STOP
  200 WRITE(6,77)
      STOP
   11 FORMAT(//' ENTER ITMAX, TOL, XLO, AND XHI  (XLO AND XHI',
     1 ' MUST BRACKET THE ROOT)'/)
   22 FORMAT(I,3E)
```

```
   33 FORMAT(/'  ITMAX = ',I3,20X,'TOL = ',G12.3,//
     1'   I',4X,'XLO',7X,'YLO',7X,' XMID',7X,'YMID',
     2 7X,'XHI',7X,'YHI'//)
   44 FORMAT(I3,6G11.3)
   55 FORMAT(//'    SORRY, FAILED TO CONVERGE IN ITMAX TRIES.'/)
   66 FORMAT(// 10X,'THE ROOT IS ',G13.4,//10X,'AND THE FUNCTION',
     1 ' EVALUATED AT THE ROOT IS ',G12.3/)
   77 FORMAT(//'  HAVE A PROBLEM!  XLO AND XHI',
     1 ' DO NOT BRACKET THE ROOT'//)
      STOP
      END
```

Discussion

BINBIS requires 322 words of core. This sample execution took approximately 0.2 sec. The execution time will depend on the complexity of the function and the number of iterations required for convergence.

It should be noted that the method used here (calculating P_1 and P_2) is equivalent to and considerably more efficient than an algorithm using logical expressions, e.g.,

   IF (Y(XLO) .LT. 0.0 .AND. Y(XMID) .GT. 0.0) GO TO 20

The following program uses the binary bisection method to find the roots of a transcendental equation.

3.2.2 BOX

Introduction

BOX calculates the energy levels and wavefunctions of a particle in a one-dimensional box with a finite potential barrier [9]. There are three parameters:

V_o, the magnitude of the potential barrier

a, the width of the well

m, the mass of the particle.

Box calculates the first four energy levels and the associated wavefunctions of the particle in the box.

Method

The Schrödinger equation for this system is

$$\frac{d^2\psi}{dx^2} + \frac{2m(E - V_o)}{\hbar^2} = 0$$

Here ψ is the wavefunction associated with the energy level E and $\hbar$ is Planck's constant divided by 2π. If $V_o = \infty$, then the eigenvalues (E) are given by

$$E_n = \frac{n^2 h^2}{8ma^2} \qquad n = 1,2,3,\ldots$$

For finite values of V_o the eigenvalues are the roots of the transcendental equation

$$\tan \frac{a\sqrt{2mE}}{\hbar} = \frac{2\sqrt{E(V_o - E)}}{2E - V_o} \qquad\qquad (3.8)$$

Figure 3.5 shows the graphical solution to the problem. The left-hand side of Eq. (3.8) is discontinuous at $n\pi/2$, where n is a positive integer. The right-hand side of Eq. (3.8) is discontinuous at $E = V_o/2$. The intersections of the two functions are the desired roots.

   Box uses the binary bisection method to iteratively approximate the first four roots of Eq. (3.8). Each root is bracketed as follows:

$$E_{lo} < E < E_{hi}$$

where E_{hi} is the ith energy level of the particle in an infinitely deep well, and E_{lo} is the value corresponding to $\pi/2$ radians less than E_{hi}. The binary bisection procedure iterates until the relative change in successive values of E is less than 0.00001.

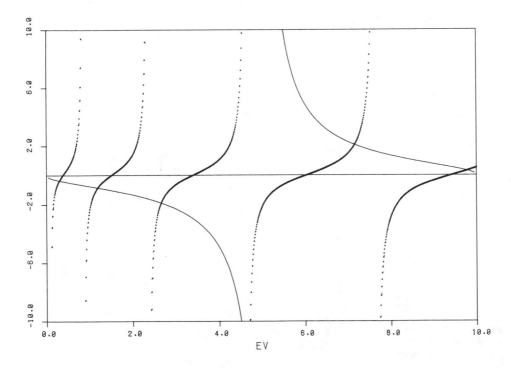

Figure 3.5. Graphical solution to the particle in a finite potential well problem (V_o = 10 eV and a = 10 Å).

The wavefunctions of the particle inside the potential well are given by [10]

$$\psi_s(x) = \cos kx \qquad \psi_a(x) = \sin kx$$

Here s and a represent symmetric $[\psi_s(x) = \psi_s(-x)]$ and antisymmetric $[\psi_a(x) = -\psi_a(-x)]$ functions. The wavefunctions with odd n are antisymmetric, those with even n are symmetric. The wavefunctions in the region $x \geq a/2$ are given by

$$\psi(x) = C^{-Kx}$$

where C is a normalizing constant and

$$K = \frac{\sqrt{2m(V_o - E)}}{\hbar}$$

The wavefunctions in the region $x < 0$ are computed from symmetry.

Sample Execution

```
ENTER WIDTH OF BOX (ANGSTROMS), HEIGHT OF BOX (IN EV),
AND MASS OF PARTICLE (ME=.0005486 AMU) ( A=0 ENDS PROGRAM)
> 10   10   0.0005486


A   =       10.0        ANGSTROMS
VO  =       10.0        EV
M   =     5.49E-04      AMU
```

I. EIGENVALUES

	FINITE BOX		INFINITE BOX
NUMBER	ENERGY IN EV	ENERGY IN ERG	ENERGY IN EV
1	0.298	4.77E-13	0.376
2	1.19	1.90E-12	1.50
3	2.65	4.25E-12	3.38
4	4.66	7.47E-12	6.02

II. EIGENVECTORS (SQUARED)

X	U1**2	U2**2	U3**2	U4**2
0.000	1.00	0.000	1.00	0.000
0.333E-08	0.991	0.342E-01	0.925	0.130
0.667E-08	0.966	0.132	0.721	0.452
0.100E-07	0.924	0.280	0.451	0.799
0.133E-07	0.867	0.459	0.196	0.991
0.167E-07	0.798	0.643	0.322E-01	0.927
0.200E-07	0.719	0.807	0.953E-02	0.642
0.233E-07	0.632	0.929	0.135	0.282
0.267E-07	0.540	0.993	0.370	0.362E-01
0.300E-07	0.447	0.989	0.645	0.311E-01

0.333E-07	0.356	0.919	0.875	0.270
0.367E-07	0.270	0.791	0.993	0.628
0.400E-07	0.191	0.623	0.962	0.920
0.433E-07	0.124	0.438	0.792	0.993
0.467E-07	0.694E-01	0.262	0.534	0.811
0.500E-07	0.298E-01	0.119	0.265	0.466
0.525E-07	0.134E-01	0.555E-01	0.132	0.258
0.550E-07	0.603E-02	0.259E-01	0.661E-01	0.143
0.575E-07	0.272E-02	0.121E-01	0.330E-01	0.790E-01
0.600E-07	0.122E-02	0.566E-02	0.165E-01	0.437E-01
0.625E-07	0.551E-03	0.265E-02	0.824E-02	0.242E-01

Listing

```
C
C
C                              BOX
C
C
C
C      THIS PROGRAM FINDS FOUR OF THE BOUND ENERGY STATES (E<V0)
C      OF A "PARTICLE IN A BOX".
C
C
C      -------------------          --------------------------
C                          :        :  A
C                          :        :  I
C                          :        :  I
C                          :        :  I   V0
C                          :        :  I
C                          :        :  I
C                          -------------------  V
C
C                       <-------A------->
C
C
C      BY SOLVING THE EQUATION:
C
C
C           2*PI*A SQRT( 2*M*E )        2*SQRT( (V0-E)*E )
C      TAN  -------------------    =    -------------------
C                    H                       2*E - V0
C
C
C
C      WHERE:
C
C           A     = WIDTH OF THE BOX (ANGSTROMS)
C           M     = MASS OF THE PARTICLE (AMU)
C           V0    = HEIGHT OF THE BARRIER (EV)
C           E     = ENERGY OF THE STATE (EV)
C           H     = PLANK'S CONSTANT
C           PI    = 3.14159265
C
C
C      THIS EQUATION IS SOLVED BY THE BINARY BISECTION METHOD
C
C      BOX ALSO CALCULATES THE EIGENVECTORS OF THESE STATES
```

```
C
C
C        THE ENERGY LEVELS FOR AN INFINITELY HIGH BOX ARE
C        GIVEN BY:
C
C                        (N*H)**2
C            E(N)  =   --------
C                        8*M*A**2
C
C
C
C        SYMBOLS:
C
C                A      = WIDTH OF THE BOX
C                V0     = HEIGHT OF THE BARRIER
C                MASS   = MASS OF THE PARTICLE
C                ANGLE  = STATEMENT FUNCT. TO FIND ARGUMENT OF THE TANGENT
C                DIFF   = STATEMENT FCT.  LEFT-HAND - RIGHT-HAND
C                         SIDE OF THE EQUATION
C                TOL    = TOLERANCE FOR CONVERGENCE
C                INF    = ENERGY OF A PARTICLE IN AN INFINITE BOX
C                OLDINF = PREVIOUS VALUE OF INF
C                ROOT   = ROOT OF THE EQUATION
C                DROOT  = LEFT - RIGHT-HAND SIDE OF EQN AT ROOT
C                LOW,HIGH = BOUNDS ON THE ROOT
C                DLOW   = LEFT - RIGHT-HAND SIDE OF EQN AT LOW
C                V2     = V0/2
C                ENERGY(4) = FIRST FOUR EIGENVALUES
C                U(4)      = THE CORRESPONDING EIGENVECTORS (SQUARED)
C                ARG(4)    = ARGUMENT FOR TRIG AND EXP FUNTIONS
C                CNST(4)   = PRE-EXPONENTIAL CONSTANT
C
C      AUTHORS:  D. L. DOERFLER AND K. J. JOHNSON
C
C
C
        REAL LOW,MASS,INF
        DIMENSION ENERGY(4),U(4),CNST(4),ARG(4)
        DATA TOL/.00001/
        ANGLE(V)=B*SQRT(V)
        DIFF(W)=(SIN(W)/COS(W))-((SQRT((V0-ROOT)*ROOT))/(ROOT-V0/2.0)
   10 WRITE(6,5)
        READ(5,7) A,V0,MASS
        IF(A.LE.0.)STOP
        WRITE(6,12) A,V0,MASS
C
C      CHECK FOR ZERO INPUT DATA
C      IF MASS = 0 OR BARRIER = 0   THERE ARE NO BOUND STATES
C      START OVER
C
        IF( MASS .LE. 0.0 .OR. V0 .LE. 0.0 ) GO TO 10
C
C      CHANGE FROM ANGSTROMS TO CM, EV TO ERGS, AMU TO GRAMS
C
        A = A * 1.0E-8
        V0 = V0 * 1.6021E-12
        MASS = MASS * 1.66039E-24
C
C      B = A * SQRT ( 2.0 * MASS ) / ( H / ( 2.0 * PI ))
```

```
C
        B = 1.341122E+27*A*SQRT(MASS)
        WRITE(6,13)
C
C       INITIALIZATION
C
        OLDINF = 0.0
        C = 2.467412/B**2
        D = (.8282E-27/MASS)*(6.6256E-27/(A**2))
        V2 = V0/2.0
C
C       BEGIN SEARCH FOR UP TO FOUR ROOTS
C
        DO 50 I=1,4
C
C       CALCULATE THE ENERGY LEVEL FOR THE INFINITE BOX
C
C       ENERGY = (N**2 H**2)/(8*MASS*A**2)
C
        INF = D * (I**2)
C
C       CALCULATE THE ENERGY THAT IS PI/2 LESS THAN THE INFINITE VALUE
C
        LOW = (2*I-1)**2*C
C
C       ARE WE BEFORE OR AFTER THE MID POINT ?
C
        IF ( LOW.GE.V2 ) GO TO 17
C
C       NOW THE ROOT IS BETWEEN LOW AND INF
C
        HIGH = INF*.999995
C
C        IGNORE THE SIGN CHANGE AT V0/2
C
        IF ( HIGH .GE. V2 ) HIGH = V2 * .999995
        GO TO 30
C
C       NOW THE ROOT IS BETWEEN LOW-PI/2 AND INF-PI/2
C
     17 HIGH = LOW * .99999
        LOW = OLDINF*1.000005
C
C       CHECK FOR FIRST LEVEL OR VERY SMALL ROOT
C
        IF ( I.EQ.1 .OR. LOW.LE.V2 ) LOW = V2 * 1.000005
C
C       CHECK FOR UNBOUND STATE  I.E., ENERGY > WALL OF BOX
C
     30 IF( LOW.GE.V0 ) GO TO 10
C
C       USE THE BISECTION METHOD TO FIND THE ROOT
C       THE ROOT IS BETWEEN HIGH AND LOW
C
C       FIND THE DIFFERENCE (LEFT-HAND SIDE - RIGHT-HAND SIDE) AT LOW
C
        ROOT=LOW
        X = ANGLE(LOW)
        DLOW = DIFF(X)
C
```

```
C       CHECK TO SEE IF HIGH IS > V0
C
        IF ( HIGH .GT. V0 ) HIGH = V0
C
C       FIND THE MID POINT OF THE INTERVAL CONTAINING THE ROOT
C
   31 ROOT = (LOW + HIGH)/2.0
C
C       FIND THE DIFFERENCE (LHS - RHS) AT  ROOT  (THE MID POINT)
C
        X = ANGLE(ROOT)
        DROOT = DIFF(X)
C
C       CHECK TO SEE IF WE HAVE FOUND A ROOT WITHIN THE TOLERANCE
C
        IF( ABS((LOW-ROOT)/ROOT) - TOL ) 40,40,33
C
C       CHECK TO SEE IF REAL ROOT IS BEFORE OR AFTER MIDDLE OF INTERVAL
C
   33 IF( DROOT*DLOW ) 36,40,34
   34 LOW = ROOT
        GO TO 31
   36 HIGH = ROOT
        GO TO 31
C
C       WE HAVE NOW FOUND THE ROOT
C
C       SAVE THE PREVIOUS VALUE OF INF
C
   40 OLDINF = INF
C
C       CHANGE FROM ERG TO EV
C
        LOW = ROOT*6.2421E11
        INF = INF*6.2421E11
        ENERGY(I)=ROOT
   50 WRITE(6,45) I,LOW,ROOT,INF
C
C       PRINT THE SQUARE OF THE FIRST 4 EIGENFUNCTIONS FOR
C
C                   0 <  X  < 1.25*A/2
C
        F=1.341122E+27*SQRT(MASS)
C
C         EIGENFUNCTION VALUES INSIDE THE BOX
C
        WRITE(6,65)
        DO 60 I=1,16
        X=(I-1)*A/30.
        DO 62 JJ=1,2
        U(2*JJ-1)=(COS(F*SQRT(ENERGY(2*JJ-1))*X))**2
   62 U(2*JJ)=(SIN(F*SQRT(ENERGY(2*JJ))*X))**2
   60 WRITE(6,66)X,(U(J),J=1,4)
C
C       EIGENFUNCTION VALUES OUTSIDE THE BOX
C
        DO 67 I=1,4
        ARG(I)=2*F*SQRT(V0-ENERGY(I))*X
```

```
          IF(ARG(I).GE.20)GO TO 68
          CNST(I)=U(I)/EXP(-ARG(I))
          GO TO 67
       68 CNST(I)=0
       67 ARG(I)=ARG(I)/X
          DO 70 I=1,5
          X=A*(.5+.025*I)
          DO 71 JJ=1,4
          ARG(JJ)=ARG(JJ)*X
          IF(ARG(JJ).GE.20)GO TO 72
          U(JJ)=CNST(JJ)*EXP(-ARG(JJ))
          GO TO 71
       72 U(JJ)=0.
       71 ARG(JJ)=ARG(JJ)/X
       70 WRITE(6,66)X,(U(J),J=1,4)
          GO TO 10
        5 FORMAT(//,' ENTER WIDTH OF BOX (ANGSTROMS), HEIGHT OF BOX (IN EV
         .),'/' AND MASS OF PARTICLE (ME=.0005486 AMU) ( A=0 ENDS',
         .' PROGRAM)',/)
        7 FORMAT(3F)
       12 FORMAT(//,' A  =',G14.3,' ANGSTROMS',/,' V0 =',G14.3,' EV',/,' M
         .  =  ',1PE10.2,'   AMU')
       13 FORMAT(//20X,'I. EIGENVALUES'//,
         .15X,'FINITE BOX',20X,'INFINITE BOX',//,' NUMBER   ENERGY',
         .7X,'ENERGY',19X,'ENERGY',/,10X,'IN EV',8X,'IN ERG',19X,
         .'IN EV',/)
       45 FORMAT(I4,G14.3,1PE12.2,13X,G14.3)
       65 FORMAT(//20X,'II. EIGENVECTORS (SQUARED)',
         .//7X,'X',7X,'U1**2',7X,'U2**2',7X,'U3**2',7X,'U4**2'/)
       66 FORMAT(5G12.3)
          END
```

<u>Discussion</u>

BOX requires 596 words of memory and executes in approximately 0.2 sec. The program can be easily modified to find more than the first four energy levels and associated wavefunctions.

<u>Problem 3.7.</u> Plot on the same graph the square of the first four wavefunctions of a particle in a finite box.

   <u>Solution.</u> See Figure 3.6.

<u>Problem 3.8.</u> Plot on the same graph the first three energy levels of an electron in a box of width 10 Å as a function of V_o.

   <u>Solution.</u> See Figure 3.7.

<u>Problem 3.9.</u> Modify BOX to calculate NMR chemical shifts and wavelengths of maximum absorption for conjugated π-electron systems [11].

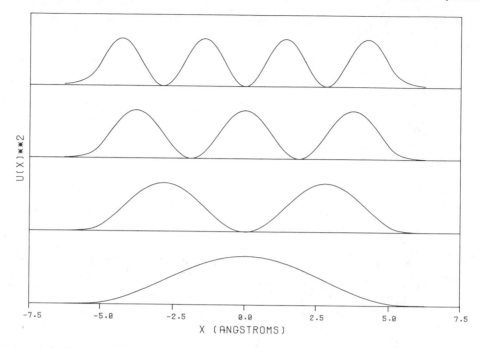

Figure 3.6. Wavefunctions for the particle in a finite box.

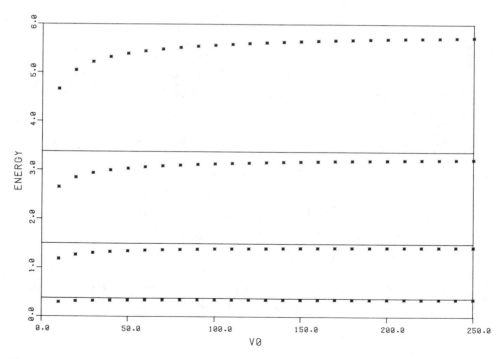

Figure 3.7. Energy eigenvalues for an electron in both infinite and finite potential wells.

3.3 THE NEWTON-RAPHSON METHOD

The Newton-Raphson method [12] is one of the most widely used iterative techniques for evaluating roots of equations. Given the function, $f(x) = 0$ and the initial approximation to the root x_0 the Newton-Raphson method is

$$x_{i+1} = x_i - \frac{f(x_i)}{f'(x_i)} \qquad (3.9)$$

Here $f(x_i)$ and $f'(x_i)$ are the function and its first derivative evaluated at $x = x_i$. The method requires an initial approximation to the root, x_0, which should be as close to the desired root as possible.

The Newton-Raphson method can be derived from a truncated Taylor series expansion of the function $f(x)$ about the initial approximation x_0,

$$f(x) = 0 = f(x_0) + f'(x_0)(x - x_0) + \frac{1}{2}f'(x_0)(x - x_0)^2 + \ldots$$

If x_0 is close to the root, then $x - x_0$ is small, and to a first approximation the quadratic and higher terms may be ignored. This gives

$$0 = f(x_0) + f'(x_0)(x - x_0)$$

or

$$x = x_0 - \frac{f(x_0)}{f'(x_0)}$$

Equation (3.9) is the iterative form of this equation.

The Newton-Raphson method is illustrated in Figure 3.8. In Figure 3.8

$$\tan \theta = \frac{f(x_0)}{x_0 - x_1} = f'(x_0)$$

or

$$x_1 = x_0 - \frac{f(x_0)}{f'(x_0)}$$

For example, consider the transcendental equation

$$\sin x = \cos x$$

Then

$$f(x) = \sin x - \cos x = 0$$

and

$$f'(x) = \cos x + \sin x$$

This function has a root at $\pi/4$ or approximately 0.78540 rad. Let 0.5 rad be the initial approximation. Then,

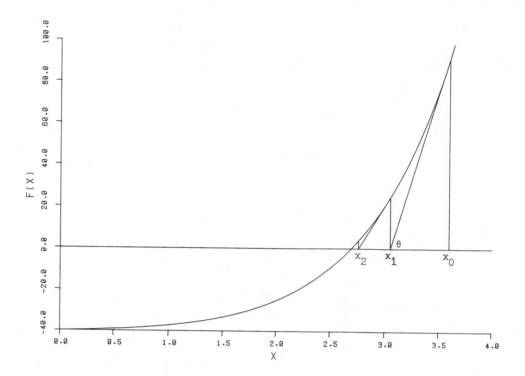

Figure 3.8 Graphical illustration of the Newton-Raphson procedure.

$$x_i = 0.5 - \frac{\sin 0.5 - \cos 0.5}{\cos 0.5 + \sin 0.5}$$

$$= 0.5 - \frac{-0.39816}{1.3570} = 0.79341$$

and

$$x_2 = 0.79341 - \frac{0.01133}{1.4142} = 0.78540$$

In this example the method converged rapidly to the root at $\pi/4$. However, the Newton-Raphson method will fail if x approaches a minimum or maximum in f(x), because at these points,

$$f'(x) = 0$$

Also, the method sometimes oscillates without converging. For example,

$$f(x) = \sqrt{x - 10}$$

Let the initial approximation to the root be 9. The Newton-Raphson procedure gives the following results.

i	0	1	2	3	$\cdots$
x_i	9	11	9	11	$\cdots$

The program in Section 3.3.1 is a general Newton-Raphson algorithm. The programs documented in Sections 3.3.2 to 3.3.4 use this method to solve aqueous and gas-phase equilibrium problems.

3.3.1 NEWTON

Introduction

This program uses the Newton-Raphson method to find a root of a function. The input to the program includes the following values:

ITMAX, the maximum number of iterations allowed

TOL, the relative convergence criterion

X, the initial approximation to the root

The program prints these values for verification purposes and a table showing the progress of the interative procedure. If the method converges, the root and the function evaluated at the root are printed.

Method

The Newton-Raphson method is

$$x_{i+1} = x_i - \frac{f(x_i)}{f'(x_i)}$$

The function and its first derivative are coded using statement functions,

```
F(X)= ...
DF(X)= ...
```

The program checks for a zero value for $f'(x_i)$. The convergence criterion is

$$\left| \frac{x_{i+1} - x_i}{x_{i+1}} \right| \leq TOL$$

The two sample executions find the two nodes in the hydrogen 3s atomic wavefunction (Figure 3.1).

Sample Execution

```
ENTER ITMAX,TOL,X0
>10  0.0001  0.0


    ITMAX =  10          TOL =     0.100E-03


  I       X               F(X)            DF(X)

  0    0.0000E+00        27.00          -27.00
  1    1.000             7.882          -12.66
  2    1.623             1.781           -7.295
  3    1.867             0.1975          -5.719
  4    1.901             0.3514E-02      -5.517
  5    1.902             0.9802E-06      -5.513


        THE ROOT IS     1.902

        AND THE FUNCTION EVALUATED AT THE ROOT IS    -0.632E-07



ENTER ITMAX,TOL,X0
>10 0.0001  8


    ITMAX =  10          TOL =     0.100E-03


  I       X               F(X)            DF(X)

  0    8.000            0.7643           0.7180
  1    6.935           -0.1622           1.019
  2    7.095           -0.3400E-02       0.9763
  3    7.098           -0.1611E-05       0.9753


        THE ROOT IS     7.098

        AND THE FUNCTION EVALUATED AT THE ROOT IS     0.000E+00
```

Listing

```
C
C                        NEWTON
C
C     PROGRAM TO FIND A ROOT OF AN EQUATION
C     USING THE NEWTON-RAPHSON METHOD
C
C     THE FUNCTION AND ITS DERIVATIVE ARE CODED
C     IN THE STATEMENT FUNCTIONS F(X) AND DF(X)
C
```

```
C    INPUT:   ITMAX  -   MAX.   OF ITERTIONS
C             TOL    -   RELATIVE CONVERGENCE TOLERANCE
C             X      -   INITIAL APPROXIMATION TO THE ROOT
C
C    AUTHOR:  K. J. JOHNSON
C
C
      F(X)=(27.-18.*X+2.*X*X)*EXP(-X/3.)
      DF(X)=(-27.+10.*X-2.*X*X/3.)*EXP(-X/3.)
C
C    INPUT
C
      WRITE(6,11)
      READ(5,22)ITMAX,TOL,X
      WRITE(6,33)ITMAX,TOL
      XOLD=X
      FX=F(X)
      DFX=DF(X)
      I=0
      WRITE(6,44)I,X,FX,DFX
C
C     ITERATION LOOP
C
      DO 10 I=1,ITMAX
      X=X-FX/DFX
      FX=F(X)
      DFX=DF(X)
      IF(ABS((X-XOLD)/X).LE.TOL)GO TO 60
      XOLD=X
   10 WRITE(6,44)I,X,FX,DFX
      WRITE(6,55)
      STOP
   60 WRITE(6,66)X,FX
      STOP
   11 FORMAT(//'  ENTER ITMAX,TOL,X0'/)
   22 FORMAT(I,2E)
   33 FORMAT(/ '     ITMAX = ',I3,10X,'TOL = ',G12.3,
     1   ///'  I',8X,'X',13X,'F(X)',10X,'DF(X)'/)
   44 FORMAT(I3,3G15.4)
   55 FORMAT(//'  SORRY, ITMAX EXCEEDED.'//)
   66 FORMAT(//10X,'THE ROOT IS ',G13.4,//10X,
     1'AND THE FUNCTION EVALUATED AT THE ROOT IS ',G12.3/)
      END
```

Discussion

NEWTON requires 256 36-bit words of core memory. The execution time depends on the number of iterations required for convergence and the number of arithmetic operations required to evaluate $f(x)$ and $f'(x)$. Each sample execution in this example took approximately 0.1 sec.

The following example illustrates the danger of divergence with the Newton-Raphson method when the iterate approaches a minimum or maximum in the function. The sine function,

$$f(x) = \sin x$$

has roots at $\pm\, n\pi$ where n = 0, 1, 2, 3, ... The following output was generated
by NEWTON with the statement function:

```
F(X)=SIN(X)
DF(X)=COS(X)
............

ENTER ITMAX,TOL,XO
>10 0.0001 4.7
```

  ITMAX = 10   TOL = 0.100E-03

I	X	F(X)	DF(X)
0	4.700	-0.9999	-0.1239E-01
1	-76.01	-0.9999	-0.1239E-01
2	-75.31	-0.5764	0.8172
3	-75.40	0.9090E-01	0.9959

    THE ROOT IS -75.40

    AND THE FUNCTION EVALUATED AT THE ROOT IS -0.252E-03

The initial approximation (4.7) is near the minimum corresponding to $x = 3\pi/2 = 4.712$.
NEWTON found the root $x = -75.40$, which is $-24\pi = 75.398$.

  The Newton-Raphson method requires the first derivative of the function. In
some cases it is useful to numerically approximate the derivative, for example,

$$f'(x_i) = \frac{f(x_i) - f(x_{i-1})}{x_i - x_{i-1}}$$

with this approximation for $f'(x)$, the algorithm becomes

$$x_{i+1} = \frac{x_{i-1}f(x_i) - x_i f(x_{i-1})}{f(x_i) - f(x_{i-1})}$$

3.3.2   HNA

Introduction

  HNA solves aqueous equilibrium problems associated with the following weak
acids and the sodium salts of the corresponding conjugate bases:

  n = 1   HA + NaA

  n = 2 H_2A + NaHA + Na_2A

  n = 3 H_3A + NaH_2A + Na_2HA + Na_3A

The entire set of acid dissociation, buffer, and hydrolysis problems involving these
species is solved by HNA. The input includes the number of acidic protons (n = 1,
2, or 3), the negative logs of the acid dissociation constants, the total concentrations of the acid and sodium salts present, and an initial approximation to the

equilibrium pH. The program prints the input data for verification purposes and a table showing the progress of the Newton-Raphson iterative procedure (Section 3.3) in finding the equilibrium $[H^+]$. If the method converges the program prints the equilibrium concentrations of all species.

<u>Method</u>

These aqueous equilibrium systems are hydrogen ion determined, i.e., when the equilibrium $[H^+]$ is known, the equilibrium concentrations of all species can be calculated. For example, for $n = 3$, the following mass action, mass balance, and charge balance equations apply:

$$H_3A \rightleftharpoons H^+ + H_2A^- \qquad K_1 = \frac{[H^+][H_2A^-]}{[H_3A]}$$

$$H_2A^- \rightleftharpoons H^+ + HA^{2-} \qquad K_2 = \frac{[H^+][HA^{2-}]}{[H_2A^-]}$$

$$HA^{2-} \rightleftharpoons H^+ + A^{3-} \qquad K_3 = \frac{[H^+][HA^{2-}]}{[HA^{2-}]}$$

$$H_2O \rightleftharpoons H^+ + OH^- \qquad K_w = [H^+][OH^-]$$

$$A_{tot} = C_1 + C_2 + C_3 + C_4 = [H_3A] + [H_2A^-] + [HA^{2-}] + [A^{3-}]$$

$$[H^+] + [Na^+] = [H_2A^-] + 2[HA^{2-}] + 3[A^{3-}] + [OH^-] \qquad (3.10)$$

Here C_1, C_2, C_3, and C_4 are the total concentrations of H_3A, NaH_2A, Na_2HA, and Na_3A, respectively. The concentrations of the four acid species in terms of A_{tot} and $[H^+]$ can be calculated using the following expressions [13]:

$$[H_3A] = f_1 A_{tot} \qquad [H_2A^-] = f_2 A_{tot}$$
$$[HA^{2-}] = f_3 A_{tot} \qquad [A^{3-}] = f_4 A_{tot}$$

where

$$f_1 = \frac{[H^+]^3}{D} \qquad f_2 = \frac{K_1[H^+]^2}{D}$$

$$f_3 = \frac{K_1 K_2 [H^+]}{D} \qquad f_4 = \frac{K_1 K_2 K_3}{D}$$

and

$$D = [H^+]^3 + K_1[H^+]^2 + K_1 K_2 [H^+] + K_1 K_2 K_3$$

The $[Na^+]$ and $[OH^-]$ are given by

$$[Na^+] = C_2 + 2C_3 + 3C_3 \qquad [OH^-] = \frac{K_w}{[H^+]}$$

The charge balance expression, Eq. (3.10), is converted by substitution into the following fifth-order polynomial in $[H^+]$:

$$f([H^+]) = a_1[H^+]^5 + a_2[H^+]^4 + a_3[H^+]^3 + a_4[H^+]^2 + a_5[H^+] + a_6$$

where

$$a_1 = 1$$
$$a_2 = K_1 + C_2 + 2C_3 + 3C_4$$
$$a_3 = K_1(K_2 - C_1 + C_3 + 2C_4) - K_w$$
$$a_4 = K_1[K_2(K_3 - 2C_1 - C_2 + C_4) - K_w]$$
$$a_5 = K_1K_2[K_3(-3C_1 - 2C_2 - C_3) - K_w]$$
$$a_6 = -K_1K_2K_3K_w$$

The polynomials associated with the HA and H_2A systems are third and fourth order, respectively. The arrangement of terms in the coefficients of the three polynomials is similar and a judicious assignment of variable names allows one algorithm to handle the three cases.

The Newton-Raphson method is used. There are three convergence criteria:

1. If $\left| ([H^+]_i - [H^+]_{i-1})/[H^+]_i \right|$ 0.0001, then convergence has occurred.

2. If $\left| ([H^+]_i - [H^+]_{i-1})/[H^+]_i \right| \geq 10000$, then the system is diverging.

3. The method is terminated after 25 iterations.

If the method converges the concentrations of the acid and its conjugate bases are calculated and printed.

Sample Executions

1. 0.001 $\underline{M}$ Nitrilotriacetic Acid - $N(CH_2COOH)_3$

ENTER NUMBER OF ACIDIC PROTONS   (0 ENDS PROGRAM)
>3

ENTER PK1,PK2 AND PK3
>2.5   2.8   10.2

ENTER THE TOTAL CONCENTRATION OF H3A, NAH2A, NA2HA AND NA3A
>0.001

FOR H3A   PK1 =   2.500   PK2 =   2.800   PK3 =   10.200

        (H3A) = 1.00E-03   (NAH2A) = 0.00E-01   (NA2HA) = 0.00E-01

        (NA3A) = 0.00E-01

ENTER THE INITIAL APPROXIMATION TO THE PH
>2

THE INITIAL APPROX. TO THE PH = 2.000

ITERATION   (H(+))   REL. CHANGE IN (H(+))

        1   7.90E-03   2.66E-01
        2   6.22E-03   2.69E-01
        3   4.89E-03   2.71E-01
        4   3.85E-03   2.73E-01
        5   3.03E-03   2.71E-01
        6   2.40E-03   2.63E-01
        7   1.93E-03   2.43E-01
        8   1.60E-03   2.04E-01
        9   1.40E-03   1.41E-01
       10   1.32E-03   6.31E-02
       11   1.31E-03   1.15E-02
       12   1.31E-03   3.52E-04
       13   1.31E-03   3.27E-07

*** CONVERGENCE ***

PH =   2.884

 (H3A) = 1.57E-04   (H2A(-)) = 3.81E-04   (HA(2-)) = 4.62E-04

 (A(3-)) = 2.23E-11

    2. 0.001 M in Phosphoric Acid and 0.001 M in Trisodium Phosphate

ENTER NUMBER OF ACIDIC PROTONS   (0 ENDS PROGRAM)
>3

ENTER PK1,PK2 AND PK3
>2.15 7.21 12.36

ENTER THE TOTAL CONCENTRATION OF H3A, NAH2A, NA2HA AND NA3A
>.001 0 0 .001

FOR H3A   PK1 =   2.150   PK2 =   7.210   PK3 =   12.360

      (H3A) = 1.00E-03 (NAH2A) = 0.00E-01 (NA2HA) = 0.00E-01

      (NA3A) = 1.00E-03

ENTER THE INITIAL APPROXIMATION TO THE PH
>7

THE INITIAL APPROX. TO THE PH = 7.000

ITERATION   (H(+))   REL. CHANGE IN (H(+))

    1   7.83E-08   2.77E-01
    2   6.66E-08   1.75E-01
    3   6.23E-08   6.93E-02
    4   6.17E-08   1.02E-02
    5   6.17E-08   2.10E-04
    6   6.17E-08   9.66E-08

*** CONVERGENCE ***

PH =   7.210

 (H3A) = 8.71E-09   (H2A(-)) = 1.00E-03   (HA(2-)) = 1.00E-03

 (A(3-)) = 7.08E-09

   Listing


```
C                              HNA
C
C
C     GENERAL HA, H2A, AND H3A WEAK ACID PROBLEM SOLVER.
C
C
C
C     THIS PROGRAM SOLVES THE ENTIRE SET OF ACID DISSOCIATION,
C     HYDROLYSIS, AND BUFFER PROBLEMS INVOLVING THE ACIDS HA, H2A,
C     AND H3A.
C
C     THE NEWTON-RAPHSON METHOD IS USED TO FIND THE EQUIL. PH
C
C  GLOSSARY:
C
C    N    -  NUMBER OF ACIDIC PROTONS
C
C  IF N = 3,
C
C    C1   -  TOTAL CONC. OF H3A
C    C2   -  TOTAL CONC. OF NAH2A
```

```
C     C3    -    TOTAL CONC. OF NA2HA
C     C4    -    TOTAL CONC. OF NA3A
C     H     -    <H(+)>
C     PK(3) -    PKA VALUES
C     H1(7) -    SUCCESSIVE POWERS OF <H(+)>
C
C     INPUT:  N,C1,C2,C3,C4,PK1,PK2,PK3, AND
C                AN INITIAL APPROXIMATION TO THE EQUIL. PH
C
C     THE ALGEBRA ASSOCIATED WITH N=1 AND N=2 IS A SPECIAL
C     CASE OF THAT FOR N=3, SO ONE FUNCTION (AND DERIVATIVE)
C     SUFFICES FOR THE THREE CASES
C
C
C            AUTHORS:  D.E.HAWKINS,JR., D.NICHOLS AND K.J.JOHNSON
C
C
      IMPLICIT REAL(K,L)
      DIMENSION K(3),H1(7),C(4),ALPHA(4)
      EQUIVALENCE(K1,K(1)),(K2,K(2)),(K3,K(3)),(C1,C(1)),(C2,C(2)),
     1(C3,C(3)),(C4,C(4))
      DATA H1(1),H1(2)/0.,1./,KW/1.E-14/
C
C     INPUT SECTION
C
    5 WRITE(6,11)
      READ(5,22)N
      GO TO(10,8,6),N
      STOP
C
C     N=3
C
    6 WRITE(6,33)
      READ(5,44)K
      WRITE(6,55)
      READ(5,44)C
      WRITE(6,66)K,C
      GO TO 15
C
C     N=2
C
    8 WRITE(6,77)
      READ(5,44)K1,K2
      WRITE(6,88)
      READ(5,44)(C(I),I=1,3)
      WRITE(6,99)K1,K2,(C(I),I=1,3)
      GO TO 12
C
C     N=1
C
   10 WRITE(6,111)
      READ(5,44)K1
      WRITE(6,122)
      READ(5,44)C1,C2
      WRITE(6,133)K1,C1,C2
      K2=0
      C3=0
   12 K3=0
      C4=0
```

```
   15 WRITE(6,144)
      READ(5,44)PH
      WRITE(6,155)PH
C
C     SET UP COEFFICIENTS OF F(H(+)) AND F'(H(+))
C     SCALE BY A FACTOR OF 1.0E7 TO AVOID UNDERFLOWS.
C
      DO 20 I=1,N
   20 K(I)=10.**(-K(I))
      H=10.**(7-PH)
      K12=K1*K2
      K123W=K12*K3*1E35*KW
      A2=1E7*(K1+C2+2*C3+3*C4)
      A3=1E14*(K1*(K2-C1+C3+2*C4)-KW)
      A4=1E21*K1*(K2*(K3-2*C1-C2+C4)-KW)
      A5=1E28*K12*(K3*(-3*C1-2*C2-C3)-KW)
      DA1=N+2
      DA2=(N+1)*A2
      DA3=N*A3
      DA4=(N-1)*A4
      DA5=(N-2)*A5
      WRITE(6,166)
C
C     THE 40 LOOP CONTAINS THE NEWTON-RAPHSON PROCEDURE.
C     TWENTY-FIVE ITERATIONS ARE ALLOWED.  THE LOOP INCLUDES BOTH A
C     CONVERGENCE AND DIVERGENCE TEST.
C
      DO 40 I=1,25
      HOLD=H
      DO 30 J=3,N+4
   30 H1(J)=H1(J-1)*H
      FH=H1(N+4)+A2*H1(N+3)+A3*H1(N+2)+A4*H1(N+1)+A5*H1(N)-K123W
      DFH=DA1*H1(N+3)+DA2*H1(N+2)+DA3*H1(N+1)+DA4*H1(N)+DA5
      H=H-FH/DFH
      H2=1E-7*H
      T=ABS((H-HOLD)/H)
      WRITE(6,177)I,H2,T
      IF(H2.LT.0)GO TO 41
      IF(T.LE.1E-4)GO TO 45
   40 IF(T.GE.1E4)GO TO 41
   41 WRITE(6,188)
      GO TO 5
C
C     THE PROGRAM HAS CONVERGED ON THE EQUILIBRIUM PH.  NOW CALCULATE
C     THE CONCENTRATION OF THE OTHER SPECIES IN THE SOLUTION
C
   45 PH=-ALOG10(H2)
      WRITE(6,211)PH
      ATOT=C(1)
      ALPHA(1)=-PH*N
      D=0
      IF(ALPHA(1).GT.-38)D=10.**ALPHA(1)
   47 DO 50 I=1,N
      ALPHA(I+1)=ALPHA(I)+PH+ALOG10(K(I))
      ATOT=ATOT+C(I+1)
      IF(ALPHA(I+1).LT.-38)GO TO 50
      D=D+10.**ALPHA(I+1)
   50 CONTINUE
```

```
        LATOT=ALOG10(ATOT)
        LD=ALOG10(D)
        DO 60 I=1,N+1
        ALPHA(I)=ALPHA(I)+LATOT-LD
        IF(ALPHA(I).LT.-38)GO TO 56
        ALPHA(I)=10.**ALPHA(I)
        GO TO 60
   56 ALPHA(I)=0
   60 CONTINUE
        GO TO(63,68,70),N
   63 WRITE(6,222)ALPHA(1),ALPHA(2)
        GO TO 5
   68 WRITE(6,233)(ALPHA(I),I=1,3)
        GO TO 5
   70 WRITE(6,244)ALPHA
        GO TO 5
   11 FORMAT(' ENTER NUMBER OF ACIDIC PROTONS   ( 0 ENDS PROGRAM )'/)
   22 FORMAT(I)
   33 FORMAT(/' ENTER PK1,PK2 AND PK3   '/)
   44 FORMAT(4F)
   55 FORMAT(/' ENTER THE TOTAL CONCENTRATION OF H3A, NAH2A, NA2HA',
      1 ' AND NA3A '/)
   66 FORMAT(/' FOR H3A  PK1 =',F9.3,6X,'PK2 =',F9.3,6X,'PK3 =',F9.3//
      1 8X,'(H3A) =',1PE9.2,'  (NAH2A) =',E9.2,'  (NA2HA) =',E9.2//7X,
      2'(NA3A) =',E9.2/)
   77 FORMAT(/' ENTER PK1 AND PK2    '/)
   88 FORMAT(/' ENTER THE TOTAL CONCENTRATIONS OF H2A,NAHA AND NA2A'/)
   99 FORMAT(/' FOR H2A  PK1 =',F9.3,5X,'PK2 =',F9.3/8X,'(H2A) =',
      1 1PE9.2,'  (NAHA) =',E9.2,'  (NA2A) =',E9.2/)
  111 FORMAT(/' ENTER PK1   '/)
  122 FORMAT(/' ENTER THE TOTAL CONCENTRATIONS OF HA AND NAA   '/)
  133 FORMAT(/' FOR HA  PK1 =',F7.3,3X,'(HA) =',1PE9.2,3X,'(NAA) =',
      1 E9.2/)
  144 FORMAT(/' ENTER THE INITIAL APPROXIMATION TO THE PH   '/)
  155 FORMAT(/' THE INITIAL APPROX. TO THE PH =',F7.3/)
  166 FORMAT(/' ITERATION',5X,'(H(+))',5X,'REL. CHANGE IN (H(+))'/)
  177 FORMAT(4X,I2,4X,1PE12.2,6X,E12.2)
  188 FORMAT(/' CONVERGENCE FAILURE.  TRY ANOTHER PH APPROXIMATION.'/)
  211 FORMAT(//' *** CONVERGENCE ***'//' PH = ',F7.3)
  222 FORMAT(/' (HA) = ',1PE9.2,10X,'(A(-)) = ',E9.2//)
  233 FORMAT(/' (H2A) = ',1PE9.2,5X,'(HA(-)) = ',E9.2,5X,'(A(2-)) = ',
      1 E9.2//)
  244 FORMAT(/' (H3A) = ',1PE9.2,5X,'(H2A(-)) = ',E9.2,5X,'(HA(2-)) = ',
      1 E9.2//' (A(3-)) = ',E9.2//)
        END
```

Discussion

   HNA requires 1617 words of memory. The execution time depends on the problem.
The execution time for both executions shown above was approximately 0.2 sec.

   A reasonable range of $[H^+]$ in this program is

$$10^{-14} \le [H^+] \le 10 \ \underline{M}$$

If the hydrogen ion concentration is less than 10^{-8} $\underline{M}$, the calculation of $[H^+]^5$
will cause an exponential underflow error in the DECSystem-10 computer. This

problem was circumvented by scaling by a factor of 10^7. The procedure is illustrated with a monoprotic acid example.

Problem 3.10. Calculate the equilibrium pH of a 1.0×10^{-6} $\underline{M}$ acetic acid solution. The K_a for acetic acid is 1.8×10^{-5}.

   Solution. The cubic equation is

$$[H^+]^3 + K_a[H^+]^2 - (C_aK_a + K_w)[H^+] - K_aK_w = 0$$

   Substituting for K_a, C_a, and K_w,

$$[H^+]^3 + 1.8 \times 10^{-5}[H^+]^2 - (1.8 \times 10^{-11} + 1.0 \times 10^{-14})[H^+] - 1.8 \times 10^{-19} = 0$$

$$[H^+]^3 + 1.8 \times 10^{-5}[H^+]^2 - 1.801 \times 10^{-11}[H^+] - 1.8 \times 10^{-19} = 0$$

   Let $x = 10^7[H^+]$, then

$$x^3 + 180x^2 - 1801x - 180 = 0$$

   The relevant root of this cubic is

$$x = 9.60$$

   Then $[H^+] = 9.60 \times 10^{-7}$ M.

   Another technique used by HNA to avoid exponential underflows is the use of logarithms. For example,

   $C = A \times B$

becomes

   $C = 10^{\log A + \log B}$

The program tests the magnitude of the exponent before computing the result.

Problem 3.11. Plot on the same graph the equilibrium pH of a polyprotic acid system as a function of the logarithm of the total concentrations of the separate species.

   Solution. See Figure 3.9.

Problem 3.12. Consider the system Na_2CO_3-H_2O. Calculate the equilibrium concentrations of all species as a function of total Na_2CO_3 concentration. Compare the results with various approximate solutions [14].

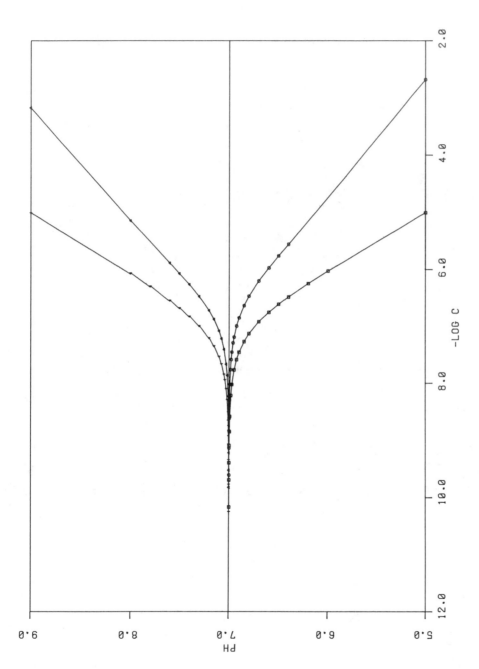

Figure 3.9 The equilibrium pH of four phosphate species as a function of log C: top curve Na_3PO_4; second curve Na_2HPO_4; third curve NaH_2PO_4; bottom curve H_3PO_4.

Problem 3.13.   Modify HNA so that equilibrium problems involving mixtures of acids
                 and bases can be solved. For example, consider a solution that is
                 a $\underline{M}$ in HCl, b $\underline{M}$ in NH_3, and c $\underline{M}$ in triethanolamine [15].

3.3.3 HNATRN

Introduction

HNATRN simulates the series of weak acid-strong base titrations involving the
weak acids HA, H_2A, and H_3A. The input includes the number of acidic protons, the
negative logarithms of the acid dissociation constants, the initial concentration
and volume of the weak acid, and the concentration of the strong base (NaOH).
HNATRN prints the input data for verification purposes, and a table containing
the percent of titration, pH, volume of base, and the concentrations of two of the
acid species. These species are [HA] and $[A^-]$ for HA, $[H_2A]$ and $[A^{2-}]$ for H_2A,
and $[H_3A]$ and $[A^{3-}]$ for H_3A.

Method

HNATRN uses the Newton-Raphson method to calculate the initial and final pH
values of the titration. The intermediate points are calculated by evaluating the
inverse function,

$$V_b = f([H^+])$$

Here V_b is the volume of titrant (NaOH) that must be added to the original H_nA
solution to give the specified $[H^+]$. The approach is similar to that used in the
EDTA program (Section 3.1.1).

1. Initial pH (0% titration). The first problem is to calculate the initial
pH of the solution. This is a special case of the chemical systems solved by the
program HNA discussed in the preceding section. For H_3A, the fifth-order polynomial
that must be solved is

$$f([H^+]) = a_1[H^+]^5 + a_2[H^+]^4 + a_3[H^+]^3 + a_4[H^+]^2 + a_5[H^+] + a_6$$

where

$$a_1 = a$$
$$a_2 = K_1$$
$$a_3 = K_1K_2 - C_aK_1 - K_w$$
$$a_4 = K_1K_2K_3 - 2C_aK_1K_2 - K_1K_w$$
$$a_5 = -3C_aK_1K_2K_3 - K_1K_2K_w$$
$$a_6 = -K_1K_2K_3K_w$$

Here C_a is the initial molar concentration of H_3A. The initial approximation to the desired root of this equation is obtained using the quadratic approximation.

$$H_3A \rightleftharpoons H^+ + H_2A^-$$
$$C_a-[H^+] \quad [H^+] \quad [H^+]$$

The approximate initial hydrogen ion concentration is

$$[H^+]_0 = \frac{-K_1 + \sqrt{K_1^2 + 4C_a}}{2}$$

The same convergence and divergence criteria used in HNA are used here. Twenty-five iterations are allowed for convergence. If the iterative procedure fails to converge, a message is printed and the program starts over.

2. Final pH. The final pH is the pH of the solution corresponding to 200% titration for HA, 300% for H_2A, and 400% for H_3A. For example, for H_3A, the following stoichiometric, mass balance, and charge balance equations apply:

$$H_3A + OH^- \rightarrow H_2O + H_2A^-$$
$$H_2A^- + OH^- \rightarrow H_2O + HA^{2-}$$
$$HA^{2-} + OH^- \rightarrow H_2O + A^{3-}$$

$$A_{tot} = [H_3A] + [H_2A^-] + [HA^{2-}] + [A^{3-}]$$

$$[H^+] + [Na^+] = [OH^-] + [H_2A^-] + 2[HA^{2-}] + 3[A^{3-}]$$

At 400% titration,

$$V_b = \frac{4C_a V_a}{C_b}$$

where V_a, V_b, C_a, and C_b refer to the volumes and concentrations of acid and base, respectively. For H_3A, the fifth-order polynomial corresponding to 400% titration is

$$f([H^+])= b_1[H^+]^5 + b_2[H^+]^4 + b_3[H^+]^3 + b_4[H^+]^2 + b_5[H^+] + b_6$$

where

$$b_1 = 1$$
$$b_2 = K_1 + 4Q$$
$$b_3 = K_1K_2 + 3K_1Q - K_w$$
$$b_4 = K_1K_2K_3 + 2K_1K_2Q - K_1K_w$$
$$b_5 = K_1K_2K_3Q - K_1K_2K_w$$

$$b_6 = -K_1 K_2 K_3 K_w$$

$$Q = \frac{C_a C_b}{4C_a + C_b}$$

At 400% of titration, there is, to a first approximation, an excess of 1 equiv $(C_a V_a)$ of OH$^-$. Then

$$[OH^-] = \frac{C_a V_a}{V_a + V_b}$$

$$V_b = \frac{4C_a V_a}{C_b}$$

so

$$[OH^-] = \frac{C_a C_b}{4V_a + V_b} = Q$$

and

$$[H^+]_0 = \frac{K_w}{Q}$$

This is the initial approximation used in the Newton-Raphson method. The same convergence and divergence criteria are used for the initial and final pH calculations.

  3. Intermediate pH values. HNATRN defines a pH increment,

$$pH_{inc} = \frac{pH_f - pH_i}{M}$$

Here pH_i and pH_f are the initial and final pH values, and M is 24 for HA, 38 for H_2A, and 58 for H_3A. The pH is incremented using the equation,

$$pH = pH_i + jpH_{inc} \qquad j = 1,2,3,\ldots$$

The volume of titrant, V_b, that must be added to reach a given intermediate pH is given by the following equation for the triprotic acid case,

$$V_b = -V_a \frac{NUM}{DEN}$$

where

$$NUM = c_1[H^+]^5 + c_2[H^+]^4 + c_3[H^+]^3 + c_4[H^+]^2 + c_5[H^+] + c_6$$

$$DEN = d_1[H^+]^5 + d_2[H^+]^4 + d_3[H^+]^3 + d_4[H^+]^2 + d_5[H^+] + d_6$$

and

$$c_1 = 1 \qquad\qquad\qquad d_1 = 1$$

$$c_2 = K_1 \qquad\qquad\qquad d_2 = K_1 + C_b$$

$$c_3 = K_1 K_2 - C_a K_1 - K_w \qquad d_3 = K_1 K_2 + C_b K_1 - K_w$$

$$c_4 = K_1 K_2 K_3 - 2 C_a K_1 K_2 - K_1 K_w \qquad d_4 = K_1 K_2 K_3 + C_b K_1 K_2 - K_1 K_w$$

$$c_5 = 3 C_a K_1 K_2 K_3 - K_1 K_2 K_w \qquad d_5 = C_b K_1 K_2 K_3 - K_1 K_2 K_w$$

$$c_6 = -K_1 K_2 K_3 K_w \qquad\qquad d_6 = -K_1 K_2 K_3 K_w$$

The polynomials associated with the HA and H_2A systems are special cases of these polynomials, and one algorithm can handle all three cases.

HNATRN generates a table containing % titration, pH, volume of base, and the concentrations of H_nA and A^{n-}. These concentrations are calculated by the subprogram SPECIE. The calling sequence is

CALL SPECIE(N,PH,HNA,AMINUS)

For example, for n = 3, $[H_3A]$ and $[A^{3-}]$ are calculated using the following equations:

$$[H_3A] = \frac{[H^+]^3}{D} \qquad \text{and} \qquad [A^{3-}] = \frac{K_1 K_2 K_3}{D}$$

where

$$D = [H^+]^3 + K_1 [H^+]^2 + K_1 K_2 [H^+] + K_1 K_2 K_3$$

On option, HNATRN also produces a teletype plot of the titration curve.

Sample Execution (Sulfurous Acid, H_2SO_3)

```
ENTER N, CA, VA, CB   ( N=0 ENDS)
> 2 .1 50 .1


ENTER  PK1, PK2
> 1.89 7.2



N = 2      CA =   0.100E+00  VA =    50.0    CB =    0.100E+00

           PK1 =    1.89   , PK2 =    7.20
```

% TITR.	PH	VOL. OF BASE	(H2A)	(A2-)
0.000	1.523	0.000	7.00E-02	6.31E-08
25.145	1.801	12.573	4.40E-02	1.43E-07
48.440	2.080	24.220	2.64E-02	3.11E-07
67.334	2.359	33.667	1.51E-02	6.44E-07

80.685	2.638	40.343	8.39E-03	1.29E-06
89.121	2.917	44.561	4.54E-03	2.52E-06
94.059	3.196	47.030	2.43E-03	4.86E-06
96.822	3.475	48.411	1.29E-03	9.32E-06
98.335	3.753	49.167	6.81E-04	1.78E-05
99.167	4.032	49.584	3.59E-04	3.39E-05
99.654	4.311	49.827	1.89E-04	6.44E-05
99.994	4.590	49.997	9.93E-05	1.22E-04
100.332	4.869	50.166	5.21E-05	2.32E-04
100.809	5.148	50.405	2.73E-05	4.37E-04
101.620	5.427	50.810	1.42E-05	8.22E-04
103.083	5.705	51.542	7.30E-06	1.53E-03
105.726	5.984	52.863	3.69E-06	2.79E-03
110.360	6.263	55.180	1.80E-06	4.93E-03
118.014	6.542	59.007	8.38E-07	8.26E-03
129.459	6.821	64.730	3.61E-07	1.28E-02
144.249	7.100	72.125	1.41E-07	1.81E-02
160.134	7.379	80.067	4.98E-08	2.31E-02
174.139	7.657	87.070	1.61E-08	2.70E-02
184.494	7.936	92.247	4.90E-09	2.97E-02
191.197	8.215	95.598	1.43E-09	3.13E-02
195.173	8.494	97.586	4.08E-10	3.22E-02
197.413	8.773	98.707	1.15E-10	3.27E-02
198.646	9.052	99.323	3.20E-11	3.30E-02
199.329	9.331	99.665	8.90E-12	3.32E-02
199.734	9.609	99.867	2.47E-12	3.32E-02
200.027	9.888	100.014	6.85E-13	3.33E-02
200.333	10.167	100.167	1.90E-13	3.33E-02
200.783	10.446	100.392	5.24E-14	3.32E-02
201.570	10.725	100.785	1.45E-14	3.31E-02
203.040	11.004	101.520	3.99E-15	3.30E-02
205.854	11.283	102.927	1.10E-15	3.27E-02
211.335	11.561	105.668	2.98E-16	3.21E-02
222.308	11.840	111.154	7.97E-17	3.10E-02
245.441	12.119	122.721	2.06E-17	2.89E-02
300.000	12.398	150.000	4.92E-18	2.50E-02

```
ENTER 1 IF YOU WANT A PLOT OF % TITR VS. PH, 0 OTHERWISE.
>1
```

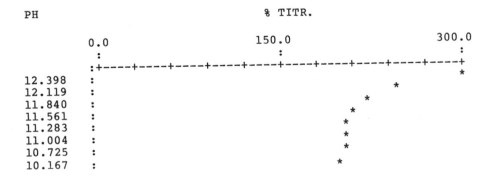

```
10.167   :                                             *
 9.888   :                                             *
 9.609   :                                             *
 9.331   :                                             *
 9.052   :                                             *
 8.773   :                                             *
 8.494   :                                             *
 8.215   :                                          *
 7.936   :                                        *
 7.657   :                                     *
 7.379   :                                  *
 7.100   :                             *
 6.821   :                          *
 6.542   :                       *
 6.263   :                   *
 5.984   :                  *
 5.705   :                *
 5.427   :                *
 5.148   :                *
 4.869   :                *
 4.590   :                *
 4.311   :                *
 4.032   :                *
 3.753   :               *
 3.475   :               *
 3.196   :               *
 2.917   :             *
 2.638   :          *
 2.359   :         *
 2.080   :       *
 1.801   :    *
 1.523   :*
```

Listing

```
C
C                              HNATRN
C
C      THIS PROGRAM SIMULATES THE WEAK ACID-STRONG BASE TITRATIONS
C      HA, H2A, AND H3A.
C
C                      K1
C      1.    HA       <====> H(+) + A(-)
C
C                      K1
C      2.    H2A      <====> H(+) + HA(-)
C
C                      K2
C            HA(-)    <====> H(+) + A(2-)
C
C                      K1
C      3.    H3A      <====> H(+) + H2A(-)
C
C                      K2
C            H2A(-)   <====> H(+) + HA(2-)
C
C                      K3
C            HA(2-)   <====> H(+) + A(3-)
```

```
C
C
C      IN ALL CASES,
C
C                   KW
C          H2O <====> H(+) + OH(-)
C
C
C
C      THE PROGRAM USES THE FOLLOWING GENERAL EQUATION:
C
C          VB = F(PH) = -VA * NUM/DEN
C WHERE
C
C          (N+2)     (N+1)                    N                         (N-1)
C     NUM=H      +K1H       +(K1K2-CAK1-KW)H +(K1K2K3-2CAK1K2-K1KW)H      +
C
C                                (N-2)
C          (-3CAK1K2K3-K1K2KW)H       -K1K2K3KW
C AND
C
C          (N+2)         (N+1)              N
C     DEN=H      +(K1+CB)H       +(K1K2+CBK1-KW)H +(K1K2K3+CBK1K2-K1KW)*
C
C          (N-1)                     (N-2)
C          H     +(CBK1K2K3-K1K2KW)H       -K1K2K3KW
C
C
C      GLOSSARY:
C
C          N              - VALUE OF 'N' IN 'HNA'
C          CA             - CONCENTRATION OF ACID
C          CB             - CONCENTRATION OF BASE
C          VA             - VOLUME OF ACID
C          VB             - VOLUME OF BASE
C          K1, K2, K3 - ACID DISSOCIATION CONSTANTS
C          KW             - WATER DISSOCIATION CONSTANT
C          H              - HYDROGEN ION CONCENTRATION
C          EVB            - VOLUME OF BASE AT 1ST EQUIVALENCE POINT -
C                           100% TITRATION
C          T              - PERCENT OF TITRATION
C
C
C
C      INPUT:
C
C          N              - INTEGER TO INDICATE WHICH TITRATION IS TO BE
C                           SIMULATED
C          CA             - INITIAL CONCENTRATION OF ACID
C          CB             - CONCENTRATION OF BASE
C          VA             - INITIAL VOLUME OF BASE
C          PK1-PKN    - -LOGS OF THE DISSOCIATION CONSTANTS
C
C      METHOD:
C
C      (1).  FIND FINAL PH (200% FOR N=1,300% FOR N=2,400% FOR N=3)
C      (2).  FIND INITIAL PH
C      (3).  INCREMENT THE PH UNIFORMLY BETWEEN THE INITIAL
C            AND THE FINAL PH
C
C      THE NEWTON-RAPHSON METHOD IS USED TO FIND THE INITIAL
C      AND FINAL PH VALUES
C
```

```
C      THE SUBROUTINE SPECIE IS CALLED TO CALCULATE THE
C      CONCENTRATIONS OF HNA AND A(N-)
C
C    AUTHORS:  B.A. SWISSHELM, D.E. HAWKINS,JR. AND K.J. JOHNSON
C
C
       IMPLICIT REAL(K,N)
       DIMENSION K(3),Hl(7),X(60),Y(60)
       COMMON CA,VA,VB,PK(3)
       EQUIVALENCE (K(1),K1),(K(2),K2),(K(3),K3)
       INTEGER KOUT(3),N,WPH(2)
       DOUBLE PRECISION B,C
       DATA Hl(1),Hl(2)/0.,1./,KOUT/'  PK1, PK2, PK3'/,KW/lE-14/,WPH/
      l'FINAL','INIT.'/
     5 WRITE(6,11)
       READ(5,22)N,CA,VA,CB
       IF(N.LE.0)STOP
       WRITE(6,33)(KOUT(I),I=1,N)
       WRITE(6,44)
       DO 6 I=1,3
     6 K(I)=0
       READ(5,55)(PK(I),I=1,N)
       WRITE(6,66)N,CA,VA,CB,(KOUT(I),PK(I),I=1,N)
       DO 10 I=1,N
    10 K(I)=10.**(-PK(I))
C
C      ASSIGN COMMONLY USED CONSTANTS
C
       K12=K1*K2
       K123W=K12*lE21*K3
       CAP=CA*CB/((N+1)*CA+CB)
C
C      ASSIGN COEFFICIENTS OF FINAL PH
C
       C2=lE7*(K1+(N+1)*CAP)
       C3=lE14*K1*(K2+N*CAP)-l.
       C4=lE21*K1*(K2*(K3+(N-1)*CAP)-KW)
       C5=lE28*K12*(K3*CAP-KW)
C
C      FIND FIRST APPROX. TO FINAL PH
C
       H=1E-7/CAP
       L=1
C
C      ASSIGN COEFFICIENTS OF DERIVATIVE
C
       NC1=N+2
    15 NC2=(N+1)*C2
       NC3=N*C3
       NC4=(N-1)*C4
       NC5=(N-2)*C5
C
C    CALCULATE THE FINAL (L=1) AND INITIAL (L=2) PH USING
C    THE NEWTON-RAPHSON METHOD.
C
       DO 30 I=1,25
       HOLD=H
       DO 20 J=3,N+4
    20 Hl(J)=Hl(J-1)*H
```

```
      FH=H1(N+4)+C2*H1(N+3)+C3*H1(N+2)+C4*H1(N+1)+C5*H1(N)-K123W
      DFH=NC1*H1(N+3)+NC2*H1(N+2)+NC3*H1(N+1)+NC4*H1(N)+NC5
      H=HOLD-FH/DFH
      IF(ABS(1-HOLD/H).LE.1E-3)GO TO(32,38),L
      IF(H.LT.0)GO TO 31
   30 IF(ABS(1-HOLD/H).GT.1E4)GO TO 31
   31 WRITE(6,88)WPH(L)
      GO TO 5
   32 PHF=-ALOG10(H)
C
C     ASSIGN COEFFICIENTS FOR INITIAL PH
C
      C2=1E7*K1
      C3=1E14*K1*(K2-CA)-1.
      C4=1E21*K1*(K2*(K3-2*CA)-KW)
      C5=-1E28*K12*(K3*3*CA+KW)
      T=0
      TF=100.*(N+1)
      VB=0
      L=2
C
C     CALCULATE FIRST APPROX. TO INITIAL PH, THEN GO BACK TO
C     NEWTON'S METHOD
C
      B=K1
      C=4*K1*CA
      IF(DABS(C/B/B).LT.1D-6)GO TO 35
      H=5D6*(DSQRT(B*B+C)-B)
      GO TO 15
   35 H=1E7*SQRT(CA*K1)
      GO TO 15
C
C     RETURN  HERE AFTER CALCULATING INITIAL PH
C
   38 PH=-ALOG10(H)
      PHOUT=PH+7
C
C     CALL SPECIE TO DETERMINE (HNA) AND (A(N-))
C
      CALL SPECIE(N,PHOUT,HNA,AMINUS)
C
C   START OF TITRATION
C    PRINT 25 (N=1), 40 (N=2), OR 60 (N=3) ENTRIES
C
      WRITE(6,77)N,N
      WRITE(6,99)T,PHOUT,VB,HNA,AMINUS
      L=(N+3)*2.5*N+13
      LP2=L+2
      X(LP2)=PHOUT
      Y(LP2)=T
      PHINC=(PHF-PH)/(L+1)
C
C     ASSIGN OTHER CONSTANTS USED IN FORMULA
C
      NC2=1E7*(K1+CB)
      NC3=1E14*K1*(K2+CB)-1.
      NC4=1E21*K1*(K2*(K3+CB)-KW)
      NC5=1E28*K12*(K3*CB-KW)
      EVB=CA*VA/CB
```

```
C
C       INCREMENT PH UNTIL 100% ABOVE LAST EQUIVALENCE POINT
C
        DO 50 I=1,L
        PH=PH+PHINC
        PHOUT=PH+7
        H=10.**(-PH)
        DO 40 J=3,N+4
     40 H1(J)=H1(J-1)*H
        NUM=H1(N+4)+C2*H1(N+3)+C3*H1(N+2)+C4*H1(N+1)+C5*H1(N)-K123W
        DEN=H1(N+4)+NC2*H1(N+3)+NC3*H1(N+2)+NC4*H1(N+1)+NC5*H1(N)-K123W
        VB=-VA*NUM/DEN
        T=100*VB/EVB
        IF(T.GE.TF)GO TO 60
        CALL SPECIE(N,PHOUT,HNA,AMINUS)
        WRITE(6,99)T,PHOUT,VB,HNA,AMINUS
        X(LP2-I)=PHOUT
        Y(LP2-I)=T
     50 CONTINUE
C
C       WRITE FINAL PH
C
     60 VB=TF*EVB/100.
        PHOUT=PHF+7
        CALL SPECIE(N,PHOUT,HNA,AMINUS)
        WRITE(6,99)TF,PHOUT,VB,HNA,AMINUS
        X(1)=PHOUT
        Y(1)=TF
C
C    IS A PLOT DESIRED?
C
        WRITE(6,101)
        READ(5,22)IPLT
        IF(IPLT.EQ.1)CALL PLOT(X,Y,LP2)
        GO TO 5
     11 FORMAT(//' ENTER N, CA, VA, CB   ( N=0 ENDS)'/)
     22 FORMAT(I,3F)
     33 FORMAT(/' ENTER',3A5)
     44 FORMAT(1X)
     55 FORMAT(3F)
     66 FORMAT(/' N =' I2,5X,'  CA  =',G12.3,'  VA  =',G12.3,'  CB  =',
      1 G12.3//11X,3(A5,' =',G12.3))
     77 FORMAT(//4X,'% TITR.',9X,'PH',6X,'VOL. OF BASE',5X,'[H',I1,
      1'A]',7X,'[A',I1,'-]'/)
     38 FORMAT(//' SORRY. THE ITERATIVE METHOD USED TO CALCULATE
      1 THE ',A5,' PH'/' DID NOT CONVERGE.'//)
     99 FORMAT(F10.3,2F13.3,4X,1P2E12.2)
    101 FORMAT(//'  ENTER 1 IF YOU WANT A PLOT OF % TITR VS. PH,',
      1 ' 0 OTHERWISE.'/)
        END
C
C
        SUBROUTINE SPECIE(N,PH,HNA,AMINUS)
C
C       THIS SUBROUTINE CALCULATES THE CONCENTRATIONS
C       OF HNA AND A(N-)
C
C
```

```
C
C    FOR N=3,
C
C    H3A = CATOT * H**3 / D
C    A(N-)    = CATOT * K1 * K2 * K3 / D
C
C    WHERE  D = H**3 + K1*H**2 + K1*K2*H + K1*K2*K3
C
C    AUTHORS:  D. E. HAWKINS,JR. AND K. J. JOHNSON
C
     DIMENSION ALPHA(4)
     COMMON CA,VA,VB,PK(3)
     REAL LATOT,LD
     ALPHA(1)=-PH*N
     D=0
     IF(ALPHA(1).GT.-38)D=10.**ALPHA(1)
     DO 10 I=1,N
     ALPHA(I+1)=ALPHA(I)+PH-PK(I)
     IF(ALPHA(I+1).LT.-38)GO TO 10
     D=D+10.**ALPHA(I+1)
 10  CONTINUE
     LATOT=ALOG10(CA*VA/(VA+VB))
     LD=ALOG10(D)
     DL=LATOT-LD
     DO 20 I=1,N+1
     ALPHA(I)=ALPHA(I)+DL
     IF(ALPHA(I).LT.-38)GO TO 15
     ALPHA(I)=10.**ALPHA(I)
     GO TO 20
 15  ALPHA(I)=0
 20  CONTINUE
     HNA=ALPHA(1)
     AMINUS=ALPHA(N+1)
     RETURN
     END
```

Discussion

HNATRN, SPECIE and PLOT require approximately 1400 36-bit words of core. The execution time for the sample shown is approximately 0.4 sec.

Scaling is used in HNATRN for the reasons discussed in the documentation of the HNA program (Section 3.3.2). The technique of solving the inverse titration function used here and in the EDTA titration simulation program (Section 3.1.1) is a noteworthy optimization feature. Clearly, HNATRN would execute considerably less efficiently if the Newton-Raphson method had to be used to calculate the root of a third-, fourth-, or fifth-order polynomial at each percentage of titration. The inverse function method also provides the bonus of concentrating the information in the region of most interest, i.e., near the equivalence point(s).

Problem 3.14. Demonstrate the dependence of acid-base titration curves on both initial concentration and the dissociation constants of the acid.

Solution. See Figure 3.10.

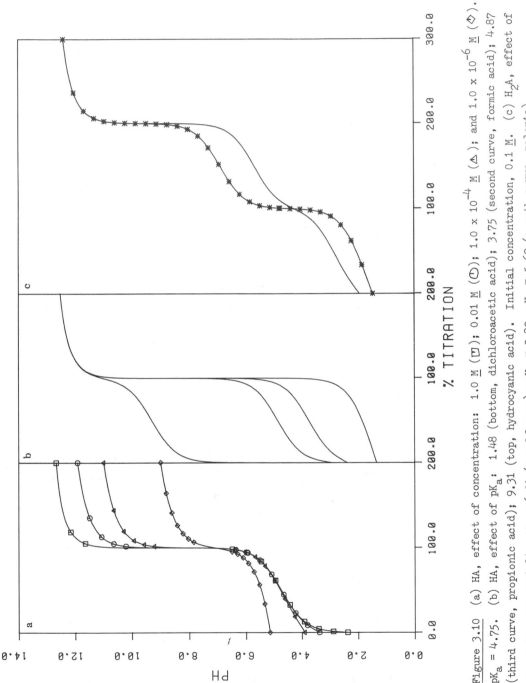

Figure 3.10 (a) HA, effect of concentration: 1.0 $\underline{M}$ ($\square$); 0.01 $\underline{M}$ ($\ominus$); 1.0 × 10^{-4} $\underline{M}$ ($\triangle$); and 1.0 × 10^{-6} $\underline{M}$ ($\diamondsuit$). pK$_a$ = 4.75. (b) HA, effect of pK$_a$: 1.48 (bottom, dichloroacetic acid); 3.75 (second curve, formic acid); 4.87 (third curve, propionic acid); 9.31 (top, hydrocyanic acid). Initial concentration, 0.1 $\underline{M}$. (c) H$_2$A, effect of K$_1$ and K$_2$: pK$_1$ = 1.81, pK$_2$ = 6.91 (*,sulfurous); pK$_1$ = 2.83, pK$_2$ = 5.69 (smooth curve, malonic).

Problem 3.16. Modify HNATRN to accommodate amino acid systems, and titrations of weak bases with strong acids.

Problem 3.17. Compare the results of HNATRN with a program that does not use the inverse titration function [16, 17].

Problem 3.18. Write a program to determine successive dissociation constants for polyprotic acids from acid-base titration data [18].

3.3.4 GASEQ

Introduction

GASEQ is concerned with the following gas-phase equilibrium system:

$$aA + bB \rightleftharpoons cC + dD \qquad K_p = \frac{p_C^c p_D^d}{p_A^a p_B^b}$$

The input includes the stoichiometric coefficients a, b, c, and d; the initial number of moles of each species, A_0, B_0, C_0, and D_0; the equilibrium constant K_p; the volume in liters V; the absolute temperature T; and an initial approximation to the equilibrium extent of reaction, in moles, x. The Newton-Raphson method is used to find the extent of the reaction at equilibrium. If the method converges the number of moles and the partial pressure of each species at equilibrium is printed, and as a check of the procedure, K_p is calculated from the equilibrium partial pressures.

Method

Let K_n be the equilibrium constant in terms of moles. Then at equilibrium,

$$aA \quad + \quad bB \quad \rightleftharpoons \quad cC \quad + \quad dD$$

moles: $A_0 - ax$ $B_0 - bx$ $C_0 + cx$ $D_0 + dx$

and

$$K_n = \frac{(C_0 + cx)^c (D_0 + dx)^d}{(A_0 - ax)^a (B_0 - bx)^b}$$

Assuming ideal gas behavior,

$$p_i = \frac{n_i RT}{V}$$

where P_i and n_i are the partial pressure and number of moles of species i, and R is the ideal gas constant. Then K_p and K_n are related by the following equation:

$$K_p = K_n \left(\frac{RT}{V}\right)^{c+d-(a+b)}$$

GASEQ calculates K_n from K_p, R, T, V, a, b, c, and d. Then the Newton-Raphson method is used to find the equilibrium extent of reaction, x, by finding the appropriate root of the function,

$$f(x) = K_n(A_0 - ax)^a(B_0 - bx)^b - (C_0 + cx)^c(D_0 + dx)^d$$

If the method converges, the equilibrium numbers of moles of the components are known.

GASEQ consists of a main program and a function subprogram,

FUNCTION FUNK(X,IPR,IGO)

Here X is x, IPR is one of the four exponents (a, b, c, or d), and IGO is an integer between 1 and 4 corresponding to one of the four species in the chemical equation. FUNK evaluates the expressions $(A_0 - ax)^a$, $(B_0 - bx)^b$, etc.

There are three convergence criteria:

Let

$$RELX = \left| \frac{x_{i+1} - x_i}{x_{i+1}} \right|$$

1. If RELX $\leq$ 0.0001, convergence is attained.
2. If RELX $\geq$ 1,000,000, the method is diverging,
3. A maximum of 20 iterations is allowed.

Sample Execution $(2NH_3 \rightleftharpoons N_2 + 3H_2)$

 FOR THE REACTION

                    P A + Q B   <======> R C +   S D
 ENTER P,Q,R AND S   (P & Q = 0 STOPS THE PROGRAM)
 >2 0 1 3

 ENTER THE INITIAL NUMBER OF MOLES OF A,B,C & D
 >1

 ENTER THE EQUIL. CONSTANT (KP), THE VOLUME (LITERS)
 AND THE ABSOLUTE TEMPERATURE
 >1.5E-6 1   298

 ENTER THE INITIAL APPROXIMATION TO THE EXTENT OF
 REACTION IN MOLES
 >.01

FOR THE GAS PHASE REACTION

   2 A + 0 B <=======>   1 C + 3 D   KP =   0.150E-05 ATM**(2)

INITIAL # OF MOLES: A0 =   1.00   B0 =   0.000

                     C0 =   0.000   AND D0 =   0.000

THE TEMP. IS   298.   K., & THE VOL. IS   1.00   LITER(S)

                         CONVERGENCE MAP

     I   X   DX   F(X)

     1   0.100E-01   0.248E-02   -0.268E-06
     2   0.752E-02   0.183E-02   -0.840E-07
     3   0.569E-02   0.130E-02   -0.259E-07
     4   0.439E-02   0.829E-03   -0.760E-08
     5   0.356E-02   0.385E-03   -0.189E-08
     6   0.318E-02   0.809E-04   -0.281E-09
     7   0.310E-02   0.303E-05   -0.972E-11
     8   0.310E-02   -0.501E-08   0.160E-13

EQUIL. # OF MOLES: AEQ =   0.994   BEQ =   0.000

                    CEQ =   0.310E-02   AND DEQ =   0.929E-02

EQUIL. PARTIAL PRESSURES: PA =   24.3   PB =   0.000

                           PC =   0.757E-01 AND PD =   0.227

THE CALCULATED KP IS   0.150E-05 ATM**(2)

```
C
C
C                  GASEQ
C
C
C
C    THIS PROGRAM SOLVES GAS PHASE EQUILIBRIUM PROBLEMS.
C
C    FOR THE REACTION
C
C         P A  +  Q B  <======>  R C  +  S D
C
C    THE FOLLOWING PARAMETERS ARE READ AS INPUT DATA:
C
C    PARAMETER(S)              DESCRIPTION
C
C    NA,NB,NC,ND               THE STOICHIOMETRIC COEFFICEICTS P,Q,R & S
C    A0,B0,C0,D0               THE INITIAL # OF MOLES OF A,B,C & D
C      CONST                   THE EQUILIBRIUM CONSTANT,KP
C        V                     THE VOLUME (LITERS)
C        T                     THE TEMPERATURE ( K)
C        X                     THE INITIAL APPROXIMATION TO THE
C                              EXTENT OF REACTION (MOLES)
C
```

Listing

```
C            THE NEWTON-RAPHSON METHOD IS USED TO FIND THE
C            EQUILIBRIUM NUMBER OF MOLES OF EACH SPECIES.
C
C
C     AUTHOR:  K. J. JOHNSON
C
      REAL KP
      COMMON NA,NB,NC,ND,A0,B0,C0,D0
C
C     INPUT
C
   10 WRITE(6,11)
      READ(5,22)NA,NB,NC,ND
      IF(NA+NB.EQ.0)STOP
      WRITE(6,33)
      READ(5,44)A0,B0,C0,D0
      WRITE(6,55)
      READ(5,44)KP,V,T
      WRITE(6,77)
      READ(5,44)X
C
C     SET CONSTANTS
C
      ITMAX=20
      CVGTOL=.0001
      DVGTOL=1.E6
      NDELT=NC+ND - (NA+NB)
      CONST0=0.08205*T/V
      CONST1=KP/(CONST0**NDELT)
C
C     PRINT DATA FOR VERIFICATION PURPOSES
C
      WRITE(6,66)NA,NB,NC,ND,KP,NDELT,A0,B0,
     1 C0,D0,T,V
      WRITE(6,88)
C
C     ITERATION LOOP
C
      DO 50 I=1,ITMAX
      FOFX=CONST1*FUNK(X,NA,1)*FUNK(X,NB,2)
     1 -FUNK(X,NC,3)*FUNK(X,ND,4)
      DFOFX=-CONST2*(NB*NB*FUNK(X,NA,1)*FUNK(X,NB-1,2)
     1 + NA*NA*FUNK(X,NB,2)*FUNK(X,NA-1,1))
     2 - (ND*ND*FUNK(X,NC,3)*FUNK(X,ND-1,4)
     3 +  NC*NC*FUNK(X,ND,4)*FUNK(X,NC-1,3))
      IF(ABS(DFOFX).LE.1.E-25)GO TO 91
      DX=FOFX/DFOFX
      WRITE(6,99)I,X,DX,FOFX
      RELCG=ABS(DX/X)
      IF(RELCG.LE.CVGTOL)GO TO 90
      IF(RELCG.GE.DVGTOL)GO TO 96
      X=X-DX
   50 CONTINUE
C
C     ERROR CONDITIONS  -  WRITE MESSAGE AND START OVER
C
      WRITE(6,95)
      GO TO 10
```

```
      91 WRITE(6,92)
         GOTO 10
      96 WRITE(6,97)
         GO TO 10
C
C     COMPUTE # OF MOLES AT EQUIL., EQUIL. PARTIAL PRESSURES,
C     AND THE EQUIL. CONSTANT  (KP)
C
      90 A=A0-NA*X
         B=B0-NB*X
         C=C0+NC*X
         D=D0+ND*X
         PA=A*CONST0
         PB=B*CONST0
         PC=C*CONST0
         PD=D*CONST0
         CALCEQ=PC**NC*PD**ND/(PA**NA*PB**NB)
         WRITE(6,98)A,B,C,D,PA,PB,PC,PD,CALCEQ,NDELT
         GO TO 10
         STOP
C
C        FORMATS
C
      11 FORMAT(//'   FOR THE REACTION'//20X,
        1 'P A  +  Q B  <=======> R C  +    S D'//,
        2 '   ENTER P,Q,R & S    ( P AND Q = 0 STOPS THE PROGRAM)'/)
      22 FORMAT(4I)
      33 FORMAT(/'   ENTER THE INITIAL NUMBER OF MOLES OF A,B,C & D'/)
      44 FORMAT(4F)
      55 FORMAT(/'   ENTER THE EQUIL. CONSTANT (KP), THE VOLUME (LITERS)',
        1 /'   AND THE ABSOLUTE TEMPERATURE   '/)
      66 FORMAT(//'   FOR THE GAS PHASE REACTION'//,2X,
        1 I2,' A  + ',I2,' B  <=======>  ',I2,' C  + ',I2,' D',
        2 5X,'KP = ',G12.3,' ATM**(',I2,')'  ,
        3 ///,'   INITIAL # OF MOLES:  A0 =',G12.3,5X,' B0 =',
        4 G12.3,//22X,' C0 =',G12.3,2X,'AND D0 =',G12.3,
        5 //'   THE TEMP. IS ',G10.3,'  K.,   & THE',
        6 ' VOL. IS ',G10.3,' LITER(S)')
      77 FORMAT(/'   ENTER THE INITIAL APPROXIMATION TO THE EXTENT OF'
        1 /'   REACTION IN MOLES '/)
      88 FORMAT(//20X,' CONVERGENCE MAP',//,'     I',10X,'X',13X,
        1 'DX',11X,'F(X)'/)
      92 FORMAT(//'   HAVE A PROBLEM DF(X) CLOSE TO ZERO'//)
      95 FORMAT(//'   CONVERGENCE FAILURE'/)
      97 FORMAT(//'   DIVERGING  --- STARTING OVER'/)
      98 FORMAT(//'   EQUIL.  OF MOLES:  AEQ =',G12.3,6X,' BEQ =',
        1 G12.3,//20X,'  CEQ =',G12.3,2X,' AND DEQ =',G12.3,
        2 //'   EQUIL. PARTIAL PRESSURES:  PA =',G12.3,4X,' PB =',
        3 G12.3,//28X,' PC =',G12.3,' AND PD =',G12.3,
        4 //'   THE CALCULATED KP IS ',G12.3,' ATM**(',I2,')',/)
      99 FORMAT(I5,3G15.3)
         END
C
C           FUNK
C
C     FUNK EVALUATES EXPRESSIONS OF THE FORM
C
C           ( A0 - NA*X ) ** NA
```

C

```
      FUNCTION FUNK(X,IPR,IGO)
      COMMON NA,NB,NC,ND,A0,B0,C0,D0
      GO TO (10,20,30,40),IGO
   10 IF(NA.LE.0)GO TO 99
      BASE=A0-NA*X
      IF(BASE)99,99,95
   20 IF(NB.LE.0)GO TO 99
      BASE=B0-NB*X
      IF(BASE)99,99,95
   30 IF(NC.LE.0)GO TO 99
      BASE=C0+NC*X
      IF(BASE)99,99,95
   40 IF(ND.LE.0)GO TO 99
      BASE=D0+ND*X
      IF(BASE)99,99,95
   95 FUNK=BASE**IPR
      RETURN
   99 FUNK=1.0
      RETURN
      END
```

Discussion

GASEQ and FUNK require approximately 800 words, respectively. The execution time for this sample execution was approximately 0.2 sec.

Problem 3.19. Calculate and plot the equilibrium partial pressures of a gas-phase system as a function of total pressure and temperature.

Solution. See Figure 3.11.

3.4 ADDITIONAL PROBLEMS

1. Write a program to calculate both real and imaginary roots of the quadratic equation.
2. (a) The equation $f(x) = x^6 - x^4 - x^3 - 1 = 0$ has one real root between 1 and 2. Find this root to seven figures.
   (b) The equation $f(x) = x^4 - 5x^3 - 12x^2 + 76x - 79 = 0$ has two real roots close to $x = 2$. Find these roots to five figures.
   (c) The equation $f(x) = \log x - \cos x$ has a real root near $x = 1$. Find this root to six figures.
   (d) Find the abscissa corresponding to the inflection point of the function $f(x) = e^{-x}\ln x$ to five figures.

   Solutions. (a) 1.403602   (b) 1.7684 and 2.2410   (c) 1.30296   (d) 2.55245

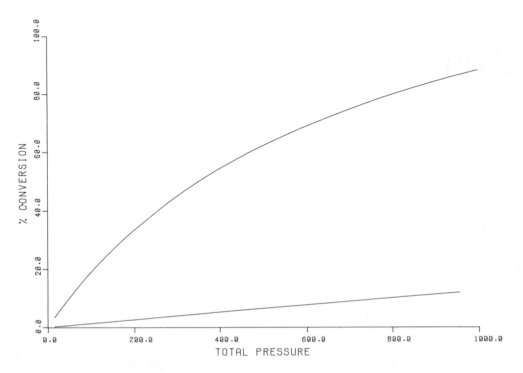

<u>Figure 3.11</u> Percent conversion of N_2 and H_2 to NH_3 as a function of total pressure
and temperature: T = 1000K (top); T = 2000K (bottom).

3. Consider the solubility equilibrium,

$$MX_s \rightleftharpoons M^{z+} + X^{z-} \qquad K_{ap} = a_{M^{z+}} a_{X^{z-}}$$

Show that the solubility of MX in a C molar 1:1 electrolyte at 25°C is a root
of the equation

$$F(S) = 2 \log S - 1.018z^2\sqrt{z^2 S + C} - \log K_{ap}$$

assuming the Debye-Hückel limiting law. Choose a series of sparingly soluble
salts and plot the solubility versus ionic strength [20].

4. Write a program to find the minimum solubility of CuCl in solution containing
HCl [21].

5. Calculate the compressibility of a real gas using the Beattie-Bridgeman
equation of state,

$$P = \frac{RT}{V} + \frac{Q}{V^2} + \frac{S}{V^3} + \frac{U}{V^4}$$

5. Continued

where Q, S, and U are temperature-dependent parameters,

$$Q = RTB_0 - A_0 - \frac{Rc}{T^2}$$

$$S = -RTB_0 b + A_0 a - \frac{RcB_0}{T^2}$$

$$U = \frac{RB_0 bc}{T^2}$$

Here A_0, B_0, a, b, and c are widely tabulated empirical constants. For example, for methane, A_0 = 2.2769, B_0 = 0.05587, a = 0.01855, b = 0.01587, and c = 1.283 x 10^4. The units are P in atmospheres, V in liters, and T in Kelvin. Compare calculated and observed compressibility values [22].

6. Compare observed and calculated compressibilities for real gases using the van der Waals, Berthelot, modified Barthelot, and Dieterici equations of state [23, 24].

7. The rate $qd\lambda$ at which electromagnetic radiation is emitted from a unit area of the surface of a black-body radiator within the wavelength λ to $d\lambda$ is given by Planck's law:

$$qd\lambda = \frac{2\pi hc^2 d\lambda}{\lambda^5 [\exp(hc/\lambda kT) - 1]} \qquad \text{ergs/cm}^2 \cdot \text{sec}$$

where c is the speed of light, 3.00 x 10^{10} cm/sec; h is Planck's constant, 6.63 x 10^{-27} erg·sec; k is Boltzmann's constant, 1.38 x 10^{-16} erg/K; and λ is the wavelength in centimeters. Calculate the wavelength at which the radiant energy is most intense as a function of temperature. This wavelength corresponds to a maximum in the preceding equation, that is, $dq/d\lambda$ = 0. Verify Wien's displacement law, $\lambda_{max} T$ = constant [25].

8. Write a program to solve equilibrium problems by the method of free energy minimization [26-28].

9. Write a program to solve an equilibrium system of interest to industrial chemistry with maximation of profit as the ultimate objective [29].

10. Write a program to calculate the fraction of enclosed electron density for an atomic orbital [30].

11. Write programs to find roots of equations using numerical methods not included in this chapter [31-33].

REFERENCES

1. J. C. Davis, Jr., "Advanced Physical Chemistry," Ronald, New York, 1965, p.
   p. 165.

2. W. J. Blaedel and V. W. Meloche, "Elementary Quantitative Analysis: Theory
   and Practice," Harper and Row, New York, 1963, pp. 578-581.

3. (a) H. Laitinen, "Chemical Analysis," McGraw-Hill, New York, 1960, p. 228ff.
   (b) A. J. Bard and D. M. King, General digital computer program for chemical
   equilibrium calculations, J. Chem. Educ., 42, 127 (1965); see also letters
   to the editor, J. Chem. Educ., 42, 624 and 625 (1965).

4. C. N. Reiley and R. W. Schmid, Chelometric titration with potentiometric end
   point detection, Anal. Chem., 30, 947 (1958).

5. "Handbook of Chemistry and Physics," Chemical Rubber Company, Cleveland,
   Vol. 45, 1964, p. A-155.

6. B. Carnahan, H. Luther, and J. Wilkes, "Applied Numerical Analysis," Wiley,
   New York, 1969, p. 203, Problem 3.30.

7. J. W. Mellor, "Higher Mathematics for Students of Chemistry and Physics,"
   4th Ed., Dover, New York, 1955, pp. 367-369.

8. B. Carnahan et al., op. cit., pp. 178-179.

9. J. C. Davis, Jr., op. cit., p. 93ff.

10. P. T. Mathews, "Introduction to Quantum Mechanics," McGraw-Hill, New York,
    1963, pp. 36-47.

11. R. B. Flewwelling and W. G. Laidlaw, Calculation of NMR shift using particle
    in a box wave functions, J. Chem. Educ., 46, 355 (1969).

12. C. Froberg, "Introduction to Numerical Analysis," Addison-Wesley, Reading,
    Mass., 1965, pp. 19-26.

13. H. Laitinen, op. cit., pp. 35-36.

14. F. S. Nakayama, Hydrolysis of sodium carbonate, J. Chem. Educ., 47, 67 (1970).

15. M. J. D. Brand, A general method for the solution of polynomial equations
    in H^+ ion concentration, J. Chem. Educ., 53, 771 (1976).

16. G. L. Breneman, A general acid-base titration curve computer program, J. Chem.
    Educ., 51, 812 (1974).

17. A. R. Emery, Computer program for the calculation of acid-base titration
    curves, J. Chem. Educ., 42, 131 (1965).

18. R. E. Jensen, et al., Determination of successive ionization constants, J. Chem.
    Educ., 47, 147 (1970).

19. C. Froberg, op. cit., p. 41.

20. H. Laitinen, op. cit., pp. 7-13.

21. J. N. Butler, "Ionic Equilibrium--A Mathematical Approach," Addison-Wesley,
    Reading, Mass., 1964, pp. 277-278.

22. B. Carnahan et al., op. cit., p. 173-177.

23. J. Ott, J. Goates, and H. Hall, Jr., Comparison of equations of state in effectively describing PVT relations, J. Chem. Educ., 48, 515 (1971).

24. R. Roskos, "Problem Solving in Physical Chemistry," West, New York, 1975, Chapter 8.

25. B. Carnahan et al., op. cit., p. 206, Problem 3.43.

26. F. van Zeggeren and S. H. Storey, "The Computation of Chemical Equilibrium," Cambridge University Press, New York, 1970.

27. S. Davidson, Iterative method of solving equilibrium problems by free-energy minimization, J. Chem. Educ., 50, 299 (1973).

28. D. R. McMillin, Determining extremums, J. Chem. Educ., 51, 496 (1974).

29. R. J. Rish, Maximizing profits in equilibrium processes, J. Chem. Educ., 52, 441 (1975).

30. I. N. Levine, Integration of the 2Pz AO, J. Chem. Educ., 49, 446 (1972).

31. B. Carnahan et al., op. cit., Chapter 3.

32. L. Kelly, "Handbook of Numerical Methods and Applications," Addison-Wesley, Reading, Mass., 1967, Chapter 6.

33. J. F. Traub, "Iterative Methods for Solutions of Equations," Prentice-Hall, Englewood Cliffs, N. J., 1964.

CHAPTER 4

SYSTEMS OF LINEAR SIMULTANEOUS EQUATIONS

A number of chemistry problems require the solution to a linear system of simultaneous equations. Among these are analysis of mixtures, multicomponent equilibrium
systems, curve smoothing, and nonlinear regression analysis. Four methods of
solving linear systems of equations are discussed in this chapter: Cramer's rule;
the Gauss-Seidel iterative method; and two elimination methods, augmented matrix
elimination and coefficient matrix inversion.

A linear system of equations has the following form:

$$a_{11}x_1 + a_{12}x_2 + a_{13}x_3 + \cdots + a_{1n}x_n = b_1$$
$$a_{21}x_1 + a_{22}x_2 + a_{23}x_3 + \cdots + a_{2n}x_n = b_2$$
$$\cdots\cdots\cdots\cdots\cdots\cdots\cdots\cdots\cdots\cdots\cdots$$
$$a_{n1}x_1 + a_{n2}x_2 + a_{n3}x_3 + \cdots + a_{nn}x_n = b_n$$

Here a_{ij} is one of n x n coefficients, b_i is one of n constants, and x_i is one of
the unknown values to be found. Introducing matrix notation,

$$AX = B$$

where

$$A = \begin{bmatrix} a_{11} & a_{12} & a_{13} & \cdots & a_{1n} \\ a_{21} & a_{22} & a_{23} & \cdots & a_{2n} \\ \cdots\cdots\cdots\cdots\cdots\cdots \\ a_{n1} & a_{n2} & a_{n3} & \cdots & a_{nn} \end{bmatrix}$$

$$B = \begin{pmatrix} b_1 \\ b_2 \\ \cdot\cdot \\ b_n \end{pmatrix} \quad \text{and} \quad X = \begin{pmatrix} x_1 \\ x_2 \\ \cdot\cdot \\ x_n \end{pmatrix}$$

The determinant of A, $|A|$, is

$$|A| = \begin{vmatrix} a_{11} & a_{12} & a_{13} & \cdots & a_{1n} \\ a_{21} & a_{22} & a_{23} & \cdots & a_{2n} \\ \cdots & \cdots & \cdots & \cdots & \cdots \\ a_{n1} & a_{n2} & a_{n3} & \cdots & a_{nn} \end{vmatrix}$$

4.1 CRAMER'S RULE

Cramer's rule [1] is

$$x_i = \frac{|D_i|}{|A|} \qquad A \neq 0$$

where D_i is the determinant of the coefficient matrix obtained when the ith column of A is replaced by B. For example, for n = 3,

$$x_1 = \frac{\begin{vmatrix} b_1 & a_{12} & a_{13} \\ b_2 & a_{22} & a_{23} \\ b_3 & a_{32} & a_{33} \end{vmatrix}}{|A|} \qquad x_2 = \frac{\begin{vmatrix} a_{11} & b_1 & a_{13} \\ a_{21} & b_2 & a_{23} \\ a_{31} & b_3 & a_{33} \end{vmatrix}}{|A|} \qquad x_3 = \frac{\begin{vmatrix} a_{11} & a_{12} & b_1 \\ a_{21} & a_{22} & b_2 \\ a_{31} & a_{32} & b_3 \end{vmatrix}}{|A|}$$

where

$$\begin{aligned} |A| &= \begin{vmatrix} a_{11} & a_{12} & a_{13} \\ a_{21} & a_{22} & a_{23} \\ a_{31} & a_{32} & a_{33} \end{vmatrix} \\ &= a_{11}a_{22}a_{33} + a_{12}a_{23}a_{31} + a_{13}a_{21}a_{32} \\ &\quad - (a_{12}a_{21}a_{33} + a_{11}a_{23}a_{32} + a_{13}a_{22}a_{31}) \end{aligned} \tag{4.1}$$

The number of factors in the determinant of order n is on the order of n!; so this method is not practical for solving large systems of equations. Also, round-off errors may accumulate if the terms in Eq. (4.1) are nearly equal in magnitude but opposite in sign. The usefulness of Cramer's rule lies in its simplicity for solving systems of low order, as illustrated in the following problem.

Problem 4.1. Use Cramer's rule to solve the linear system

$$\begin{aligned} 3x_1 + x_2 - x_3 &= 2 \\ x_1 + 2x_2 - x_3 &= 2 \\ x_1 - x_2 + 4x_3 &= 11 \end{aligned}$$

Solution.

$$|A| = \begin{vmatrix} 3 & 1 & -1 \\ 1 & 2 & -1 \\ 1 & -1 & 4 \end{vmatrix} = 19$$

Then

$$x_1 = \frac{\begin{vmatrix} 2 & 1 & -1 \\ 2 & 2 & -1 \\ 11 & -1 & 4 \end{vmatrix}}{19} \qquad x_2 = \frac{\begin{vmatrix} 3 & 2 & -1 \\ 1 & 2 & -1 \\ 1 & 11 & 4 \end{vmatrix}}{19} \qquad x_3 = \frac{\begin{vmatrix} 3 & 1 & 2 \\ 1 & -1 & 11 \\ 1 & -1 & 11 \end{vmatrix}}{19}$$

$$x_1 = 1 \qquad x_2 = 2 \qquad x_3 = 3$$

4.2 THE GAUSS-SEIDEL METHOD

The Gauss-Seidel method [2, 3] is an iterative method for solving linear systems of equations. The procedure is illustrated below for a system of order 3. First an initial approximation to the solution is required. Let this initial approximation be

$$X = \begin{pmatrix} 0 \\ 0 \\ 0 \end{pmatrix}$$

that is, $x_1 = x_2 = x_3 = 0$. The linear system is

$$a_{11}x_1 + a_{12}x_2 + a_{13}x_3 = b_1$$
$$a_{21}x_1 + a_{22}x_2 + a_{23}x_3 = b_2$$
$$a_{31}x_1 + a_{32}x_2 + a_{33}x_3 = b_3$$

First iteration:

$$x_1^{(1)} = \frac{b_1 - 0 - 0}{a_{11}}$$

$$x_2^{(1)} = \frac{b_2 - a_{21}x_1^{(1)} - 0}{a_{22}}$$

$$x_3^{(1)} = \frac{b_3 - a_{31}x_1^{(1)} - a_{32}x_2^{(1)}}{a_{33}}$$

This method assumes that a_{11}, a_{22}, and a_{33} are nonzero. Notice that the Gauss-Seidel method uses the most recently calculated values for x_i.

Second iteration:

$$x_1^{(2)} = \frac{b_1 - a_{12}x_2^{(1)} - a_{13}x_3^{(1)}}{a_{11}}$$

$$x_2^{(2)} = \frac{b_2 - a_{21}x_1^{(2)} - a_{23}x_3^{(1)}}{a_{22}}$$

$$x_3^{(2)} = \frac{b_3 - a_{31}x_1^{(2)} - a_{32}x_2^{(2)}}{a_{33}}$$

After m iterations:

$$x_1^{(m+1)} = \frac{b_1 - a_{12}x_2^{(m)} - a_{13}x_3^{(m)}}{a_{11}}$$

$$x_2^{(m+1)} = \frac{b_2 - a_{21}x_1^{(m+1)} - a_{23}x_3^{(m)}}{a_{22}}$$

$$x_3^{(m+1)} = \frac{b_3 - a_{31}x_1^{(m+1)} - a_{32}x_2^{(m+1)}}{a_{33}}$$

For a linear system of order n,

$$x_i^{(m+1)} = \frac{b_i - \sum_{j=1}^{i=1} a_{ij}x_j^{(m+1)} - \sum_{j=i+1}^{n} a_{ij}x_j^{(m)}}{a_{ii}}$$

with $a_{ii} \neq 0$.

Problem 4.2. Use the Gauss-Seidel method to find the solution to the following linear system of equations:

$$30x_1 + x_2 - x_3 = 29$$
$$x_1 + 20x_2 - x_3 = 38$$
$$x_1 + x_2 + 10x_3 = 33$$

Solution. Assume $x_1^0 = x_2^0 = x_3^0 = 0$

First iteration:

$$x_1^{(1)} = \frac{29}{30} = 0.967$$

$$x_2^{(1)} = \frac{38 - 0.967}{20} = 1.85$$

$$x_3^{(1)} = \frac{33 - 0.967 - 1.85}{10} = 3.02$$

Second iteration:

$$x_1^{(2)} = \frac{29 - 1.85 + 3.02}{30} = 1.01$$

$$x_2^{(2)} = \frac{38 - 1.01 + 3.02}{20} = 2.02$$

$$x_3^{(2)} = \frac{33 - 1.01 - 2.02}{10} = 3.00$$

Third iteration:

$$x_1^{(3)} = \frac{29 - 2.02 - 3.00}{30} = 0.999$$

$$x_2^{(3)} = \frac{38 - 0.999 + 3.00}{20} = 2.00$$

$$x_3^{(3)} = \frac{33 - 0.999 - 2.00}{10} = 3.00$$

The method is converging on the solution

$$x_1 = 1 \qquad x_2 = 2 \qquad x_3 = 3$$

The n x n coefficient matrix A is said to be diagonally dominant if

$$|a_{ii}| \geq \sum_{\substack{j=1 \\ j \neq i}}^{n} |a_{ij}|$$

For example, the coefficient matrix in the preceding example is diagonally dominant,

$$30 > 1 + 1 \qquad 20 > 1 + 1 \qquad 10 > 1 + 1$$

The Gauss-Siedel method will converge if the coefficient matrix is diagonally dominant [2]. The rate of convergence depends on the diagonal dominance of the coefficient matrix and the proximity of the initial approximations to the solution vector.

An example in chemistry that involves a diagonally dominant matrix is the analysis of a mixture by spectrophotometry.

Problem 4.3. Determine the molar concentrations of a five-component mixture in solution from the following spectrophotometric data:

Wavelength	Molar absorptivity of component					Observed total absorbance
	1	2	3	4	5	
1	100	10	2	1	0.5	0.1135
2	12	120	10	5	0.9	0.2218
3	30	30	90	10	2	0.2700
4	1	4	18	140	24	0.2992
5	2	4	8	16	120	0.1350

Assume the optical path length is unity, and that the solvent does not absorb at these wavelengths.

Solution. Assuming that Beer's law is obeyed, then at a given wavelength, i,

$$A_{tot_i} = \sum_{j=1}^{5} \epsilon_{ij} c_j$$

A_{tot_i} is the total observed absorbance at λ_i, ϵ_{ij} is the molar absorptivity of the jth component at λ_i, and c_j is the molar concentration of the jth component in the mixture. The length of the optical cell is assumed to be unity. This is a linear system of simultaneous equations,

$$AX = B$$

where $a_{ij} = \epsilon_{ij}$, $x_i = c_i$, and $b_i = A_{tot_i}$. The coefficient matrix is diagonally dominant. This problem is solved in the sample execution of GAUSEI (Section 4.2.1).

4.2.1 GAUSEI

Introduction

GAUSEI solves linear systems of simultaneous equations using the iterative Gauss-Seidel method. The input to the program includes the following: the order of the system, the maximum number of iterations allowed before the procedure is terminated, the convergence and divergence criteria, the coefficient matrix, the constant vector, and an initial approximation to the solution vector. GAUSEI prints the input data for verification purposes, a convergence map, and, if convergence is attained, the solution vector and the difference vector AX - B.

Method

The iterative Gauss-Seidel method is used. There are three convergence criteria. Let

$$RELSUM = \sum_{i=1}^{n} \left| \frac{x_i^m - x_i^{m-1}}{x_i^m} \right| \qquad (x_i^m \neq 0)$$

Here n is the order of the linear system and x_i^m and x_i^{m-1} are the values of x_i at iterations m and m - 1. If RELSUM is less than the convergence criterion, the solution has been found. On the other hand, if RELSUM is greater than the divergence tolerance or if the maximum number of iterations has been exceeded, then the method is not converging and the execution is terminated.

GAUSEI also calculates the difference vector D,

AX = B

  D = AX - B

The elements of D should be zero, assuming no round-off errors.

Sample Execution (See Problem 4.3)


```
 ENTER N,ITMAX,CVGTOL AND DVGTOL     N < 2 STOPS THE PROGRAM.
>5 10 .0001 1000

 ENTER THE COEFFICIENT MATRIX BY ROWS
>100 10 2 1 .5
>12 120 10 5 .9
>30 30 90 10 2
>1 4 18 140 24
>2 4 8 16 120

 ENTER THE CONSTANT VECTOR
>.1135 .2218 .27 .2992 .1350

 ENTER THE INITIAL APPROXIMATION TO THE SOLUTION VECTOR
> 0 0 0 0 0
```

        N = 5   ITMAX = 10

      CVGTOL =   0.100E-03   DVGTOL =   0.100E+04

FOR AX = B, A =

100.	10.0	2.00	1.00	0.500
12.0	120.	10.0	5.00	0.900
30.0	30.0	90.0	10.0	2.00
1.00	4.00	18.0	140.	24.0
2.00	4.00	8.00	16.0	120.

AND B =   0.114   0.222   0.270   0.299   0.135

```
                           CONVERGENCE MAP

ITN                       ( X(I),   I = 1,2,3,...,N )

  0    0.000         0.000        0.000        0.000        0.000
  1    0.113E-02     0.173E-02    0.204E-02    0.182E-02    0.670E-03
  2    0.899E-03     0.151E-02    0.198E-02    0.172E-02    0.699E-03
  3    0.924E-03     0.151E-02    0.198E-02    0.171E-02    0.699E-03
  4    0.923E-03     0.151E-02    0.198E-02    0.171E-02    0.699E-03
  5    0.923E-03     0.151E-02    0.198E-02    0.171E-02    0.699E-03
  6    0.923E-03     0.151E-02    0.198E-02    0.171E-02    0.699E-03

       THE SOLUTION IS:

           I         X(I)        A(I,J)*X(J)-B(J)
                                  J=1,2,3,...,N

           1       0.9233E-03      -0.40E-07
           2       0.1514E-02       0.00E+00
           3       0.1982E-02       0.37E-08
           4       0.1713E-02       0.00E+00
           5       0.6987E-03       0.00E+00
```

Listing

```
C
C              GAUSEI
C
C  GAUSEI SOLVES LINEAR SYSTEMS OF SIMULTANEOUS
C  EQUATIONS USING THE GAUSS-SEIDEL METHOD
C
C  INPUT:  N   -   ORDER OF THE LINEAR SYSTEM
C       ITMAX -   MAXIMUM   OF ITERATIONS
C       CVGTOL -  RELATIVE CONVERGENCE CRITERION
C       DVGTOL -  RELATIVE DIVERGENCE CRITERION
C         A   -   COEFFICIENT MATRIX
C         B   -   CONSTANT VECTOR
C         X   -   INITIAL APPROXIMATION TO SOLUTION VECTOR
C
C   AUTHOR:  K. J. JOHNSON
C
      DIMENSION A(10,10),B(10),X(10)
C
C   INPUT
C
   10 WRITE(6,11)
      READ(5,22)N,ITMAX,CVGTOL,DVGTOL
      IF(N.LE.1)STOP
      WRITE(6,33)
      DO 20 I=1,N
      READ(5,44)(A(I,J),J=1,N)
   20 IF(A(I,I).EQ.0.0)GO TO 200
      WRITE(6,55)
      READ(5,44)(B(I),I=1,N)
      WRITE(6,66)
      READ(5,44)(X(I),I=1,N)
      WRITE(6,77)N,ITMAX,CVGTOL,DVGTOL
```

```
          DO 30 I=1,N
    30 WRITE(6,88)(A(I,J),J=1,N)
          WRITE(6,89)(B(I),I=1,N)
          ITN=0
          WRITE(6,99)ITN,(X(I),I=1,N)
C
C    THE OUTERMOST LOOP (100) CONTROLS THE ITERATION COUNTER (ITN)
C
          DO 100 ITN=1,ITMAX
          SUMDIF=0.0
C
C      THE INNER LOOP (80) CONTAINS THE GAUSS-SEIDEL ALGORITHM
C
          DO 80 I=1,N
          SUM=0.0
          IM1=I-1
          DO 40 J=1,IM1
          IF(J.GT.IM1)GO TO 50
    40 SUM=SUM+A(I,J)*X(J)
    50 IP1=I+1
          IF(IP1.GT.N)GO TO 70
          DO 60 J=IP1,N
    60 SUM=SUM+A(I,J)*X(J)
    70 XOLD=X(I)
          X(I)=(B(I)-SUM)/A(I,I)
          IF(X(I).EQ.0.)GO TO 80
          SUMDIF=SUMDIF+ABS((X(I)-XOLD)/X(I))
    80 CONTINUE
          WRITE(6,101)ITN,(X(I),I=1,N)
C
C      TEST FOR CONVERGENCE AND DIVERGENCE
C
          IF(SUMDIF.LE.CVGTOL)GO TO 300
          IF(SUMDIF.GE.DVGTOL)GO TO 400
   100 CONTINUE
C
C    FAILED TO CONVERGE IN ITMAX TRIES
C
          WRITE(6,155)
          GO TO 10
   200 WRITE(6,133)
          GO TO 10
   300 WRITE(6,111)
C
C      PRINT THE SOLUTION AND TEST AX-B=0 ?
C
          DO 330 I=1,N
          SUM=0.
          DO 320 J=1,N
   320 SUM=SUM+A(I,J)*X(J)
          DIFF=SUM-B(I)
   330 WRITE(6,122)I,X(I),DIFF
          GO TO 10
   400 WRITE(6,144)
          GO TO 10
    11 FORMAT(//'  ENTER N,ITMAX,CVGTOL AND DVGTOL',
      1'    N < 2 STOPS THE PROGRAM.'/)
    22 FORMAT(2I,2E)
    33 FORMAT(/'  ENTER THE COEFFICIENT MATRIX BY ROWS'/)
    44 FORMAT(10E)
```

```
 55 FORMAT(/'   ENTER THE CONSTANT VECTOR'/)
 66 FORMAT(/'   ENTER THE INITIAL APPROXIMATION TO THE SOLUTION',
   1' VECTOR'/)
 77 FORMAT(///10X,'N = ',I2,10X,'ITMAX = ',I2,//10X,'CVGTOL = ',
   1 G12.3,10X,'DVGTOL = ',G12.3//,' FOR AX = B,  A ='//)
 88 FORMAT(5G12.3,/5X,5G12.3)
 89 FORMAT(//'   AND B = ',5G12.3,/10X,5G12.3)
 99 FORMAT(//,25X,'CONVERGENCE MAP'//,
   1' ITN',15X,'( X(I),   I = 1,2,3,...,N )',
   2//I3,5G12.3,/5X,5G12.3)
101 FORMAT(I3,5G12.3,/5X,5G12.3)
111 FORMAT(//'     THE SOLUTION IS:'//,
   1   10X,'I',8X,'X(I)',5X,'A(I,J)*X(J)-B(J)'
   2/30X,'J=1,2,3,...,N'/)
122 FORMAT(9X,I2,2G15.4)
133 FORMAT(//'  A(I,I) CANNOT BE 0.0'/)
144 FORMAT(//'  THE SOLUTION IS DIVERGING. EXECUTION TERMINATED.'//)
155 FORMAT(//'  CONVERGENCE FAILURE.  EXECUTION TERMINATED.'//)
    END
```

Discussion

GAUSEI requires 714 words of core. This sample execution took 0.2 sec.
The solution to Problem 4.3 is

$[A] = 9.2 \times 10^{-4}$ $\underline{M}$ $[B] = 1.5 \times 10^{-3}$ $\underline{M}$

$[C] = 2.0 \times 10^{-3}$ $\underline{M}$ $[D] = 1.7 \times 10^{-3}$ $\underline{M}$

$[E] = 7.0 \times 10^{-4}$ $\underline{M}$

The coefficient matrix in this example is diagonally dominant and the convergence is rapid. The Gauss-Seidel method will often converge in linear systems
that are not diagonally dominant.

The Gauss-Seidel method is iterative and therefore may require more execution
time than the elimination methods considered in the following sections. One
advantage of this method is that only one row of the linear system is needed in
core at a time. The elimination methods require that the entire coefficient
matrix reside in core. Therefore systems of linear equations of order 100 can be
handled using the Gauss-Seidel method without using in excess of 10,000 words of
core memory. This is particularly useful in the iterative solution of partial
differential equations, for example, nuclear reactor neutron flux problems [3].

4.3 ELIMINATION METHODS

In this section the elimination methods of Gauss and Jordan [4] are considered.
Both involve a series of elementary row operations on the augmented matrix $(A|B)$.
For an n x n linear system, $(A|B)$ is the n x (n + 1) matrix

$$(A|B) = \begin{bmatrix} a_{11} & a_{12} & a_{13} & \cdots & a_{1n} & b_1 \\ a_{21} & a_{22} & a_{23} & \cdots & a_{2n} & b_2 \\ \cdots\cdots\cdots\cdots\cdots\cdots\cdots\cdots \\ a_{n1} & a_{n2} & a_{n3} & \cdots & a_{nn} & b_n \end{bmatrix}$$

The three elementary row operations are defined as follows:

$E_1(i,j)$, interchange rows i and j.

$E_2(c,j)$, multiply row j by the constant c.

$E_3(c,i,j)$, multiply row i by the constant c and add the result to row j.

Gauss' method involves a series of elementary row operations designed to reduce the augmented matrix $(A|B)$ to upper triangular form, for example, for n = 3,

$$\begin{bmatrix} a_{11} & a_{12} & a_{13} & b_1 \\ a_{21} & a_{22} & a_{23} & b_2 \\ a_{31} & a_{32} & a_{33} & b_3 \end{bmatrix} \rightarrow \rightarrow \rightarrow \begin{bmatrix} f_{11} & f_{12} & f_{13} & g_1 \\ 0 & f_{22} & f_{23} & g_2 \\ 0 & 0 & f_{33} & g_3 \end{bmatrix}$$

The solution is then obtained by back substitution:

$$x_3 = \frac{g_3}{f_{33}}$$

$$x_2 = \frac{g_2 - f_{23}x_3}{f_{22}}$$

$$x_1 = \frac{g_1 - f_{12}x_2 - f_{13}x_3}{f_{11}}$$

The procedure is illustrated in the following problem.

Problem 4.3. Solve the linear system

$$2x_1 - x_2 + 2x_3 = 6$$
$$-2x_1 - x_2 + x_3 = -1$$
$$-x_1 + x_2 + 2x_3 = 7$$

Solution. Here the augmented matrix is

$$(A|B) = \begin{bmatrix} 2 & -1 & 2 & 6 \\ -2 & -1 & 1 & -1 \\ -1 & 1 & 2 & 7 \end{bmatrix}$$

First, elements a_{21} and a_{31} are eliminated as follows:

$$
\begin{bmatrix}
2 & -1 & 2 & 6 \\
-2 & -1 & 1 & -1 \\
-1 & 1 & 2 & 7
\end{bmatrix}
\xrightarrow{E_3(0.5,1,3)}
\begin{bmatrix}
2 & -1 & 2 & 6 \\
-2 & -1 & 1 & -1 \\
0 & 0.5 & 3 & 10
\end{bmatrix}
\xrightarrow{E_3(1,1,2)}
\begin{bmatrix}
2 & -1 & 2 & 6 \\
0 & -2 & 3 & 5 \\
0 & 0.5 & 3 & 10
\end{bmatrix}
$$

Then, element a_{32} is eliminated giving an upper triangular matrix,

$$
\begin{bmatrix}
2 & -1 & 2 & 6 \\
0 & -2 & 3 & 5 \\
0 & 0.5 & 3 & 10
\end{bmatrix}
\xrightarrow{E_3(0.25,2,3)}
\begin{bmatrix}
2 & -1 & 2 & 6 \\
0 & -2 & 3 & 5 \\
0 & 0 & 15/4 & 45/4
\end{bmatrix}
$$

or

$$
\begin{aligned}
2x_1 - x_2 + 2x_3 &= 6 \\
-2x_2 + 3x_3 &= 5 \\
\tfrac{15}{4}x_3 &= \tfrac{45}{4}
\end{aligned}
$$

Finally, solving by back substitution,

$$
\begin{aligned}
x_3 &= 3 \\
x_2 &= \frac{5 - 3(3)}{2} = 2 \\
x_3 &= \frac{6 + 2 - 6}{2} = 1
\end{aligned}
$$

The Gauss-Jordan method involves elimination of all off-diagonal elements, for example, for $n = 3$,

$$
\begin{bmatrix}
a_{11} & a_{12} & a_{13} & b_1 \\
a_{21} & a_{22} & a_{23} & b_2 \\
a_{31} & a_{32} & a_{33} & b_3
\end{bmatrix}
\rightarrow \rightarrow \rightarrow
\begin{bmatrix}
f_1 & 0 & 0 & g_1 \\
0 & f_2 & 0 & g_2 \\
0 & 0 & f_3 & g_3
\end{bmatrix}
$$

Then the solution vector is simply

$$
x_i = \frac{g_i}{f_i} \qquad i = 1, 2, 3
$$

Both methods require division by an element in the coefficient matrix called the pivot element. A computer program designed to solve linear systems of equations by either of these methods should check for a zero or minute pivot element. A pivot element of zero indicates that the coefficient matrix is singular, that is, $|A| = 0$, and therefore no unique solution exists. A very small pivot element can cause significant round-off errors as the following example illustrates. The linear system

$$0.0003x_1 + 3.0000x_2 = 2.0001$$
$$1.0000x_1 + 1.0000x_2 = 1.0000$$

has the exact solution $x_1 = 1/3$ and $x_2 = 2/3$. If 0.0003 is used as the pivot element to eliminate a_{21}, one obtains (to four decimal places)

$$0.0003x_1 + 3.0000x_2 = 2.0001$$
$$-9999x_2 = -6666$$

So

$$x_2 = \frac{6666}{9999} = 0.6667$$

and

$$x_1 = \frac{2.0001 - 3.0000x_2}{0.0003} = 0.0$$

However, if the equations are simply reversed and 1.0000 is used as the pivot element, then

$$1.0000x_1 + 1.0000x_2 = 1.0000$$
$$2.9999x_2 = 1.9998$$

So

$$x_2 = \frac{1.9998}{2.9999} = 0.6667$$

and

$$x_1 = 1.0000 - x_2 = 0.3333$$

A computer program should search the matrix for the largest (absolute value) eligible element and use it for the pivot element. The following program, a Gauss-Jordan routine, uses this technique.

4.3.1 GAUJOR

Introduction

GAUJOR solves linear systems of simultaneous equations using the Gauss-Jordan method. The input to the program includes the order of the system of equations N, the coefficient matrix A, and the constant vector B. The program prints the coefficient matrix and constant vector for verification purposes and then the solution vector X. GAUJOR also prints the elements of the difference vector AX - B, which should be zero within round-off error.

Method

The Gauss-Jordan method of complete elimination of the augmented matrix $(A|B)$ is described in Section 4.3 GAUJOR consists of a main program and a subroutine.

The calling sequence for the subroutine is

    CALL GAUJOR(A,X,N,ISTOP)

Here ISTOP is set to 1 if a pivot element is less than 1×10^{-35}. GAUJOR uses two work vectors, LOC(N) and IPIV(N). One column at a time is eliminated. The pivot row is the row in which the maximum eligible pivot element is found. A given row can contain the pivot element only once. IPIV is used to indicate whether a given row has already been used as a pivot row. LOC is used to store the row number of each pivot row. The solution vector is calculated using the equation

    $X(I)=A(L,NP1)/A(L,I)$ $I = 1, 2, \ldots, N$

where

    $L = LOC(I)$ and $NP1 = N + 1$

Consider, for example, the following linear system,

$$x_1 + 3x_2 + 2x_3 = 11$$
$$2x_1 + x_2 + x_3 = 9$$
$$4x_1 - x_2 - 2x_3 = 8$$

In the elimination procedure used here, row 3 would be the first pivot row, since 4 is the largest element in column 1. Therefore LOC(1) = 3. Similarly, LOC(2) = 1 and LOC(3) = 2. After elimination, the augmented matrix is

$$\begin{bmatrix} 0 & 3\ 1/4 & 0 & 6\ 1/2 \\ 0 & 0 & 11/13 & 11/13 \\ 4 & 0 & 0 & 12 \end{bmatrix}$$

The solution is

$$x_1 = \frac{12}{4} = 3$$
$$x_2 = \frac{6\ 1/2}{3\ 1/4} = 2$$
$$x_3 = \frac{11/13}{11/13} = 1$$

Sample Execution (See Problem 4.3.)
  ENTER N
>5

 ENTER THE COEFFICIENT MATRIX BY ROWS
>100 10 2 1 .5
>12 120 10 5 .9
>30 30 90 10 2
>1 4 18 140 24
>2 4 8 16 120

```
   ENTER THE CONSTANT VECTOR
>.1135 .2218 .27 .2992 .1350

 N  =   5

THE COEFFICIENT MATRIX IS:

    100.         10.0        2.00        1.00        0.500
    12.0         120.        10.0        5.00        0.900
    30.0         30.0        90.0        10.0        2.00
    1.00         4.00        18.0        140.        24.0
    2.00         4.00        8.00        16.0        120.

THE CONSTANT VECTOR IS:

  0.114        0.222        0.270        0.299        0.135

THE SOLUTION TO THE LINEAR SYSTEM IS:

    I           X(I)         A(I,J)*X(J)-A(I,N+1)
                                J=1,2,...,N

    1         0.9233E-03        0.00
    2         0.1514E-02        0.00
    3         0.1982E-02        0.37E-08
    4         0.1713E-02        0.00
    5         0.6987E-03        0.00
```

Listing

```
C
C
C            GAUJOR
C
C        THIS PROGRAM SOLVES SYSTEMS OF LINEAR
C        SIMULTANEOUS EQUATIONS BY THE GAUSS-JORDAN METHOD
C
C
C
C       INPUT:
C
C          N         -  ORDER OF THE LINEAR SYSTEM
C          A(N,N)    -  COEFFICIENT MATRX
C          A(N,N+1)  -  CONSTANT VECTOR
C
C       OUTPUT:
C
C          N,A(N,N+1)       FOR VERIFICATION PURPOSES
C             D             DETERMINANT OF A
C             X             SOLUTION TO THE LINEAR SYSTEM
C
C     I/O IS HANDLED BY THE MAIN PROGRAM
C     THE SOLUTION IS FOUND BY THE SUBROUTINE GAUJOR
C
C     USAGE:   CALL GAUJOR(A,X,N,ISTOP)
C
C        ISTOP = 1 IF THE COEFFICIENT MATRIX IS SINGULAR
C           ( PIVOT ELEMENT < 1.0E-35  )
```

```
C      AUTHOR:  K. J. JOHNSON
C
C
       DIMENSION A(10,11),STORA(10,11),X(10)
C
C      INPUT
C
  5    WRITE(6,11)
       READ(5,22)N
       IF(N.LE.0)STOP
       NP1=N+1
       WRITE(6,33)
       DO 10 I=1,N
 10    READ(5,44)(A(I,J),J=1,N)
       WRITE(6,55)
       READ(5,44)(A(I,NP1),I=1,N)
C
C   STORE A(N,NP1) IN STORA(N,NP1) FOR
C   SUBSEQUENT VERIFICATION OF SOLUTION
C
       DO 15 I=1,N
       DO 15 J=1,NP1
 15    STORA(I,J)=A(I,J)
       WRITE(6,66)N
       DO 20 I=1,N
 20    WRITE(6,77)(A(I,J),J=1,N)
       WRITE(6,88)(A(I,NP1),I=1,N)
C
C      CALL THE GAUSS-JORDAN ROUTINE
       CALL GAUJOR(A,X,N,ISTOP)
C
       IF(ISTOP.EQ.1)GO TO 222
C
C      PRINT THE SOLUTION AND THE DIFFERENCE VECTOR
C
       WRITE(6,100)
       DO 30 I=1,N
       SUM=0.
       DO 25 J=1,N
 25    SUM=SUM+STORA(I,J)*X(J)
       DIFF=SUM-STORA(I,NP1)
 30    WRITE(6,111)I,X(I),DIFF
       GO TO 5
 222   WRITE(6,223)
        GO TO 5
       STOP
 11    FORMAT(//'    ENTER N   '/)
 22    FORMAT(I)
 33    FORMAT(//'  ENTER THE COEFFICIENT MATRIX BY ROWS    '/)
 44    FORMAT(10F)
 55    FORMAT(//'    ENTER THE CONSTANT VECTOR     '/)
 66    FORMAT(/'  N =   ',I2,//'  THE COEFFICIENT MATRIX IS:'//)
 77    FORMAT(6G12.3)
 88    FORMAT(//'  THE CONSTANT VECTOR IS:'//6G12.3)
 100   FORMAT(//'  THE SOLUTION TO THE LINEAR SYSTEM IS:'//
      1 5X,'I',8X,'X(I)',5X,'  A(I,J)*X(J)-A(I,N+1)',
      2 /30X,'J=1,2,...,N'//)
 111   FORMAT(I6,2X,G14.4,5X,G14.2)
 223   FORMAT(//'  PROGRAM FAILURE - PIVOT ELEMENT < 10**(-35)'/)
       END
```

```
C
C            GAUJOR
C
C      GAUJOR SOLVES LINEAR SYSTEMS OF SIMULTANEOUS
C      EQUATIONS USING THE GAUSS-JORDAN METHOD
C
C      AUTHOR:  K. J. JOHNSON
C
C
       SUBROUTINE GAUJOR(A,X,N,ISTOP)
       DIMENSION A(10,11),X(10),LOC(10),IPIV(10)
       NP1=N+1
       ISTOP=0
       DO 10 I=1,N
   10  IPIV(I)=0
C
C    MAIN LOOP.   ELIMINATE ONE COLUMN AT A TIME
C
       DO 100 I=1,N
       IP1=I+1
C
C    FIND MAXIMUM ELEMENT IN I-TH COLUMN
C
       AMAX=0.
       DO 20 K=1,N
       ELEMNT=ABS(A(K,I))
       IF(ELEMNT.LE.AMAX)GO TO 20
C
C    HAS THIS ROW PREVIOUSLY BEEN USED AS A PIVOT?
C
       IF(IPIV(K).EQ.1)GO TO 20
       LOC(I)=K
       AMAX=ELEMNT
   20  CONTINUE
C
C     IS THE COEFFICIENT MATRIX SINGULAR?
C
       IF(AMAX.LE.1.0E-35)GO TO 950
       L=LOC(I)
       IPIV(L)=1
C
C    PERFORM ELIMINATION; L IS PIVOT ROW, A(L,I) IS PIVOT ELEMENT
C
       DO 50 J=1,N
       IF(J.EQ.L) GO TO 50
       F=-A(J,I)/A(L,I)
       DO 40 K=IP1,NP1
   40  A(J,K)=A(J,K)+F*A(L,K)
   50  CONTINUE
  100  CONTINUE
C
C      CALCULATE THE SOLUTION VECTOR, X
C
       DO 200 I=1,N
       L=LOC(I)
  200  X(I)=A(L,NP1)/A(L,I)
       RETURN
  950  ISTOP=1
       RETURN
       END
```

Discussion

GAUJOR, both main program and subroutine, requires approximately 750 words of core. This sample execution took approximately 0.2 sec.

4.3.2 Matrix Inversion

Matrix inversion is probably the most widely used method for solving linear systems of equations. Let A, B, and C be n x n matrices. The matrix product,

$$C = AB$$

is defined as follows:

$$C_{ij} = \sum_{k=1}^{n} A_{ik}B_{kj} \qquad i,j = 1,2,3,\dots,n$$

The identity matrix is the n x n matrix

$$I_{ij} = 1.0 \qquad \text{if } i = j$$

$$I_{ij} = 0.0 \qquad \text{if } i \neq j$$

The inverse of the matrix A, denoted A^{-1}, is the matrix for which

$$AA^{-1} = A^{-1}A = I$$

The linear system of equations,

$$AX = B$$

is readily solved if the inverse of the coefficient matrix is known,

$$AX = B$$
$$A^{-1}AX = A^{-1}B$$
$$IX = A^{-1}B$$
$$X = A^{-1}B$$

or

$$x_i = \sum_{k=1}^{n} A_{ik}^{-1}B_k$$

The inverse of a matrix can be determined by the elimination procedure [5]. Consider the n x 2n augmented matrix $(A|I)$. A series of elementary row operations is performed to convert the left half of this augmented matrix to the identity,

$$(A|I) \quad \rightarrow\rightarrow\rightarrow \quad (I|B)$$

It can be shown that B is A^{-1}.

Problem 4.4. Find the inverse of the matrix

$$A = \begin{bmatrix} 1 & 2 & -1 \\ 2 & 1 & -3 \\ -3 & 1 & 2 \end{bmatrix}$$

Solution. The 3 x 3 augmented matrix $(A|I)$ is converted to $(I|A^{-1})$ by the following set of elementary row operations,

$$
\begin{bmatrix}
1 & 2 & -1 & 1 & 0 & 0 \\
2 & 1 & -3 & 0 & 1 & 0 \\
-3 & -1 & 2 & 0 & 0 & 1
\end{bmatrix}
$$

1. $E_3(-2,1,2)$
2. $E_3(3,1,3)$
3. $E_2(-1/3,2)$
4. $E_3(-2,2,1)$
5. $E_3(-5,2,3)$
6. $E_2(-3/8,3)$
7. $E_3(5/3,3,1)$
8. $E_3(1/3,3,2)$

$$
\longrightarrow
\begin{bmatrix}
1 & 0 & 0 & -1/8 & -3/8 & -5/8 \\
0 & 1 & 0 & 5/8 & -1/8 & 1/8 \\
0 & 0 & 1 & 1/8 & -5/8 & -3/8
\end{bmatrix}
$$

The result is

$$A^{-1} = \begin{bmatrix} -1/8 & -3/8 & -5/8 \\ 5/8 & -1/8 & 1/8 \\ 1/8 & -5/8 & -3/8 \end{bmatrix}$$

This can be verified by showing that

$$A^{-1}A = AA^{-1} = I$$

Problem 4.5. Use the solution to Problem 4.4 to solve the following linear system of equations.

$$
\begin{aligned}
x_1 + 2x_2 - x_3 &= 2 \\
2x_1 + x_2 - 3x_3 &= -5 \\
-3x_1 - x_2 - 2x_3 &= 1
\end{aligned}
$$

Solution. $X = A^{-1}B$

$$X = \begin{bmatrix} -1/8 & -3/8 & -5/8 \\ 5/8 & -1/8 & 1/8 \\ 1/8 & -5/8 & -3/8 \end{bmatrix} \begin{pmatrix} 2 \\ -5 \\ 1 \end{pmatrix} = \begin{pmatrix} 1 \\ 2 \\ 3 \end{pmatrix}$$

The solution is $x_1 = 1$, $x_2 = 2$, and $x_3 = 3$.

The following program solves linear systems by coefficient matrix inversion.

4.3.3 LINSYS

Introduction

LINSYS solves systems of linear simultaneous equations by coefficient matrix inversion. The input to the program includes the order of the system N, the coefficient matrix A, and the constant vector B. LINSYS prints A and B for verification purposes, the determinant of A, an error parameter which tests the equality $AA^{-1} = I$, the solution vector X, and the difference vector AX - B.

Method

LINSYS consists of a main program and a subroutine, MATINV. The calling sequence is

CALL MATINV(A,N,D)

Here D is the determinant of the coefficient matrix A. MATINV uses the elimination method discussed above. The accuracy of the inverse is tested by the error parameter DIFF, which is the difference in absolute value between the order of the system, N, and the sum of the absolute values of all elements in the matrix product AA^{-1},

$$DIFF = N - \sum_{i,j=1}^{N} |I_{ij}|$$

where

$$I_{ij} = \sum_{k=1}^{n} A_{ik}^{-1} A_{kj}$$

The solution vector is $X = A^{-1}B$. The solution vector is tested by computing the difference vector AX - B. DIFF and the elements of this difference vector should be zero within round-off error.

MATINV executes the following sequence of operations (see Listing). A work matrix IPV(N,3) is dimensioned. The first two columns of IPV hold the row and column indices of the N pivot elements. The third column of IPV is a flag which is set to 1 if a given column has already provided a pivot element. The 100 loop directs the elimination procedure. One row is eliminated at a time. The 25 loop finds the largest available element in the matrix for the pivot element. Next the singularity test is performed. If the pivot element is less than 1×10^{-35} in absolute value, the determinant is set to zero and control is returned to the main program. If the pivot element does not appear on the diagonal, it is placed there by the 30 loop. The determinant is updated, and the pivot row is normalized in the 40 loop. The 50 and 55 loops eliminate the elements in the pivot column, retaining the desired elements of A^{-1}. The 200 and 250 loops reorder the rows and columns of A and multiply the accumulated determinant by the correct power of -1.

For example, consider the matrix inverted in Problem 4.4. At the termination of the 100 loop, IPV and A would have the following values,

$$IPV = \begin{bmatrix} 2 & 3 & 1 \\ 1 & 2 & 1 \\ 1 & 1 & 1 \end{bmatrix} \qquad A^{-1} = \begin{bmatrix} -5/8 & -1/8 & -3/8 \\ 1/8 & 5/8 & -1/8 \\ -3/8 & 1/8 & -5/8 \end{bmatrix}$$

The 200 and 250 loops use IPV to swap rows and columns so that the correct inverse is returned,

$$A^{-1} = \begin{bmatrix} -1/8 & -3/8 & -5/8 \\ 5/8 & -1/8 & 1/8 \\ 1/8 & -5/8 & -3/8 \end{bmatrix}$$

Sample Execution (See Problem 4.3.)

```
ENTER N      ( N<2 ENDS )
>5

ENTER THE COEFFICIENT MATRIX BY ROWS
>100 10 2 1 .5
>12 120 10 5 .9
>30 30 90 10 2
>1 4 18 140 24
>2 4 8 16 120

ENTER THE CONSTANT VECTOR
>.1135 .2218 .27 .2992 .1350
```

THE COEFFICIENT MATRIX IS:

100.	10.0	2.00	1.00	0.500
12.0	120.	10.0	5.00	0.900
30.0	30.0	90.0	10.0	2.00
1.00	4.00	18.0	140.	24.0
2.00	4.00	8.00	16.0	120.

THE CONSTANT VECTOR IS:

0.114	0.222	0.270	0.299	0.135

THE DETERMINANT OF THE COEFFICIENT MATRIX IS   0.1677E+11

AND THE SCALAR DIFFERENCE   ABS(A*A(-1) - N) IS   0.0000

THE SOLUTION TO THE LINEAR SYSTEM IS:

I	X(I)	A(I,J)*X(J)-B(I) J=1,2,...,N
1	0.9233E-03	0.93E-09
2	0.1514E-02	0.00E+00
3	0.1982E-02	0.37E-08
4	0.1713E-02	-0.37E-08
5	0.6987E-03	0.19E-08

Listing

```
C
C           LINSYS
C
C        THIS PROGRAM SOLVES SYSTEMS OF LINEAR
C        SIMULTANEOUS EQUATIONS BY MATRIX INVERSION
C
C                      AX=B
C             A(-1)*A*X=A(-1)*B
C                      X=A(-1)*B
C
C     INPUT:
C
C        N    -  ORDER OF THE LINEAR SYSTEM
C        A    -  COEFFICIENT MATRIX
C        B    -  CONSTANT VECTOR
C
C     OUTPUT:
C
C        N,A & B        FOR VERIFICATION PURPOSES
C         D             DETERMINANT OF A
C         X             SOLUTION TO THE LINEAR SYSTEM
C
C   THE MAIN PROGRAM HANDLES I/O
C   THE COEFFICIENT MATRIX IS INVERTED
C   BY THE SUBROUTINE MATINV,
C
C      CALL MATINV(A,N,D)
C
C        THE INVERSE IS RETURNED IN A
C        IF THE COEFFICIENT MATRIX IS SINGULAR
C        (PIVOT < 1.0E-35) THE DETERMINANT IS SET TO 0.0
C
C
C   AUTHORS:  S. B. LEVITT AND K. J. JOHNSON
C
C
      DIMENSION A(10,10),STORA(10,10),B(10),X(10)
C
C      INPUT
C
  5   WRITE(6,11)
      READ(5,22)N
      IF(N.LT.2)STOP
      WRITE(6,33)
      DO 10 I=1,N
 10   READ(5,44)(A(I,J),J=1,N)
      WRITE(6,55)
      READ(5,44)(B(I),I=1,N)
      WRITE(6,66)
      DO 20 I=1,N
 20   WRITE(6,77)(A(I,J),J=1,N)
      WRITE(6,88)(B(I),I=1,N)
C
C   SAVE THE COEFFICIENT MATRIX
C
      DO 25 I=1,N
      DO 25 J=1,N
 25   STORA(I,J)=A(I,J)
```

```
C
C          INVERT THE COEFFICIENT MATRIX
C
      CALL MATINV(A,N,D)
C
C   IS THE COEFFICIENT MATRIX SINGULAR?
C
      IF(D.EQ.0)GO TO 90
C
C      PRINT THE DETERMINANT   AND TEST A(-1)
C      IS A*A(-1) = I ?
C
      SUMTOT=0.
      DO 30 I=1,N
      DO 30 J=1,N
      SUM=0.
      DO 35 K=1,N
   35 SUM=SUM+STORA(I,K)*A(K,J)
   30 SUMTOT=SUMTOT+ABS(SUM)
      DIFF=ABS(SUMTOT-N)
      WRITE(6,99)D,DIFF
C
C    EVALUATE AND PRINT THE SOLUTION AND DIFFERENCE VECTORS
C
      DO 40 I=1,N
      X(I)=0.
      DO 50 J=1,N
   50 X(I)=X(I)+A(I,J)*B(J)
   40 CONTINUE
      DO 60 I=1,N
      SUM=0.
      DO 65 J=1,N
   65 SUM=SUM+STORA(I,J)*X(J)
      DIFF=SUM-B(I)
   60 WRITE(6,101)I,X(I),DIFF
C
C      START OVER
C
      GO TO 5
   90 WRITE(6,133)
      GO TO 5
      STOP
   11 FORMAT(//'  ENTER N    ( N<2 ENDS )'/)
   22 FORMAT(I)
   33 FORMAT(//'  ENTER THE COEFFICIENT MATRIX BY ROWS    '/)
   44 FORMAT(10E)
   55 FORMAT(//'  ENTER THE CONSTANT VECTOR     '/)
   66 FORMAT(/'  THE COEFFICIENT MATRIX IS:'//)
   77 FORMAT(6G12.3)
   88 FORMAT(//'  THE CONSTANT VECTOR IS:'//6G12.3)
   99 FORMAT(/'  THE DETERMINANT OF THE COEFFICIENT MATRIX IS',
     1 G14.4//,'  AND THE SCALAR DIFFERENCE  ABS( A*A(-1) - N ) IS'
     2 G14.4//,'  THE SOLUTION TO THE LINEAR SYSTEM IS:'//
     3 5X,'I',8X,'X(I)',5X,'A(I,J)*X(J)-B(I)',
     4 /24X,'J=1,2,...,N'/)
  101 FORMAT(I6,2X,G14.4,G12.2)
  133 FORMAT(//'  THE COEFFICIENT MATRIX IS SINGULAR.'/)
      END
```

```
C                   MATINV
C
C          THIS SUBROUTINE INVERTS REAL,SQUARE MATRICES
C
C     USAGE:      CALL MATINV(A,N,D)
C
C          WHERE      A IS THE MATRIX (NXN)
C                     N IS THE ORDER OF A
C             AND     D IS THE DETERMINANT OF A
C
C          NOTE:  A IS REPLACED BY A(-1)
C
C     AUTHORS:  S. B. LEVITT AND K. J. JOHNSON
C
C
       SUBROUTINE MATINV(A,N,D)
       DIMENSION A(10,10),IPV(10,3)
       D=1.0
       TOL=1.0E-35
       DO 10 J=1,N
   10  IPV(J,3)=0
C
C   MAIN LOOP, ELIMINATE ONE ROW AT A TIME
C
       DO 100 I=1,N
C
C          SEARCH REMAINING MATRIX FOR MAXIMUM ELEMENT (PIVOT);
C          TEST FOR A SINGULAR COEFFICIENT MATRIX
C
       AMAX=0.0
       DO 25 J=1,N
       IF(IPV(J,3).EQ.1)GO TO 25
       DO 20 K=1,N
       IF(IPV(K,3).EQ.1)GO TO 20
       IF(AMAX.GE.ABS(A(J,K)))GO TO 20
       IROW=J
       ICOLUM=K
       AMAX=ABS(A(J,K))
   20  CONTINUE
   25  CONTINUE
       IF(AMAX.LE.TOL)GO TO 300
       IPV(ICOLUM,3)=1
       IPV(I,1)=IROW
       IPV(I,2)=ICOLUM
C
C     INTERCHANGE ROWS (IF NECESSARY) TO PUT PIVOT ELEMENT ON DIAGONAL
C
       IF(IROW.EQ.ICOLUM)GO TO 35
       DO 30 L=1,N
       SWAP=A(IROW,L)
       A(IROW,L)=A(ICOLUM,L)
   30  A(ICOLUM,L)=SWAP
C
C          UPDATE DETERMINANT AND NORMALIZE PIVOT ROW
C
   35  PIVOT=A(ICOLUM,ICOLUM)
       D=D*PIVOT
       A(ICOLUM,ICOLUM)=1.0
```

```
      DO 40 L=1,N
   40 A(ICOLUM,L)=A(ICOLUM,L)/PIVOT
C
C          ELIMINATE ICOLUM RETAINING INVERSE ELEMENTS
C
      DO 55 L1=1,N
      IF(L1.EQ.ICOLUM)GO TO 55
      T=A(L1,ICOLUM)
      A(L1,ICOLUM)=0.0
      DO 50 L=1,N
   50 A(L1,L)=A(L1,L)-A(ICOLUM,L)*T
   55 CONTINUE
  100 CONTINUE
C
C          INTERCHANGE COLUMNS AND MODIFY DETERMINANT
C
      NSWAP=0
      DO 250 I=1,N
      L=N-I+1
      IF(IPV(L,1).EQ.IPV(L,2))GO TO 250
      JROW=IPV(L,1)
      JCOLUM=IPV(L,2)
      NSWAP=NSWAP+1
      DO 200 K=1,N
      SWAP=A(K,JROW)
      A(K,JROW)=A(K,JCOLUM)
      A(K,JCOLUM)=SWAP
  200 CONTINUE
  250 CONTINUE
      D=D*((-1)**NSWAP)
      RETURN
  300 D=0.0
      RETURN
      END
```

Discussion

LINSYS (including MATINV) requires 926 words of memory. This sample execution took 0.2 sec. A double-precision version of LINSYS generated the following results:

$$DIFF = 0.2D\text{-}20 \qquad AX\text{-}B = \begin{pmatrix} 0.20D\text{-}20 \\ 0.10D\text{-}23 \\ 0.D0D\text{-}22 \end{pmatrix}$$

4.4 NONLINEAR SYSTEMS OF EQUATIONS

The system of equations

$$F(x,y) = 0 \qquad G(x,y) = 0$$

is nonlinear if x and/or y occur with exponents other than unity; for example,

$$X^2 + X - Y^2 - 0.15 = 0 \tag{4.2}$$

$$X^2 - Y + Y^2 + 0.17 = 0 \tag{4.3}$$

This system has the solution x = 0.2 and y = 0.3. Two methods of solving nonlinear simultaneous equations are discussed in this section, the method of successive approximations [6, 7] and the extension of the Newton-Raphson method (Section 3.3) to systems of nonlinear equations [8].

The method of successive approximations involves deriving iteration formulas in which one of the roots is expressed in terms of the others. For example, for a system of order 2,

$$F(x,y) = 0 \qquad G(x,y) = 0$$

equations of the following form are derived,

$$x_{i+1} = P(x_i, y_i) \qquad y_{i+1} = Q(x_{i+1}, y_i)$$

Initial approximations, x_0 and y_0, are required to start the procedure.

Problem 4.6.   Derive iteration formulas and use them to find the solution to Eqs. (4.2) and (4.3). Use as initial approximations $x = 0.15$ and $y = 0.35$.

Solution.   The iteration formulas are

$$x_{i+1} = 0.15 + y_i^2 - x_i^2$$

$$y_{i+1} = 0.17 + x_{i+1}^2 + y_i^2$$

i	x_i	y_i
0	0.1500	0.3500
1	0.2500	0.3550
2	0.2135	0.3416
3	0.2211	0.3356
4	0.2137	0.3283
5	0.2121	0.3228
....	....	....
20	0.2002	0.3004

The procedure is slowly converging to the solution $x = 0.2$ and $y = 0.3$.

Problem 4.7.   Consider the chemical equilibria associated with the $PbCl_2$-$NaCl$-H_2O system [9, 10]:

$$PbCl_2 \rightleftharpoons Pb^{2+} + 2Cl^- \qquad K_{sp} = 1.6 \times 10^{-5}$$

$$Pb^{2+} + Cl^- \rightleftharpoons PbCl^+ \qquad \beta_1 = 7.6$$

$$Pb^{2+} + 2Cl^- \rightleftharpoons PbCl_2(aq) \qquad \beta_2 = 30.9$$

$$Pb^{2+} + 3Cl^- \rightleftharpoons PbCl_3^- \qquad \beta_3 = 12.3$$

$$Pb^{2+} + 4Cl^- \rightleftharpoons PbCl_4^{2-} \qquad \beta_4 = 8.7$$

Calculate and plot the solubility of $PbCl_2$ and the equilibrium concentrations of all species as a function of the molar concentration of NaCl. Assume ideal solution behavior and that all other chemical equilibria can be safely ignored.

Solution. Let S be the molar solubility of $PbCl_2$ and C be the molar concentration of NaCl. The mass balance and charge balance equations are:

$$Pb_{tot} = S = [Pb^{2+}] + [PbCl^+] + [PbCl_2] + [PbCl_3^-] + [PbCl_4^{2-}]$$

$$Cl_{tot} = C + 2S = [Cl^-] + [PbCl^+] + 2[PbCl_2] + 3[PbCl_3^-] + 4[PbCl_4^{2-}]$$

$$[Na^+] + 2[Pb^{2+}] + [PbCl^+] = [Cl^-] + [PbCl_3^-] + 2[PbCl_4^{2-}]$$

The system is solved by the following program.

4.4.1 ITERAT

Introduction

ITERAT uses the method of iterative equations to solve Problem 4.7. The input includes the maximum number of iterations, the convergence criterion, the total concentration of NaCl, and initial approximations to the equilibrium solubility and the equilibrium Cl^- concentration. ITERAT prints the input values for verification purposes and a tabulation of the iterative procedure. If the method converges, the program prints the equilibrium concentrations of all species and tests the mass balance and charge balance equations.

Method

The mass balance equations can be rearranged to give the following iteration equations:

$$[Cl^-] = C + 2S - [PbCl^-] + 2[PbCl_2] + 3[PbCl_3^-] + 4[PbCl_4^{2-}]$$

$$= C + 2S - [Pb^{2+}]\left\{\beta_1[Cl^-] + 2\beta_2[Cl^-]^2 + 3\beta_3[Cl^-]^3 + 4\beta_4[Cl^-]^4\right\}$$

$$= C + 2S - K_{sp}\left\{\frac{\beta_1}{[Cl^-]} + 2\beta_2 + 3\beta_3[Cl^-] + 4\beta_4[Cl^-]^2\right\} \qquad (4.4)$$

$$S = [Pb^{2+}] + [PbCl^+] + [PbCl_2] + [PbCl_3^-] + [PbCl_4^{2-}]$$

$$= [Pb^{2+}] \left\{ 1 + \beta_1[Cl^-] + \beta_2[Cl^-]^2 + \beta_3[Cl^-]^3 + \beta_4[Cl^-]^4 \right\}$$

$$= K_{sp} \left\{ \frac{1}{[Cl^-]^2} + \frac{\beta_1}{[Cl^-]} + \beta_2 + \beta_3[Cl^-] + \beta_4[Cl^-]^2 \right\} \qquad (4.5)$$

ITERAT reads C, the concentration of NaCl, and initial approximations to S, the solubility of $PbCl_2$, and X, the equilibrium $[Cl^-]$. The initial approximations are used in Eq. (4.4) to obtain a better approximation for X, and then Eq. (4.5) is evaluated to get a better approximation to S. This cycle continues until the method converges.

Sample Execution (C = 0.020 M)

```
ENTER ITMAX,CVGTOL,CNACL,SO,CLO
> 30 0.00001 0.02 0.013 0.04

TOTAL NACL =    0.200D-01 MOLAR

  ITN          SOLUBILITY              <CL(-)>

   0         0.13000D-01            0.40000D-01
   1         0.12495D-01            0.41947D-01
   2         0.12946D-01            0.41077D-01
   3         0.12510D-01            0.41917D-01
   4         0.12931D-01            0.41105D-01
   5         0.12524D-01            0.41889D-01
   6         0.12917D-01            0.41131D-01
   7         0.12537D-01            0.41863D-01
   8         0.12904D-01            0.41155D-01
   9         0.12550D-01            0.41839D-01
  10         0.12721D-01            0.41509D-01
  11         0.12723D-01            0.41503D-01
  12         0.12723D-01            0.41502D-01
  13         0.12723D-01            0.41501D-01
  14         0.12723D-01            0.41501D-01

  CONVERGENCE!

  SOLUBILITY =     0.127D-01           <CL(-)> =     0.415D-01

     <PB(2+)> =    0.929D-02          <PBCL(+)> =    0.293D-02

      <PBCL2> =    0.494D-03          <PBCL3(-)> =   0.817D-05

  PBCL4(2-)> =     0.240D-06

TOTAL LEAD BALANCE:    -0.339D-20

TOTAL CHLORINE BALANCE:    0.326D-12

CHARGE BALANCE:    0.326D-12
```

Listing

```
C
C                 ITERAT
C
C  PROGRAM TO SOLVE THE PBCL2-NACL-H2O SYSTEM
C  BY THE METHOD OF ITERATION FORMULAS
C
C  GLOSSARY:
C
C  ITMAX         -  MAXIMUM NUMBER OF ITERATIONS
C  CVGTOL        -  CONVERGENCE TOLERANCE
C  CNACL         -  TOTAL NACL CONCENTRATION
C  SO            -  INITIAL APPROXIMATION TO THE EQUIL. SOLUBILITY
C  CLO           -  INITIAL APPROX. TO THE EQUIL. CL(-) CONC.
C
C    AUTHOR:   K. J. JOHNSON
C
C
       IMPLICIT DOUBLE PRECISION(A-Z)
       INTEGER I,ITN,ITMAX
       DATA B1,B2,B3,B4,KSP/7.6D0,30.9D0,12.3D0,
      1 8.7D0,1.6D-5/
    1  WRITE(6,11)
       READ(5,22)ITMAX,CVGTOL,CNACL,SO,CLO
       IF(ITMAX.EQ.0)STOP
       ITN=0
       S=SO
       CL=CLO
       WRITE(6,33)CNACL,ITN,SO,CLO
C
C    ITERATION LOOP
C
       DO 10 ITN=1,ITMAX
       CL=CNACL + 2.*S - KSP*(B1/CL +2.*B2 + CL*(3.*B3
      1 + 4.*B4*CL))
       S=KSP*(1./CL *( 1./CL + B1) + B2 + CL*(B3 + B4*CL) )
       RELSUM=DABS((S-SO)/S) + DABS((CL-CLO)/CL)
       IF(RELSUM.LE.CVGTOL)GO TO 20
       IF(ITN.LT.10)GO TO 8
       S=(S+SO)/2.
       CL=(CL+CLO)/2.
    8  SO=S
       CLO=CL
   10  WRITE(6,44)ITN,S,CL
C
C    CONVERGENCE FAILURE
C
       WRITE(6,55)
       GO TO 1
   20  CONTINUE
C
C    CONVERGENCE.  CALCULATE ALL SPECIES AND TEST THE
C    MASS AND CHARGE BALANCE EQUATIONS
C
       CL2=CL*CL
       PB=KSP/CL2
       PBCL=B1*PB*CL
       PBCL2=B2*PB*CL2
       PBCL3=B3*PB*CL2*CL
```

```
      PBCL4=B4*PB*CL2**2
      F1=S-(PB+PBCL +PBCL2+PBCL3+PBCL4)
      F2=CNACL+2.*S-(CL+PBCL+2.*PBCL2+3.*PBCL3+4.*PBCL4)
      F3=CNACL + 2.*PB + PBCL -(CL + PBCL3 + 2.*PBCL4)
      WRITE(6,66)S, CL,PB,PBCL,PBCL2,PBCL3,PBCL4,F1,F2,F3
      GO TO 1
C
   11 FORMAT(//'  ENTER ITMAX,CVGTOL,CNACL,SO,CLO'/)
   22 FORMAT(I,4F)
   33 FORMAT(//'  TOTAL NACL = ',G12.3, ' MOLAR',///'  ITN',
     110X,'SOLUBILITY',12X,'<CL(-)>'//,
     2I5,2G20.5)
   44 FORMAT(I5,2G20.5)
   55 FORMAT(//'  SORRY. THE METHOD FAILED TO CONVERGE.'//)
   66 FORMAT(//'  CONVERGENCE!',//'    SOLUBILITY = ',
     1 G12.3, 10X, '<CL(-)> = ',G12.3,//6X,'<PB(2+)> = ',G12.3,
     2 7X,' <PBCL(+)> = ',G12.3,//7X,'<PBCL2> = ',G12.3,
     3 7X,' <PBCL3(-)> = ',G12.3,//4X,'PBCL4(2-)> = ',
     4  G12.3,///'  TOTAL LEAD BALANCE:',G12.3,//,
     5 '  TOTAL CHLORINE BALANCE:',G12.3//,
     6 '  CHARGE BALANCE:',G12.3//)
      END
```

Discussion

ITERAT requires 389 words of core memory. This sample execution took 0.2 sec. Figure 4.1 was prepared from data provided by ITERAT.

The rate of convergence can be relatively slow using iterative formulas. One device that often improves the rate of convergence is to use average values:

$$X_{i+1} = \frac{X_i + X_{i-1}}{2}$$

This technique is used in ITERAT if the procedure fails to converge by the tenth iteration. Double precision is used because both equations are susceptible to round-off error.

The second method for solving systems of nonlinear equations considered here is an extension of the Newton-Raphson method for finding the roots of a nonlinear equation (Section 3.3). Consider a nonlinear system of order 2,

$$F(x,y) = 0 \qquad G(x,y) = 0$$

Let x_0 and y_0 be initial approximations to the solutions. Now expand these functions about the initial approximations using a Taylor series, dropping all terms of order greater than one. This gives

$$F(x,y) = 0 = F_0 + F_{x_0}\Delta x + F_{y_0}\Delta y$$

$$G(x,y) = 0 = G_0 + G_{x_0}\Delta x + G_{y_0}\Delta y$$

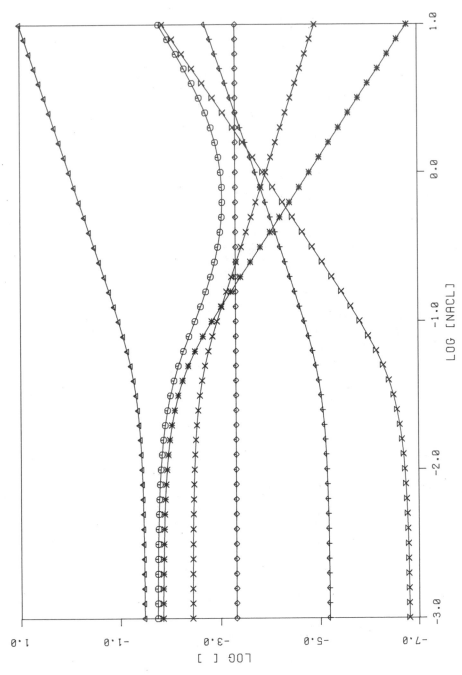

Figure 4.1 Solubility of $PbCl_2$ and the equilibrium concentrations of all species as a function of total NaCl: solubility = "$\bigcirc$"; $[Cl^-]$ = "$\triangle$"; $[Pb^{2+}]$ = "$*$"; $[PbCl^+]$ = "$\times$"; $[PbCl_2]$ = "$\diamond$"; $[PbCl_3^-]$ = "$\oplus$"; and $[PbCl_4^{2-}]$ = "$\times\!\!\!\!\!\diagdown$"

Here

$$F_0 = F(x_0, y_0) \qquad\qquad G_0 = G(x_0, y_0)$$

$$F_{x_0} = \left|\frac{\partial F(x,y)}{\partial x}\right|_0 \qquad G_{x_0} = \left|\frac{\partial G(x,y)}{\partial x}\right|_0$$

$$F_{y_0} = \left|\frac{\partial F(x,y)}{\partial y}\right|_0 \qquad G_{y_0} = \left|\frac{\partial G(x,y)}{\partial y}\right|_0$$

and

$$\Delta x = x - x_0 \qquad \Delta y = y - y_0$$

The partial derivatives F_{x_0}, G_{y_0}, etc., can all be calculated if $F(x,y)$ and $G(x,y)$ are differentiable. The nonlinear system has been approximated by a linear system of equations in the correction terms Δx, and Δy,

$$F_{x_0} \Delta x + F_{y_0} \Delta y = -F_0$$

$$G_{x_0} \Delta x + G_{y_0} \Delta y = -G_0$$

The solution to this linear system of equations gives Δx and Δy. If the procedure converges, Δx and Δy provide better approximations to the desired solution,

$$x = x_0 + \Delta x$$

$$y = y_0 + \Delta y$$

These new values become the initial approximations for the next iteration, and the entire procedure is continued until a convergence criterion is satisfied.

Problem 4.8. Solve Problem 4.6 using the Newton-Raphson method. Use as initial
            approximations $x = 0.15$ and $y = 0.35$.

Solution. $F(x,y) = x^2 + x - y^2 - 0.15$

$G(x,y) = x^2 - y + y^2 + 0.17$

$F_x = 2x + 1$

$F_y = -2y$

$G_x = 2x$

$G_y = 2y - 1$

Using Cramer's rule (Section 4.1),

$$\Delta x = \frac{G_0 F_{y_0} - F_0 G_{y_0}}{D}$$

$$\Delta y = \frac{F_0 G_{x_0} - F_{x_0} G_0}{D}$$

where

$$D = F_{x_0} G_{y_0} - G_{x_0} F_{y_0}$$

First iteration:

$x_0 = 0.15$ $y_0 = 0.35$

$F(0.15, 0.35) = -0.1000$ $G(0.15, 0.35) = -0.0350$

$F_{x_0} = 1.3000$ $G_{x_0} = 0.3000$

$F_{y_0} = -0.7000$ $G_{y_0} = -0.3000$

$D = -0.1800$

$\Delta x = 0.0306$ $\Delta y = -0.0861$

$x = 0.1806$ $y = 0.2639$

Second iteration:

$x_0 = 0.1806$ $y_0 = 0.2639$

$F(0.1806, 0.2639) = -0.0065$ $G(0.1806, 0.2639) = 0.0083$

$F_{x_0} = 1.3611$ $G_{x_0} = 0.3611$

$F_{y_0} = -0.5278$ $G_{y_0} = -0.4722$

$D = -0.4522$

$\Delta x = 0.0165$ $\Delta y = 0.0303$

$x = 0.1971$ $y = 0.2942$

Third iteration:

$x_0 = 0.1971$ $y_0 = 0.2942$

$F(0.1971, 0.2942) = -0.0006$ $G(0.1971, 0.2942) = 0.0012$

$F_{x_0} = 1.3941$ $G_{x_0} = 0.3941$

$F_{y_0} = -0.5884$ $G_{y_0} = -0.4116$

$D = -0.3419$

$\Delta x = 0.0028$ $\Delta y = 0.0056$

$x = 0.1999$ $y = 0.2998$

The method is rapidly converging on the solution $x = 0.2$ and $y = 0.3$.

Problem 4.8. Consider the following aqueous equilibrium system:

$$M(ClO_4)_n + NaX + H_2O$$

Here the metal ion M^{n+} forms m complexes with the ligand X^-. For example, if $n = 2$ and $m = 6$,

$$M^{2+} + X^- \rightleftharpoons MX^+ \qquad\qquad K_1$$

$$MX^+ + X^- \rightleftharpoons MX_2 \qquad\qquad K_2$$

$$\cdots\cdots\cdots\cdots\cdots\cdots\cdots\cdots\cdots\cdots$$

$$MX_5^{3-} + X^- \rightleftharpoons MX_6^{4-} \qquad\qquad K_6$$

Also, X^- is the conjugate base of the weak acid HX. Calculate and plot the concentrations of all species at equilibrium as a function of both total metal and total ligand concentrations. This problem is solved by the following program.

4.4.2 COMPLX

Introduction

COMPLX solves the general complexation equilibrium problem posed in Problem 4.8. With $m = 6$ there are 11 species at equilibrium: M^{2+}, MX^+, MX_2, MX_3^-, MX_4^{2-}, MX_5^{3-}, MX_6^{4-}, H^+, OH^-, X^-, and HX. Eleven equations are required to solve the system. The first six are the mass action equations of the form

$$K_i = \frac{[MX_i^{2-i}]}{[MX_{i-1}^{3-i}][X^-]} \qquad i = 1,2,3,\ldots,6$$

The remaining five are

$$HX \rightleftharpoons H^+ + X^- \qquad\qquad K_a$$

$$H_2O \rightleftharpoons H^+ + OH^- \qquad\qquad K_w$$

$$M_{tot} = [M^{2+}] + \sum_{i=1}^{6} [MX_i^{2-i}]$$

$$X_{tot} = [X^-] + [HX] + \sum_{i=1}^{6} i[MX_i^{2-i}]$$

$$[H^+] + [Na^+] - [OH^-] - [ClO_4^-] - [X^-] + 2[M^{2+}] + \sum_{i=1}^{6} (2 - i)[MX^{2-i}] = 0$$

It is assumed here that this system of equations adequately describes the chemical system and that additional reactions and activity effects can be safely ignored. The problem to be solved is the following: Given C_m mol of $M(ClO_4)_n$ and C_x mol of NaX in 1.0 liter of H_2O, what are the concentrations of all 11 species at equilibrium?

The input includes the following: C_m and C_x (CM and CX); the ionic charge of the metal ion, n (N); the coordination number of the metal, m (M); the pK_a of the weak acid (PKA); the logarithms of the stepwise formation constants of the complexes (C(6)); and the initial approximations to the equilibrium concentrations of the three species X^-, M^{n+}, and H^+. COMPLX prints these values for verification purposes and a convergence map showing the progress of the iterative method. If the system converges, the concentrations of all species are printed.

Method

Let the coordination number be 6. The 11 x 11 system can be reduced by appropriate substitutions to a system of order 3 in the unknowns $[M^{n+}]$, $[X^-]$, and $[H^+]$. Using overall formation constants derived from the stepwise formation constants,

$$[MX_i^{n-i}] = \beta_i [M^{n+}][X^-]^i$$

where

$$\beta_i = K_1 K_2 \cdots K_i$$

Let the unknowns $[X^-]$, $[M^{n+}]$, and $[H^+]$ be represented by x, y, and z, respectively. The three equations are the mass balance and charge balance equations,

$$F(x,y,z) = x + \frac{xz}{K_a} + \sum_{i=1}^{6} i\beta_i yx^i - C_x = 0$$

$$G(x,y,z) = y + \sum_{i=1}^{6} \beta_i yx^i - C_m = 0$$

$$H(x,y,z) = z + C_x - \frac{K_w}{z} - nC_m - x + ny + \sum_{i=1}^{m} (n-i)\beta_i yx^i = 0$$

The Newton-Raphson method requires the nine partial derivatives F_x, F_y, F_z, G_x, G_y, G_z, H_x, H_y, and H_z. These partial derivatives and the three functions are coded in the subroutine EVAL. The matrix inversion routine MATINV (Section 4.3.2) is used to compute the vector of correction terms,

$$D = A^{-1}B$$

where

$$D = \begin{pmatrix} \Delta_x \\ \Delta_y \\ \Delta_z \end{pmatrix} \qquad B = \begin{pmatrix} -F_0 \\ -G_0 \\ -H_0 \end{pmatrix} \qquad A = \begin{bmatrix} F_{x_0} & F_{y_0} & F_{z_0} \\ G_{x_0} & G_{y_0} & G_{z_0} \\ H_{x_0} & H_{y_0} & H_{z_0} \end{bmatrix}$$

COMPLX uses three convergence criteria. Let RELSUM be the relative changes in the absolute values of the three roots,

$$\text{RELSUM} = \left| \frac{\Delta_x}{x} \right| + \left| \frac{\Delta_y}{y} \right| + \left| \frac{\Delta_z}{z} \right|$$

If RELSUM is less than 0.001, the method has converged. If RELSUM is greater than 1000, then the method is diverging and the procedure is terminated. If the method fails to converge in 20 iterations, the program terminates.

Sample Execution $(Cd(ClO_4)_2 + NaCN + H_2O)$

THIS PROGRAM SOLVES THE AQUEOUS EQUILIBRIUM SYSTEM:

   M(CLO4)N + NAX + H2O

 ENTER THE INITIAL CONCENTRATIONS OF M(CLO4)N AND NAX.
 >.001 .01

 ENTER THE CHARGE AND COORD. NUMBER OF THE METAL.
 >2 4

 ENTER THE PKA OF THE ACID HX.
 >9.4

 ENTER LOG(K(I)), I=1,4. K(I) IS THE STEPWISE FORMATION
 CONSTANT FOR THE SPECIES MXI(2-I).
 >5.48 5.14 4.56 3.58

 FINALLY, ENTER THE INITIAL APPROXIMATIONS TO THE
 EQUILIBRIUM CONCENTRATIONS OF X(-), M(2+) AND H(+).
 >.009 1E-7 1E-11

 INITIAL CONCENTRATIONS: M(CLO4)2 = 1.00D-03
                         NAX   = 1.00D-02

 INITIAL APPROXIMATIONS TO FINAL CONCENTRATIONS:

                    X(-) = 9.00D-03
                    M(2+) = 1.00D-07
                    H(+) = 1.00D-11

                    PKA FOR HX = 9.40

AND FOR SPECIES: MX (+)   MX2   MX3(-)   MX4(2-)
        LOG(K):   5.480   5.140   4.560   3.580

ITN   SPECIES   X(I)   REL. CHANGE

 1
        X(-)   9.00D-03   -3.08D-05
        M(2+)   1.22D-11   -1.00D+00
        H(+)   1.63D-11   6.31D-01
 2
        X(-)   7.63D-03   -1.52D-01
        M(2+)   7.44D-12   -3.93D-01
        H(+)   2.13D-11   3.06D-01

```
3
        X(-)            6.41D-03           -1.59D-01
        M(2+)           4.75D-12           -3.61D-01
        H(+)            2.44D-11            1.44D-01
4
        X(-)            5.83D-03           -9.14D-02
        M(2+)           1.82D-12           -6.17D-01
        H(+)            2.60D-11            6.69D-02
5
        X(-)            5.68D-03           -2.52D-02
        M(2+)           3.26D-13           -8.21D-01
        H(+)            2.65D-11            1.79D-02
6
        X(-)            5.67D-03           -2.06D-03
        M(2+)           1.62D-13           -5.02D-01
        H(+)            2.65D-11            1.42D-03
7
        X(-)            5.67D-03           -1.56D-05
        M(2+)           1.61D-13           -8.26D-03
        H(+)            2.65D-11            1.03D-05
```

THE SYSTEM HAS CONVERGED. THE SOLUTION IS:

I	SPECIES	X(I)	F(I)
1	MX (+)	2.75D-10	
2	MX2	2.15D-07	
3	MX3(-)	4.43D-05	
4	MX4(2-)	9.55D-04	
5	HX	3.77D-04	
6	X(-)	5.67D-03	2.05D-09
7	M(2+)	1.61D-13	5.17D-10
8	H(+)	2.65D-11	-1.02D-09
9	OH(-)	3.77D-04	

<u>Listing</u>

```
C
C                       COMPLX
C
C          THIS PROGRAM USES THE NEWTON-RAPHSON METHOD TO
C          SOLVE A SYSTEM OF NONLINEAR SIMULTANEOUS EQUATIONS.
C
C          IN THIS EXAMPLE THE SYSTEM M(CLO4)N-NAX IS SOLVED,
C          WHERE M(N+) COMPLEXES X(-).
C
C          THE SIX TO ELEVEN NONLINEAR EQUATIONS ARE REDUCED
C          TO THREE, IN [X(-)], [M(N+)], AND [H(+)].
C
C          THE EQUATIONS ARE STORED IN F(3).
C          THE PARTIAL DERVATIVES ARE STORED IN G(3,3).
C          F AND G ARE COMPUTED IN THE SUBROUTINE EVAL.
C
C          ORDER OF INPUT DATA:
C
C              CM   = INITIAL CONCENTRATION OF M(N+).
C              CX   = INITIAL CONCENTRATION OF X(-).
```

```
C                N    = CHARGE ON METAL (0 < N & N < 5)
C                M    = COORDINATION NO. OF THE METAL (0<M & M<7).
C               PKA   = PKA OF THE ACID, HX.
C       (C(J), J=1,M) = LOGS OF STEPWISE FORMATION CONSTANTS.
C               X(1)  = INITIAL APPROXIMATION TO THE EQUILIBRIUM
C                       CONCENTRATION OF X(-).
C               X(2)  = INITIAL APPROXIMATION TO THE EQUILIBRIUM
C                       CONCENTRATION OF M(N+).
C               X(3)  = INITIAL APPROXIMATION TO THE EQUILIBRIUM
C                       CONCENTRATION OF H(+).
C
C
C          GLOSSARY
C
C          F(3)        - NONLINEAR EQUATIONS.
C          G(3,3)      -  COEFF. MATRIX OF PARTIAL DERIVATIVES
C          X(3)        - ROOTS OF THE NONLINEAR SYSTEM.
C          SPECES(11) - SYMBOLS FOR THE IONIC SPECIES AND COMPLEXES
C         CHARGE(10)  - IONIC CHARGES OF SPECIES
C          BETA(J)     -  PRODUCT OF THE ANTILOG OF C(K), K=1,J.
C
C
C          THE SUBROUTINE MATINV INVERTS THE COEFFICIENT MATRIX.
C
C
C     AUTHORS:  R. E. SHULIK AND K. J. JOHNSON
C
      IMPLICIT DOUBLE PRECISION (A-H,O-Z)
      INTEGER SPECES(11),CHARGE(10)
      DOUBLE PRECISION KW,KA
      COMMON G(3,3),BETA(6),C(6),X(3),F(3),CM,CX,KA,KW,N,M
C
C     INITIALIZATION
C
      DATA SPECES/'MX','MX2','MX3','MX4','MX5','MX6','HX','X(-)','M',
     1'H(+)','OH(-)'/
      DATA CHARGE/'(5-)','(4-)','(3-)','(2-)','(-)','    ','(+)',
     1 '(2+)','(3+)','(4+)'/
      DATA G(2,3),KW/0.0D0,1.0D-14/
      DATA CVGTOL,DVGTOL/1.0D-2,1.0D3/
C
C          INPUT
C
      WRITE(6,11)
      READ(5,55) CM,CX
      WRITE(6,33)
      READ(5,22) N,M
      NP6=N+6
      WRITE(6,44)
      READ(5,55) PKA
      WRITE(6,66) M,N
      READ(5,55) (C(J), J=1,M)
      WRITE(6,77) N
      READ(5,55) X(1),X(2),X(3)
C
C          INPUT PRINTED OUT FOR VERIFICATION.
C
      WRITE(6,88) N,CM,CX,X(1),N,X(2),X(3),PKA,(SPECES(J),
     1 CHARGE(NP6-J),J=1,M)
      WRITE(6,166) (C(J), J=1,M)
C
C          CONVERT C(L) TO BETA(L), L=1,M
C
```

```
      BETA(1)=10.D0**C(1)
      IF(M.EQ.1)GO TO 21
      DO 20 J=2,M
   20 BETA(J)=BETA(J-1)*10.D0**C(J)
C
C          PKA IS CONVERTED TO KA.
C
   21 KA=10.D0**(-PKA)
      WRITE(6,199)
C
C          MAIN LOOP.  NEWTON-RAPHSON ITERATIVE PROCEDURE.
C
      DO 100 ITN=1,10
C
C          EVALUATE F AND G
C
      CALL EVAL
      DO 40 K=1,3
   40 F(K)=-F(K)
C
C          CALL MATRIX INVERSION ROUTINE.
C
      CALL MATINV(G,3,D)
      IF(D.EQ.0.D0)GO TO 330
C
C          EVALUATE THE CORRECTIONS
C
      WRITE(6,244)ITN
      ZUM=0.0D0
      DO 70 I=1,3
      Z=0.0D0
      DO 50 J=1,3
   50 Z=Z+G(I,J)*F(J)
      ZZ=Z/X(I)
      ZUM=ZUM+DABS(ZZ)
      X(I)=X(I)+Z
      II=I+7
      IF(I.NE.2)WRITE(6,277) SPECES(II),X(I),ZZ
   70 IF(I.EQ.2)WRITE(6,288) SPECES(9),CHARGE(N+6),X(2),ZZ
C
C          TEST FOR CONVERGENCE
C
      IF(ZUM.LE.CVGTOL)GO TO 105
C
C          TEST FOR DIVERGENCE
C
  100 IF(ZUM.GT.DVGTOL)GO TO 115
      WRITE(6,111)
      GO TO 125
  105 CALL EVAL
      WRITE(6,122)
      GO TO 125
  115 WRITE(6,133)
C
C          OUTPUT FINAL VALUES
C
  125 WRITE(6,144)
C
C   CALCULATE AND PRINT MXL, L = 1,2,...,M
C
```

```
      Y=X(2)
      XLIG=X(1)
      DO 130 L=1,M
      Y=10.0D0**C(L)*Y*XLIG
  130 WRITE(6,177)L,SPECES(L),CHARGE(NP6-L),Y
C
C  CALCULATE AND PRINT [HX],[X(-)],[M(N+)],[H(+)], AND [OH(-)]
C
      Y=X(1)*X(3)/KA
      L=M+1
      WRITE(6,255) L,SPECES(7),Y
      L=L+1
      WRITE(6,255) L,SPECES(8),X(1),F(1)
      L=L+1
      WRITE(6,266) L,SPECES(9),CHARGE(N+6),X(2),F(2)
      L=L+1
      WRITE(6,255) L,SPECES(10),X(3),F(3)
      L=L+1
      Y=KW/X(3)
      WRITE(6,255) L,SPECES(11),Y
      STOP
  330 WRITE(6,333)
      STOP
C
C   FORMATS
C
   11 FORMAT('  THIS PROGRAM SOLVES THE AQUEOUS EQUILIBRIUM SYSTEM:',
     1//5X,'M(CLO4)N + NAX + H2O',//,'  ENTER THE',
     2 ' INITIAL CONCENTRATIONS OF M(CLO4)N AND NAX. ',/)
   22 FORMAT(2I)
   33 FORMAT('  ENTER THE CHARGE AND COORD. NUMBER OF THE METAL.'/)
   44 FORMAT('  ENTER THE PKA OF THE ACID HX. ',/)
   55 FORMAT(6E)
   66 FORMAT('  ENTER LOG(K(I)), I=1,',I1,'. K(I) IS THE STEPWISE
     1 FORMATION',/,'  CONSTANT FOR THE SPECIES MXI(',I1,'-I).'/)
   77 FORMAT('  FINALLY, ENTER THE INITIAL APPROXIMATIONS TO THE '
     1,/,'  EQUILIBRIUM CONCENTRATIONS OF X(-), M(',I1,'+) AND',
     2' H(+)/)
   88 FORMAT('  INITIAL CONCENTRATIONS: M(CLO4)',I1,' =',1PD9.2,/
     1 28X,' NAX   =',1PD9.2//'  INITIAL APPROXIMATIONS TO FINAL ',
     2'CONCENTRATIONS:'//,21X,'X(-)  = ',1PD9.2,/,21X,'M(',I1,'+) = ',
     31PD9.2,/20X,' H(+)  = ',1PD9.2,//,20X,' PKA FOR HX =',0PF6.2,//,
     4 ' AND FOR SPECIES:  ',6(2X,A3,A4))
  111 FORMAT('     ITMAX EXCEEDED. THE BEST VALUES ARE:',/)
  122 FORMAT(//' THE SYSTEM HAS CONVERGED.  THE SOLUTION IS:'/)
  133 FORMAT('   THE SYSTEM IS DIVERGING. THE BEST VALUES ARE:'/)
  144 FORMAT(/T9,'I',2X,'SPECIES',T36,'X(I)',T51,'F(I)'/)
  166 FORMAT(T10,' LOG(K):',6(3X,F6.3))
  177 FORMAT(T5,I5,2X,A3,A4,T25,2(1PD16.2))
  199 FORMAT(/' ITN    SPECIES',8X,'X(I)',8X,'REL. CHANGE'/)
  244 FORMAT(I3)
  255 FORMAT(T5,I5,2X,A5,T25,2(1PD16.2))
  266 FORMAT(T5,I5,2X,A1,A4,T25,2(1PD16.2))
  277 FORMAT(T9,A5,T14,2(1PD16.2))
  288 FORMAT(T9,A1,A4,T14,2(1PD16.2))
  333 FORMAT(//'  HAVE A SINGULAR COEFFICIENT MATRIX.'//)
      END
```

```
C                   EVAL
C
C
C              THIS SUBROUTINE EVALUATES THE FUNCTIONS AND THEIR PARTIAL
C              DERVATIVES.
C
C              F(1)  IS THE MASS BALANCE EQUATION FOR X(-).
C              F(2)  IS THE MASS BALANCE EQUATION FOR M(N+).
C              F(3)  IS THE CHARGE BALANCE EQUATION.
C
C              G(1,1)=DF(1)/DX(1)   G(1,2)=DF(1)/DX(2)   G(1,3)=DF(1)/DX(3)
C              G(2,1)=DF(2)/DX(1)   G(2,2)=DF(2)/DX(2)   G(2,1)=DF(2)/DX(3)
C              G(3,1)=DF(3)/DX(1)   G(3,2)=DF(3)/DX(2)   G(3,3)=DF(3)/DX(3)
C
       SUBROUTINE EVAL
       IMPLICIT DOUBLE PRECISION (A-H,O-Z)
       DOUBLE PRECISION KW,KA
       COMMON G(3,3),BETA(6),C(6),X(3),F(3),CM,CX,KA,KW,N,M
       DO 10 I=1,3
       F(I)=0.0D0
       G(I,1)=0.0D0
   10  G(I,2)=0.0D0
       T1=1.
       DO 20 I=1,M
       T1=T1*X(1)
       T2=T1*BETA(I)
       G(1,1)=G(1,1)+T2*I*I
       G(1,2)=G(1,2)+T2*I
       G(2,2)=G(2,2)+T2
       G(3,1)=G(3,1)+I*(N-I)*T2
   20  G(3,2)=G(3,2)+(N-I)*T2
       G(1,1)=X(2)*G(1,1)/X(1)+X(3)/KA+1.
       G(1,3)=X(1)/KA
       G(2,1)=X(2)*G(1,2)/X(1)
       G(2,2)=G(2,2)+1.
       G(3,1)=X(2)*G(3,1)/X(1)-1.
       G(3,2)=G(3,2)+N
       G(3,3)=KW/(X(3)*X(3))+1.
       F(1)=X(2)*G(1,2)+X(3)*G(1,3)+X(1)-CX
       F(2)=X(2)*G(2,2)-CM
       F(3)=X(2)*G(3,2)+X(3)+CX-(X(1)+N*CM+KW/X(3))
       RETURN
       END
```

Discussion

The program lengths in 36-bit words are main program, 870 words; EVAL, 79 words; and MATINV, 303 words. COMPLX requires approximately 3k words of memory. This sample execution took approximately 0.3 sec.

Figure 4.2 was plotted using data provided by COMPLX.

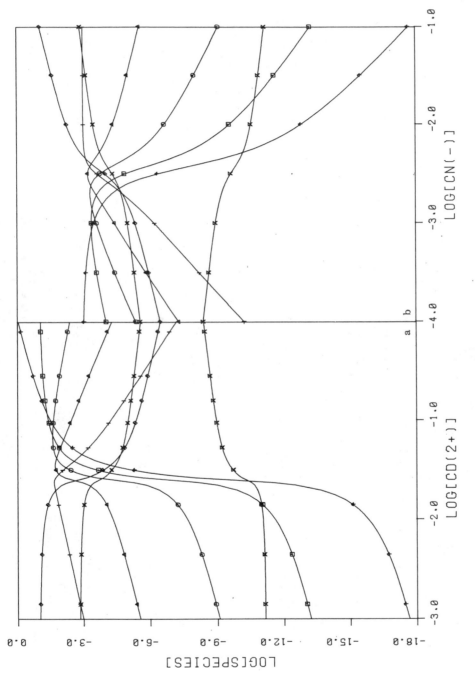

Figure 4.2   (a) Solution of the $Cd(ClO_4)_2$-NaCN-H_2O system as a function of total $Cd(NO_3)_2$; total cyanide = 0.1 M: Cd^{2+} = "▲"; $CdCN^+$ = "□"; $Cd(CN)_2$ = "⊙"; $Cd(CN)_3^-$ = "⊙"; $Cd(CN)_4^{2-}$ = "⊣"; CN^- = "◇"; HCN = "✕"; and H^+ = "✕."
(b) Solution of the $Cd(ClO_4)_2$-NaCN-H_2O system as a function of total NaCN; total cadmium = 0.001 M.

4.5 ADDITIONAL PROBLEMS

1. Solve the following linear systems of equations:

   (a)   x + 2y - 12z + 8v = 27

          5x + 44y + 7z - 2v = 4

         -3x + 7y + 9z + 5v = 11

          6x - 12y - 9z + 3v = 49

   (b)   2x + 10y - 6z + 4u + 8v = 8

         -3x - 12y - 9z + 6u + 3v = 3

          -x +   y - 35z + 15u + 18v = 29

          4x + 18y   + 4u + 14v = -2

          5x + 26y - 19z + 25u + 36v = 23

   Solutions [11]:

      (a) x = 3, y = -2, z = 1, v = 5

      (b) x = 0, y = 1, z = -1, u = 2, v = -2

2. Determine the concentrations of the four compounds in the following mixture
    assuming Beer's law [12].

Molar Absorptivity (liters/mol·cm)

Wave-length	p-Xylene	m-Xylene	o-Xylene	Ethyl-benzene	Absorbance
12.5	1.502	0.0514	0	0.0408	0.1013
13.0	0.0261	1.1516	0	0.0820	0.09943
13.4	0.0342	0.0355	2.532	0.2933	0.2194
14.3	0.0340	0.0684	0	0.3470	0.03396

Assume an optical path length of 1.0 cm.

3. Mass spectroscopy can be used to determine the concentrations of species in a
    mixture if observed peak height is a linear function of the partial pressure
    [13-15].

   $$H_i = \sum_{j=1}^{n} S_{ij} P_j$$

Here H_i is the observed peak height, n is the number of gases in the mixture,
S_{ij} is the sensitivity of the jth gas at a particular m/e value, measured using
a sample of the pure gas, and P_j is the partial pressure of the jth gas.
Calculate the mole fraction of each gas in the following mixtures:

(a) A mixture of hydrocarbons [13]:

Sensitivities

m/e	Ethylcyclo-pentane	Cyclo-hexane	Cyclo-heptane	Methylcyclo-hexane	Peak height of mixture
69	121.0	22.4	27.1	23.0	87.6
83	9.35	4.61	20.7	100.0	58.8
84	1.38	74.9	1.30	6.57	47.2
98	20.2	0.0	32.8	43.8	100.0

Solution. Ethylcyclopentane, 0.25006; cyclohexane, 0.25039; cycloheptane, 0.25018; and methylcyclohexane, 0.24938.

(b) A mixture of hydrogen and hydrocarbons [14]:

Sensitivities

m/e	H_2	CH_4	C_2H_4	C_2H_6	C_3H_6	C_3H_8	$n-C_5H_{12}$	Peak height of mixture
2	16.87	0.165	0.2019	0.3170	0.234	0.182	0.110	17.1
16	0	27.7	0.862	0.062	0.0730	0.131	0.120	65.1
26	0	0	22.35	13.05	4.420	6.001	3.043	186.0
30	0	0	0	11.28	0.0	1.110	0.3710	82.7
40	0	0	0	0	9.850	1.684	2.108	84.2
44	0	0	0	0	0.2990	15.98	2.107	63.7
72	0	0	0	0	0	0	4.670	119.7

The total pressure is 39.9 μm of Hg.

(c) A mixture of alcohols [15]:

Peak Heights

m/e	CH_3OH	C_2H_5OH	C_3H_7OH	$(CH_3)_2CHOH$	Unknown mixture
15	42.72	10.72	4.03	12.51	17.26
19	0.53	3.46	0.98	7.82	5.10
27	0.0	22.53	16.51	16.53	22.92
29	62.75	22.53	15.95	10.75	29.75
31	100.00	100.00	100.00	7.01	100.00
32	75.37	2.34	3.51	0.0	13.43
39	0	0	4.85	5.99	5.08

Table (continued)

m/e	CH_3OH	C_2H_5OH	C_3H_7OH	$(CH_3)_2CHOH$	Unknown Mixture
43	0	8.59	4.03	18.15	12.88
45	0	39.60	4.98	100.00	59.89
46	0	18.52	0	0	6.03
59	0	0	10.13	3.95	6.80

Sensitivities: 0.138, 0.282, 0.416, 0.369.

4. Solve the nonlinear system of equations

$$x - 0.1y^2 + 0.05z^2 = 0.7$$

$$y + 0.3x^2 - 0.1xz = 0.5$$

$$z + 0.4y^2 - 0.1xy = 1.2$$

Solution [16]. $x = 0.660$, $y = 0.441$, $z = 1.093$.

5. Write a program to balance chemical reactions using matrix methods [17, 18].

6. Write a program to solve complex kinetic systems by the Laplace transform method [19].

7. The principal reactions in the production of synthesis gas by partial oxidation of methane with oxygen are [20]

$$CH_4 + \frac{1}{2} O_2 \rightleftharpoons CO + 2H_2$$

$$CH_4 + H_2O \rightleftharpoons CO + 3H_2$$

$$H_2 + CO_2 \rightleftharpoons CO + H_2O$$

Determine the O_2/CH_4 reactant ratio that will produce an adiabatic equilibrium temperature of 2200°F at an operating pressure of 20 atm when the reactant gases are preheated to an entering temperature of 1000°F.

8. Solve the following aqueous equilibrium problems, i.e., calculate the molar concentrations of all species at equilibrium as a function of total concentrations of the relevant species:

(a) $FeNH_4(SO_4)_2 + H_2C_2O_4$ [21].

(b) $Ni^{2+} + NH_3 + EDTA$ [22].

(c) $Ni^{2+} + CN^- + NH_3 + Ag^+ + I^-$ [23].

(d) The mixed ligand system Cu^{2+} + alanine + valine [24].

9. Use the pit-mapping method of Sileen [25] to determine the stability constants of tin-pyrocatechol violet complexes [26].

10. Solve the gas-phase equilibrium problem involving

    (a) N_2, O_2, NO, and NO_2 as a function of temperature and total pressure [27].

    (b) CH_4, H_2O, CO_2, CO, and H_2O as a function of temperature [28].

    (c) C_3H_8, O_2, and N_2 [29].

    See Ref. 30 for additional problems and methods for solving equilibrium problems.

11. Predict the structure of boron hydrides using the "styx" formulas [31-33].

REFERENCES

1. C. Froberg, "Introduction to Numerical Analysis," Addison-Wesley, Reading, Mass., 1965, p. 73.

2. B. Carnahan, H. Luther, and J. Wilkes, "Applied Numerical Methods," Wiley, New York, 1965, pp. 299-307.

3. R. S. Varga, "Matrix Iterative Analysis," Prentice-Hall, Englewood Cliffs, N. J., 1962, pp. 56-61.

4. B. Carnahan et al., op. cit., pp. 270-281.

5. B. Carnahan et al., ibid., pp. 270-296.

6. B. Carnahan et al., ibid., pp. 308-318.

7. T. R. Dickson, "The Computer and Chemistry," Freeman, San Francisco, 1968, pp. 144-147.

8. B. Carnahan et al., op. cit., pp. 319-329.

9. J. Butler, "Ionic Equilibrium," Addison-Wesley, Reading, Mass., 1964, pp. 261-266.

10. J. Bjerrum, G. Schwarzenbach, and L. Sileen, "Stability Constants," Parts I and II, Special Publications Numbers 6 and 7 of the Chemical Society, London, 1958.

11. C. Froberg, op. cit., pp. 92-93.

12. T. R. Dickson, op. cit., p. 150.

13. T. L. Isenhour and P. C. Jurs, "Computer Programming for Chemistry," Allyn and Bacon, Boston, 1972, pp. 265-269.

14. B. Carnahan et al., op. cit., pp. 331-332.

15. T. R. Dickson, op. cit., pp. 151-153.

16. C. Froberg, op. cit., p. 93.

17. V. L. Fabishak, Matrix method for balancing chemical equations, Chemistry, 40, 18 (1967).

18. J. P. Brown et al., A computer program for balancing chemical equations, J. Chem. Educ., 49, 754 (1972).

19. E. McLaughlin and R. W. Rozett, The kinetics of four components linked by equilibria, J. Chem. Educ., 49, 482 (1972).

20. B. Carnahan et al., op. cit., pp. 321-329.

21. J. W. Swinnerton and W. W. Miller, Use of a digital computer to solve a complex chemical equilibrium, J. Chem. Educ., 36, 485 (1959).

22. A. J. Bard and D. M. King, General digital computer program for chemical equilibrium calculations, J. Chem. Educ., 42, 127 (1965); see also letters to the editor, J. Chem. Educ., 42, 624 and 625 (1965).

23. S. A. Carrano et al., A demonstration and exercise in simultaneous equilibria, J. Chem. Educ., 43, 603 (1966).

24. Ting-Po I and G. H. Nancollas, EQUIL -- A general computational method for the calculation of solution equilibria, Anal. Chem., 44, 1940 (1972).

25. L. G. Sileen, High-speed computer as a supplement to graphical methods, Acta Chem. Scand., 18, 1085 (1964).

26. W. Wakley and L. Varga, Stability constants of tin pyrocatechol violet complexes from computer analysis of absorption spectra, Anal. Chem., 44, 169 (1972).

27. E. Schonfeld, Computer calculated concentrations on the reactions of nitrogen and oxygen, J. Chem. Educ., 45, 173 (1968).

28. E. E. Stone, J. Chem. Educ., 43, 241 (1966).

29. D. H. Yean and J. R. Riter, Jr., Exact thermodynamics of propane combustion, J. Chem. Educ., 51, 505 (1974).

30. F. van Zeggeren and S. H. Storey, "The Computation of Chemical Equilibria," Cambridge University Press, New York, 1970.

31. W. Lipscomb, "Boron Hydrides," Benjamin, New York, 1963, pp. 43-49.

32. T. R. Dickson, op. cit., pp. 188-189.

33. D. Jones and E. Carberry, A computer evaluation of the equations of balance for boron hydride species, J. Chem. Educ., 49, 681 (1972).

CHAPTER 5

REGRESSION ANALYSIS

Regression analysis, or curve fitting, is frequently used in chemistry to represent experimental data or to determine values for parameters of a mathematical model of a chemical system. For example, an analytical chemist might represent experimental calibration data using a quadratic equation,

$$y = a_0 + a_1 x + a_2 x^2$$

Here x is the independent variable, x is the dependent variable, and a_0, a_1, and a_2 are the coefficients of the quadratic function. Given a table of experimental x and y values the problem is to determine the set of coefficients that define the quadratic function that fits the experimental data as closely as possible.

The usual procedure is to minimize the sum of the squares of the residuals R_i,

$$R_i = y_i - F(x_i)$$

where

$$F(x_i) = a_0 + a_1 x_i + a_2 x_i^2$$

The residuals are illustrated in Figure 5.1. This procedure is known as the least-squares method. This subject has been extensively treated in the literature, both in books [1-11] and in journal articles [12-56]. Each ordinate, y_i, may have a known associated error, for example, a standard deviation,

$$y_i \pm \sigma_i$$

where

$$\sigma_i^2 = \frac{\sum_{j=1}^{m} (\bar{y} - y_j)^2}{m - 1}$$

Here $\bar{y}$ is the average value of m observed values,

$$\bar{y} = \frac{\sum_{j=1}^{m} y_j}{m}$$

and σ_i is the standard deviation of this average value. If the distribution of error in the observed ordinates follows the normal (Gaussian) error curve [57],

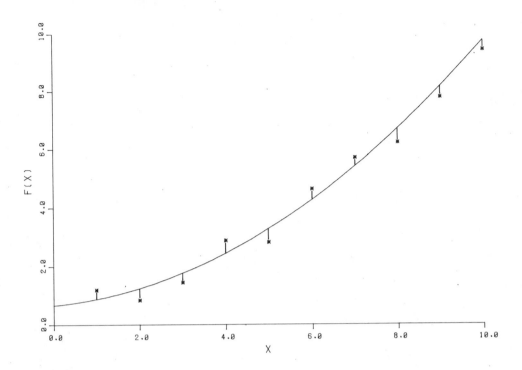

<u>Figure 5.1</u> Quadratic fit to a set of data points. The residuals are indicated by
the short vertical lines.

then relative weighting factors, approximated by

$$\omega_i = \frac{1}{\sigma_i^2}$$

can be assigned in computing the minimum in the residual function

$$R_i = \omega_i [y_i - F(x_i)]^2$$

This method of weighting is referred to as <u>instrumental</u> <u>weighting</u>. When standard
deviations are not available, it may be useful to use <u>statistical</u> <u>weighting</u> [58].

$$\omega_i = \frac{1}{y_i}$$

The weighting factor is frequently ignored completely by assuming unit weighting,

$$\omega_i = 1.0 \qquad (i = 1,2,3,\ldots)$$

5.1 LINEAR REGRESSION ANALYSIS

One of the most frequently used computer applications in chemistry is linear regression analysis. Here the "fitting function" is the straight line

$$y_i = a_0 + a_1 x_i$$

The least-squares procedure is to minimize the sum of the weighted residual as follows:

$$R = \sum_{i=1}^{N} \omega_i [y_i - (a_0 + a_1 x_i)]^2$$

$$\frac{\partial R}{\partial a_0} = 0 = 2 \sum_{i=1}^{N} \omega_i [y_i - (a_0 + a_1 x_i)](-1)$$

$$\frac{\partial R}{\partial a_1} = 0 = 2 \sum_{i=1}^{N} \omega_i [y_i - (a_0 + a_1 x_i)] (-x_i)$$

These equations define the following linear system of order 2:

$$\sum_{i=1}^{N} \omega_i a_0 + \sum_{i=1}^{N} \omega_i x_i a_1 = \sum_{i=1}^{N} \omega_i y_i$$

$$\sum_{i=1}^{N} \omega_i x_i a_0 + \sum_{i=1}^{N} \omega_i x_i^2 a_1 = \sum_{i=1}^{N} \omega_i x_i y_i$$

Here N is the number of data points. Solving by Cramer's rule (Section 4.1),

$$a_0 = \frac{\sum_{i=1}^{N} \omega_i y_i \sum_{i=1}^{N} \omega_i x_i^2 - \sum_{i=1}^{N} \omega_i x_i y_i \sum_{i=1}^{N} \omega_i x_i}{D}$$

$$a_1 = \frac{\sum_{i=1}^{N} \omega_i \sum_{i=1}^{N} \omega_i x_i y_i - \sum_{i=1}^{N} \omega_i x_i \sum_{i=1}^{N} \omega_i y_i}{D}$$

where

$$D = \sum_{i=1}^{N} \omega_i \sum_{i=1}^{N} \omega_i x_i^2 - \left(\sum_{i=1}^{N} \omega_i x_i \right)^2 \tag{5.1}$$

The variance of the fit is given by [59]

$$\sigma^2 = \frac{R}{N - 2} \tag{5.2}$$

and the standard deviations of the intercept (σ_{a_0}) and slope (σ_{a_1}) are

$$\sigma_{a_0}^2 = \frac{\sigma^2 \sum_{i=1}^{N} \omega_i x_i^2}{D}$$

$$\sigma_{a_1}^2 = \frac{\sigma^2 \sum_{i=1}^{N} \omega_i}{D}$$

The correlation coefficient is [60]

$$\rho = \left| \frac{N \sum_{i=1}^{N} \omega_i x_i y_i - \sum_{i=1}^{N} \omega_i x_i \sum_{i=1}^{N} \omega_i y_i}{\left[\left\{ N \sum_{i=1}^{N} \omega_i x_i^2 - \left(\sum_{i=1}^{N} \omega_i x_i \right)^2 \right\} \left\{ N \sum_{i=1}^{N} \omega_i y_i^2 - \left(\sum_{i=1}^{N} \omega_i y_i \right)^2 \right\} \right]^{1/2}} \right|$$

The correlation coefficient has the value 1.0 for a perfect linear correlation between x and y and 0 for no linear correlation. If unit weighting is used,

$$\sum_{i=1}^{N} \omega_i = N$$

Problem 5.1. Perform a linear regression analysis on the following calibration
             data [61]:

x	4.0	8.0	12.5	16.0	20.0	25.0	31.0	35.0	40.0	40.0
y	3.7	7.8	12.1	15.6	19.8	24.5	31.1	35.5	39.4	39.5

Calculate the intercept, slope, standard deviations in the intercept
and slope, the variance of the fit, and the correlation coefficient.
Prepare a table of calculated and observed values and the differences
between them. Plot the calculated linear function and the observed
points on the same graph.

Solution. Assume unit weighting.

$$N = 10 \qquad\qquad \sum_{i=1}^{N} \omega_i = 10$$

$$\sum_{i=1}^{N} \omega_i x_i = 232.5 \qquad\qquad \sum_{i=1}^{N} \omega_i y_i = 229.0$$

$$\sum_{i=1}^{N} \omega_i x_i y_i = 6884.65 \qquad\qquad \sum_{i=1}^{N} \omega_i x^2 = 6974.25$$

$$\sum_{i=1}^{N} \omega_i y_i^2 = 6796.66$$

$$D = 10(6974.25) - 232.5^2 = 15686.25$$

$$a_0 = \frac{229.0(6974.25) - 6884.65(232.5)}{15686.25} = -0.2281$$

$$a_1 = \frac{10(6884.65) - 232.5(229.0)}{15686.25} = 0.9948$$

$$\sigma^2 = 0.04273$$

$$\sigma_{a_0} = \left[\frac{0.04273(6974.25)}{15686.25}\right]^{1/2} = 0.0138$$

$$\sigma_{a_1} = \left[\frac{0.04273(10)}{15686.25}\right]^{1/2} = 0.000522$$

$$\rho = \left| \frac{10(6884.65) - 232.5(229)}{([10(6974.25) - 232.5^2][10(6796.66) - (229.0^2)])^{1/2}} \right| = 0.9999$$

The table and figure are included as part of the documentation of the following program.

5.1.1 LINEAR

Introduction

LINEAR performs linear regression analysis. The input includes the number of data points (N), a weighting option (IWT), and the data (x_i, y_i; i = 1,2,3,...,N). Three weighting options are provided:

1. Unit weighting (IWT = 1), $\omega_i = 1.0$

2. Statistical weighting (IWT = 0), $\omega_i = 1/y_i$

3. Instrumental weighting (IWT = -1), $\omega_i = 1/\sigma_i^2$

Here σ_i is the measured standard deviation in the dependent variable y_i. The weights are assigned by the program if either unit or statistical weighting is used. LINEAR calculates the intercept and slope of the regression line, the standard deviations of these parameters, the variance of the fit, the correlation coefficient, and a table containing observed and calculated ordinates and the corresponding differences.

Method

LINEAR uses Cramer's rule (Section 4.1) to find the regression coefficients. Double precision is used to minimize round-off errors. Four tests for zero divisors are made:

1. If statistical weighting is used, $y_i \neq 0$.

2. If instrumental weighting is used, $\sigma_i \neq 0$.

3. $D \neq 0$ [Eq. (5.1)].

4. $N - 2 \neq 0$ [Eq. (5.2)].

LINEAR first reads N and IWT. If IWT is 0 or 1, the data, $(x_i$ and y_i, i = 1,2, 3,...,N), are read and the weights are assigned. If IWT = -1, then LINEAR reads x_i, y_i, and σ_i, and the weights are assigned the values $1/\sigma_i^2$. The required sums are accumulated in the 100 loop, the regression parameters are calculated, and the results including a table showing the agreement between observed and calculated ordinates are printed.

Sample Execution

```
 ENTER N        ( N < 3 STOPS PGM. )
>10

 ENTER THE INTEGER CORRESPONDING TO THE DESIRED
 WEIGHTING OPTION AS FOLLOWS:

     1    FOR UNIT WEIGHTING          (W(I)=1.0)
     0    FOR STATISTICAL WEIGHTING   (W(I)=1./Y(I) )
    -1    FOR INSTRUMENTAL WEIGHTING  (W(I)=1./SIGMA(I)**2)

>1

 ENTER 10 DATA PAIRS: (X(I),Y(I), I=1 TO 10)
>4 3.7
>8 7.8
>12.5 12.1
>16 15.6
>20 19.8
>25 24.5
>31 31.1
>36 35.5
>40 39.4
>40 39.5

        UNIT WEIGHTING, W(I)=1.0

  FOR   Y = A + BX,  A =   -2.281D-01 +/-   1.38D-01

                     B =    9.948D-01 +/-  5.22D-03

   THE VARIANCE OF THE FIT IS    4.273D-02

   AND THE CORRELATION COEFFICIENT IS    9.999D-01
```

I	X(I)	W(I)	Y(I)	A+B*X(I)	DIFF
1	4.000	1.000	3.700	3.751	-0.5094E-01
2	8.000	1.000	7.800	7.730	0.7004E-01
3	12.50	1.000	12.10	12.21	-0.1064
4	16.00	1.000	15.60	15.69	-0.8801E-01
5	20.00	1.000	19.80	19.67	0.1330
6	25.00	1.000	24.50	24.64	-0.1408
7	31.00	1.000	31.10	30.61	0.4906
8	36.00	1.000	35.50	35.58	-0.8315E-01
9	40.00	1.000	39.40	39.56	-0.1622
10	40.00	1.000	39.50	39.56	-0.6217E-01

Listing

```
C
C                         LINEAR
C
C     PROGRAM TO FIT Y(X) TO   Y = A + B*X
C
C     INPUT SPECIFICATIONS:
C
C         N - THE NUMBER OF DATA POINTS    ( N .LE. 40 )
C
C         IWT - THE WEIGHTING OPTION:
C
C                 +1 FOR UNIT WEIGHTING   (W(I)=1.0,I=1,N)
C                  0 FOR STATISTICAL WEIGHTING  ( W(I)=1./Y(I) )
C                 +1 FOR UNIT WEIGHTING   (W(I)=1.0,I=1,N)
C                  0 FOR STATISTICAL WEIGHTING  ( W(I)=1./Y(I) )
C                 -1 FOR INSTRUMENTAL WEIGHTING ( W(I)=1./SIGMA(I)**2 )
C                     WHERE SIGMA(I) IS THE STD. DEV. OF Y(I)
C
C         (Y(I), X(I), I = 1,N ) - THE DATA POINTS
C
C         SIGMA(I) IS ALSO READ WITH INSTRUMENTAL WEIGHTING
C
C         A VALUE OF N < OR = 2 TERMINATES THE PROGRAM
C
C   AUTHOR:  K. J. JOHNSON
C
C
      IMPLICIT DOUBLE PRECISION (S)
      DIMENSION Y(50), X(50),W(50)
C
C    INPUT
C
   10 WRITE(6,6)
   12 READ(5,11)N
      IF(N.LE.2) STOP
      WRITE(6,22)
      READ(5,11)IWT
      IF(IWT.LT.0)GO TO 50
C
C   HAVE UNIT OR STATISTICAL WEIGHTING
C
      WRITE(6,33)N,N
      DO 20 I=1,N
      READ(5,55)X(I),Y(I)
      W(I)=1.
```

```
         IF(IWT.EQ.1)GO TO 20
         IF(Y(I).EQ.0.)GO TO 300
         W(I)=1./Y(I)
   20 CONTINUE
         GO TO 70
   50 CONTINUE
C
C    HAVE INSTRUMENTAL WEIGHTING
C
         WRITE(6,44) N,N
         DO 30 I = 1,N
         READ(5,55) X(I),Y(I),W(I)
         IF(W(I).EQ.0.0)W(I)=1.0
   30 W(I)=1./(W(I)*W(I))
C
C    COMPUTE THE APPROPRIATE SUMS
C
   70 SUM=0.
         SUMX=0.
         SUMY=0.
         SUMXY=0.
         SUMXX=0.
         SUMYY=0.
         DO 100 I=1,N
         SUM=SUM+W(I)
         SUMX=SUMX+X(I)*W(I)
         SUMY=SUMY+Y(I)*W(I)
         SUMXY=SUMXY+X(I)*Y(I)*W(I)
         SUMXX=SUMXX+X(I)*X(I)*W(I)
  100 SUMYY=SUMYY+Y(I)*Y(I)*W(I)
C
C    CALCULATE THE REGRESSION PARAMETERS
C
         SDELT=SUM*SUMXX-SUMX*SUMX
         SA=(SUMXX*SUMY-SUMX*SUMXY)/SDELT
         SB=(SUM*SUMXY-SUMX*SUMY)/SDELT
         SRHO=DABS((SUM*SUMXY-SUMX*SUMY)/
        1DSQRT(DABS(SDELT*(SUM*SUMYY-SUMY*SUMY))))
         SIGMA=(SUMYY-SA*SUMY-SB*SUMXY)/(N-2.)
         SIGMA=SIGMA/(N-2.0)
         STDA=DSQRT(SIGMA*SUMXX/SDELT)
         STDB=DSQRT(SIGMA*SUM/SDELT)
C
C    OUTPUT
C
         IF(IWT.EQ.-1)WRITE(6,56)
         IF(IWT.EQ.0) WRITE(6,57)
         IF(IWT.EQ.1) WRITE(6,58)
         WRITE(6,66)SA,STDA,SB,STDB,SIGMA,SRHO
         WRITE(6,77)
         DO 200 I = 1,N
         Z = SA + SB * X(I)
         DIFF = Y(I) - Z
  200 WRITE(6,88) I,X(I),W(I),Y(I),Z,DIFF
         GO TO 10
  300 WRITE(6,99)
         GO TO 10
    6 FORMAT(//'  ENTER N      ( N < 3 STOPS PGM. )'/)
   11 FORMAT(I)
```

```
   22 FORMAT(/'  ENTER THE INTEGER CORRESPONDING TO THE DESIRED',/
      1 '  WEIGHTING OPTION AS FOLLOWS:'//,
      2 '         1    FOR UNIT WEIGHTING          (W(I)=1.0)',/,
      3 '         0    FOR STATISTICAL WEIGHTING   (W(I)=1./Y(I) )'/,
      4 '        -1    FOR INSTRUMENTAL WEIGHTING (W(I)=1./SIGMA(I)**2)'/)
   33 FORMAT(/,' ENTER ',I2,' DATA PAIRS: (X(I),Y(I), I=1 TO ',I2,')'/)
   44 FORMAT(/' ENTER ',I2,' DATA TRIPLETS: (X(I),Y(I),W(I), I=1,',
      1  I2,')'/)
      FORMAT(3F)
   56 FORMAT(//'          INSTRUMENTAL WEIGHTING, W(I)=1./(SIGMA(I)**2)')
   57 FORMAT(//'          STATISTICAL WEIGHTING, W(I)=1./Y(I)')
    8 FORMAT(//'          UNIT WEIGHTING, W(I)=1.0')
   66 FORMAT(//'  FOR    Y = A + BX,   A = ',1PD12.3,
      1 ' +/- ',1PD10.2,//21X,'B = ',1PD12.3,' +/-',1PD10.2,
      2 //'  THE VARIANCE OF THE FIT IS ',1PD12.3,
      3 //'  AND THE CORRELATION COEFFICIENT IS ',1PD12.3//)
   77 FORMAT(//'  I',5X,'X(I)',9X,'W(I)',9X,'Y(т)',
      17X,'A+B*X(I)',7X,'DIFF'/)
   88 FORMAT(I3,5G13.4)
   99 FORMAT(//'  TERMINAL ERROR!  ATTEMPT TO DIVIDE BY 0. ',
      1 '  STARTING OVER.'/)
      END
```

<u>Discussion</u>

LINEAR requires 1354 words of core memory. This sample execution took approximately 0.2 sec. The linear fit is illustrated in Figure 5.2.

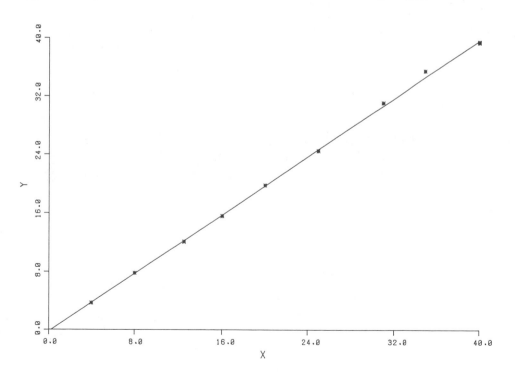

Figure 5.2 LINEAR's solution to Problem 5.1.

An optimization feature is worthy of note. A separate loop is not required to compute the variance of the fit [Eq. (5.2)]. The calculation can be done in one statement by completing the square and using terms in the expression already computed.

$$\sigma^2 = \frac{R}{N-2}$$

$$= \frac{\sum\limits_{i=1}^{N} \omega_i [y_i - (a_0 + a_1 x_i)]^2}{N-2}$$

$$= \frac{\sum\limits_{i=1}^{N} \omega_i y_i^2 - a_0 \sum\limits_{i=1}^{N} y_i - a_1 \sum\limits_{i=1}^{N} x_i y_i}{N-2}$$

5.2 POLYNOMIAL REGRESSION ANALYSIS

There are a number of reasons why scientists and engineers represent data with polynomials,

$$F(x) = a_0 + a_1 x + a_2 x^2 + \cdots + a_m x^m$$

If the fitting function closely approximates the data, then it may be used for interpolation, differentiation, integration, and with extreme caution, extrapolation. The coefficients of the polynomial, a_0, a_1, ..., a_m, are determined by minimization of the weighted residual function,

$$R = \sum\limits_{i=1}^{N} \omega_i [y_i - F(x_i)]^2$$

Here N is the number of data points (x_i and y_i, i = 1,2,3,...,N). This is done by taking partial derivatives of the residual function with respect to each of the coefficients of the polynomial,

$$\frac{\partial R}{\partial a_j} = 0 \qquad (j = 0,1,2,...,m)$$

This results in a linear system of equations of order m + 1:

$$CA = B$$

where C is the coefficient matrix of order n + 1:

$$
C = \begin{bmatrix}
\sum_{i=1}^{N} \omega_i & \sum_{i=1}^{N} \omega_i x_i & \cdots & \sum_{i=1}^{N} \omega_i x_i^{m} \\[2ex]
\sum_{i=1}^{N} \omega_i x_i & \sum_{i=1}^{N} \omega_i x_i^{2} & \cdots & \sum_{i=1}^{N} \omega_i x_i^{m+1} \\[2ex]
\sum_{i=1}^{N} \omega_i x_i^{2} & \sum_{i=1}^{N} \omega_i x_i^{3} & \cdots & \sum_{i=1}^{N} \omega_i x_i^{m+2} \\[2ex]
\cdots\cdots\cdots & & & \\[1ex]
\sum_{i=1}^{N} \omega_i x_i^{m} & \sum_{i=1}^{N} \omega_i x_i^{m+1} & \cdots & \sum_{i=1}^{N} \omega_i x_i^{2m}
\end{bmatrix}
$$

The constant vector B and the unknown vector A are

$$
B = \begin{pmatrix}
\sum_{i=1}^{N} \omega_i y_i \\[2ex]
\sum_{i=1}^{N} \omega_i x_i y_i \\[2ex]
\sum_{i=1}^{N} \omega_i x_i^{2} y_i \\[2ex]
\cdots\cdots \\[1ex]
\sum_{i=1}^{N} \omega_i x_i^{m} y_i
\end{pmatrix}
\qquad
A = \begin{pmatrix}
a_0 \\[2ex]
a_1 \\[2ex]
a_2 \\[2ex]
\cdots \\[1ex]
a_m
\end{pmatrix}
$$

A is determined by inverting the coefficient matrix,

$$A = C^{-1}B$$

The variance of the fit and the standard deviations of the coefficients are given by the following equations [62, 63]:

$$\sigma^2 = \frac{R}{N - m - 1}$$

$$\sigma_{a_j^2}^2 = \sigma^2 C_{jj}^{-1} \qquad (j = 1,2,3,\ldots,m)$$

Here C_{jj}^{-1} is the jth diagonal element of the inverse of the coefficient matrix.

Problem 5.2. Given the following data [63],

x	y	x	y
0.23	5.64	5.36	16.21
1.01	7.83	5.51	14.57
2.29	17.04	6.36	0.78
2.87	21.38	6.84	-7.64
4.15	24.56	7.0	-12.52

and assuming unit weighting, find the coefficients, variance of the
fit, and the standard deviations of the coefficients for polynomial
fitting functions with m = 2, 3, 4, 5, and 6. Plot the results for
the fitting function that gives the minimum value of the variance.

Solution. This problem is solved in the sample execution of the following
program.

5.2.1 POLREG

Introduction

POLREG performs polynomial regression analysis. The fitting function is

$$F(x) = a_0 + a_1 x + a_2 x^2 + \cdots + a_m x^m$$

The input includes N, the number of data points; m, the order of the polynomial;
a weighting option; and the data (x_i and y_i, i = 1,2,3,...,N). The weighting
option is the same as that used in LINEAR (Section 5.1.1). The program prints the
following: the determinant of the coefficient matrix; a table containing x_i, ω_i,
y_i, $F(x_i)$, and the difference $y_i - F(x_i)$; the variance of the fit; and the m + 1
regression coefficients with the corresponding standard deviations.

Method

The coefficient matrix is of order m + 1. However, it contains only 2m + 1
unique elements. For example, let m = 3 and IWT = 1. The 4 x 4 coefficient
matrix, with unit weighting, is

$$C = \begin{bmatrix} N & \sum_{i=1}^{N} x_i & \sum_{i=1}^{N} x_i^2 & \sum_{i=1}^{N} x_i^3 \\ \sum_{i=1}^{N} x_i & \sum_{i=1}^{N} x_i^2 & \sum_{i=1}^{N} x_i^3 & \sum_{i=1}^{N} x_i^4 \\ \sum_{i=1}^{N} x_i^2 & \sum_{i=1}^{N} x_i^3 & \sum_{i=1}^{N} x_i^4 & \sum_{i=1}^{N} x_i^5 \\ \sum_{i=1}^{N} x_i^3 & \sum_{i=1}^{N} x_i^4 & \sum_{i=1}^{N} x_i^5 & \sum_{i=1}^{N} x_i^6 \end{bmatrix}$$

There are seven, 2x3+1, unique elements in C: C_{11}, C_{12}, C_{13}, C_{14}, C_{24}, C_{34}, and C_{44}.
POLREG computes only these elements and assigns the rest. The diagonal elements
and the remaining elements in the upper-right triangle,

$$C_{ij} \qquad i \leq j$$

are assigned using

$$C_{ij} = C_{i-1,j+1}$$

For example, $C_{22} = C_{13}$, $C_{23} = C_{14}$, and $C_{33} = C_{24}$. The elements in the lower-left triangle are assigned by symmetry,

$$C_{ij} = C_{ji} \qquad i > j$$

For example, $C_{21} = C_{12}$, $C_{32} = C_{23}$, etc.

POLREG uses double-precision variable names to minimize round-off errors. The coefficient matrix is inverted using a double-precision version of MATINV (Section 4.3.3).

Sample Execution (See Problem 5.2)

```
ENTER N, THE NUMBER OF DATA POINTS AND
M,THE DEGREE OF THE POLYNOMIAL
 NOTE: M < 10 AND N > M+1
>10 2

 ENTER THE INTEGER CORRESPONDING TO THE DESIRED
 WEIGHTING OPTION AS FOLLOWS:

     1    FOR UNIT WEIGHTING          (W(I)=1.0)
     0    FOR STATISTICAL WEIGHTING   (W(I)=1./Y(I) )
    -1    FOR INSTRUMENTAL WEIGHTING  (W(I)=1./SIGMA(I)**2)
>1

ENTER  10 DATA PAIRS: (X(I),Y(I), I=1 TO  10)
>0.23 5.64
>1.01 7.83
>2.29 17.04
>2.87 21.38
>4.15 24.56
>5.36 16.21
>5.51 14.57
>6.36 0.78
>6.84 -7.64
>7.00 -12.52
```

THE DETERMINANT OF A IS   9.25D+04

I	X(I)	W(I)	Y(I)	YCALC	DIFF
1	2.30D-01	1.00D+00	5.640D+00	1.103D+00	4.54D+00
2	1.01D+00	1.00D+00	7.830D+00	1.115D+01	-3.32D+00
3	2.29D+00	1.00D+00	1.704D+01	2.126D+01	-4.22D+00
4	2.87D+00	1.00D+00	2.138D+01	2.324D+01	-1.86D+00
5	4.15D+00	1.00D+00	2.456D+01	2.185D+01	2.71D+00
6	5.36D+00	1.00D+00	1.621D+01	1.326D+01	2.95D+00
7	5.51D+00	1.00D+00	1.457D+01	1.170D+01	2.87D+00
8	6.36D+00	1.00D+00	7.800D-01	8.244D-01	-4.44D-02
9	6.84D+00	1.00D+00	-7.640D+00	-6.861D+00	-7.79D-01
10	7.00D+00	1.00D+00	-1.252D+01	-9.669D+00	-2.85D+00

```
     THE VARIANCE OF THE FIT =        1.23D+01

        FOR A0 + A1*X + ...

        A( 0)=   -2.420D+00     +/-           3.22D+00
        A( 1)=    1.587D+01     +/-           2.06D+00
        A( 2)=   -2.415D+00     +/-           2.67D-01
```

2. Cubic Fit (Input and table omitted.)

```
     THE VARIANCE OF THE FIT =        1.50D+00

         FOR A0 + A1*X + ...

        A( 0)=    3.687D+00     +/-           1.41D+00
        A( 1)=    4.239D+00     +/-           1.78D+00
        A( 2)=    1.635D+00     +/-           5.73D-01
        A( 3)=   -3.681D-01     +/-           5.14D-02
```

3. Quartic fit.

```
     THE VARIANCE OF THE FIT =        2.57D-01

         FOR A0 + A1*X + ...

        A( 0)=    6.316D+00     +/-           7.57D-01
        A( 1)=   -4.230D+00     +/-           1.71D+00
        A( 2)=    6.955D+00     +/-           1.00D+00
        A( 3)=   -1.493D+00     +/-           2.07D-01
        A( 4)=    7.608D-02     +/-           1.39D-02
```

4. Fifth-order fit.

```
     THE VARIANCE OF THE FIT =        3.19D-01

        FOR A0 + A1*X + ...

        A( 0)=    6.216D+00     +/-           1.04D+00
        A( 1)=   -3.789D+00     +/-           3.30D+00
        A( 2)=    6.523D+00     +/-           2.86D+00
        A( 3)=   -1.332D+00     +/-           1.01D+00
        A( 4)=    5.089D-02     +/-           1.54D-01
        A( 5)=    1.402D-03     +/-           8.55D-03
```

5. Sixth-order fit.

```
     THE VARIANCE OF THE FIT =        3.50D-01

         FOR A0 + A1*X + ...

        A( 0)=    5.475D+00     +/-           1.43D+00
        A( 1)=    4.717D-01     +/-           6.33D+00
        A( 2)=    5.959D-01     +/-           7.96D+00
        A( 3)=    1.945D+00     +/-           4.22D+00
        A( 4)=   -7.955D-01     +/-           1.07D+00
        A( 5)=    1.039D-01     +/-           1.28D-01
        A( 6)=   -4.707D-03     +/-           5.86D-03
```

Listing

```
C
C                      POLREG
C
C        POLREG FITS DATA TO THE POLYNOMIAL
C
C            Y(I)= A0 + A1*X(I) +A2*X(I)**2 +...+AM*X(I)**M
C
C        BY THE METHOD OF LEAST SQUARES.
C
C        INPUT SPECIFICATIONS:
C
C            N      NUMBER OF DATA POINTS
C            M      DEGREE OF POLYNOMIAL
C
C          X(I),Y(I)     CO-ORDINATES OF DATA POINTS
C
C          IWT - THE WEIGHTING OPTION:
C
C                +1 FOR UNIT WEIGHTING   (W(I)=1.0,I=1,N)
C                 0 FOR STATISTICAL WEIGHTING  ( W(I)=1./Y(I) )
C                -1 FOR INSTRUMENTAL WEIGHTING ( W(I)=1./SIGMA(I)**2 )
C                    WHERE SIGMA(I) IS THE STD. DEV. OF Y(I)
C
C     POLREG CALLS THE SUBROUTINE MATINV(A,N,D)
C
C
C        WHERE A IS THE COEFFICIENT MATRIX
C              N IS THE ORDER OF A (M+1)
C        AND   D IS THE DETERMINANT OF THE MATRIX A
C
C            A(-1) IS RETURNED IN A
C
C     AUTHORS: J. A. BUCHINO AND K. J. JOHNSON
C
      IMPLICIT DOUBLE PRECISION (A-H,O-Z)
      DIMENSION X(100),Y(100),W(100),A(11,11),B(11),ALFA(11)
      DATA W,A,B,ALFA,SMSQ/100*1.0D0,144*0.0D0/
C
C     INPUT
C
      WRITE(6,6)
      READ(5,11)N,M
      MP1=M+1
      IF(N.LE.2 .OR. N.LE. MP1 .OR. M .GT. 10 )STOP
      WRITE(6,22)
      READ(5,11)IWT
      IF(IWT.LT.0)GO TO 25
C
C   HAVE UNIT OR STATISTICAL WEIGHTING
C
      WRITE(6,33)N,N
      DO 10 I=1,N
      READ(5,55)X(I),Y(I)
      IF(IWT.EQ.1)GO TO 10
      IF(Y(I).EQ.0.0D0)GO TO 200
      W(I)=1./Y(I)
   10 CONTINUE
```

```
          GO TO 30
     25 CONTINUE
C
C      HAVE INSTRUMENTAL WEIGHTING
C
          WRITE(6,44) N,N
          DO 28 I = 1,N
          READ(5,55) X(I),Y(I),W(I)
          IF(W(I).EQ.0.0D0)W(I)=1.0D0
     28 W(I)=1./(W(I)*W(I))
C
C      AX=B  -    CALCULATE B
C
     30 DO 40 I=1,N
          B(1)=B(1)+W(I)*Y(I)
          XYVAL=Y(I)
          DO 40 J=2,MP1
          XYVAL=XYVAL*X(I)
     40 B(J)=B(J)+XYVAL*W(I)
C
C        CALCULATE THE 2M+1 UNIQUE ELEMENTS OF A. THIS INCLUDES THE
C        UNIQUE OFF DIAGONAL ELEMENTS AND THE FIRST AND LAST DIAGONAL
C        ELEMENTS (A(1,1) AND A(M+1,M+1)) OF A.
C
C
          DO 50 I=1,N
          A(1,1)=A(1,1)+W(I)
          XVAL=1.0D0
          DO 50 J=2,2*M+1
          XVAL=XVAL*X(I)
          IOVER=J-MP1
          IF(IOVER.GT.0)GO TO 45
          A(1,J)=A(1,J)+W(I)*XVAL
          GO TO 50
     45 A(IOVER+1,MP1)=A(IOVER+1,MP1)+W(I)*XVAL
     50 CONTINUE
C
C        ASSIGN THE REMAINING UPPER RIGHT TRIANGLE (INCLUDING
C        DIAGONAL ELEMENTS)
C
          IF(M.LE.1.)GO TO 65
          DO 60 I=2,M
          DO 60 J=I,M
     60 A(I,J)=A(I-1,J+1)
C
C        ASSIGN SYMMETRIC ELEMENTS OF A
C
     65 DO 70 J=1,M
          DO 70 I=J+1,MP1
     70 A(I,J)=A(J,I)
C
C        INVERT THE COEFFICIENT MATRIX
C
          WRITE(21,211)((A(I,J),J=1,MP1),I=1,MP1)
    211 FORMAT(6D12.3)
          CALL MATINV(A,MP1,D)
          WRITE(6,66)D
          IF(D.EQ.0.)GO TO 98
```

```
C
C        FIND THE COEFFICIENT VECTOR, ALFA.
C
         DO 80 I=1,MP1
         DO 80 J=1,MP1
   80    ALFA(I)=ALFA(I)+A(I,J)*B(J)
C
C        CALCULATE THE VARIANCE AND PRINT THE TABLE
C
         DO 100 I=1,N
         YCALC=ALFA(1)
         XVAL=1.
         DO 90 J=2,MP1
         XVAL=XVAL*X(I)
   90    YCALC=YCALC+ALFA(J)*XVAL
         DIFF=Y(I)-YCALC
         WRITE(6,144)I,X(I),W(I),Y(I),YCALC,DIFF
  100    SMSQ=SMSQ+DABS(W(I))*DIFF*DIFF
         VAR=SMSQ/(N-MP1)
         SIGMA=DSQRT(VAR)
         WRITE(6,99)VAR
         DO 110 J=1,MP1
         SIGALF=SIGMA*DSQRT(DABS(A(J,J)))
         I=J-1
         SGLALF=ALFA(J)
  110    WRITE(6,77)(I,SGLALF,SIGALF)
         STOP
   98    WRITE(6,88)
         STOP
  200    WRITE(6,188)
    6    FORMAT(///' ENTER N, THE NUMBER OF DATA POINTS AND'
        1/' M,THE DEGREE OF THE POLYNOMIAL    ',/,
        2 '   NOTE: M < 10 AND N > M+1'//)
   11    FORMAT(2I)
   22    FORMAT(/'  ENTER THE INTEGER CORRESPONDING TO THE DESIRED',/
        1 ' WEIGHTING OPTION AS FOLLOWS:'//,
        2 '      1    FOR UNIT WEIGHTING          (W(I)=1.0)',/,
        3 '      0    FOR STATISTICAL WEIGHTING   (W(I)=1./Y(I) )'/,
        4 '     -1    FOR INSTRUMENTAL WEIGHTING (W(I)=1./SIGMA(I)**2)'/)
   33    FORMAT(/,' ENTER ',I3,' DATA PAIRS: (X(I),Y(I), I=1 TO ',I3,
        1 ')'/)
   44    FORMAT(/' ENTER ',I2,' DATA TRIPLETS: (X(I),Y(I),W(I), I=1,',
        1 I2,')'/)
   55    FORMAT(3D)
   66    FORMAT(//' THE DETERMINANT OF A IS ',1PD11.2//
        1 '  I',8X,'X(I)',8X,'W(I)',8X,'Y(I)',9X,'YCALC',
        2 8X,'DIFF'/)
   77    FORMAT ('  A(',I2,')=',1PD12.3,'    +/-    ',1PD12.2)
   88    FORMAT(//' THE COEFFICIENT MATRIX IS SINGULAR.'//)
   99    FORMAT(//' THE VARIANCE OF THE FIT = ',1PD12.2/,
        1 /'  FOR A0 + A1*X + ...',/)
  144    FORMAT(I3,2(1PD13.2),2(1PD13.3),1PD13.2)
  188    FORMAT(/'  TERMINAL ERROR.  Y(I) CANNOT BE 0.0 ',
        1 '  USING STATISTICAL WEIGHTING'//)
  199    FORMAT(/' N IS LESS THAN OR EQUAL TO M+1.  THE VARIANCE OF',
        1/' THE FIT AND THE STANDARD DEVIATION CANNOT BE CALCULATED.'/)
         END
```

Discussion

POLREG (including MATINV) requires approximately 2000 words of core. The sample execution corresponding to m = 4 took approximately 0.4 sec.

Note that the variance of the fit reaches a minimum at m = 4.

m	2	3	4	5	6
σ^2	12.3	1.5	0.26	0.32	0.35

The output corresponding to the quartic fit is

THE DETERMINANT OF A IS   6.98D+10

I	X(I)	W(I)	Y(I)	YCALC	DIFF
1	2.30D-01	1.00D+00	5.640D+00	5.693D+00	-5.27D-02
2	1.01D+00	1.00D+00	7.830D+00	7.679D+00	1.51D-01
3	2.29D+00	1.00D+00	1.704D+01	1.726D+01	-2.23D-01
4	2.87D+00	1.00D+00	2.138D+01	2.133D+01	5.24D-02
5	4.15D+00	1.00D+00	2.456D+01	2.439D+01	1.67D-01
6	5.36D+00	1.00D+00	1.621D+01	1.633D+01	-1.21D-01
7	5.51D+00	1.00D+00	1.457D+01	1.452D+01	5.21D-02
8	6.36D+00	1.00D+00	7.800D-01	1.108D+00	-3.28D-01
9	6.84D+00	1.00D+00	-7.640D+00	-8.501D+00	8.61D-01
10	7.00D+00	1.00D+00	-1.252D+01	-1.196D+01	-5.59D-01

THE VARIANCE OF THE FIT =   2.57D-01

FOR A0 + A1*X + ...

```
A( 0)=    6.316D+00    +/-        7.57D-01
A( 1)=   -4.230D+00    +/-        1.71D+00
A( 2)=    6.955D+00    +/-        1.00D+00
A( 3)=   -1.493D+00    +/-        2.07D-01
A( 4)=    7.608D-02    +/-        1.39D-02
```

These results are plotted in Figure 5.3. The effect of statistical rather than unit weighting on the regression analysis for m = 4 is shown in the following:

THE DETERMINANT OF A IS   -1.87D+06

I	X(I)	W(I)	Y(I)	YCALC	DIFF
1	2.30D-01	1.77D-01	5.640D+00	5.731D+00	-9.08D-02
2	1.01D+00	1.28D-01	7.830D+00	7.532D+00	2.98D-01
3	2.29D+00	5.87D-02	1.704D+01	1.746D+01	-4.21D-01
4	2.87D+00	4.68D-02	2.138D+01	2.163D+01	-2.54D-01
5	4.15D+00	4.07D-02	2.456D+01	2.449D+01	6.60D-02
6	5.36D+00	6.17D-02	1.621D+01	1.593D+01	2.77D-01
7	5.51D+00	6.86D-02	1.457D+01	1.408D+01	4.89D-01
8	6.36D+00	1.28D+00	7.800D-01	8.115D-01	-3.15D-02
9	6.84D+00	-1.31D-01	-7.640D+00	-8.276D+00	6.36D-01
10	7.00D+00	-7.99D-02	-1.252D+01	-1.146D+01	-1.06D+00

```
THE VARIANCE OF THE FIT =       3.84D-02

FOR A0 + A1X + ...

A( 0)=    6.549D+00    +/-        7.13D-01
A( 1)=   -5.266D+00    +/-        1.71D+00
A( 2)=    7.810D+00    +/-        9.49D-01
A( 3)=   -1.710D+00    +/-        1.76D-01
A( 4)=    9.275D-02    +/-        1.01D-02
```

The improvement in the variance of the fit, 0.038 for statistical weighting compared to 0.26 for unit weighting, is due to the inclusion of the weighting factors, all but one less than unity, in the calculation of the residual function.

The danger of round-off error in polynomial regression is illustrated by the test polynomial

$$y = 1 - x + x^2 - x^3 + x^4 - x^5 + x^6 - x^7 + x^8 - x^9$$

The following program was used to generate the "data" to test POLREG.

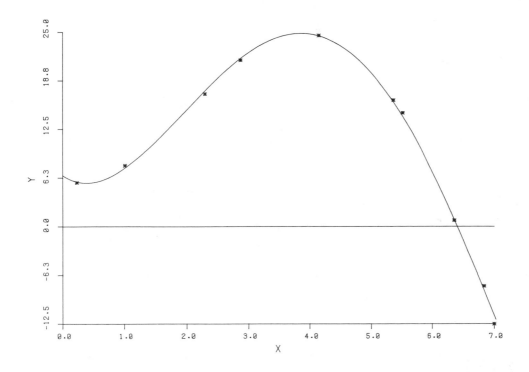

Figure 5.3. POLREG's solution to Problem 5.2. The smooth curve is the quartic fit.

```
        WRITE(26,11)X,Y
100 CONTINUE
 11 FORMAT(2D30.16)
    END
    DOUBLE PRECISION X,Y
    DO 100 I=1,100
    X=I*0.01D0
    Y=1.D0-X+X**2-X**3+X**4-X**5+X**6-X**7+X**8-X**9
```

The data, 100 X-Y pairs, were written on file FOR26.DAT. A segment of the output
follows:

```
0.1000000000000000D-01              0.9900990099009901D+00
0.2000000000000000D-01              0.9803921568627451D+00
0.3000000000000000D-01              0.9708737864077664D+00
0.4000000000000000D-01              0.9615384615384515D+00
                 :                                   :
                 :                                   :
                 :                                   :
0.9600000000000000D+00              0.1710037571471933D+00
0.9700000000000000D+00              0.1332872452310009D+00
0.9800000000000000D+00              0.9238747126891571D-01
0.9900000000000000D+00              0.4804920853828920D-01
0.1000000000000000D+01              0.0000000000000000D+00
```

POLREG returned the following results (input omitted):

```
  I        X(I)          W(I)          Y(I)         YCALC          DIFF

  1      1.00D-02      1.00D+00      9.901D-01      9.901D-01      3.02D-13
  2      2.00D-02      1.00D+00      9.804D-01      9.804D-01     -1.68D-13
  3      3.00D-02      1.00D+00      9.709D-01      9.709D-01     -5.17D-13
  4      4.00D-02      1.00D+00      9.615D-01      9.615D-01     -7.93D-13
  5      5.00D-02      1.00D+00      9.524D-01      9.524D-01     -1.04D-12
                                        :
                                        :  ( Rest of Output Omitted )

 95      9.50D-01      1.00D+00      2.058D-01      2.058D-01      8.38D-08
 96      9.60D-01      1.00D+00      1.710D-01      1.710D-01      9.02D-08
 97      9.70D-01      1.00D+00      1.333D-01      1.333D-01      9.71D-08
 98      9.80D-01      1.00D+00      9.239D-02      9.239D-02      1.04D-07
 99      9.90D-01      1.00D+00      4.805D-02      4.805D-02      1.12D-07
100      1.00D+00      1.00D+00      3.725D-08     -8.315D-08      1.20D-07
```

THE VARIANCE OF THE FIT =   1.15D-15

FOR A0 + A1*X + ...

```
A( 0)=    1.000D+00   +/-        4.44D-08
A( 1)=   -1.000D+00   +/-        2.36D-06
A( 2)=    1.000D+00   +/-        4.16D-05
A( 3)=   -1.000D+00   +/-        3.43D-04
A( 4)=    1.000D+00   +/-        1.55D-03
A( 5)=   -1.000D+00   +/-        4.13D-03
A( 6)=    1.000D+00   +/-        6.64D-03
A( 7)=   -1.000D+00   +/-        6.35D-03
A( 8)=    1.000D+00   +/-        3.32D-03
A( 9)=   -1.000D+00   +/-        7.29D-04
```

A single-precision version of POLREG returned the following results:

THE DETERMINANT OF A IS   -1.62D-28

I	X(I)	W(I)	Y(I)	YCALC	DIFF
1	1.00E-02	1.00E+00	9.901E-01	9.930E-01	-2.85E-03
2	2.00E-02	1.00E+00	9.804E-01	9.819E-01	-1.53E-03
3	3.00E-02	1.00E+00	9.709E-01	9.714E-01	-4.99E-04
4	4.00E-02	1.00E+00	9.615E-01	9.613E-01	2.87E-04
5	5.00E-02	1.00E+00	9.524E-01	9.515E-01	8.60E-04
		$\vdots$			
95	9.50E-01	1.00E+00	2.058E-01	-1.258E+00	1.46E+00
96	9.60E-01	1.00E+00	1.710E-01	-1.385E+00	1.56E+00
97	9.70E-01	1.00E+00	1.333E-01	-1.519E+00	1.65E+00
98	9.80E-01	1.00E+00	9.239E-02	-1.662E+00	1.75E+00
99	9.90E-01	1.00E+00	4.805E-02	-1.814E+00	1.86E+00
100	1.00E+00	1.00E+00	3.725E-08	-1.975E+00	1.97E+00

THE VARIANCE OF THE FIT =   3.66D-01

FOR A0 + A1*X + ...

A(0)=   1.005D+00   +/-   8.13D-01
A(1)=   -1.183D+00   +/-   2.97D+01
A(2)=   2.859D+00   +/-   3.04D+02
A(3)=   -8.750D+00   +/-   1.28D+03
A(4)=   1.606D+01   +/-   2.38D+03
A(5)=   -1.306D+01   +/-   1.04D+03
A(6)=   -2.531D+00   +/-   1.31D+03
A(7)=   5.031D+00   +/-   1.49D+03
A(8)=   1.219D+00   +/-   1.10D+03
A(9)=   -2.625D+00   +/-   4.40D+02

The single-precision version of POLREG read the data file using a "2E" rather than a "2D" format. This, plus the round-off errors that accumulate with the inversion of the 10 x 10 coefficient matrix, accounts for the poor fit. This example under-scores the requirement that extreme caution be used in regression analysis.

The method of polynomial regression discussed above sometimes fails because of problems associated with a nearly singular coefficient matrix. An alternative approach is to use orthogonal polynomials [64-66]. Let $P_i(x)$ and $P_j(x)$ be two polynomials in x. They are orthogonal if

$$\sum_{k=1}^{N} P_i(x_k)P_j(x_k) = 0 \qquad \text{for } i \neq j$$

The fitting function can be expressed either as a sum of polynomials,

$$F(x) = \sum_{j=0}^{m} a_j P_j(x) = a_0 P_0(x) + a_1 P_1(x) + \cdots + a_m P_m(x)$$

or by combining the terms in the polynomials, as a power series in x,

$$F(x) = C_0 + C_1 x + C_2 x^2 + \cdots + C_m x^m$$

The fact that the polynomials of the fitting function are orthogonal conveniently eliminates the off-diagonal elements of the coefficient matrix. Forsythe [64] has shown that a useful set of orthogonal polynomials is generated by the recursion formula

$$P_m(x) = (x - u_m)P_{m-1}(x) - v_{m-1}P_{m-2}(x)$$

with

$$P_{-1}(x) = 0 \qquad \text{and} \qquad P_0(x) = 1$$

then

$$P_1(x) = x - u_1$$

$$P_2(x) = (x - u_2)P_1(x) - v_1$$

$$P_3(x) = (x - u_3)P_2(x) - v_2 P_1(x)$$

$$\cdots\cdots\cdots\cdots\cdots\cdots\cdots\cdots\cdots$$

The coefficients u and v are calculated using the following equations:

$$u_m = \frac{\sum\limits_{i=1}^{N} x_i [P_{m-1}(x_i)]^2}{D_{m-1}}$$

$$v_{m-1} = \frac{\sum\limits_{i=1}^{N} x_i P_{m-1}(x_i) P_{m-2}(x_i)}{D_{m-2}}$$

where

$$D_m = \sum_{i=1}^{N} [P_m(x_i)]^2$$

Here all sums are from i = 1 to N, the number of data points. Given data (x_i and y_i, i = 1,2,3,...,N), these equations are used to define m polynomials, $P_j(x)$ (j = 1,2,...,m). The regression equation is

$$F(x) = \sum_{j=0}^{m} a_j P_j(x)$$

where

$$a_j = \frac{\sum\limits_{i=1}^{N} y_i P_j(x_i)}{D_j}$$

<u>Problem 5.3.</u> Fit the following data [65] to a quadratic function (m = 2). Assume
unit weighting.

x	0	1	2	3
y	-2	0	0	-2

<u>Solution.</u> n = 1

$$D_0 = \sum_{i=1}^{4} [P_0(x_i)]^2$$

$$u_1 = \frac{\sum\limits_{i=1}^{4} x_i [P_0(x_i)]^2}{D_0} = \frac{0 + 1 + 2 + 3}{4} = \frac{3}{2}$$

$$P_1(x) = x - \frac{3}{2}$$

m = 2

$$D_1 = \sum_{i=1}^{4} [P_1(x_i)]^2 = \sum_{i=1}^{4} (x_i - \frac{3}{2})^2$$

$$= (0 - \frac{3}{2})^2 + (1 - \frac{3}{2})^2 + (2 - \frac{3}{2})^2 + (3 - \frac{3}{2})^2 = 5$$

$$u_2 = \frac{\sum\limits_{i=1}^{4} x_i [P_1(x_i)]^2}{D_1} = \frac{\sum\limits_{i=1}^{4} x_i (x_i - \frac{3}{2})^2}{5} =$$

$$= \frac{0 + (1 - \frac{3}{2})^2 + 2(2 - \frac{3}{2})^2 + 3(3 - \frac{3}{2})^2}{5} = \frac{3}{2}$$

$$v_1 = \frac{\sum\limits_{i=1}^{4} x_i P_1(x_i) P_0(x_i)}{D_0} = \frac{\sum\limits_{i=1}^{4} x_i (x_i - \frac{3}{2})}{4}$$

$$= \frac{0 + (1 - \frac{3}{2}) + 2(2 - \frac{3}{2}) + 3(3 - \frac{3}{2})}{4} = \frac{5}{4}$$

$$P_2(x) = (x - u_2) P_1(x) - v_1 = (x - \frac{3}{2})(x - \frac{3}{2}) - \frac{5}{4}$$

$$= x^2 - 3x + 1$$

$$D_2 = \sum_{i=1}^{4} [P_2(x_i)]^2 = \sum_{i=1}^{4} (x_i^2 - 3x_i + 1)^2$$

$$= (0 - 3 + 1)^2 + (1 - 3 + 1)^2 + (4 - 12 + 1)^2 + (27 - 9 + 1)^2 = 4$$

$$a_0 = \frac{\sum_{i=1}^{4} y_i P_0(x_i)}{D_0} = \frac{-2 + 0 + 0 - 2}{4} = -1$$

$$a_1 = \frac{\sum_{i=1}^{4} y_i P_1(x_i)}{D_1} = \frac{\sum_{i=1}^{4} y_i (x_i - \frac{3}{2})}{D_1}$$

$$= \frac{-2(0 - \frac{3}{2}) + 0 + 0 - 2(3 - \frac{3}{2})}{5} = 0$$

$$a_2 = \frac{\sum_{i=1}^{4} y_i P_2(x_i)}{D_2} = \frac{\sum_{i=1}^{4} y_i (x_i^2 - 3x_i + 1)}{D_2}$$

$$= -2(0 - 0 + 1) + 0 + 0 - 2(9 - 9 + 1) = -1$$

Finally,

$$F(x) = a_0 P_u(x) + a_1 P_1(x) + a_2 P_2(x)$$

$$= -1(1) + 0(x - \frac{3}{2}) + (-1)(x^2 - 3x + 1) = -2 + 3x - x^2$$

It is a relatively straightforward procedure to calculate the coefficients u_i, v_i, and a_i. What is usually desired however is not a regression equation expressed as a sum of polynomials, but rather a power series in x. The coefficients of the power series can be calculated as follows:

$$F(x) = \sum_{k=0}^{m} c_{km} x^k$$

$$= c_{0m} + c_{1m} x + c_{2m} x^2 + \cdots + c_{mm} x^m$$

where

$$c_{km} = \sum_{j=k}^{m} a_j b_{kj}$$

and

$$b_{kj} \begin{cases} = 0 & \text{for } k < 0 \text{ or } k > j \\ = 1 & \text{for } k = j \\ = b_{k-1, j-1} - u_j b_{k, j-1} - v_{j-1} b_{k, j-2} & \text{for } 0 \le k < j \end{cases}$$

<u>Problem 5.4.</u> Use the values obtained in the solution to Problem 5.3 to fit the data to a quadratic function in x.

<u>Solution.</u>

$$F(x) = \sum_{k=0}^{2} c_{k2}{}^{k}$$

$$= c_{02} + c_{12}x + c_{22}x^{2}$$

Here

$$c_{02} = \sum_{j=0}^{2} a_{j}b_{0j}$$

$$= a_{0}b_{00} + a_{1}b_{01} + a_{2}b_{02}$$

$$c_{12} = \sum_{j=1}^{2} a_{j}b_{1j} = a_{1}b_{11} + a_{2}b_{12}$$

$$c_{22} = \sum_{j=2}^{2} a_{j}b_{2j} = a_{2}b_{22}$$

But

$$b_{00} = b_{11} = b_{22} = 1$$

and

$$b_{01} = b_{-10} - u_{1}b_{00} - v_{-1}b_{0-1}$$

$$= 0 - \frac{3}{2} - 0 = -\frac{3}{2}$$

$$b_{03} = b_{-11} - u_{2}b_{01} - v_{1}b_{00}$$

$$= 0 - \frac{3}{2}(-\frac{3}{2}) - \frac{5}{4}(1) = 1$$

$$b_{12} = b_{01} - u_{2}b_{11} - v_{1}b_{10}$$

$$= -\frac{3}{2} - \frac{3}{2} - 0 = -3$$

Therefore

$$c_{02} = a_{0}b_{00} + a_{1}b_{01} + a_{2}b_{02} = -1 + 0 + (-1) = -2$$

$$c_{12} = a_{1}b_{11} + a_{2}b_{12} = 0 + 3 = 3$$

and

$$c_{22} = a_{2}b_{22} = -1$$

Finally,

$$F(x) = \sum_{k=1}^{2} c_{km}x^k = c_{02} + c_{12}x + c_{22}x^2 = -2 + 3x - x^2$$

The following program uses this technique to fit data to polynomials in x.

Problem 5.5. Fit the following data [66] to a quadratic function using the Forsythe method.

x	y	x	y
0	-0.89	55	1.22
5	-0.69	60	1.45
10	-0.53	65	1.68
15	-0.34	70	1.88
20	-0.15	75	2.10
25	0.02	80	2.31
30	0.20	85	2.54
35	0.42	90	2.78
40	0.61	95	3.00
45	0.82	100	3.22

Solution. This problem is solved in the sample execution of the following program.

5.2.2 FORSY

Introduction

FORSY performs polynomial regression analysis using the Forsythe method discussed above. The input includes N, the number of data points; m', the order of the fitting polynomial; and the data (x_i and y_i, i = 1,2,...N). The output includes the coefficients of the fitting polynomial, the variance of the fit, a table containing x_i, y_i, and F_i, the values for y calculated using the fitting function, and the difference between each observed and calculated ordinate.

Method

The Forsythe method is described above. Three function subprograms are used: GETP(K,X), BCALC(M,K), and GETB(M,K). GETP(K,X) evaluates the polynomials,
GETP(K,X) evaluates the polynomials,

$$P_m(x) = (x - u_m)P_{m-1}(x) - v_{m-1}P_{m-2}(x)$$

BCALC(M,K) and GETB(M,K) are used to calculate the values of B(m,k). GETB assigns B(m,k) as follows:

$$B(m,k) = 0 \qquad \text{if } k < 0 \quad \text{or} \quad k > m$$
$$B(m,k) = 1 \qquad \text{if } k = m$$

If $0 \le k < m$, then

$$B(m,k) = B(m-1,k-1) - u_m B(m-1,k) - v_{m-1} B(m-2,k)$$

The values of $B(m,k)$, where $0 \le k < m$, are stored in the vector B. They are retrieved using the algorithm

$$B(m,k) = B(\text{INDEX})$$

where

$$\text{INDEX} = m + km' - \sum_{k=1}^{k} k$$

$$= m + \frac{k(2m' - k - 1)}{2}$$

Here m' is the order of the fitting polynomial, and m is an integer. For example, if m' is 4, the following values of $B(m,k)$ will be stored in B: $B(0,1)$, $B(0,2)$, $B(0,3)$, $B(0,4)$, $B(1,2)$, $B(1,3)$, $B(1,4)$, $B(2,3)$, $B(2,4)$, and $B(3,4)$. The retrieving algorithm assigns INDEX as follows:

k	0	0	0	0	1	1	1	2	2	3
m	1	2	3	4	2	3	4	3	4	4
INDX	1	2	3	4	5	6	7	8	9	10

The dimensionality of B is $m'(m' + 1)/2$. In FORSY, the maximum order is 10, therefore B is dimensioned 55.

FORSY first reads the input data. The 20 loop calculates the values u_m, v_m, and D_m. The 30 loop calculates the coefficients a_m, which gives the regression equation in terms of the orthogonal polynomials,

$$F(x) = \sum_{m=0}^{m'} a_m P_m(x)$$

The 40 loop calculates and stores in B the coefficients $B(m,k)$. The 50 loop calculates the desired coefficients of the power series. The 60 loop generates the table of results and computes the variance of the fit,

$$\sigma^2 = \frac{\sum_{i=1}^{N} [y_i - F(x_i)]^2}{N - m' - 1}$$

Sample Execution

ENTER N AND M
>21 2

```
    ENTER  X(I),Y(I), I = 1,2,...,N
>0 -.89
>5 -.69
>10 -.53
>15 -.34
>20 -.15
>25 .02
>30 .2
>35 .42
>40 .61
>45 .82
>50 1.03
>55 1.22
>60 1.45
>65 1.68
>70 1.88
>75 2.1
>80 2.31
>85 2.54
>90 2.78
>95 3
>100 3.22
```

FOR Y = C(0) + C(1)*X + C(2)*X**2 + ... + C(M)*X**M,

I	C(I)
0	-8.862D-01
1	3.524D-02
2	5.979D-05

I	X(I)	Y(I)	YCALC	DIFF
1	0.000D-01	-8.900D-01	-8.862D-01	-3.755D-03
2	5.000D+00	-6.900D-01	-7.086D-01	1.855D-02
3	1.000D+01	-5.300D-01	-5.279D-01	-2.128D-03
4	1.500D+01	-3.400D-01	-3.442D-01	4.202D-03
5	2.000D+01	-1.500D-01	-1.575D-01	7.542D-03
6	2.500D+01	2.000D-02	3.211D-02	-1.211D-02
7	3.000D+01	2.000D-01	2.247D-01	-2.475D-02
8	3.500D+01	4.200D-01	4.204D-01	-3.740D-04
9	4.000D+01	6.100D-01	6.190D-01	-8.991D-03
10	4.500D+01	8.200D-01	8.206D-01	-5.983D-04
11	5.000D+01	1.030D+00	1.025D+00	4.805D-03
12	5.500D+01	1.220D+00	1.233D+00	-1.278D-02
13	6.000D+01	1.450D+00	1.443D+00	6.645D-03
14	6.500D+01	1.680D+00	1.657D+00	2.308D-02
15	7.000D+01	1.880D+00	1.873D+00	6.527D-03
16	7.500D+01	2.100D+00	2.093D+00	6.984D-03
17	8.000D+01	2.310D+00	2.316D+00	-5.549D-03
18	8.500D+01	2.540D+00	2.541D+00	-1.071D-03
19	9.000D+01	2.780D+00	2.770D+00	1.042D-02
20	9.500D+01	3.000D+00	3.001D+00	-1.083D-03
21	1.000D+02	3.220D+00	3.236D+00	-1.557D-02

    THE VARIANCE OF THE FIT IS   1.398D-04

ENTER ANOTHER CHOICE FOR M IF DESIRED.
   (M .LE. 0 STOPS THE PROGRAM).
>0

   Listing

```
C
C                         FORSY
C
C   FORSY PERFORMS POLYNOMIAL REGRESSION ANALYSIS
C   USING THE FORSYTHE METHOD
C
C     AUTHOR:  K. J. JOHNSON
C
C
      IMPLICIT DOUBLE PRECISION (A-H,O-Z)
      DIMENSION X(100),Y(100),U(10),V(10),A(0/10),C(0/10),
     1  B(0/55),D(0/10)
      COMMON U,V,B,MAXM
C
C   INPUT
C
      WRITE(6,11)
      READ(5,22)N,MAXM
      IF(N.EQ.0 .OR. MAXM.GT.(N-1).OR.MAXM.GT.10)STOP
      WRITE(6,33)
      DO 10 I=1,N
   10 READ(5,44)X(I),Y(I)
      D(0)=N
      M0=1
C
C   LOOP TO CALC. U(M),V(M),AND D(M)
C
   15 DO 20 M=M0,MAXM
      SUM=0.D0
      DO 21 I=1,N
   21 SUM=SUM+X(I)*GETP(M-1,X(I))**2
      U(M)=SUM/D(M-1)
      SUM=0.D0
      DO 24 I=1,N
   24 SUM=SUM+X(I)*GETP(M,X(I))*GETP(M-1,X(I))
      V(M)=SUM/D(M-1)
      SUM=0.D0
      DO 28 I=1,N
   28 SUM=SUM+GETP(M,X(I))**2
      D(M)=SUM
   20 CONTINUE
C
C   LOOP TO CALCULATE A(M), COEFFICIENTS OF POLYNOMIALS
C
      DO 30 M=0,MAXM
      SUM=0.D0
      DO 35 I=1,N
   35 SUM=SUM+Y(I)*GETP(M,X(I))
   30 A(M)=SUM/D(M)
C
C   CALCULATE ELEMENTS OF B
```

```
      B(0)=1.
      DO 40 K=0,MAXM-1
      KP1=K+1
      DO 40 M=KP1,MAXM
      INDX=M+K*(2*MAXM-K-1)/2
      B(INDX)=BCALC(M,K)
   40 CONTINUE
C
C   CALCULATE THE COEFFICIENTS OF THE POLYNOMIAL
C
      WRITE(6,66)
      DO 50 K=0,MAXM
      C(K)=A(K)
      KP1=K+1
      IF(KP1.GT.MAXM)GO TO 50
      DO 58 M=KP1,MAXM
      INDX=M+K*(2*MAXM-K-1)/2
      C(K)=C(K)+A(M)*B(INDX)
   58 CONTINUE
   50 WRITE(6,77)K,C(K)
C
C   PRINT RESULTS AND COMPUTE VARIANCE
C
      WRITE(6,88)
      SUM=0.D0
      DO 60 I=1,N
      YCALC=C(0)
      XFCTR=1.
      DO 65 J=1,MAXM
      XFCTR=XFCTR*X(I)
   65 YCALC=YCALC+C(J)*XFCTR
      DIFF=Y(I)-YCALC
      SUM=SUM+DIFF*DIFF
   60 WRITE(6,99)I,X(I),Y(I),YCALC,DIFF
      VAR=SUM/(N-MAXM-1)
      WRITE(6,101)VAR
C
C   CYCLE ON OPTION
C
      WRITE(6,111)
      M0=MAXM
      READ(5,22)MAXM
      IF(MAXM.LE.0 .OR. MAXM.GT.(N-1) .OR.MAXM.GT.10)STOP
      GO TO 15
      STOP
C
C   FORMATS
C
   11 FORMAT(//'  ENTER N AND M'/)
   22 FORMAT(2I)
   33 FORMAT(/'  ENTER  X(I),Y(I), I = 1,2,...,N'/)
   44 FORMAT(2D)
   66 FORMAT(//'  FOR Y = C(0) + C(1)*X + C(2)*X**2 + ... + ',
     1 'C(M)*X**M,',//11X,'I',11X,'C(I)'/)
   77 FORMAT(10X,I2,5X,1PD13.3)
   88 FORMAT(//'  I',6X,'X(I)',9X,'Y(I)',8X,'YCALC',8X,'DIFF'/)
   99 FORMAT(I3,4(1PD13.3))
  101 FORMAT(//'   THE VARIANCE OF THE FIT IS ',1PD13.3)
  111 FORMAT(//'  ENTER ANOTHER CHOICE FOR M IF DESIRED.',
     1/'   (M .LE. 0 STOPS THE PROGRAM).'/)
```

```fortran
        STOP
        END
C
C           GETP
C
        DOUBLE PRECISION FUNCTION GETP(M,X)
        DOUBLE PRECISION U,V,P
        DIMENSION U(10),V(10),P(0/10)
        COMMON U,V
        GETP=1.D0
        IF(M.EQ.0)RETURN
        P(1)=X-U(1)
        P(2)=(X-U(2))*P(1)-V(1)
        IF(M.LE.2)GO TO 20
        DO 10 I=3,M-1
   10   P(I)=(X-U(I))*P(I-1)-V(I-1)*P(I-2)
        GETP=(X-U(M))*P(M-1)-V(M-1)*P(M-2)
        RETURN
   20   GETP=P(M)
        RETURN
        END
C
C       BCALC
C
C
C
C   BCALC AND GETB CALCULATE THE VALUE OF B(K,M)
C
C   B(K,M) = B(K-1,M-1)-U(M)*B(K,M-1)-V(M-1)*B(K,M-2)
C
C
C
C   GETB ASSIGNS A VALUE TO B(K,M) AS FOLLOWS:
C
C      B(K,M) = 0    IF   K < 0   OR   K > M
C             = 1    IF   K = M
C
C    OTHERWISE, B(K,M) IS RETRIEVED FROM THE VECTOR B(0/55)
C    USING THE RETRIEVING FUNCTION, INDX=M+K*(2*MAXM-K-1)/2
C
        DOUBLE PRECISION FUNCTION BCALC(M,K)
        DOUBLE PRECISION U,V
        DIMENSION U(10),V(10)
        COMMON U,V,B,MAXM
        BCALC=GETB(M-1,K-1)-U(M)*GETB(M-1,K)-V(M-1)*GETB(M-2,K)
        RETURN
        END
C
C       GETB
C
        DOUBLE PRECISION FUNCTION GETB(M,K)
        DOUBLE PRECISION U,V,B
        DIMENSION U(10),V(10),B(0/55)
        COMMON U,V,B,MAXM
        GETB=0.D0
        IF(K.LT.0)RETURN
        IF(K.GT.M)RETURN
        IF(K.EQ.M)GO TO 10
        INDX=M+K*(2*MAXM-K-1)/2
        GETB=B(INDX)
        RETURN
   10   GETB=1.D0
        RETURN
        END
```

Discussion

FORSY, including the three subroutines, requires approximately 1200 words of core. This sample execution took 0.2 sec.

FORSY contains a nonstandard Fortran feature provided by Digital Equipment Corporation's Fortran-10 compiler,

DIMENSION $A(i/j)$

Here i and j are integers which may be negative or zero. The use of the feature allows for more readable code when variables with zero or negative subscripts are encountered, for example, $D(0)$, $B(0,1)$, etc.

There are two advantages of the Forsythe method compared to that used by POLREG. Problems associated with coefficient matrix inversion are eliminated. Also, the amount of work that must be done to determine the variance using a polynomial of order $m + 1$ having already determined the variance with a polynomial of order m, is considerably more efficient using the Forsythe method.

The application of FORSY to the data in Problem 5.3 using a quartic fitting function is shown in the following:

```
ENTER N AND M
> 10 4

 ENTER  X(I),Y(I), I = 1,2,...,N
>0.23 5.64
>1.01 7.83
>2.29 17.04
>2.87 21.38
>4.15 24.56
>5.36 16.21
>5.51 14.57
>6.36 0.78
>6.84 -7.64
>7.00 -12.52
```

FOR Y = C(0) + C(1)*X + C(2)*X**2 + ... + C(M)*X**M,

I	C(I)
0	6.316D+00
1	−4.230D+00
2	6.955D+00
3	−1.493D+00
4	7.608D−02

I	X(I)	Y(I)	YCALC	DIFF
1	2.300D−01	5.640D+00	5.693D+00	−5.272D−02
2	1.010D+00	7.830D+00	7.679D+00	1.513D−01
3	2.290D+00	1.704D+01	1.726D+01	−2.227D−01

```
 2     1.010D+00      7.830D+00      7.679D+00      1.513D-01
 3     2.290D+00      1.704D+01      1.726D+01     -2.227D-01
 4     2.870D+00      2.138D+01      2.133D+01      5.242D-02
 5     4.150D+00      2.456D+01      2.439D+01      1.669D-01
 6     5.360D+00      1.621D+01      1.633D+01     -1.206D-01
 7     5.510D+00      1.457D+01      1.452D+01      5.206D-02
 8     6.360D+00      7.800D-01      1.108D+00     -3.284D-01
 9     6.840D+00     -7.640D+00     -8.501D+00      8.612D-01
10     7.000D+00     -1.252D+01     -1.196D+01     -5.595D-01
```

THE VARIANCE OF THE FIT IS   2.571D-01

The performance of FORSY on the test data used in the discussion of POLREG is shown in the following (input omitted):

FOR Y = C(0) + C(1)*X + C(2)*X**2 + ... + C(M)*X**M,

```
      I              C(I)

      0           1.000D+00
      1          -1.000D+00
      2           1.000D+00
      3          -9.999D-01
      4           1.000D+00
      5          -1.002D+00
      6           1.006D+00
      7          -1.007D+00
      8           1.005D+00
      9          -1.001D+00
```

```
  I     X(I)          Y(I)          YCALC          DIFF

  1    1.000D-02      9.901D-01      9.901D-01      1.022D-08
  2    2.000D-02      9.804D-01      9.804D-01      9.571D-10
  3    3.000D-02      9.709D-01      9.709D-01     -5.422D-09
  4    4.000D-02      9.615D-01      9.615D-01     -9.356D-09
  5    5.000D-02      9.524D-01      9.524D-01     -1.128D-08
              :
              :       (Rest of Output Omitted)
              :
 95    9.500D-01      2.058D-01      2.058D-01     -4.646D-07
 96    9.600D-01      1.710D-01      1.710D-01     -5.005D-07
 97    9.700D-01      1.333D-01      1.333D-01     -5.350D-07
 98    9.800D-01      9.239D-02      9.239D-02     -5.664D-07
 99    9.900D-01      4.805D-02      4.805D-02     -5.922D-07
100    1.000D+00      3.725D-08      6.466D-07     -6.093D-07
```

THE VARIANCE OF THE FIT IS   3.420D-14

5.3 CURVE SMOOTHING

In some experiments it is difficult to achieve a satisfactory signal-to-noise ratio. Figure 5.4 (a) shows a simulation of a spectrum which consists of two overlapping Gaussian bands,

$$F(x) = a \exp[-b(x-c)^2] + d \exp[-f(x-g)^2] \qquad (5.3)$$

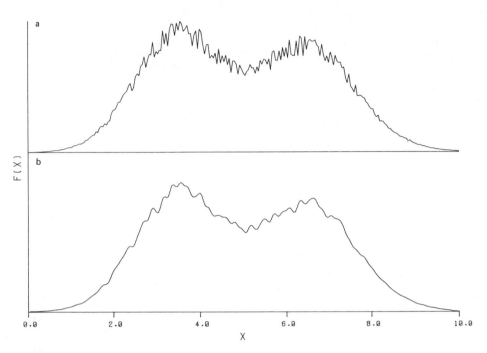

Figure 5.4   (a) Simulated spectrum consisting of two overlapping Gaussian bands.
(b) Results of "numerical filtering" by the SMOOTH routine (see below).

The "noisy" spectrum was generated by superimposing a uniformly distributed $\pm$ 10%
random error on the calculated points. The program that generated the spectrum in
Figure 5.4(a) is listed below.

```
C
C       PROGRAM TO SIMULATE A SPECTRUM CONSISTING OF
C       TWO OVERLAPPING GAUSSIAN BANDS.
C
C       Y(I,1)  -  SPECTRUM
C       Y(I,2)  -  SPECTRUM WITH +/- 10% RANDOM NOISE SUPERIMPOSED
C
        DIMENSION X(400),Y(400,2)
        DATA A,B,C,D,E,F/0.9,0.5,3.5,0.8,0.4,6.5/
        GAUSS(A1,A2,A3,Z) = A1*EXP(-A2*(Z-A3)**2)
        DO 10 I=1,200
        J=I+200
        X(I)=I*0.04
        X(J)=J*0.04
        Y(I,1) = GAUSS(A,B,C,X(I)) + GAUSS(D,E,F,X(I))
        Y(J,1) = GAUSS(A,B,C,X(J)) + GAUSS(D,E,F,X(J))
        Y(I,2) = (0.9+0.2*RAN(XXX))*Y(I,1)
        Y(J,2) = (0.9+0.2*RAN(XXX))*Y(J,1)
     10 WRITE(6,11) X(I),Y(I,1),Y(I,2),X(J),Y(J,1),Y(J,2)
     11 FORMAT(6G12.3)
        END
```

   Numerical techniques are available [67-70] to filter some of the random noise
in experimental data. Figure 5.5 shows a seven-point segment of this spectrum.
In Figure 5.5 the numerical values assigned to the abscissa are the integers -3,
-2, -1, 0, 1, 2, and 3. The smooth curve is the quartic function,

$$F(x) = a_0 + a_1 x + a_2 x^2 + a_3 x^3 + a_4 x^4$$

This quartic function could be obtained using the polynomial regression analysis
methods discussed in Section 5.2. The fourth point has the abscissa value 0. For
this point,

$$F(x) = F(0) = a_0$$

a_0 is the numerically filtered value corresponding to the fourth point in the
original seven-point segment. The value a_0 for a given seven-point segment can
be calculated in closed-form. The least-squares coefficient matrix with unit
weighting for a quartic fit is

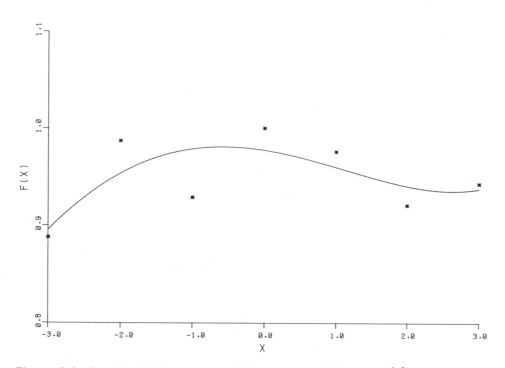

Figure 5.5   Quartic fit to a seven-point segment of Figure 5.4(a).

$$
\begin{bmatrix}
7 & \sum_{i=1}^{7} x_i & \sum_{i=1}^{7} x_i^2 & \sum_{i=1}^{7} x_i^3 & \sum_{i=1}^{7} x_i^4 \\[2ex]
\sum_{i=1}^{7} x_i & \sum_{i=1}^{7} x_i^2 & \sum_{i=1}^{7} x_i^3 & \sum_{i=1}^{7} x_i^4 & \sum_{i=1}^{7} x_i^5 \\[2ex]
\sum_{i=1}^{7} x_i^2 & \sum_{i=1}^{7} x_i^3 & \sum_{i=1}^{7} x_i^4 & \sum_{i=1}^{7} x_i^5 & \sum_{i=1}^{7} x_i^6 \\[2ex]
\sum_{i=1}^{7} x_i^3 & \sum_{i=1}^{7} x_i^4 & \sum_{i=1}^{7} x_i^5 & \sum_{i=1}^{7} x_i^6 & \sum_{i=1}^{7} x_i^7 \\[2ex]
\sum_{i=1}^{7} x_i^4 & \sum_{i=1}^{7} x_i^5 & \sum_{i=1}^{7} x_i^6 & \sum_{i=1}^{7} x_i^7 & \sum_{i=1}^{7} x_i^8
\end{bmatrix}
$$

Here all sums extend from 1 to 7. Because of the choice of abscissa,

$$
\sum_{i=1}^{7} x_i = (-3) + (-2) + (-1) + 0 + 1 + 2 + 3 = 0
$$

$$
\sum_{i=1}^{7} x_i^3 = \sum_{i=1}^{7} x_i^5 = \sum_{i=1}^{7} x_i^7 = 0
$$

$$
\sum_{i=1}^{7} x_i^2 = 28
$$

$$
\sum_{i=1}^{7} x_i^4 = 196
$$

$$
\sum_{i=1}^{7} x_i^8 = 13636
$$

$$
\sum_{i=1}^{7} x_i^6 = 1588
$$

The augmented matrix is

$$
\begin{bmatrix}
7 & 0 & 28 & 0 & 196 & \sum_{i=1}^{7} y_i \\[2ex]
0 & 28 & 0 & 196 & 0 & \sum_{i=1}^{7} y_i x_i \\[2ex]
28 & 0 & 196 & 0 & 1588 & \sum_{i=1}^{7} y_i x_i^2 \\[2ex]
0 & 196 & 0 & 1588 & 0 & \sum_{i=1}^{7} y_i x_i^3 \\[2ex]
196 & 0 & 1588 & 0 & 13636 & \sum_{i=1}^{7} y_i x_i^4
\end{bmatrix}
$$

a_0 can be calculated using elimination methods (Section 4.3). The result is

$$a_0 = \frac{150912(196T_1 - 28T_2) - 38416T_3 + 311248T_2}{89921664}$$

where

$$T_1 = \sum_{i=1}^{7} y_i$$

$$T_2 = \sum_{i=1}^{7} y_i x_i^2$$

$$T_3 = \sum_{i=1}^{7} y_i x_i^4$$

The curve-smoothing technique discussed here is to consider the first seven-point segment of the spectrum and replace the fourth point by the value calculated for a_0; then to consider points 2 through 8 of the original spectrum and replace the fifth value by the corresponding value for a_0; then to consider points 3 through 9, etc. The spectrum is smoothed, one point at a time. The first and last sets of three points are not smoothed. The entire procedure can be repeated, if necessary, to achieve further smoothing. Figure 5.4(b) is the result of 10 applications of this procedure to the simulated spectrum shown in Figure 5.4(a).

A quantitative measure of the effectiveness of this numerical smoothing procedure can be obtained by defining an arbitrary relative noise parameter, NOISE,

$$\text{NOISE} = \sum_{i=1}^{N} \left| \frac{y_{ni} - y_{si}}{y_{si}} \right|$$

Here N is the number of points in the spectrum, y_{ni} is the ith value in the noisy spectrum, and y_{si} is the corresponding value in the smooth spectrum. The effect of repetitive smoothing should be reflected by a decreasing value for NOISE. The following results were obtained in this spectral simulation:

Iteration no.	1	2	3	4	5	6	7	8	9	10
NOISE	13.7	4.5	1.9	1.1	0.77	0.60	0.50	0.42	0.37	0.32

Figure 5.4(b) is the result of 10 iterations of the routine SMOOTH, which is listed below.

```
C     SMOOTH
C
C   THIS SUBROUTINE USES A 7-POINT QUARTIC
C   ALGORITHM TO SMOOTH A SPECTRUM
C
C      YN   -  ORIGINAL SPECTRUM
C      YS   -  SMOOTHED SPECTRUM
C    DIFF   -  NOISE PARAMETER
C
C    AUTHOR:   K. J. JOHNSON
C
       SUBROUTINE SMOOTH(YN,YS,N,DIFF)
       DIMENSION X2(7),X4(7),YN(100),YS(100),SUM(3)
       DATA X2/9.,4.,1.,0.,1.,4.,9./
       DATA X4/81.,16.,1.,0.,1.,16.,81./
       DIFF=0.
C
C     THE FIRST AND LAST SETS OF THREE
C     POINTS ARE NOT SMOOTHED
C
       DO 10 I=1,3
       YS(I)=YN(I)
   10  YS(N-I+1)=YN(N-I+1)
       DO 100 I=4,N-4
       DO 20 KJJ=1,3
   20  SUM(KJJ)=0.
       DO 50 J=1,7
       Y=YN(I+J-4)
       SUM(1)=SUM(1)+Y
       SUM(2)=SUM(2)+Y*X2(J)
   50  SUM(3)=SUM(3)+Y*X4(J)
       YS(I)=0.56709957*SUM(1)-0.2651515*SUM(2)+0.02272727*SUM(3)
       IF(YS(I). EQ. 0.) GO TO 100
       DIFF=DIFF+ ABS( ( YN(I)-YS(I) ) /YS(I) )
  100  CONTINUE
       RETURN
       END
```

5.4 NONLINEAR REGRESSION ANALYSIS

A significant component of chemical research is concerned with the hypothesis of mathematical models to represent experimental data. These mathematical models are usually represented by a function of a single independent variable and a set of adjustable parameters,

$$y = F(x;\ a_1, a_2, a_3, \ldots, a_p)$$

Here y and x are the observed dependent and independent variables, and a_1, a_2, ..., a_p are the parameters to be determined. The objective in nonlinear regression analysis is to find the set of parameters that gives the best possible fit to experimental values. For example, Eq. (5.3) is the "fitting function" for the spectral data in Figure 5.4. In this case the fitting function has six parameters. If the function is linear in the parameters, for example,

$$F(x) = a_1 + a_2 x + \frac{a_3}{x} + a_4 \exp(-x^2)$$

then the methods discussed in Section 5.3 can be used to find the set of parameters that gives the best fit in the least squares sense. Frequently, however, the function is not linear in the parameters. For example, Eq. (5.3) is not linear in the six parameters.

The method of nonlinear regression analysis considered here involves the minimization of the weighted residual function,

$$R = \sum_{i=1}^{N} \omega_i [y_i - F(x_i)]^2$$

Here ω_i is the weight assigned the observed datum (x_i, y_i), and $F(x_i)$ is the value of the fitting function evaluated at x_i. The minimization procedure described above for linear and polynomial regression,

$$\frac{\partial R}{\partial a_j} = 0 \qquad (j = 1, 2, \ldots, p)$$

would give a system of p nonlinear simultaneous equations. One approach to solving the problem is to approximate the fitting function by a truncated Taylor series expansion [71]. Given a set of initial approximations to the parameters,

$$a_{j0} \qquad j = 1,2,3,\ldots,p$$

then the fitting function,

$$F_i = F(x_i; a_1, a_2, \ldots, a_p)$$

becomes

$$F_i = F_{i0} + \frac{\partial F_i}{\partial a_1}\bigg|_0 (a_1 - x_{10}) + \frac{\partial F_i}{\partial a_2}\bigg|_0 (a_2 - a_{20}) + \cdots + \frac{\partial F_i}{\partial a_p}\bigg|_0 (a_p - a_{p0})$$

Here F_{i0} is the function evaluated at x_i, a_{10}, a_{20}, etc., and $(\partial F_i / \partial F_j)_0$ is the derivative of the function with respect to a_j, evaluated at x_i, a_{10}, a_{20}, etc. This procedure of linerization by a truncated Taylor series expansion about initial approximations is similar to the method discussed in Section 4.4 for solving systems of nonlinear simultaneous equations.

To simplify notation, let p = 2 and let

$$M_i = \frac{\partial F_i}{\partial a_1}\bigg|_0 \qquad N_i = \frac{\partial F_i}{\partial a_2}\bigg|_0$$

$$\Delta a_1 = a_1 - a_{10} \qquad \Delta a_2 = a_2 - a_{20}$$

Then

$$F_i = F_{i0} + M_i \Delta a_1 + N_i \Delta a_2$$

The residual function, assuming unit weighting, is

$$R = \sum_{i=1}^{N} (y_i - F_i)^2$$

$$= \sum_{i=1}^{N} [y_i - (F_{i0} + M_i \; \Delta a_1 + N_i \; \Delta a_2)]^2$$

This function is linear in the correction terms Δa_1, and Δa_2. Minimizing the residual function with respect to these correction terms,

$$\partial \frac{\partial R}{\Delta a_1} = 0 = 2 \sum_{i=1}^{N} [y_i - (F_{i0} + M_i \Delta a_1 + N_i \; \Delta a_2)](-M_i)$$

$$\partial \frac{\partial R}{\Delta a_2} = 0 = 2 \sum_{i=1}^{N} [y_i - (F_{i0} + M_i \Delta a_1 + N_i \; \Delta a_2)](-N_i)$$

This is rearranged to the following linear system of equations:

$$\sum_{i=1}^{N} M_i^2 \; \Delta a_1 + \sum_{i=1}^{N} M_i N_i \; \Delta a_2 = \sum_{i=1}^{N} (y_i - F_{i0}) M_i$$

$$\sum_{i=1}^{N} M_i N_i \; \Delta a_1 + \sum_{i=1}^{N} N_i^2 \; \Delta a_2 = \sum_{i=1}^{N} (y_i - F_{i0}) N_i$$

These equations can be solved for the correction terms, Δa_1 and Δa_2. Then, if the procedure is converging, better approximations to the parameters have been found:

$$a_1 = a_{10} + \Delta a_1 \qquad\qquad a_2 = a_{20} + \Delta a_2$$

Extending this treatment to p parameters, an iterative procedure is suggested:

$$a_j^{m+1} = a_j^m + \Delta a_j^m \qquad j = 1,2,3,\ldots,p$$

The method is said to converge when

$$\sum_{j=1}^{p} \left| \frac{\Delta a_j}{a_j} \right| < \varepsilon \qquad\qquad a_j \neq 0$$

Here ε is an arbitrary convergence tolerance. Let the p x p linear system of equations be represented by

$$PA=Z$$

Here P is the p x p coefficient matrix of partial derivatives,

$$P_{rs} = \sum_{i=1}^{N} \left. \frac{\partial F_i}{\partial a_r} \right|_0 \left. \frac{\partial F_i}{\partial a_s} \right|_0$$

A is the p x 1 unknown vector of correction terms,

$$A = \begin{pmatrix} a_1 \\ a_2 \\ .. \\ a_p \end{pmatrix}$$

and Z is the p x 1 constant vector. The rth element of Z is

$$Z_r = \sum_{i=1}^{N} (y_i - F_{i0}) \left. \frac{\partial F_i}{\partial a_r} \right|_0$$

Note that the coefficient matrix is symmetric. The solution to the system can be obtained by matrix inversion (Section 4.4.2),

$$A = P^{-1}Z$$

The variance of the fit is

$$\sigma^2 = \frac{\sum_{i=1}^{N} {}_i[y_i - F(x_i; a_1, a_2, \ldots, a_p)]^2}{N - p - 1}$$

and the standard deviations of the parameters are

$$\sigma_{a_r} = [\sigma^2 P^{-1}_{rr}]^{1/2} \qquad r = 1, 2, 3, \ldots, p$$

Here P^{-1}_{rr} is the rth diagonal element in the inverse of the coefficient matrix.

In summary, the procedure for nonlinear regression is as follows: Given a set of data (x_i and y_i, i = 1,2,3,...,N), an appropriate fitting function, F(x; $a_1, a_2, \ldots, a_p$), and a set of initial approximations to the p parameters ($a_{10}, a_{20}, \ldots, a_{p0}$), then the following iterative procedure is established:

1. Evaluate the function and the p partial derivatives,

$$F_{i0} = F(x_i; a_{10}, a_{20}, \ldots, a_{p0}) \qquad \left. \frac{\partial F_i}{\partial a_r} \right|_0 \qquad r = 1, 2, \ldots, p$$

2. Assign the values to the coefficient matrix P and the constant vector Z.

3. Invert the coefficient matrix, and calculate the correction terms, $A = P^{-1}Z$, where $A_j = \Delta a_j = a_j - a_{j0}$.

4. Test for convergence. If the test fails, use the current values for the parameters as initial approximations, and repeat the process.

The following program provides an example of this technique of nonlinear curve fitting.

5.4.1 NONLIN

Introduction

NONLIN performs nonlinear regression analysis. The objective is to find a set
of parameters, a_1, a_2, ..., a_p, that minimize the residual function,

$$R = \sum_{i=1}^{N} \omega_i [y_i - F(x_i;\ a_1, a_2, \ldots, a_p)]^2$$

Here y_i and x_i are the observed dependent and independent variables and $F(x_i, a_1,$
$a_2, \ldots, a_p)$ is the fitting function that is nonlinear in the parameters.

NONLIN is documented using the following example from chemical kinetics,

$$A \underset{k_4}{\overset{k_1}{\rightleftharpoons}} B \underset{k_3}{\overset{k_2}{\rightleftharpoons}} C \qquad (a_0 = 1,\ b_0 = c_0 = 0)$$

This kinetics system is solved in closed form (Section 2.2). For the purposes of
this documentation, the dependent variable is the concentration of the intermediate
species, B, and the independent variable is time. There are four parameters, k_1,
k_2, k_3, and k_4. The function,

$$B(t;\ k_1, k_2, k_3, k_4)$$

is [72]

$$B(t) = T_4 + T_5\ \exp^{-\lambda_2 t} + T_6\ \exp^{-\lambda_3 t}$$

where

$$T_4 = \frac{k_1 k_3}{\lambda_2 \lambda_3}$$

$$T_5 = \frac{k_1(k_3 - \lambda_2)}{\lambda_2(\lambda_2 - \lambda_3)}$$

$$T_6 = \frac{k_1(\lambda_3 - k_3)}{\lambda_3(\lambda_2 - \lambda_3)}$$

$$\lambda_2 = \frac{P + Q}{2}$$

$$\lambda_3 = \frac{P - Q}{2}$$

with

$$P = k_1 + k_2 + k_3 + k_4$$

and

$$Q = P^2 - 4(k_1 k_2 + k_4 k_3 + k_1 k_3)$$

This is a function for which the parameters are manifestly nonlinear.

NONLIN reads the following values: M, the number of parameters; N, the number of data points; the data (X_i and Y_i, i = 1,2,3,...,N); three convergence criteria; and the initial approximations to the parameters. NONLIN prints the input data for verification purposes and a convergence map showing the progress of the iterative procedure. If the method converges, the variance of the fit, the values of the parameters, and the associated standard deviations are printed, and optionally a table containing X_i, Y_i, F_i, and $Y_i - F_i$.

Method

The numerical procedure used by NONLIN is that described above. NONLIN consists of a main program and four subroutines, VALUE, FUNCTN, CLOSE, and MATINV.

MAIN: This is the control program. It contains the iteration loop, calculates the elements of the coefficient matrix (P) and the constant vector (Z) from the partial derivatives (G) and the function values (F) and tests for convergence.

VALUE: This is the input subroutine. For purposes of this documentation, the "input data" are calculated in VALUE.

FUNCTN: This subroutine calculates the values of the fitting function and the partial derivatives.

CLOSE: This is the output subroutine. CLOSE calculates the variance of the fit, the standard deviation, and, on option, the table.

MATINV: The matrix inversion routine (Section 4.4.2).

Three convergence criteria are used. Let

$$\text{RELSUM} = \sum_{j=1}^{p} \left| \frac{a_j^{(m+1)} - a_j^{(m)}}{a_{a_j}^{(m+1)}} \right|$$

If RELSUM is less than the convergence criterion, the method has converged. If RELSUM exceeds the divergence criterion, then the method is diverging and the procedure is terminated. Finally, if the method fails to converge in a specified number of iterations, the procedure is terminated.

In this sample execution, internally generated data are used to test the procedure. The parameters used in generating the data are

$k_1 = 1.0$ $k_2 = 0.5$ $k_3 = 0.25$ $k_4 = 0.5$

N = 100   and $0 \leq t \leq 15$ (arbitrary units)

Then, the following initial approximations are given:

$k_{10} = 1.3$ $k_{20} = 0.65$ $k_{30} = 0.175$ $k_{40} = 0.36$

In the second part of the sample execution, a random $\pm$ 5% error is superimposed on the generated "data" to simulate a set of experimental data.

Sample Execution

1. Internally generated data.

ENTER N AND M
>100 4

ENTER THE INITIAL APPROXIMATIONS TO THE CONSTANTS
>.8 .6 .2 .75

ENTER MAXIMUM NO. OF ITERATIONS AND CONVERGENCE AND
DIVERGENCE TOLERANCES
>10 1E-3 1E4

N = 100   M =   4

K1 =   0.800   K2 =   0.600   K3 =   0.200   K4 =   0.750

ITMAX = 10   CVGTOL =   0.100E-02   DVGTOL =   0.100E+05

CONVERGENCE MAP

ITERATION	I	PARAMETER(I)	CORRECTION(I)
1			
	1	1.00	0.201
	2	0.475	-0.125
	3	0.257	0.566E-01
	4	0.486	-0.264
2			
	1	1.00	-0.856E-03
	2	0.499	0.244E-01
	3	0.250	-0.646E-02
	4	0.500	0.143E-01
3			
	1	1.00	0.871E-04
	2	0.500	0.784E-03
	3	0.250	-0.162E-03
	4	0.500	-0.380E-03
4			
	1	1.00	-0.330E-07
	2	0.500	-0.612E-07
	3	0.250	-0.248E-06
	4	0.500	-0.787E-07

CONVERGENCE ATTAINED

THE VARIANCE OF THE FIT IS   0.124E-16

I	PARAMETER(I)	STD. DEV.(I)
1	1.00	0.103E-07
2	0.500	0.141E-07
3	0.250	0.448E-08
4	0.500	0.316E-07

ENTER 1 IF A TABLE OF Y VALUES (OBSERVED AND
CALCULATED) IS DESIRED, 0 OTHERWISE
>1

I	X(I)	Y(I)	F(I)	Y(I)-F(I)
1	0.150	0.129	0.129	-0.149E-07
2	0.300	0.224	0.224	0.000E+00
3	0.450	0.293	0.293	0.373E-08
4	0.600	0.342	0.342	0.373E-08
5	0.750	0.376	0.376	0.000E+00
		⋮	(Rest of Table Omitted)	
96	14.4	0.286	0.286	0.745E-08
97	14.6	0.286	0.286	0.745E-08
98	14.7	0.286	0.286	0.745E-08
99	14.9	0.286	0.286	0.373E-08
100	15.0	0.286	0.286	0.373E-08

2. _Internally generated data with ± 5% error superimposed._

ITERATION	I	PARAMETER(I)	CORRECTION(I)
1			
	1	0.963	0.163
	2	0.569	-0.310E-01
	3	0.280	0.800E-01
	4	0.312	-0.438
2			
	1	0.962	-0.130E-02
	2	0.571	0.206E-02
	3	0.273	-0.744E-02
	4	0.371	0.597E-01
3			
	1	0.963	0.868E-03
	2	0.571	0.292E-03
	3	0.273	0.250E-04
	4	0.374	0.229E-02
4			
	1	0.963	0.489E-04
	2	0.571	-0.742E-04
	3	0.273	-0.167E-04
	4	0.374	0.174E-03

CONVERGENCE ATTAINED

THE VARIANCE OF THE FIT IS   0.827E-04

I	PARAMETER(I)	STD. DEV.(I)
1	0.963	0.256E-01
2	0.571	0.447E-01
3	0.273	0.125E-01
4	0.374	0.895E-01

ENTER 1 IF A TABLE OF Y VALUES (OBSERVED AND
CALCULATED) IS DESIRED, 0 OTHERWISE
>1

I	X(I)	Y(I)	F(I)	Y(I)-F(I)
1	0.150	0.125	0.125	0.806E-04
2	0.300	0.230	0.219	0.110E-01
3	0.450	0.296	0.287	0.928E-02
4	0.600	0.336	0.336	-0.359E-03
5	0.750	0.361	0.371	-0.973E-02
6	0.900	0.386	0.395	-0.824E-02
7	1.05	0.395	0.410	-0.147E-01
8	1.20	0.430	0.418	0.118E-01
9	1.35	0.431	0.422	0.819E-02
10	1.50	0.406	0.423	-0.172E-01
		:		
		:	(Rest of table omitted)	
		:		
90	13.5	0.285	0.287	-0.237E-02
91	13.6	0.273	0.287	-0.137E-01
92	13.8	0.300	0.287	0.128E-01
93	14.0	0.280	0.287	-0.740E-02
94	14.1	0.292	0.287	0.509E-02
95	14.3	0.276	0.287	-0.112E-01
96	14.4	0.299	0.287	0.119E-01
97	14.6	0.296	0.287	0.924E-02
98	14.7	0.300	0.287	0.131E-01
99	14.9	0.299	0.287	0.122E-01
100	15.0	0.297	0.287	0.103E-01

Listing

```
C
C
C                         NONLIN
C
C
C     THIS PROGRAM PERFORMS NONLINEAR LEAST SQUARES ANALYSIS.
C
C
C     PROGRAM ARRANGEMENT:
C           MAIN   - CONTROL, MAIN ITERATION LOOP
C           VALUE  - INPUT
C           FUNCTN - CONTAINS THE FUNCTION AND ITS DERIVATIVES.
C           CLOSE  - OUTPUT
C           MATINV - INVERSION OF COEFFICIENT MATRIX
C
C
```

```
C     GLOSSARY:
C
C          N      = NUMBER OF DATA POINTS
C          M      = NUMBER OF PARAMETERS
C          X(N)   = INDEPENDENT VARIABLE
C          Y(N)   = DEPENDENT VARIABLE (OBSERVED)
C          F(N)   = DEPENDENT VARIABLE (CALCULATED)
C          A(M)   = PARAMETERS
C          W(N)   = WEIGHTING FACTORS
C          G(M,N) = PARTIAL DERIVATIVES
C          DELA   = CHANGE IN A(M)
C          P(M,M) = COEFFICIENT MATRIX
C          Z(M)   = CONSTANT VECTOR
C          ITMAX  = MAXIMUM NUMBER OF ITERATIONS
C          CVGTOL = CONVERGENCE TOLERANCE
C          DVGTOL = DIVERGENCE TOLERANCE
C          H      = RELAXATION FACTOR
C          P(M,M) = COEFFICIENT MATRIX
C          Z(M)   = CONSTANT VECTOR
C          ITMAX  = MAXIMUM NUMBER OF ITERATIONS
C          CVGTOL = CONVERGENCE TOLERANCE
C          DVGTOL = DIVERGENCE TOLERANCE
C          H      = RELAXATION FACTOR
C
C
C     AUTHORS:  B. A. SWISSHELM, D. E. HAWKINS, JR., AND K. J. JOHNSON
C
C

      DIMENSION Y(100),F(100),A(10),W(100),
     1P(10,10),Z(10),X(100),G(10,100)
      H=1.0
C
C     INPUT
C
      CALL VALUE(N,M,ITMAX,CVGTOL,DVGTOL,A,X,Y,W)
      WRITE(6,11)
C
C     ITERATION LOOP
C
      DO 100 ICOUNT=1,ITMAX
      WRITE(6,22) ICOUNT
C
C     CALL FUNCTN TO SET UP P MATRIX
C
      CALL FUNCTN(N,M,A,X,F,G)
C
C     ZERO THE Z VECTOR AND P MATRIX
C
      DO 25 I=1,M
      Z(I)=0.0
      DO 25 J=1,M
   25 P(I,J)=0.0
C
C     FORM THE DIAGONAL AND LOWER TRIANGLE OF P MATRIX
C
      DO 30 I=1,N
      DO 30 J=1,M
      DO 30 L=1,J
   30 P(J,L)=P(J,L)+W(I)*G(J,I)*G(L,I)
```

```
C
C      FORM UPPER TRIANGLE OF P MATRIX
C
       DO 40 I=2,M
       L=I-1
       DO 40 J=1,L
    40 P(J,I)=P(I,J)
C
C      FORM THE Z VECTOR
C
       DO 50 I=1,N
       DIF=Y(I)-F(I)
       DO 50 J=1,M
    50 Z(J)=Z(J)+W(I)*DIF*G(J,I)
C
C      CALL MATINV TO INVERT P MATRIX
C
       CALL MATINV(P,M,D)
       IF(D.EQ.0.0) GO TO 200
       TEST=0.0
C
C      TEST FOR CONVERGENCE
C
       DO 60 I=1,M
       DELA=0.0D+0
       DO 65 J=1,M
    65 DELA=DELA+P(I,J)*Z(J)
       TEST=TEST+ABS(DELA/A(I))
C
C      FORM NEW A VALUES
C
       A(I)=A(I)+H*DELA
    60 WRITE(6,33)I,A(I),DELA
C
C      TEST FOR CONVERGENCE
C
       IF(TEST.LE.CVGTOL) GO TO 70
C
C      TEST FOR DIVERGENCE
C
   100 IF(TEST.GT.DVGTOL) GO TO 80
C
C      CONVERGENCE FAILURE
C
       WRITE(6,44)
       STOP
C
C      CONVERGENCE ATTAINED
C
C      CALL CLOSE TO OUTPUT FINAL VALUES
C
    70 CALL CLOSE(N,M,A,X,Y,F,P,W)
       STOP
C
C       DIVERGENCE
C
    80 WRITE(6,55)
       STOP
C
C      SINGULAR MATRIX
```

```
C
  200 WRITE(6,66)
      STOP
   11 FORMAT(/,T20,' CONVERGENCE MAP'//' ITERATION   ',T15,'I',
     1T19,'PARAMETER(I)',T35,'CORRECTION(I)'/)
   22 FORMAT(T6,I2)
   33 FORMAT(T15,I1,T19,G12.3,T35,G12.3)
   44 FORMAT(/' SORRY,ITMAX EXCEEDED'//)
   55 FORMAT(/,' THE SYSTEM IS DIVERGING',/)
   66 FORMAT(' THE COEFFICIENT MATRIX IS SINGULAR')
      END
C
C
C     SUBROUTINE VALUE
C
C
C     THIS ROUTINE IMPUTS N, M, (X(I),Y(I),W(I),I=1,N), THE INITIAL
C     APPROXIMATIONS, (A(I),I=1,M), ITMAX, CVGTOL, AND DVGTOL.
C
C
C
C     FOR THE KINETIC SYSTEM
C
C                K1          K2
C           A <=====> B <=====> C
C                K4          K3
C
C     THE VARIABLES ARE:
C
C           A(1) = K1
C           A(2) = K2
C           A(3) = K3
C           A(4) = K4
C           X(I) = TIME
C           Y(I) = OBSERVED CONCENTRATION OF B
C           F(I) = CALCULATED CONCENTRATION OF B
C
C     THE INITIAL CONCENTRATIONS ARE: A=1, B=C=0.
C
C
      SUBROUTINE VALUE(N,M,ITMAX,CVGTOL,DVGTOL,A,X,Y,W)
      IMPLICIT REAL(K,L)
      DIMENSION Y(100),W(100),A(10),X(100)
      WRITE(6,11)
      READ(5,22)N,M
      WRITE(6,33)
      READ(5,44)(A(I),I=1,M)
      WRITE(6,55)
      READ(5,66) ITMAX,CVGTOL,DVGTOL
C
C     THE FOLLOWING LOOP GENERATES DATA TO TEST THE PROGRAM
C
C           K1 = 1.0     K2 = 0.5
C
C           K4 = 0.5     K3 = 0.25
C
C
C     T VARIES FROM 0. TO 15.
C
```

```
C        ASSIGN CONSTANTS
C
         K1=1.
         K2=.5
         K3=.25
         K4=.5
C
         P=K1+K2+K3+K4
         DS=P*P-4*(K1*K2+K3*K4+K1*K3)
         Q=SQRT(DS)
         L2=(P+Q)/2
         L3=(P-Q)/2
         TR1=K1*K3/L2/L3
         TR2=K1*(K3-L2)/L2/(L2-L3)
         TR3=K1*(L3-K3)/L3/(L2-L3)
         DO 10 I=1,100
         T=I*0.15
         X(I)=T
      10 Y(I)=TR1+TR2*EXP(-L2*T)+TR3*EXP(-L3*T)
C
C        ASSUME UNIT WEIGHTING
C
         DO 50 I=1,N
      50 W(I)=1.
         WRITE(6,88) N,M,(A(I),I=1,M),ITMAX,CVGTOL,DVGTOL
         RETURN
      11 FORMAT(/' ENTER N AND M '/)
      22 FORMAT(2I)
      33 FORMAT(/' ENTER THE INITIAL APPROXIMATIONS TO THE CONSTANTS'/)
      44 FORMAT(10F)
      55 FORMAT(/,' ENTER MAXIMUM NO. OF ITERATIONS AND CONVERGENCE AND '
        1/' DIVERGENCE TOLERANCES ',/)
      66 FORMAT(I,2F)
      77 FORMAT(2E12.6)
      88 FORMAT(/' N =',I4,10X,'M =',I3//' K1 =',G12.3,3X,'K2 =',G12.3,
        13X'K3 =',G12.3,3X,'K4 =',G12.3//' ITMAX =',I3,5X,'CVGTOL =',
        2G12.3,5X,'DVGTOL =',G12.3/)
         END
C
C        SUBROUTINE FUNCTN
C
C        THIS ROUTINE EVALUATES THE FUNCTION AND ITS M DERIVATIVES.
C        THE FUNCTION VALUES ARE STORED IN F(M) AND THE DERIVATIVES IN
C        G(M,N).
C
C
         SUBROUTINE FUNCTN(N,M,A,X,F,G)
         IMPLICIT REAL (K,L)
         DIMENSION A(10),G(10,100),X(100),F(100)
         REAL NUMER
C
C        FORM REPEATEDLY USED CONSTANTS
C
         K1=A(1)
         K2=A(2)
         K3=A(3)
         K4=A(4)
C
         P=K1+K2+K3+K4
         Q=SQRT(P*P-4*(K1*K2+K3*K4+K1*K3))
```

```
      L2=(P+Q)/2
      L3=(P-Q)/2
      QUE1=(K1-K2-K3+K4)/Q
      QUE2=(-K1+K2+K3+K4)/Q
      QUE3=-QUE1
      QUE4=(K1+K2-K3+K4)/Q
C
      K1H=K1/2
      K1XK3=K1*K3
      K3ML3=K3-L3
      L2MK3=L2-K3
      L2PL3=L2+L3
      L2ML3=L2-L3
      L2XL3=L2*L3
      L2ML3C=L2ML3*L2PL3
      L2TRM2=L2ML3*L2ML3-2*L2XL3
      DEN=L2XL3*L2ML3
      DENSX2=DEN*DEN*2
      CNST11=K3*(L2ML3+K1*QUE1)
      CNST12=K3*L3-L2XL3
      CNST13=L2XL3-K3*L2
      CNST21=K1XK3*QUE2
      CNST31=K1*(L2ML3+K3*QUE3)
      CNST41=K1XK3*QUE4
      CNSTN1=L2ML3*K1XK3
      CNSTN2=K1*L3*L2MK3
      CNSTN3=K1*L2*K3ML3
C
C     SET UP PARTIAL DERIVATIVES AND EVALUATE F(I)
C
      DO 10 I=1,N
      T=X(I)
      TTERM1=T*L2MK3
      TTERM2=T*K3ML3
      L2TRM1=L2*(TTERM2+1)
      L3TERM=L3*(TTERM1-1)
      EL2=EXP(-L2*T)
      EL3=EXP(-L3*T)
      NUMER=CNSTN1-EL2*CNSTN2-EL3*CNSTN3
      F(I)=NUMER/DEN
      DER1P2=NUMER*(L2ML3C-L2TRM2*QUE1)/DENSX2
      DER2P2=NUMER*(L2ML3C-L2TRM2*QUE2)/DENSX2
      DER3P2=NUMER*(L2ML3C-L2TRM2*QUE3)/DENSX2
      DER4P2=NUMER*(L2ML3C-L2TRM2*QUE4)/DENSX2
C
C     EVALUATE THE PARTIAL DERIVATIVES
C
      G(1,I)=(CNST11+EL2*(CNST12+K1H*(L3TERM-L2MK3+(L3TERM+L2MK3)
     1*QUE1))+EL3*(CNST13+K1H*(L2TRM1-K3ML3-(L2TRM1+K3ML3)*QUE1)))/DEN
     2-DER1P2
C
      G(2,I)=(CNST21+K1H*(EL2*(L3TERM-L2MK3+(L3TERM+L2MK3)*QUE2)
     1+EL3*(L2TRM1-K3ML3-(L2TRM1+K3ML3)*QUE2)))/DEN-DER2P2
C
      G(3,I)=(CNST31+K1H*(EL2*(L3*(TTERM1+1)-L2MK3+(L3TERM+L2MK3)*QUE3)
     1+EL3*(L2*(TTERM2-1)-K3ML3-(L2TRM1+K3ML3)*QUE3)))/DEN-DER3P2
C
      G(4,I)=(CNST41+K1H*(EL2*(L3TERM-L2MK3+(L3TERM+L2MK3)*QUE4)+
     1EL3*(L2TRM1-K3ML3-(L2TRM1+K3ML3)*QUE4)))/DEN-DER4P2
   10 CONTINUE
      RETURN
      END
```

```
C
C      SUBROUTINE CLOSE
C
C
C      IF THE SYSTEM CONVERGES, THIS ROUTINE COMPUTES AND PRINTS THE
C      VARIANCE OF THE FIT AND THE STANDARD DEVIATIONS OF THE
C      PARAMETERS.  ON OPTION, IT ALSO PRINTS A TABLE CONTAINING THE
C      OBSERVED DATA, (X(I),Y(I),I=1,N), THE FUCTION EVALUATED AT EACH
C      X(I), AND THE DIFFERENCES, Y(I)-F(I).
C
C
       SUBROUTINE CLOSE(N,M,A,X,Y,F,P,W)
       DIMENSION Y(N),X(N),W(N),A(M),F(N),P(10,10),G(10,100)
       WRITE(6,11)
C
C      CALCULATE OUTPUT VALUES OF F(I)
C
       CALL FUNCTN(N,M,A,X,F,G)
C
C      CALCULATE THE VARIANCE OF THE FIT
C
       S=0.0
       DO 5 I=1,N
     5 S=S+W(I)*(Y(I)-F(I))**2
       S=S/(N-M-1.0)
       WRITE(6,22) S
       WRITE(6,33)
C
C      CALCULATE THE STANDARD DEVIATION OF EACH PARAMETER
C
       DO 20 J=1,M
       SIGMA=SQRT(ABS(S*P(J,J)))
    20 WRITE(6,44) J,A(J),SIGMA
C
C      DETERMINE IF A TABLE OF VALUES IS DESIRED
C
C      ITABLE=0 STOPS PROGRAM
C      ITABLE>0 CAUSES TABLE OF OBSERVED AND CALCULATED Y(I)
C               TO BE PRINTED
C
       WRITE(6,55)
       READ(5,66)ITABLE
       IF(ITABLE.LE.0) RETURN
C
C      OUTPUT TABLE OF Y(I)
C
       WRITE(6,77)
       DO 10 I=1,N
       DIFF=Y(I)-F(I)
       WRITE(6,88)I,X(I),Y(I),F(I),DIFF
    10 CONTINUE
```

```
   RETURN
11 FORMAT(/,T11,' CONVERGENCE ATTAINED')
22 FORMAT(/,' THE VARIANCE OF THE FIT IS',G12.3)
 3 FORMAT(//,T15,'I',T19,'PARAMETER(I)',T35,'STD. DEV.(I)',/)
44 FORMAT(T15,I1,T19,G12.3,T35,G12.3)
55 FORMAT(/,' ENTER 1 IF A TABLE OF Y VALUES (OBSERVED AND '
  1/' CALCULATED) IS DESIRED, 0 OTHERWISE',/)
66 FORMAT(I1)
77 FORMAT(//,'    I',T10,'X(I)',T29,'Y(I)',T45,'F(I)',T60,
  1'Y(I)-F(I)',/)
88 FORMAT(I5,T7,G12.3,T24,G12.3,T40,G12.3,T58,G12.3)
   END
```

Discussion

The memory requirements for NONLIN are main program, 1847; VALUE, 344, FUNCTN, 393; CLOSE, 1338; and MATINV, 274. This totals 4095 36-bit words. Each sample execution took approximately 1 sec. The results of the fit to the data with superimposed random noise are shown in Figure 5.6.

NONLIN must be modified for every application. The main program does not have to be changed unless the number of parameters exceeds 10. VALUE and FUNCTN have to be extensively rewritten to accommodate the input and the calculation of the function and the corresponding partial derivatives. For example, if Eq. (5.3) were the fitting function, the following assignment statements would be used in FUNCTN:

$$F_i \leftarrow F(x_i;\ a_1, a_2, \ldots, a_6)$$

$$G_{ij} \leftarrow \left. \frac{\partial F_i}{\partial a_j} \right|_0$$

For Eq. (5.3),

```
    F(I) = A1*EXP(-A2*(X(I)-A3)**2) + A4*EXP(-A5*(X(I)-A6)**2)

    G(1,I) = EXP(-A2*(X(I)-A3)**2)

    G(2,I) = -A1*(X(I)-A3)**2*EXP(-A2*(X(I)-A3)**2)

    G(3,I) = -2*A1*A2*(X(I)-A3)*EXP(-A2*(X(I)-A3)**2)

    G(4,I) = EXP(-A5*(X(I)-A6)**2)

    G(5,I) = -A4*(X(I)-A6)**2*EXP(-A5*(X(I)-A6)**2)

    G(6,I) = -2*A4*A5*(X(I)-A6)*EXP(-A5*X(I)-A6)**2
```

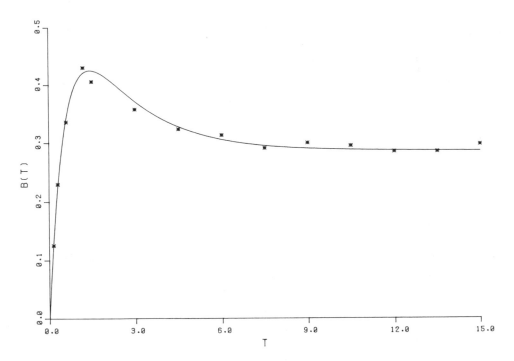

Figure 5.6 Fit of the simulated kinetic data with random noise superimposed. The smooth curve is the fitting function.

5.5 ADDITIONAL PROBLEMS

1. Calculate the heat of vaporization of H_2O and chlorobenzene from the following data [73, 74], assuming that the heat of vaporization is constant over this temperature range.

	Equilibrium Vapor Pressure (torr)	
T ($^{\circ}$C)	H_2O	C_5H_5Cl
0	4.579	
10	9.209	4.85
20	17.535	8.78
30	31.824	15.49
40	55.324	26.02
50	92.51	41.97
60	149.38	65.53
70	233.7	97.92
80	335.1	144.78
90	525.76	208.37
100	760.00	292.86

2. Calculate the first-order rate constant for the decomposition of N_2O_5 from the following data at 44.9°C [74]:

Time (min)	0	10	20	30	40	50	60	70
$P_{N_2O_5}$ (torr)	348.5	248	185	142	106	79.0	56.5	43.4

Time (min)	80	90	100	120	140	160
$P_{N_2O_5}$	32.8	25.3	19.1	10.5	5.12	2.99

3. Calculate the first-order rate constant for the hydrolysis of t-butyl bromide,

$$(CH_3)_3CBr + H_2O \longrightarrow (CH_3)_3COH + H^+ + Br^-$$

The initial concentration of t-butyl bromide is 0.1039 M, and the solvent is a 10% water-90% acetone mixture [75, 76].

T = 25°C		T = 50°C	
Time (hr)	$[(CH_3)_3CBr]$ (M)	Time (hr)	$[(CH_3)_3CBr]$ (M)
0	0.1039		
3.15	0.0896	0	0.1056
4.10	0.0859	9	0.0961
6.20	0.0776	18	0.0856
8.20	0.701	27	0.0767
10.0	0.0639	40	0.0645
13.5	0.0529	54	0.0536
18.3	0.0353	72	0.0432
26.0	0.0270	105	0.0270
30.8	0.0207	135	0.0174
37.3	0.0152	180	0.0089
43.8	0.0101		

Solution. k = 1.44 x 10^{-5} sec(-1); half-life 13.4 hr.

4. Calculate the second-order rate constant for the reaction of isobutylbromide and sodium ethoxide from the following data in ethyl alcohol at 95.15°C. The initial concentration of C_4H_9Br is 0.0535 M, and that of NaOEt is 0.0792 M [77, 78].

Time (min)	2.5	5.0	7.5	10.0	13.0	17.0
[Products]	0.0029	0.0056	0.0077	0.0105	0.0136	0.0152

Time (min)	20.0	30.0	40.0	50.0	60.0	70.0	90.0
[Products]	0.0200	0.0247	0.0281	0.0305	0.0325	0.0356	0.0396

Solution. k = 0.33 liter/mol·min

5. The hydrolysis of ethylnitrobenzoate by hydroxide,

$$NO_2C_6H_4COOC_2H_5 + OH^- \longrightarrow NO_2C_6H_4COOH + C_2H_5OH$$

proceeds as follows at 15°C when the initial concentrations of both reactants are 0.05 M [79, 80]:

Time (sec)	120	180	240	330	530	600
% hydrolyzed	32.95	41.75	48.8	58.05	69.0	70.35

Calculate the second-order rate constant.

Solution. k = 0.8 liter/mol·sec

6. The reaction $2NO + 2H_2 \quad N_2 + 2H_2O$ was studied with equimolar quantities of NO and H_2 at various initial pressures [79]:

Initial pressure (torr)	345	340.5	375	288	251	243	202
Half-life (min)	81	102	95	140	180	176	224

Estimate the overall order of the reaction.

Solution. Approximately 2.5.

7. The rate constant for the reaction,

$$CH_3I + C_2H_5ONa \longrightarrow CH_3OC_2H_5 + NaI$$

in ethyl alcohol shows the following temperature dependence [81, 82]:

T (°C)	0	6	12	18	24	30
$k_2 \times 10^5$ (liter/mol·sec)	5.6	11.8	24.5	48.8	100	208

Assuming Arrhenius behavior, calculate the activation energy and the pre-exponential factor.

Solution. E_a = 19.5 kcal; A = 2.4 x 10^{11}

8. The following results were found for the dielectric constant (ϵ) of gaseous sulfur dioxide at 1 atm as a function of temperature [83]:

T ($^\circ$K)	267.6	297.2	336.9	443.8
ϵ	1.009918	1.008120	1.005477	1.003199

Estimate the dipole moment of SO_2, assuming ideal gas behavior.

Solution. 1.67 debye

9. The following values were found for the dielectric constant (ϵ) and the density (ρ) of isopropyl cyanide as a function of mole fraction in benzene at 25°C [83, 84]:

x	0.00301	0.00523	0.00956	0.01301	0.01834	0.02517
ϵ	2.326	2.366	2.422	2.502	2.598	2.718
ρ	0.87326	0.87301	0.87260	0.87226	0.87121	0.87108

For pure C_3H_7NC, $\rho = 0.76572$, refractive index, $n_0 = 1.3712$; for pure benzene, $\rho = 0.87345$, $n_0 = 1.5106$. Calculate the dipole moment of isopropyl cyanide.

Solution. 2.65 debye.

10. The following data have been published [85, 86] for the tar and nicotine content (in milligrams) of several brands of king-size filter and regular cigarettes:

Filter		Regular	
Tar	Nicotine	Tar	Nicotine
8.3	0.32	32.4	1.69
12.3	0.46	33.0	1.75
18.8	1.10	34.1	1.48
22.9	1.32	34.8	1.89
23.1	1.26	36.7	1.73
24.0	1.44	37.2	2.11
27.3	1.42	38.5	2.35
30.0	1.96	41.1	2.45
35.9	2.23	41.5	1.97
41.6	2.20	43.4	2.64

Perform linear regression, $y = a + bx$, on each of these sets of data, with y the nicotine content and x the tar content. Test the hypothesis that B is the same for both cases.

11. Fit the following equilibrium data for the reaction

$$N_2 + 3H_2 \rightleftharpoons 2NH_3$$

to the function $K_p = a + bP + cP^2$, given the following data at $500°C$ [87, 88]:

P (atm)	10	20	50	100	300	600
K_p	0.00381	0.00386	0.00388	0.00402	0.00498	0.00651

12. Write a program to calculate partial molal volumes of solutes. The program should read weight percent-density data, derive an expression relating density (ρ) to molality (m), for example,

$$\rho = a_0 + a_1 m + a_2 m^2 + a_3 m^3$$

The partial molal volume is obtained by extrapolating the apparent molal volume to infinite dilution [89].

13. Consider the following heat capacity data for CO_2 (in calories per mole per degree) as a function of temperature [90]:

T (°K)	300	310	320	330	340	
c_p	9.01	9.09	9.15	9.25	9.34	

T (°K)	350	360	370	380	390	400
c_p	9.41	9.49	9.57	9.65	9.74	9.80

14. The following data were obtained for the ammine complexes of Hg^{2+} [91, 92]:

$-\log[NH_3]$	9.23	8.94	8.63	8.535	8.34	8.09	7.33
$\bar{n}$	0.441	0.655	0.969	1.290	1.62	1.92	1.98

$-\log[NH_3]$	2.604	1.941	1.522	1.120	0.807	0.477	0.195
$\bar{n}$	2.017	2.173	2.238	2.638	3.132	3.502	3.656

Here $\bar{n}$ is the average number of bound NH_3 molecules,

$$\bar{n} = \frac{C_a - [NH_3]}{C_m} = \frac{\sum_{i=1}^{N} i[Hg(NH_3)_i^{2+}]}{C_m}$$

Here C_a is the total concentration of NH_3, C_m is the total concentration of Hg^{2+}, and n is the coordination number of Hg^{2+}. Assume that Hg^{2+} forms four ammine complexes, $[Hg(NH_3)_i^{2+}]$, where $i = 1,2,3,4$. Devise both linear and nonlinear least-squares schemes to determine the stepwise formation constants.

15. Calculate the bond distances in carbonyl sulfide from moment of inertia data obtained from microwave spectroscopy [36].

16. Write a program to determine the formation constants of complexes from experimental data [20-22, 25, 27, 33, 37, 49, 52].

17. Write a program to determine the equilibrium points in titration systems [15-17, 31].

18. Write a program to determine the equivalence points and pK values for titrations of polyprotic acids [17].

19. Write programs to compare the resolution of spectra using Lorentzian, Gaussian, and the log-normal distribution functions [26, 29, 38].

20. Write a program to extract ion-pair formation constants from kinetic data [32].

21. Explore alternative approaches to nonlinear regression analysis, notably, the gradiant and Marquardt method [47, 93].

REFERENCES

1. J. R. Green and D. Hargerison, "Statistical Treatment of Experimental Data," Elsevier, New York, 1977.
2. C. L. Lawson and R. J. Hanson, "Solving Least Squares Problems," Prentice-Hall, Englewood Cliffs, N. J., 1974.
3. H. L. Youmans, "Statistics for Chemistry," Merrill, Columbus, Ohio, 1973.
4. D. F. DeTar (ed.), "Computer Programs for Chemists," 3rd Ed., Benjamin, New York, 1972.
5. P. R. Bevington, "Data Reduction and Error Analysis for the Physical Sciences," McGraw-Hill, New York, 1969, Chapter 6.
6. P. D. Clark, B. R. Craven, and R. C. L. Bosworth, "The Handling of Chemical Data," Pergamon Press, New York, 1968.
7. N. Draper and H. Smith, "Applied Regression Analysis," Wiley, New York, 1966.
8. W. E. Deming, "Statistical Adjustment of Data," Wiley, New York, 1943; reprinted by Dover, New York, 1964.
9. W. C. Hamilton, "Statistics in Physical Sciences," Ronald, New York, 1964.
10. P. G. Guest, "Numerical Methods of Curve Fitting," Cambridge University Press, London, 1961.
11. W. J. Youden, "Statistical Methods for Chemists," Wiley, New York, 1951, p. 41ff.

12. S. R. Goode, Computerized curve-fitting to determine the equivalence point in spectrophotometric titrations, Anal. Chem., 49, 1408 (1977).

13. D. E. Metzler, et al., Digital analysis of electronic absorption spectra, Anal. Chem., 49, 864A (1977).

14. G. M. Ridder and D. W. Margerum, Simultaneous kinetic analysis of multicomponent mixtures, Anal. Chem., 49, 2090 (1977).

15. M. Meloun and J. Cermak, Multiparametric curve fitting. I. Computer-assisted evaluation of chelatometric titrations with metallochromic indicators, Talanta, 23, 15 (1976).

16. J. G. McCullough and L. Meites, Titrimetric applications of multiparametric curve fitting: locations of end points in amperometric, conductometric, spectrophotometric and similar titrations, Anal. Chem., 47, 1081 (1975).

17. T. N. Briggs and J. E. Stuehr, Simultaneous potentiometric determination of precise equivalence points and pK values of two and three pK systems, Anal. Chem., 47, 1916 (1975).

18. I. L. Larsen and J. J. Wagner, Linear instrument calibration with statistical applications, J. Chem. Educ., 52, 215 (1975).

19. G. F. Pollinow and M. J. VanderWielen, A superior curve fitting program for functions of two variables, J. Chem. Educ., 52, 543 (1975).

20. H. F. DeBrabander et al., Polynuclear complex formation between lead(II) and 2-mercaptoethanol and 3-mercapto-1,2-propanediol, J. Coord. Chem., 4, 887 (1974).

21. W. A. E. McBryde, Spectrophotometric determination of equilibrium constants in solution, Talanta, 21, 979 (1974).

22. H. S. Dunsmore and D. Midgley, Computer calculation of the composition of equilibrium mixtures in solution, Anal. Chim. Acta, 72, 121 (1974).

23. D. E. Sands, Weighting factors in least squares, J. Chem. Educ., 51, 473 (1974).

24. S. D. Christian, et al., Linear least squares analysis, J. Chem. Educ., 51, 475 (1974).

25. K. R. Magnell, Formation constants of multi-specie equilibria from spectrophotometric data using small computer systems, J. Chem. Educ., 50, 619 (1973).

26. E. R. Price et al., Computer simulation of electronic absorption spectra, J. Chem. Educ., 50, 177 (1973).

27. W. Likussar, Computer approach to the continuous variations method for spectrophotometric determination of extraction and formation constants, Anal. Chem., 45, 1926 (1973).

28. B. Musulin, Least squares: linear and nonlinear, J. Chem. Educ., 50, 79 (1973).

29. D. B. Siano, The log-normal distribution function, J. Chem. Educ., 49, 755 (1972).

30. T. J. MacDonald et al., Computer evaluation of titrations by Gran's method, J. Chem. Educ., 49, 200 (1972).

31. W. J. Kozarek and Q. Fernando, Location of the equivalence points in potentio-
     metric titrations, J. Chem. Educ., 49, 202 (1972).

32. P. K. Cosgrove et al., Ion association and ionic reaction rates, J. Chem. Educ.,
     48, 626 (1971).

33. F. J. C. Rossotti et al., The use of electronic computing techniques in the
     calculation of stability constants, J. Inorg. Nucl. Chem., 33, 2051 (1971).

34. G. F. Pollnow, Least squares fit for two forms of the Clausius-Clapyron
     equation, J. Chem. Educ., 48, 518 (1971).

35. J. L. Dye and V. A. D. Nicely, A general purpose curve fitting program for
     class and research use, J. Chem. Educ., 48, 443 (1971).

36. H. Kim, Computer programming in the physical chemistry laboratory, J. Chem.
     Educ., 47, 120 (1970).

37. P. J. Lingane and Z. Z. Hugus, Jr., Normal equations for the Gaussian least-
     squares refinement of formation constants with simultaneous adjustment of the
     spectra of the absorbing species, Inorg. Chem., 9, 757 (1970).

38. A. H. Anderson et al., Computer analysis of unresolved non-Gaussian gas
     chromatograms by curve fitting, Anal. Chem., 42, 434 (1970).

39. H. Sato and K. Momoki, Least-squares curve fitting method for end-point
     detection of chelatometric titrations with metal indicators, Anal. Chem., 42,
     1477 (1970).

40. B. G. Willis et al., Simultaneous kinetic determination of mixtures by on-line
     regression analysis, Anal. Chem., 42, 1350 (1970).

41. R. C. Williams and J. W. Taylor, Computer calculation of first-order rate
     constants, J. Chem. Educ., 47, 129 (1970).

42. T. P. Kohman, Least-squares fitting of data with large errors, J. Chem. Educ.,
     47, 657 (1970).

43. M. Bader, A computerized physical chemistry experiment, J. Chem. Educ., 46,
     206 (1969).

44. B. Musulin, A computer program for chemical kinetics by titration, J. Chem.
     Educ., 46, 109 (1969).

45. L. E. Erickson, A computer program for the analysis of the N_2O_4 dissociation
     equilibrium, J. Chem. Educ., 46, 383 (1969).

46. F. A. Settle and E. Talley, A computer program for calculation of dipole
     moments from oscillometric data, J. Chem. Educ., 46, 744 (1969).

47. J. C. Becsey et al., Nonlinear least-squares methods -- a direct grid search
     approach, J. Chem. Educ., 45, 728 (1968).

48. D. A. Brandreth, Some practical aspects of curve fitting, J. Chem. Educ., 45,
     657 (1968).

49. R. W. Ramette, Equilibrium constants from spectrophotometric data, J. Chem.
     Educ., 44, 647 (1967).

50. K. P. Anderson and R. L. Snow, A relative deviation least-squares method of data treatment, J. Chem. Educ., 44, 356 (1967).

51. K. Negano and D. E. Metzler, Machine computation of equilibrium constants and plotting of spectra of individual ionic species in the pyridoxal-alanine system, J. Amer. Chem. Soc., 89, 2891 (1967).

52. L. Petrakis, Spectral line shapes, J. Chem. Educ., 44, 432 (1967).

53. E. D. Smith and D. M. Matthews, Least-squares regression lines, J. Chem. Educ., 44, 757 (1967).

54. D. F. DeTar, Significant figures and correlation of d parameters, J. Chem. Educ., 44, 757 (1967).

55. W. E. Wentworth, Rigorous least-squares adjustment -- applications to some nonlinear equations, J. Chem. Educ., 42, 96,162 (1965).

56. W. B. White et al., Chemical equilibrium in complex mixtures, J. Chem. Phys., 8, 751 (1958).

57. P. R. Bevington, op. cit., p. 43ff.

58. P. R. Bevington, ibid., p. 106ff.

59. P. R. Bevington, ibid., p. 113ff.

60. P. R. Bevington, ibid., p. 119ff.

61. W. J. Youden, op. cit., p. 41.

62. P. R. Bevington, op. cit., Chapter 8.

63. B. Carnahan, et al., "Applied Numerical Methods," Wiley, New York, 1969, p. 573ff.

64. G. E. Forsythe, Generation and use of orthogonal polynomials for data fitting with a digital computer, J. Soc. Indust. Appl. Math., 5, 74 (1957).

65. L. G. Kelly, "Handbook of Numerical Methods and Applications," Addison-Wesley, Reading, Mass., 1967, p. 68 ff.

66. B. Carnahan et al., op. cit., pp. 574-575.

67. A. Savitsky and M. J. F. Gulay, Smoothing and differentiation of data by simplified least squares, Anal. Chem., 36, 1627 (1964).

68. C. Allen Bush, Fourier method for digital data smoothing in circular dichroism spectrometry, Anal. Chem., 46, 890 (1974).

69. R. Bracewell, "The Fourier Transform and Its Applications," McGraw-Hill, New York, 1965.

70. J. W. Hayes et al., Some observations on digital smoothing of electroanalytical data based on the Fourier transformation, Anal. Chem., 44, 943 (1972).

71. P. R. Bevington, op cit., p. 232ff.

72. A. A. Frost and R. G. Pearson, "Kinetics and Mechanism," Wiley, New York, 1965, pp. 175-176.

73. R. Weast (ed.), "Handbook of Chemistry and Physics," 56th Ed., Chemical Rubber Co., Cleveland, 1975, p. D180.

74. T. R. Dickson, "The Computer and Chemistry," Freeman, San Francisco, 1968, pp. 159-160.

75. G. M. Barrow, "Physical Chemistry," McGraw-Hill, New York, 1966, p. 466.

76. L. C. Bateman et al., Mechanism of substitution at a saturated carbon atom. Part XIX, J. Chem. Soc., 1940, 960.

77. G. M. Barrow, op. cit., p. 474.

78. I. Dostrovsky and E. R. Hughes, Mechanism of substitution at a saturated carbon atom. Part XXVI, J. Chem. Soc., 1946, 157.

79. W. J. Moore, "Physical Chemistry," Prentice-Hall, Englewood Cliffs, N. J., 3rd Ed., 1963, p. 316.

80. W. B. S. Newling and C. N. Hinshelwood, The kinetics of the acid and alkaline hydrolysis of esters, J. Chem. Soc., 1936, 1357.

81. G. M. Barrow, op. cit., p. 486.

82. W. Heckt and M. Conrad, Z. Physik. Chem., 3, 450 (1889).

83. W. J. Moore, op. cit., p. 606.

84. M. T. Rogers, The electric moments of some aliphatic fluorides, cyanides, and amines, J. Amer. Chem. Soc., 69, 457 (1947).

85. B. Carnahan et al., op. cit., p. 590.

86. Time, 89, 51 (1967).

87. I. Klotz, "Chemical Thermodynamics," Benjamin, New York, 1964, p. 23ff.

88. A. T. Larson and R. L. Dodge, The ammonia equilibrium, J. Amer. Chem. Soc., 45, 2918 (1923); and The ammonia equilibrium at high temperatures, J. Amer. Chem. Soc., 45, 367 (1924).

89. I. Klotz, op. cit., p. 241ff.

90. T. R. Dickson, op. cit., p. 166.

91. J. N. Butler, "Ionic equilibrium," Addison-Wesley, Reading, Mass., 1964, p. 334.

92. N. Bjerrum, "Metal Ammine Formation in Aqueous Solution," Haase and Son, Copenhagen, 1957.

93. P. R. Bevington, op. cit. p. 235.

CHAPTER 6

NUMERICAL INTEGRATION

Several numerical methods are used to approximate definite integrals,

$$I = \int_a^b F(x)\ dx$$

Numerical techniques are required to solve integrals for which closed-form solutions are not available. Most of the methods discussed in this chapter involve approximating the integral with a finite sum,

$$I \simeq \sum_{i=1}^n c_i F(x_i)$$

Here c_i is a constant, and x_i is a value in the range $a \leqslant x \leqslant b$.

The numerical methods and associated computer programs in this chapter will enable the reader to approximate such integrals as

$$\int_a^b \frac{x^4 e^x\ dx}{(e^x - 1)^2}$$

This integral is encountered in solid-state chemistry (see Section 6.6, Problem 9).

6.1 TRAPEZOIDAL RULE

The area of a trapezoid with parallel sides labeled c and d is $(h/2)(c + d)$, where h is the base of the trapezoid. The trapezoidal rule [1] involves dividing up the interval (a,b) into n trapezoids and summing the resultant subareas. This method is illustrated in Figure 6.1. The integral is approximated by the following summation:

$$I = \int_a^b F(x)\ dx \simeq \sum_{i=1}^n a_i$$

where a_i is the area of the ith trapezoid.

$$I \simeq \frac{h}{2}[F(a) + F(a + h)] + \frac{h}{2}[F(a + h) + F(a + 2h)] + \cdots + \frac{h}{2}[F(a + (n - 1)h + F(b)]$$

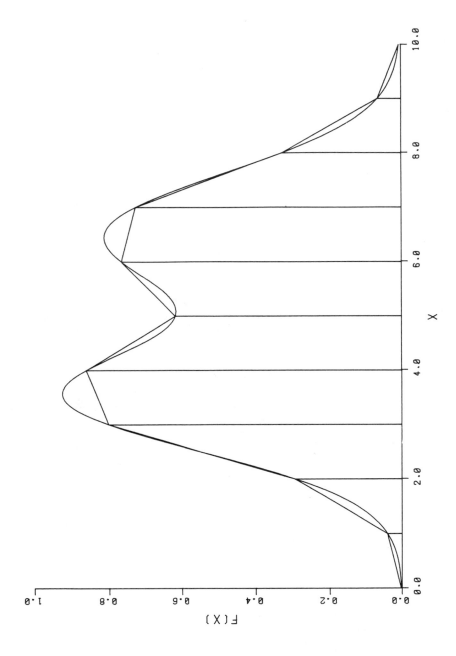

Figure 6.1 Approximating the area under a curve by a set of trapezoids.

$$I \simeq \frac{h}{2}[F(a) + F(b)] + h \sum_{i=1}^{n-1} F(a + ih) \qquad (6.1)$$

where

$$h = \frac{b - a}{n} \qquad (6.2)$$

Here n is the number of trapezoids used to approximate the area (integral).

Problem 6.1.   Use the trapezoidal rule to approximate the integral

$$\int_0^1 x^2 \, dx$$

Solution.   The analytical solution is 1/3.   Let n = 1, then

$$h = \frac{1 - 0}{1} = 1$$

and

$$I = \frac{1}{2}[F(0) + F(1)] = \frac{1}{2}(0 + 1) = 0.5$$

Next try n = 2.

$$h = \frac{1 - 0}{2} = \frac{1}{2}$$

and

$$I = \frac{1}{4}[F(0) + F(1)] + \frac{1}{2}F(\tfrac{1}{2}) = \frac{1}{4}(0 + 1) + \frac{1}{2}(\tfrac{1}{4}) = \frac{3}{8} = 0.375$$

For n = 3,

$$h = \frac{1}{4}$$

and

$$I = \frac{1}{8}[F(0) + F(1)] + \frac{1}{4} \sum_{i=1}^{3} F(a + ih)$$

$$= \frac{1}{8}(0 + 1) + \frac{1}{4}[F(\tfrac{1}{4}) + F(\tfrac{1}{2}) + F(\tfrac{3}{4})] = \frac{22}{64} = 0.344$$

The approximation improves with decreasing h:

n	h	I
8	0.125	0.334
16	0.0625	0.334
64	0.0156	0.333

The usual approach with a trapezoidal rule algorithm is to approximate the desired integral iteratively. The following program uses this approach.

6.1.1 TRAP

Introduction

TRAP approximates the value of a definite integral using an iterative trapezoidal rule algorithm. The algorithm is

$$I = \int_a^b F(x)\ dx \simeq \frac{h}{2}[F(a) + F(b)] + h \sum_{i=1}^{n-1} F(a + ih)$$

The input includes the integration limits a and b, the maximum number of iterations allowed, and a convergence tolerance. The function, $F(x)$, is coded in an arithmetic statement function. The program prints the input data for verification purposes and an iteration map containing the decreasing step size and the corresponding value of the integral.

Method

The trapezoidal rule is described above. Let $I(h)$ and $I(h/2)$ be the successive approximations using the step sizes h and $h/2$, respectively. The convergence test is

$$\left| \frac{I(h/2) - I(h)}{I(h/2)} \right| \quad TOL$$

TOL is the relative convergence criterion.

The algorithm is tested using the integral

$$\int_0^1 x\ \ln(1 + x) = \frac{1}{4}$$

Sample Execution

```
ENTER ITMAX,A,B,TOL
> 10   0   1   1E-5
```

```
        A =      0.000E+00           B =       1.00

   ITMAX =   10              AND TOL =      0.100E-04
```

ITN.	H	AREA
1	0.50000	0.27465
2	0.25000	0.25620
3	0.12500	0.25155
4	0.62500E-01	0.25039
5	0.31250E-01	0.25010
6	0.15625E-01	0.25002
7	0.78125E-02	0.25001
8	0.39063E-02	0.25000
9	0.19531E-02	0.25000

THE AREA IS   0.2500004

Listing

```
C
C              TRAP
C
C      ITERATIVE TRAPEZOIDAL RULE ROUTINE
C
C      TRAP INTEGRATES F(X) BETWEEN A AND B
C
C      F(X) IS AN ARITHMETIC STATEMENT FUNCTION
C
C      INPUT:  ITMAX   -  MAXIMUM NUMBER OF ITERATIONS
C              A,B     -  LOWER AND UPPER LIMITS OF INTEGRATION
C              TOL     -  CONVERGENCE TOLERANCE
C
C      TEST INTEGRAL:      1
C                          INT ( X*LN(1+X) )  =  1/4
C                          0
C
C      AUTHOR:  K. J. JOHNSON
C
       F(X)=X*ALOG(1.+X)
       OLD=1.E35
       WRITE(6,11)
       READ(5,22)ITMAX,A,B,TOL
       WRITE(6,33)A,B,ITMAX,TOL



       ENDS=(F(A)+F(B))/2.
       WRITE(6,44)
       H=B-A
       SUM=0.
C
C      ITERATIVE PROCEDURE.  CALCULATE F(X) ONLY WHERE NECESSARY
C
       DO 10 ITN=1,ITMAX
       H=H/2.
       M=2**ITN - 1
       DO 20 I=1,M,2
```

```
 20 SUM=SUM + F(A+I*H)
    AREA=(ENDS + SUM)*H
    WRITE(6,55)ITN,H,AREA
    IF(ABS((AREA-OLD)/AREA).LT.TOL)GO TO 99
    OLD=AREA
 10 CONTINUE
    WRITE(6,66) AREA
    STOP
 99 WRITE(6,77)AREA
    STOP
 11 FORMAT(//'  ENTER ITMAX,A,B,TOL ')
 22 FORMAT(I,3E)
 33 FORMAT(//5X,'    A = ',G12.3,5X,'    B = ',G12.3,
   1//,'     ITMAX = ',I3,10X,'  AND TOL = ',G12.3)
 44 FORMAT(//'  ITN.  ',10X,'H',16X,'AREA'/)
 55 FORMAT(I5,2G20.5)
 66 FORMAT(//'  SORRY, ITMAX EXCEEDED.'//,
   1 '  THE BEST APPROXIMATION TO THE AREA IS ',1PE15.7/)
 77 FORMAT(//'  THE AREA IS ',G15.7/)
    END
```

Discussion

TRAP requires approximately 275 words of core. This sample execution took approximately 0.2 sec.

TRAP contains an optimization feature worthy of note. Let i be the iteration counter. The step size is decreased by successive powers of 2,

$$h = \frac{b - a}{2^i} \qquad i = 1,2,3,\ldots$$

Then, from Eq. (6.2), $n = 2^i$. For example, with $a = 0$ and $b = 1$, notice the pattern of function evaluations:

j	n	h	$\sum\limits_{i=1}^{n-1} F(a + ih)$
1	2	1/2	$F(1/2)$
2	4	1/4	$F(1/4) + F(1/2) + F(3/4)$
3	8	1/8	$F(1/8) + F(1/4) + F(3/8) + F(1/2) + F(5/8) + F(3/4) + F(7/8)$
⋮	⋮	⋮	⋮

Note the redundancy in function evaluations in this iterative process. Now consider
two iterative trapezoidal algorithms.

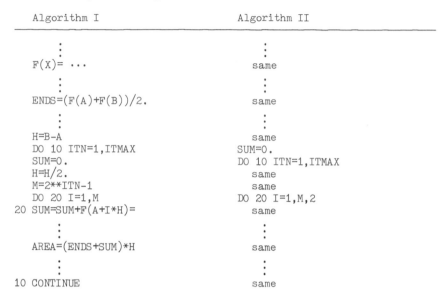

```
            Algorithm I                    Algorithm II
     _____

                 ⋮                            ⋮
            F(X)= ...                          same

                 ⋮                            ⋮
            ENDS=(F(A)+F(B))/2.                same

                 ⋮                            ⋮
            H=B-A                              same
            DO 10 ITN=1,ITMAX             SUM=0.
            SUM=0.                        DO 10 ITN=1,ITMAX
            H=H/2.                             same
            M=2**ITN-1                         same
            DO 20 I=1,M                   DO 20 I=1,M,2
         20 SUM=SUM+F(A+I*H)=                  same

                 ⋮                            ⋮
            AREA=(ENDS+SUM)*H                  same

                 ⋮                            ⋮
         10 CONTINUE                           same
```

Algorithm I recalculates ordinate positions, algorithm II does not. The simple
expedient of initializing SUM outside the 10 loop, and incrementing the 20 loop
by 2 makes algorithm II approximately twice as fast as algorithm I. The number
of times $F(X)$ is evaluated after 10 iterations is

$$\text{Algorithm I:} \qquad 2 + \sum_{j=1}^{10} (2^j - 1) = 2038$$

$$\text{Algorithm II:} \qquad 2 + \sum_{j=1}^{10} 2^{(j-1)} = 1025$$

TRAP uses algorithm II.

The trapezoidal rule, a two-point algorithm, uses a linear function to approximate a segment of the function to be integrated. It is particularly useful for
integrating experimental data collected at nonuniform intervals. The main advantage
of the trapezoidal rule is its simplicity. However, for applications in which
several significant digits of accuracy are required the trapezoidal rule might
require excessive amounts of computer time.

6.2 SIMPSON'S RULE

Simpson's rule [2] is a three-point algorithm,

$$I \simeq \frac{h}{3}[F(a) + 4F(a + h) + F(b)] \tag{6.3}$$

where $h = (b - a)/2$. Here a parabola is used to approximate the function, $F(x)$, over the interval (a,b). If the interval is divided into $2n$ segments, with step size

$$h = \frac{b - a}{2n}$$

then Eq. (6.3) becomes

$$I \simeq \frac{h}{3}[F(a) + 4F(a + h) + 2F(a + 2h) + 4F(a + 3h)$$
$$+ 2F(a + 4h) + \cdots + 2F(a + (2n - 2)h) + 4F(a + (2n - 1)h) + F(b)]$$

Problem 6.2. Use Simpson's rule to approximate the integral

$$\int_0^1 x^4 \, dx$$

Solution. Let $n = 1$. Then $h = 1/2$, and

$$I \approx \frac{1}{6}[F(0) + 4F(\tfrac{1}{2}) + F(1)]$$

$$= \frac{1}{6}[(0) + 4(\tfrac{1}{16}) + 1]$$

$$= \frac{5}{24} = 0.2083$$

Now try $n = 2$. Then $h = 1/4$, and

$$I \approx \frac{1}{12}[F(0) + 4F(\tfrac{1}{4}) + 2F(\tfrac{1}{2}) + 4F(\tfrac{3}{4}) + F(1)]$$

$$= \frac{1}{12}[0 + 4(\tfrac{1}{256}) + 2(\tfrac{1}{16}) + 4(\tfrac{81}{256}) + 1]$$

$$= \frac{154}{768} = 0.2005$$

The exact value is 0.200.

The following program is an iterative Simpson's rule routine.

6.2.1 SIMP

Introduction

SIMP approximates the value of a definite integral using an iterative Simpson's rule algorithm. The first iteration uses the equation

$$I = \int_{a}^{b} F(x)\ dx \simeq \frac{h}{3}[F(a) + 4F(a + h) + F(b)]$$

where

$$h = \frac{b - a}{2}$$

The input includes the limits a and b, the maximum number of iterations allowed, and a tolerance. The function, F(x), is coded in an arithmetic statement function. The program prints the input data for verification purposes and a table of the iteratively decreasing step size h and the corresponding approximation to the integral.

Method

Simpson's rule is discussed above. SIMP uses the same iterative procedure as the trapezoidal rule routine (Section 6.1).

Sample Execution

```
ENTER ITMAX,A,B,TOL
> 10   0   1   1E-5


          A =     0.000E+00           B =      1.00

      ITMAX =   10             AND TOL =    0.100E-04


   ITN.            H                  AREA

      1          0.5000           0.2506796
      2          0.2500           0.2500503
      3          0.1250           0.2500033
      4          0.6250E-01       0.2500002
      5          0.3125E-01       0.2500000


      THE AREA IS    0.2500000
```

Listing

```
C
C                     SIMP
C
C    SIMP INTEGRATES F(X) BETWEEN A AND B
C    USING AN ITERATIVE SIMPSON'S RULE
C
C    F(X) IS CODED IN THE STATEMENT FUNCTION F(X)
C
C    INPUT:  ITMAX,A,B,TOL   (I,3F)
C
C    AUTHOR: K. J. JOHNSON
```

```
C
      REAL INTGRL
      F(X)=X*ALOG(1.0+X)
      DATA N,EVEN/1,0.0/
      WRITE(6,11)
      READ(5,22)ITMAX,A,B,TOL
      WRITE(6,33)A,B,ITMAX,TOL
      H=(B-A)/2.
      ENDS=F(A)+F(B)
      ODD=4.*F(A+H)
      OLD=(ENDS+ODD)*H/3.
      WRITE(6,44)
      WRITE(6,55)N,H,OLD
C
C     ITERATIVE PROCEDURE   (DOES NOT RECALCULATE F(X) )
C
      DO 10 IT=2,ITMAX
      EVEN=EVEN+ODD/2.
      N=2*N
      H=H/2.
      M=2*N-1
      ODD=0.
      DO 20 J=1,M,2
   20 ODD=ODD+F(A+J*H)
      ODD=4.*ODD
      INTGRL=(ENDS+EVEN+ODD)*H/3.
      WRITE(6,55)IT,H,INTGRL
      IF(ABS((INTGRL-OLD)/INTGRL).LT.TOL)GO TO 99
      OLD=INTGRL
   10 CONTINUE
      WRITE(6,66)
      STOP
   99 WRITE(6,77)INTGRL
      STOP
   11 FORMAT(//'  ENTER ITMAX,A,B,TOL    ')
   22 FORMAT(I,3E)
   33 FORMAT(//5X,'    A = ',G12.3,5X,'    B = ',G12.3,//,
     1 '     ITMAX = ',I3,10X,'  AND TOL = ',G12.3)
   44 FORMAT(//'  ITN.    ',7X,'H',14X,'AREA'/)
   55 FORMAT(I6,5X,G12.4,2X,G16.7)
   66 FORMAT(//10X,'SORRY, ITMAX EXCEEDED'/)
   77 FORMAT(//5X,'THE AREA IS ',G15.7/)
      END
```

Discussion

SIMP requires approximately 300 words of core memory. This sample execution took approximately 0.1 sec. Note that SIMP contains the same optimization feature discussed above with TRAP. The pattern of function values needed with j, the iteration number, a = 0 and b = 1, is indicated below:

j	h	Function values required
1	1/2	$F(0) + 4F(1/2) + F(1)$
2	1/4	$F(0) + 4F(1/4) + 2F(1/2) + 4F(3/4) + F(1)$
3	1/8	$F(0) + 4F(1/8) + 2F(1/4) + 4F(3/8) + 2F(1/2)$
		$+ 4F(5/8) + 2F(3/4) + 4F(7/8) + F(1)$
$\vdots$	$\vdots$	$\vdots$

Note that three terms arise:

1. $F(a) + F(b)$
2. $4F(a + ih)$ $i = 1,3,5,7,\ldots$
3. $2F(a + jh)$ $j = 2,4,6,8,\ldots$

The proper arrangement of the double loop guarantees that there is no redundancy in function evaluations.

Simpson's rule is a three-point formula,

$$\int_a^b F(x)\ dx \simeq c_1 F(a) + c_2 F(a + h) + c_3 F(b)$$

The coefficients c_1, c_2, and c_3 are chosen so that the finite sum approximation gives the exact integral for the three functions, $F(x) = 1$, $F(x) = x$, and $F(x) = x^2$. The coefficients are derived as follows:

$$h = \frac{b - a}{2}$$

so

$$b = a + 2h$$

$$\int_a^{a+2h} dx = x \Big|_a^{a+2h} = 2h = c_1 + c_2 + c_3$$

$$\int_a^{a+2h} x\ dx = \frac{x^2}{2} \Big|_a^{a+2h} = 2h^2 + 2ah = c_1 a + c_2(a + h) + c_3(a + 2h)$$

$$\int_a^{a+2h} x^2\ dx = \frac{x^3}{3} \Big|_a^{a+2h} = 2ah + 4ah^2 + \frac{8}{3}h^3 = c_1 a^2 + c_2(a + h)^2 + c_3(a + 2h)^2$$

These three equations can be rearranged to give

$$c_1 + c_2 + c_3 = 2h$$
$$c_2 + 2c_3 = 2h$$
$$c_2 + 4c_3 = \frac{8h}{3}$$

The solution to this system of equations is

$$c_1 = c_3 = \frac{h}{3} \qquad \text{and} \qquad c_2 = \frac{4h}{3}$$

So

$$\int_a^b F(x)\ dx \approx \frac{h}{3}\left[F(a) + 4F(a + h) + F(b)\right]$$

Simpson's rule was derived to calculate the integral of quadratic functions exactly. A "fringe benefit" of this procedure is that the algorithm also returns the exact integral of cubic functions,

$$\int_0^1 x^3 \, dx = \frac{1}{6}[F(0) + 4F(\tfrac{1}{2}) + F(1)]$$
$$\equiv \frac{1}{6}[0 + \frac{4}{8} + 1]$$
$$= \frac{12}{48} = \frac{1}{4}$$

The Cotes numbers are derived by this technique.

6.3 COTES NUMBERS

Cotes numbers [3] are the coefficients, A_j, in the following algorithm:

$$\int_0^{n-h} F(x) \, dx \simeq \frac{nh}{N} \sum_{j=0}^{n} A_j F(jh) \tag{6.4}$$

The coefficients of the first six Cotes formulas are given in Table 6.1. For example, consider the seven-point algorithm (n = 6):

$$\int_0^{6h} F(x) \, dx \simeq \frac{6h}{840}[41F(0) + 216F(h) + 27F(2h)$$
$$+ 272F(3h) + 27F(4h) + 216F(5h) + 41F(6h)]$$

Problem 6.3. Approximate the following integral using the seven-point formula.

$$\int_0^{1.2} e^x \, dx$$

Solution. $$\int_0^{1.2} e^x \, dx \simeq \frac{6(0.2)}{840} [41e^0 + 216e^{0.2}$$
$$+ 27e^{0.4} + 272e^{0.6} + 27e^{0.8}$$
$$+ 216e^{1.0} + 41e^{1.2} = 2.320116928$$

Analytically,
$$e^{1.2} - e^0 = 2.3201169227$$

Table 6.1 Table of Cotes Numbers

n	N	A_1	A_2	A_3	A_4	A_5	A_6	A_7
1	2	1	1			(Trapezoidal rule)		
2	6	1	4	1		(Simpson's rule)		
3	8	1	3	3	1			
4	90	7	32	12	32	7		
5	288	19	75	50	50	75	19	
6	840	41	216	27	272	27	216	41

Error functions have been calculated for the Cotes formulas [3]. For example, the error in the seven-point formula is less than

$$6.4 \times 10^{-10} (nh)^9 F^{VIII}(x)$$

Here $F^{VIII}(x)$ is the maximum value of the eighth derivative of $F(x)$. The error in the solution to Problem 6.3 should be less than

$$6.4 \times 10^{-10} (1.2)^9 e^{1.2} = 11 \times 10^{-9}$$

The observed error is 5×10^{-9}.

Equation (6.4) can be transformed to correspond to the interval (a,b) as follows

$$\int_a^b F(x)\, dx = \int_a^{a+nh} F(x)\, dx \approx \frac{nh}{N} \sum_{j=0}^n A_j F(a + jh)$$

Problem 6.4. Write a program to approximate the integral

$$\int_a^b F(x)\, dx$$

using the seven-point Cotes formula. Test the routine with the same integral used in TRAP and SIMP.

Solution.

```
C
C     SEVEN-POINT COTES FORMULA
C
      IMPLICIT DOUBLE PRECISION (A-Z)
      INTEGER I
      DIMENSION C(4)
      DATA C/41.D0,216.D0,27.D0,272.D0/
      F(X)=X*DLOG(1.D0+X)
      WRITE(6,11)
      READ(5,22)A,B
      AREA=0.0D0
      H=(B-A)/6.0D0
      DO 10 I=1,3
   10 AREA=AREA + C(I)*( F(A+(I-1)*H) + F( A + (7-I)*H) )
      AREA=(B-A)*(AREA + C(4)*F(A+3*H))/840.D0
   20 WRITE(6,33) A,B,AREA
   11 FORMAT(//'  ENTER A AND B')
   22 FORMAT(2D)
   33 FORMAT(//'  A = ',1PD10.2, '    B = ',1PD10.2,
     1//'    THE AREA IS ',1PD20.12)
      END
```

The sample execution of this program is given next.

```
 ENTER A AND B
>0 1

  A =    0.00D+00    B =    1.00D+00

   THE AREA IS    2.500001882166D-01
```

6.4 GAUSSIAN QUADRATURE

Cotes formulas use equidistant abscissa values and coefficients calculated to obtain
as high an accuracy as possible. Gaussian quadrature formulas [4-6] are derived by
parameterizing both the coefficients and the abscissas. This procedure significantly
increases the accuracy of the finite sum approximation to an integral.

Gaussian quadrature formulas are of the form

$$\int_a^b w(x)F(x) \ dx \simeq \sum_{j=1}^n A_j F(x_j) \tag{6.5}$$

Here $w(x)$ is a weighting function, A_j are the coefficients, and x_j are the abscissas.
The weights and abscissas are derived using orthogonal polynomials. Four of the
more frequently used polynomials used in Gaussian quadrature are given in Table 6.2.

6.4.1 Legendre Polynomials

The first few Legendre polynomials are [7]

$$P_0(x) = 1$$

$$P_1(x) = x$$

$$P_2(x) = \tfrac{1}{2}(3x^2 - 1)$$

$$P_3(x) = \tfrac{1}{2}(5x^3 - 3x)$$

$$P_4(x) = \tfrac{1}{8}(35x^4 - 30x^2 + 3)$$

The Legendre polynomials are generated by the following recursion formula

$$P_n(x) = (\tfrac{2n - 1}{n})P_{n-1}(x) - (\tfrac{n - 1}{n})P_{n-2}(x)$$

These polynomials are orthogonal on the interval $(-1,1)$, that is,

$$\int_{-1}^1 P_m(x)P_n(x) = \begin{cases} 0 & \text{if } n \text{ is} \neq m \\ \dfrac{2}{2N + 1} & \text{if } n = m \end{cases}$$

Table 6.2 Orthogonal Polynomials

Polynomial	a	b	$w(x)$
Legendre	-1	1	1
Laguerre	0		e^{-x}
Hermite	-		e^{-x^2}
Chebyshev	-1	1	$1/\sqrt{1 - x^2}$

The roots of $P_n(x)$ are real, distinct (nondegenerate), and in the interval $(-1,1)$. An n-point Gauss-Legendre quadrature formula has the form

$$\int_{-1}^{1} F(x)\ dx \simeq \sum_{j=1}^{n} A_j F(x_j)$$

The abscissas are the roots of $P_n(x)$, and the corresponding coefficients are

$$A_j = \frac{2}{[P_n'(x_j)]^2 (1 - x_j)^2}$$

The n-point Gauss-Legendre quadrature formula calculates integrals exactly for polynomials of order $2n - 1$ or less.

Problem 6.5. Derive the three-point Gaussian quadrature formula and use it to integrate the three functions, x^4, x^5, and x^6 from $x = -1$ to $x = 1$.

Solution. $P_3(x) = \frac{1}{2}(5x^3 - 3x)$

The roots of $P_3(x)$ are $\pm\sqrt{0.6}$ and 0. Let $x_1 = -\sqrt{0.6}$, $x_2 = 0$, and $x_3 = \sqrt{0.6}$. Then

$$A_1 = \frac{2}{[P_3'(x_j)]^2 (1 - x_1)^2} = -\frac{5}{9}$$

Similarly, $A_2 = 8/9$ and $A_3 = 5/9$. Then the Gauss-Legendre quadrature formula is

$$\int_{-1}^{1} F(x)\ dx \simeq \frac{5}{9}F(-\sqrt{0.6}) + \frac{8}{9}F(0) + \frac{5}{9}F(\sqrt{0.6})$$

For $F(x) = x^4$,

$$\int_{-1}^{1} x^4\ dx \quad \frac{5}{9}(-\sqrt{0.6})^4 + 0 + \frac{5}{9}(\sqrt{0.6})^2$$

$$= \frac{2(5)(0.6)^2}{9} = \frac{2}{5}$$

For $F(x) = x^5$,

$$\int_{-1}^{1} x^5\ dx \quad \frac{5}{9}(-\sqrt{0.6})^5 + 0 + \frac{5}{9}(\sqrt{0.6})^5 = \frac{5}{9}(0) = 0$$

For $F(x) = x^6$,

$$\int_{-1}^{1} x^6\ dx \quad \frac{5}{9}(-\sqrt{0.6})^6 + \frac{8}{9}(0) + \frac{5}{9}(\sqrt{0.6})^6 = \frac{2(5)(0.6)^3}{9} = \frac{6}{25} = 0.24$$

Analytically, the three integrals are 2/5, 0, and 2/7 (0.2857).

The abscissas and coefficients for the first six Gauss-Legendre quadrature formulas are given in Table 6.3 [5]. Abscissas and coefficients accurate to 30 significant figures and for n up to 512 have been calculated [8]. Equation (6.5) can be transformed to the interval (a,b), resulting in the following approximation [9],

$$\int_a^b F(x)\ dx \simeq \frac{b-a}{2} \sum_{j=1}^n a_j F\left\{\frac{a+b}{2} + \frac{b-a}{2}\ x_j\right\} \tag{6.6}$$

Problem 6.6. Use the three-point Gauss-Legendre quadrature formula to approximate the integral

$$\int_0^1 x^5\ dx$$

Solution. a = 0, b = 1, and using Eq. (6.6),

$$\int_0^1 x^5\ dx \quad \frac{1}{2} \sum_{j=1}^3 A_j F\left\{\frac{1}{2} + \frac{x_j}{2}\right\}$$

$$= \frac{1}{2}\left\{\frac{5}{9}\left(\frac{1}{2} - \frac{\sqrt{0.6}}{2}\right)^5 + \frac{8}{9}\left(\frac{1}{2} + 0\right)^5 + \frac{5}{9}\left(\frac{1}{2} + \frac{\sqrt{0.6}}{2}\right)^5\right\}$$

$$= \frac{1}{2}\left\{\frac{5}{9}(0.00002) + \frac{1}{36} + \frac{5}{9}(0.54998)\right\}$$

$$= 0.16667$$

The analytical solution is 1/6, or 0.166666... .

Table 6.3 Abscissas and Coefficients for Gauss-Legendre Quadrature

n	x_j	a_j
2	±0.57735 02691 89626	1.00000 00000 00000
3	±0.77459 66692 41483 0.00000 00000 00000	0.55555 55555 55556 0.88888 88888 88889
4	±0.86113 63115 94053 ±0.33998 10435 84856	0.34785 48451 37454 0.65214 51548 62546
5	±0.90617 98459 38664 ±0.53846 93101 05683 0.00000 00000 00000	0.23692 68850 56189 0.47862 86704 99366 0.56888 88888 88889
6	±0.93246 95142 03152 ±0.66120 93864 66265 ±0.23861 91860 83197	0.17132 44923 79170 0.36076 15730 48139 0.46791 39345 72691

Problem 6.7.   Write a program to approximate the integral of a function using an
               iterative six-point Gauss-Legendre quadrature formula. Test the
               routine with the same integral used in the previous integration
               programs.

     Solution. See the following program.

6.4.2   GAULEG

Introduction

     GAULEG uses an iterative six-point Gauss-Legendre quadrature formula to
approximate the integral

$$\int_a^b F(x)\ dx$$

The input to the program includes the integration limits a and b, the maximum number
of iterations, and the convergence criterion. The function $F(x)$ is coded in an
arithmetic statement function. GAULEG prints the input data for verification
purposes and the progress of the iterative Gauss-Legendre algorithm.

Method

     The following approximation is used

$$\int_a^b F(x)\ dx \simeq \frac{b-a}{2} \sum_{j=1}^{6} A_j F\left\{\frac{a+b}{2} + \frac{b-a}{2}\,x_j\right\}$$

where

$A_1 = A_4 = 0.171324492379170$ $x_1 = -x_4 = 0.932469514203152$

$A_2 = A_5 = 0.360761573048139$ $x_2 = -x_5 = 0.661209386466265$

$A_3 = A_6 = 0.467913934572691$ $x_3 = -x_6 = 0.238619186083197$

     In the first iteration GAULEG approximates the integral using this approximation.
Then the interval (a,b) is reduced by a factor of 2 and two summations are performed.
If the relative change in the resultant approximation to the integral is greater
than the convergence criterion, the interval is reduced by a factor of 3, and three
summations are performed. The procedure is repeated until convergence is attained
or ITMAX is exceeded.

Sample Execution

```
 ENTER ITMAX,A,B,& TOL
> 10   0   1   1D-5
```

```
      A =     0.000D+00       B =     0.100D+01

      ITMAX =  10     AND TOL =    0.100D-04


      ITN      H               INTEGRAL

       1   0.100D+01      0.2499999999D+00
       2   0.500D+00      0.2500000000D+00


      THE AREA IS     0.2499999999999203D+00
```

Listing

```
C
C                       GAULEG
C
C     ITERATIVE SIX-POINT GAUSS-LEGENDRE INTEGRATION ROUTINE
C
C
C     THE FUNCTION TO BE INTEGRATED IS CODED IN THE
C     ARITHMETIC STATEMENT FUNCTION, FUNC(X)
C
C      INPUT:
C
C      ITMAX    -    MAXIMUM NUMBER OF ITERATIONS
C      A        -    LOWER INTEGRATION LIMIT
C      B        -    UPPER INTEGRATION LIMIT
C      TOL      -    RELATIVE CONVERGENCE CRITERION
C
C
C   AUTHOR:  K. J. JOHNSON
C
C
       IMPLICIT DOUBLE PRECISION (A-H,O-Z)
       DIMENSION X(3),ALF(3)
C
C   INITIALIZE ABSCISSAS AND COEFFICIENTS
C
       DATA X/0.932469514203152D0,0.661209386466265D0,
      1 0.238619186083197D0/
       DATA ALF/0.171324492379170D0,0.360761573048139D0,
      1 0.467913934572691D0/
       FUNC(X)=X*DLOG(1.0D0+X)
C
C   INPUT
C
       WRITE(6,11)
       READ(5,22)ITMAX,A,B,TOL
       WRITE(6,33)A,B,ITMAX,TOL
C
C   SET CONSTANTS
C
       AO=A
       BO=B
       OLD=1.0D35
C
C   OUTERMOST LOOP (100) CONTROLS ITERATION
C
```

```
      DO 100 ITN=1,ITMAX
      H=(BO-AO)/ITN
      AREA=0.0D0
      A=AO-H
C
C   MIDDLE LOOP (50) COMPUTES AREA WITH H
C
      DO 50 I=1,ITN
      A=A+H
      B=A+H
      C=(B-A)/2.D0
      D=(A+B)/2.D0
C
C   INNERMOST LOOP (10) USES THE 6-POINT ALGORITHM
C
      DO 10 J=1,3
      T=C*X(J)
      ARG1=-T+D
      ARG2=T+D
   10 AREA=AREA+ALF(J)*(FUNC(ARG1)+FUNC(ARG2))
   50 CONTINUE
      AREA=AREA*C
      WRITE(6,66)ITN,H,AREA
      IF(DABS((AREA-OLD)/AREA).LT.TOL)GO TO 200
      OLD=AREA
  100 CONTINUE
      WRITE(6,99) AREA
      STOP
  200 WRITE(6,88)AREA
      STOP
   11 FORMAT(//'  ENTER ITMAX,A,B,& TOL')
   22 FORMAT(I,3D)
   33 FORMAT(//'  A = ',G12.3,6X,'B =',G12.3,
     1//'  ITMAX = ',I3,5X,'AND TOL =',D12.3///,
     2 2X,'ITN',6X,'H',11X,'INTEGRAL'/)
   66 FORMAT(I4,D12.3,D20.10)
   88 FORMAT(//'  THE AREA IS ',D25.16//)
   99 FORMAT(//'  THE TOLERANCE HAS NOT BEEN MET.'/,
     1 '  THE BEST APPROXIMATION TO THE AREA IS',D25.16//)
      END
```

Discussion

GAULEG requires 465 words of memory. This sample execution took approximately 0.1 sec. GAULEG was tested using the following family of integrals:

$$\int_0^1 x^n \, dx = \frac{1}{m+1}$$

The following results were obtained:

n	Area
1	0.5000000000000000D+00
2	0.3333333333333333D+00
3	0.2500000000000000D+00
4	0.2000000000000000D+00
5	0.1666666666666667D+00

n	Area
6	0.1428571428571429D+00
7	0.1250000000000000D+00
8	0.1111111111111111D+00
9	0.1000000000000000D+00
10	0.9090909090909092D-01
11	0.8333333333333335D-01
12	0.7692307692290740D-01
13	0.7142857142853653D-01
14	0.6666666666650605D-01
15	0.6249999999940620D-01

A six-point formula should be exact for functions of order 11 $(2n - 1)$ or less. Analytically,

$$\frac{1}{11} = 0.09090909\cdots \qquad\qquad \frac{1}{12} = 0.08333333\cdots$$

$$\frac{1}{13} = 0.076923076923077 \qquad\qquad \frac{1}{14} = 0.071428571428571$$

$$\frac{1}{15} = 0.066666666666667 \qquad\qquad \frac{1}{16} = 0.06250000\cdots$$

6.4.3 Gauss-Laguerre Quadrature

The Laguerre polynomials $L_n(x)$ are orthogonal over the interval $(0, \ )$, with the weighting function

$$w(x) = e^{-x}$$

That is,

$$\int_0^\infty e^{-x} L_n(x) L_m(x)\ dx = \begin{cases} 0 & \text{if } m \ne n \\ \text{constant} & \text{if } m = n \end{cases}$$

The first few Laguerre polynomials are [7]

$$L_0(x) = 1$$
$$L_1(x) = -x + 1$$
$$L_2(x) = x^2 - 4x + 2$$
$$L_3(x) = -x^3 + 9x^2 - 18x + 6$$

The Laguerre polynomials are generated by the recursion formula

$$L_n(x) = (2n - x - 1)L_{n-1}(x) - (n - 1)^2 L_{n-2}(x)$$

The Gauss-Laguerre quadrature formula is

$$\int_0^\infty e^{-x} F(x)\ dx \simeq \sum_{j=1}^n A_j F(x_j) \qquad\qquad\qquad (6.7)$$

Here x_j are the roots of $L_n(x)$, and

$$A_j = \frac{(n!)^2}{x_j[L_n'(x_j)]^2}$$

A table containing values for x_j and the corresponding A_j are given in Table 6.4 [7]. Abscissas and coefficients accurate to 30 significant figures and for n up to 68 have been calculated [10].

Table 6.4 Abscissas and Coefficients for Gauss-Laguerre Quadrature

n	x_j	A_j
2	0.585786437627	0.853553390593
	3.414213562373	0.146446609407
3	0.415774556783	0.711093009929
	2.294280360279	0.278517733569
	6.289945082937	0.010389256502
4	0.322547689619	0.603154104342
	1.745761101158	0.357418692438
	4.536620296921	0.0388879085150
	9.395070912301	0.000539294705561
5	0.263560319718	0.521755610583
	1.413403059107	0.398666811083
	3.596425771041	0.0759424496817
	7.085810005859	0.00361175867992
	12.640800844276	0.0000233699723858

Problem 6.8. Write a Gauss-Laguerre quadrature routine using the five-point formula. Test the algorithm with the following polynomial functions:

$$\int_0^\infty e^{-x} x^n \, dx \qquad n = 1, 2, 3, \ldots, 10$$

Solution.

```
C
C     GAUSS-LAGUERRE QUADRATURE - 5 POINT FORMULA
C
      IMPLICIT DOUBLE PRECISION (A-Z)
      INTEGER I,JJ
      DIMENSION X(5),A(5)
      DATA X/0.263560319718D0,1.413403059107D0,3.596425771041D0,
     1 7.0858100005859D0,12.640800844276D0/
      DATA A/0.521755610583D0,0.398666811083D0,0.759424496817D-1,
     1 0.361175867992D-2,0.233699723858D-4/
C
C     TEST FUNCTION:   F(X) = X**I,   I=1,2,3,....
C
      F(Z)=Z**JJ
      DO 20 JJ=1,10
      AREA=0.D0
```

```
      DO 10 I=1,5
      Z=X(I)
      T=A(I)*F(Z)
  10  AREA=AREA+T
  20  WRITE(6,22)JJ, AREA
  22  FORMAT(/'  JJ = ',I3,10X,  'AREA = ',1PD20.13)
      END
```

The sample execution of this routine follows.

        JJ = 1   AREA = 9.9999999998086D-01

        JJ = 2   AREA = 1.9999999997302D+00

        JJ = 3   AREA = 5.9999999971318D+00

        JJ = 4   AREA = 2.3999999972898D+01

        JJ = 5   AREA = 1.1999999975994D+02

        JJ = 6   AREA = 7.1999999795878D+02

        JJ = 7   AREA = 5.0399999831263D+03

        JJ = 8   AREA = 4.0319999863363D+04

        JJ = 9   AREA = 3.6287999891088D+05

        JJ = 10   AREA = 3.6143999914264D+06

The analytical solution to the integral is [11]

$$\int_0^\infty e^{-x} x^n \, dx = n!$$

n	1	2	3	4	5	6	7	8	9	10
n!	1	2	6	24	120	720	5040	40320	362,880	3,628,800

The algorithm gives the correct result for polynomial functions of order 9 ($2n - 1$) or less. The relative error for the 10th-order polynomial is approximately 0.4%.

Equation (6.7) can be transformed to include a finite lower limit as follows [12]:

$$\int_a^\infty e^{-x} F(x) \, dx \simeq e^{-a} \sum_{j=1}^n A_j F(ax_j)$$

6.4.4 Gauss-Hermite Quadrature

The Hermite polynomials $H_n(x)$ are orthogonal over the interval $(-\infty, \infty)$, with weighting function,

$$w(x) = e^{-x^2}$$

$$\int_{-\infty}^\infty e^{-x^2} H_m(x) H_n(x) \, dx = \begin{cases} 0 & \text{if } n \neq m \\ \text{constant} & \text{if } n = m \end{cases}$$

The first few Hermite polynomials are [7]

$$H_0(x) = 1$$

$$H_1(x) = 2x$$

$$H_2(x) = 4x^2 - 2$$

$$H_3(x) = 8x^3 - 12x$$

The recursion formula for the Hermite polynomials is

$$H_n(x) = 2xH_{n-1}(x) - 2(n - 1)H_{n-2}(x)$$

The abscissas are the roots of $H_n(x)$ and the coefficients are

$$A_j = \frac{2^{n+1}n! \sqrt{\pi}}{[H_n'(x_j)]^2}$$

Table 6.5 contains some abscissas and coefficients for Gauss-Hermite quadrature formulas [5]. Abscissas and coefficients accurate to 30 significant figures up to n = 136 have been calculated [13].

Table 6.5. Abscissas and Coefficients for Gauss-Hermite Quadrature

n	x_j	A_j
2	±0.707106781187	0.886226925453
3	0	1.181635900604
	±1.224744871392	0.295408975151
4	±0.524647623275	0.804914090006
	±1.650680123886	0.081312835447
5	0	0.945308720483
	±0.958572464614	0.393619323152
	±2.020182870456	0.019953242059
6	±0.436077411928	0.724629595224
	±1.335849074014	0.157067320323
	±2.350604973674	0.00453000990551

Problem 6.9. Write a Gauss-Hermite quadrature routine using the six-point formula. Test the routine with the following polynomial functions:

$$\int_{-\infty}^{\infty} e^{-x^2} F(x)\ dx$$

where

$$F(x) = x^{2n}, \qquad n = 1,2,3,\ldots$$

Solution.

```
C
C     GAUSS-HERMITE QUADRATURE - 6 POINT FORMULA
C
```

```
      IMPLICIT DOUBLE PRECISION (A-Z)
      INTEGER I,N
      DIMENSION X(3),A(3)
      DATA X/0.436077411928D0,1.335849074014D0,2.350604973674D0/
      DATA A/0.724629595224D0,0.157067320323D0,0.453000990551D-2/
C
C   TEST FUNCTION:  F(X) = X**(2*N),      N=1,2,3,...
C
      F(Z)=Z**(2*N)
      DO 20 N=1,6
      JJ=2*N
      AREA=0.D0
      DO 10 I=1,3
      Z=X(I)
      T=A(I)*(F(Z)+F(-Z))
   10 AREA=AREA+T
   20 WRITE(6,22)N,AREA
   22 FORMAT(/' N = ',I3,10X,' AREA = ',1PD20.13)
      END
```

The sample execution of this program follows.

N =	1	AREA =	8.8622692545385D-01
N =	2	AREA =	1.3293403881810D+00
N =	3	AREA =	3.3233509704504D+00
N =	4	AREA =	1.1631728396564D+01
N =	5	AREA =	5.2342777784486D+01
N =	6	AREA =	2.6794517199181D+02

The analytical solution to this integral is [11]

$$\int_{-\infty}^{\infty} x^{2n} e^{-x^2}\, dx \qquad n = 1,2,3,\ldots$$

$$= \frac{1(3)(5)\cdots(2n - 1)\ \sqrt{\pi}}{2^n}$$

The results are correct for the polynomials of order 10 or less. The error in the 12th-order polynomial is approximately 7%.

6.4.5   Gauss-Chebyshev Quadrature

The Chebyshev polynomials $T_n(x)$ are orthogonal on the interval $(-1,1)$, with respect to the weighting function,

$$w(x) = \frac{1}{\sqrt{1 - x^2}}$$

$$\int_{-1}^{1} \frac{T_m(x)T_n(x)\ dx}{\sqrt{1 - x^2}} = \begin{cases} 0 & \text{if } n \neq m \\ \text{constant} & \text{if } n = m \end{cases}$$

The first few Chebyshev polynomials are [7]

$$T_0(x) = 1$$

$$T_1(x) = x$$

$$T_2(x) = 2x^2 - 1$$

$$T_3(x) = 4x^3 - 3x$$

The general recursion relation is

$$T_n(x) = 2xT_{n-1}(x) - T_{n-2}(x)$$

The Gauss-Chebyshev quadrature formula is

$$\int_{-1}^{1} \frac{F(x)\ dx}{\sqrt{1 - x^2}} \simeq \sum_{j=1}^{N} A_j F(x_j)$$

where

$$A_j = \frac{\pi}{n} \qquad x_j = \cos \frac{(2j - 1)\pi}{2n}$$

Problem 6.10. Transform the Gauss-Chebyshev quadrature formula to the integral
interval (a,b) with the unit weighting function, and write a program
to test the algorithm using the integral

$$\int_{-1}^{1} \frac{T_n(x)T_n(x)\ dx}{\sqrt{1 - x^2}} = \frac{\pi}{2} \qquad (m = n \neq 0)$$

Use the fifth Chebyshev polynomial,

$$T_5(x) = 16x^5 - 20x^3 + 5x$$

Also compare the results of Gauss-Chebyshev quadrature and Gauss-
Legendre quadrature using the integral in previous examples.

$$\int_{0}^{1} x \ln(1 + x) = \frac{1}{4}$$

Solution. The transformed equation is [14]

$$\int_{a}^{b} F(x)\ dx \simeq \frac{(b - a)}{2n} \sum_{i=1}^{n} F(x_i) \sqrt{1 - z_1^2}$$

where

$$x_i = \frac{z_i(b - a) + a + b}{2} \qquad z_i = \cos \frac{(2i - 1)}{2n}$$

The algorithm to test the procedure is given next.

```
C
C     ITERATIVE GAUSS-CHEBYSHEV QUADRATURE
C
      IMPLICIT DOUBLE PRECISION (A-Z)
      INTEGER I,N,ITMAX,ITN
      DATA PI/3.14159265359D0/
      F(XI)=(16.*XI**5-20.*XI**3+5*XI)**2/(DSQRT(1.0D0-XI*XI))
      WRITE(6,11)
      READ(5,22)ITMAX,A,B,TOL
      WRITE(6,23)A,B,TOL
      BMA=B-A
      AREA0=1.0D35
      N=1
      DO 100 ITN=1,ITMAX
      N=N*2
      CONST=PI/(2.0D0*FLOAT(N))
      AREA=0.0D0
      DO 10 I=1,N
      ZI=DCOS( FLOAT( 2*I-1 )*CONST )
      XI=( ZI*BMA + A + B )/2.0D0
   10 AREA=AREA + F(XI)*DSQRT( 1.0D0-ZI*ZI )
      AREA=AREA*BMA*CONST
      WRITE(6,33) N,AREA
      IF(ABS((AREA-AREA0)/AREA) .LE. TOL) GO TO 200
  100 AREA0=AREA
      WRITE(6,44)
      STOP
  200 WRITE(6,55)AREA
   11 FORMAT(//'  ENTER ITMAX, A, B, AND TOL')
   22 FORMAT(I,3D)
   23 FORMAT(//'   A= ',1PD10.2,5X,'B = ',1PD10.2,5X,
     1 'TOL = ',1PD10.2///,'     N',9X,'AREA'/)
   33 FORMAT(I5,1PD20.10)
   44 FORMAT(//'  SORRY, ITMAX EXCEEDED'//)
   55 FORMAT(//10X,'THE AREA IS ',1PD20.12)
      END
```

The sample execution of this program follows.

```
 ENTER ITMAX, A, B, AND TOL
> 10  -1  1  1.0D-12


 A =  -1.00D+00     B =   1.00D+00     TOL =   1.00D-12


     N          AREA

     2    1.5707963268D+00
     4    1.5707963268D+00


        THE AREA IS   1.570796326794D+00
```

The modification required to integrate other functions is simply the replacement of the arithmetic statement function, for example,

F(XI)=XI*DLOG(1.0D0+XI)

The sample execution of the program with this change follows.

ENTER ITMAX, A, B, AND TOL
> 10   0   1   .00001

A =   0.00D+00   B =   1.00D+00   TOL =   1.00D-05

N	AREA
2	3.0364113470D-01
4	2.5957434163D-01
8	2.5226402103D-01
16	2.5055899015D-01
32	2.5013932274D-01
64	2.5003480434D-01
128	2.5000869944D-01
256	2.5000217476D-01
512	2.5000054368D-01

THE AREA IS   2.500005436830D-01

6.5   THE MONTE-CARLO METHOD

For some applications the proceding methods fail because of numerical difficulties, for example, singularities in the function, or because they are too costly in computer time. The Monte-Carlo method uses random numbers to roughly approximate integrals. For example, consider the integral

$$\int_0^1 \sqrt{1 - x^2} \ dx = \frac{\pi}{4}$$

This is one-fourth of the unit circle. Given a random number generator that produces uniformly distributed random numbers between 0 and 1, one can generate a set of random coordinates,

$$(x_i, y_i) \qquad i = 1,2,3,\ldots,N$$

The procedure can be described as a simulated dart game (Monte Carlo) in which the target is the area under the curve and the dart board is the unit square (see Figure 6.2). The random point (x_i', y_i) is a "hit" if

$$y_i \leq \sqrt{1 - x_i^2}$$

That is, if y_i falls on or under the curve.

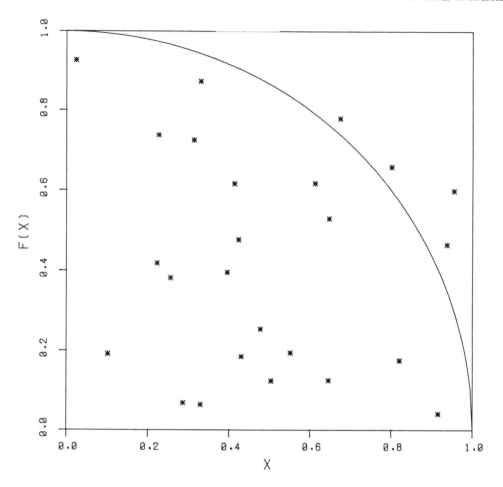

<u>Figure 6.2</u> Illustration of the Monte Carlo technique.

The area under the curve is approximated by

$$\int_0^1 \sqrt{1 - x^2} \, dx \simeq \frac{\text{No. of hits}}{N} \ A_{tot}$$

Here A_{tot} is the enclosing area, which for the unit square is unity, and N is the
number of random coordinates used.

The following program uses the Monte Carlo method to approximate the area of
the unit circle $(3.14159\cdots)$ and also tests the distribution of the random numbers
generated by the DECSystem-10 pseudo-random number routine (RAN).

6.5.1 MONCAR

Introduction

MONCAR demonstrates the Monte Carlo method of numerical integration by approximating the integral

$$\int_0^1 \sqrt{1 - x^2}\ dx = \frac{\pi}{4}$$

MONCAR reads N, the number of random numbers, which are assumed to be uniformly distributed between 0 and 1. The program acquires 2N random numbers, approximates π, calculates the average of the random numbers, and calculates the distribution of the first digits in the random numbers. A perfect uniform-distribution random number generator would have an average of 0.50, and an observed probability distribution,

$$p(0) = p(1) = p(2) = \cdots = p(9) = 0.10$$

Here $p(i)$ is the probability that the first digit in the random number is the integer i.

Method

The method is illustrated with 25 random numbers in Figure 6.2. The fraction of random points that fall on or under the curve approximates the area,

AREA - NHITS*ATOT/NTOT

where NHITS is the number of points that fall on or under the curve; ATOT is the total area here simply the area of the unit square (1.0); and NTOT is the number of random points. From Figure 6.2, NTOT = 25 and NHITS = 21. The calculated area is 21/25, or 0.84. The correct area is $\pi/4$, or 0.7854.

The average of the random numbers is calculated using.

$$AVG = \frac{\sum_{i=1}^{NTOT} (XR + YR)}{2NTOT} \ -$$

The approximation to the area should improve with the number of random points used.

Sample Execution.

ENTER N
>100

THE DISTRIBUTION OF FIRST DIGITS FOR 200 RANDOM NUMBERS IS:

0	1	2	3	4	5	6	7	8	9
16	22	25	26	29	21	16	13	18	14

THE AVERAGE IS   0.46431

AND THE CALCULATED VALUE OF PI IS   3.4400

ENTER N
>400

THE DISTRIBUTION OF FIRST DIGITS FOR   1000 RANDOM NUMBERS IS:

0	1	2	3	4	5	6	7	8	9
101	88	114	127	100	95	103	83	101	88

THE AVERAGE IS   0.48547

AND THE CALCULATED VALUE OF PI IS   3.1920

ENTER N
>2000

THE DISTRIBUTION OF FIRST DIGITS FOR   5000 RANDOM NUMBERS IS:

0	1	2	3	4	5	6	7	8	9
458	466	567	524	502	480	522	497	505	479

THE AVERAGE IS   0.50020

AND THE CALCULATED VALUE OF PI IS   3.1552

ENTER N
>2500

THE DISTRIBUTION OF FIRST DIGITS FOR   10000 RANDOM NUMBERS IS:

0	1	2	3	4	5	6	7	8	9
946	991	1084	1021	1017	957	1000	990	1019	975

THE AVERAGE IS   0.49945

AND THE CALCULATED VALUE OF PI IS   3.1320


```
      Listing
C
C                      MONCAR
C
C   MONCAR APPROXIMATES PI BY USING THE MONTE-CARLO METHOD
C   TO INTEGRATE THE UPPER-RIGHT QUADRANT OF THE UNIT CIRCLE.
C   MONCAR ALSO CHECKS THE RANDOM NUMBER GENERATOR BY
C   COMPUTING THE AVERAGE AND DISTRIBUTION OF N RANDOM NUMBERS
C
```

```
C
C     GLOSSARY
C
C         N   -   NUMBER OF RANDOM NUMBERS
C         XR  -   RANDOM VALUE ALONG THE X-AXIS
C         YR  -   RANDOM VALUE ALONG THE Y-AXIS
C       NUMB  -   THE DISTRIBUTION OF FIRST DIGITS IN XR,YR
C
C
C     AUTHOR:   K. J. JOHNSON
C
      DIMENSION NUMB(0/9)
      DATA SUM,NTOT,HITS,NUMB,ATOT/13*0.,1./
C
C        STATEMENT FUNCTION FOR THE INTEGRAL TO BE APPROXIMATED
C
      F(X)=SQRT(1.-X*X)
C
C     START THE RANDOM NUMBER GENERATOR WITH THE CURRENT TIME
C
      CALL TIME(A,B)
      CALL SETRAN(B)
C
C
C     INPUT
C
   10 WRITE(6,11)
      READ(5,22)N
      IF(N.LE.0)STOP
      DO 20 I=1,N
      XR=RAN(XR)
      YR=RAN(YR)
      SUM=SUM+XR+YR
      IXR=IFIX(XR*10.)
      IYR=IFIX(YR*10.)
      NUMB(IXR)=NUMB(IXR)+1
      NUMB(IYR)=NUMB(IYR)+1
C
C     IF YR FALLS ON OR UNDER THE FUNCTION, SCORE A HIT
C
   20 IF(YR.LE.F(XR)) HITS=HITS+1.
      NTOT=NTOT+N
      NTOT2=2*NTOT
      AVG=SUM/NTOT2
      AREA=ATOT*HITS/NTOT
      PI=4.*AREA
      WRITE(6,33)NTOT2,(I,I=0,9),(NUMB(I),I=0,9),AVG,PI
      GO TO 10
   11 FORMAT(//'   ENTER N'/)
   22 FORMAT(I)
   33 FORMAT(//'   THE DISTRIBUTION OF FIRST DIGITS FOR',I7,
     1 ' RANDOM NUMBERS IS:'//,
     3 10I6/1X,10I6//5X,'THE AVERAGE IS ',G12.5,
     4 //5X,'AND THE CALCULATED VALUE OF PI IS ',G12.5//)
      END
```

Discussion

MONCAR requires 72 words of core. This sample execution took approximately 0.6 sec.

The Monte Carlo method used to approximate the integrals of numerically complex functions for which other methods are not appropriate [15]. For more accurate work, a subroutine to generate more uniformly distributed random numbers is available [16]. Additional applications of the Monte Carlo technique in chemistry are discussed in Section 7.6.

6.6 ADDITIONAL PROBLEMS

1. Evaluate the following integrals:

(a) $\displaystyle\int_0^1 x^x \, dx$

(b) $\displaystyle\int_0^{\pi/2} \frac{\cos x \, dx}{1 - x}$

(c) $\displaystyle\int_0^{\pi/2} (1 - \frac{\sin^2 x}{4}) dx$

(d) $\displaystyle\int_0^1 \exp(-\exp(-x)) dx$

(e) $\displaystyle\int_0^1 \frac{\sin^{3/2} x}{x^2} = dx$

(f) $\displaystyle\int_0^\infty \frac{e^{-x^2}}{1 + x^2}$

(g) $\displaystyle\int_0^\infty e^{-x} \ln (x) \, dx$

(h) $\displaystyle\int_0^\infty e^{-\frac{x + 1}{x}} \, dx$

Solutions [17]. (a) 0.7834 (b) 0.6736 (c) 1.4675 (d) 0.54003 (e) 1.9049
                     (f) 0.6716 (g) -0.57722 (h) 0.2797

2. The error function erf(x) is defined by the following integrals

$$\text{erf}(x) = \frac{2}{\sqrt{\pi}} \int_0^\infty e^{-t^2} \, dt = (1 - \frac{2}{\sqrt{\pi}}) \int_x^\infty e^{-t^2} \, dt$$

Evaluate erf (x) using both of these formulas. Compare the results with tabulated values, for example, erf(0.5) ≃ 0.5205 [18].

3. Evaluate the integral

$$\int_{-\infty}^\infty \frac{x^2}{1 + x^4} \, dx$$

Solution [18]. $\dfrac{\pi}{2 \sin(\pi/4)}$.

4. Evaluate the gamma function [18],

$$\Gamma(a) = \int_0^\infty e^{-x} x^{a-1} \, dx$$

Compare the results with tabulated values, for example, $\Gamma(1.0) = 0.88623$.

5. Evaluate the exponential integral,

$$E(a) = \int_a^\infty \frac{e^{-x}}{x} \, dx$$

Compare the results with tabulated values, for example, $E(1) = 1.89512$ [19].

6. The fugacity f (atm) of a gas at a pressure P (atm) and a specified temperature is given by

$$\ln \frac{f}{P} = \int_0^P \frac{z-1}{P} \, dP$$

where z is the compressibility function $(x = PV/RT)$. Evaluate the fugacity as a function of pressure using the following data [20].

P	Compressibility CH_4 $(-70^\circ C)$	Compressibility NH_3 $(200^\circ C)$	P	Compressibility CH_4 $(-70^\circ C)$	Compressibility NH_3 $(200^\circ C)$
1	0.9940	0.9975	140	0.4753	
10	0.9370	0.9805	160	0.5252	
20	0.8683	0.9611	180	0.5752	
30	0.7928	0.9418	200	0.6246	0.5505
40	0.7034	0.9219	250	0.7468	
50	0.5936	0.9020	300	0.8663	0.4615
60	0.4515	0.8821	400	1.0980	0.4948
80	0.3429	0.8411	500	1.3236	0.5567
100	0.3767	0.8008	600	1.5409	0.6212
120	0.4259		800	1.9626	0.7545

7. Write a program to approximate a periodic function by a Fourier expansion,

$$G(x) \quad \sum_{m=0}^N C_m \cos mx + \sum_{m=0}^N D_m \sin mx$$

where

$$C_m = \frac{1}{\pi} \int_{-\pi}^\pi G(x) \cos mx \, dx$$

and

$$D_m = \frac{1}{\pi} \int_{-\pi}^{\pi} G(x) \sin mx\, dx$$

Solution. See Ref. 21-23.

8. The following values for the heat capacity of nitromethane as a function of absolute temperature have been observed [24]:

T	C_o	T	C_o
15	0.89	120	13.56
20	2.07	140	14.45
30	4.59	160	15.31
40	6.90	180	16.19
50	8.53	200	17.08
60	9.76	220	17.98
70	10.70	240	18.88
80	11.47	260	25.01
90	12.10	280	25.17
100	12.62	300	25.35

The melting point is 244.7 K, and the heat of fusion is 2319 cal/mole. The vapor pressure of the liquid at 298 K is 36.66 torr. The heat of vaporization at 298 K is 9147 cal/mol. Calculate the third-law entropy of nitromethane gas at 298 K and 1 atm.

Solution. 65.7 cal/mole· K [24].

9. A binary mixture is separated by gas chromatography. Two clearly separated peaks appear on the gas chromatogram [25].

Time	Intensity	Time	Intensity
40	3	130	4
53	17	250	0.5
64	40	258	6
75	69	266	14
86	90	274	21
97	68	282	13
108	43	290	5
119	19	298	0
130	4		

Estimate the relative amounts of the two substances.

Solution. Approximately 8:1.

10. The theoretical heat capacity of a solid is given by the Debye-Einstein equation,

$$C_v = \frac{9Nk}{x_0^3} \int_0^{x_0} \frac{x^4 e^x}{(e^x - 1)^2}$$

where $x = h\nu/kT$, $x_0 = h\nu_m/kT = T_0/T$, with ν_m the maximum vibration frequency, T_0 the Debye temperature, N Avogadro's number, k Boltzman's constant, and T the absolute temperature. Some values for T_0 (in K) are Al, 398; Na, 159; Cu, 315; Ag, 215; Fe, 420; KCl, 227; NaCl, 281; and AgCl, 183. Evaluate the heat capacities of these materials as a function of temperature and compare them to experimental values [26].

11. The fraction of electromagnetic radiation emitted by a black-body radiator between the wavelengths λ_1 and λ_2 is given by

$$f_{\lambda_1 \lambda_2} = \frac{8\pi hc}{7.56 \times 10^{-15} T^4} \int_{\lambda_1}^{\lambda_2} \frac{d\lambda}{\lambda^5 \exp(hc/\lambda kT - 1)}$$

Here h is planck's constant, c is the speed of light, T is the absolute temperature, and λ is wavelength. Calculate $f_{\lambda_1 \lambda_2}$ for various regions of the electromagnetic spectrum as a function of temperature [26, 27].

12. An early application of computers in quantum chemistry was to calculate the ground-state energy of H_3 and H_3^+ [28]. Write algorithms to verify the tabulated integrals. Compare this problem to a more recent one [29].

13. Calculate the electronic energy levels of the ions isoelectronic with Li, that is, Be^+, B^{2+}, C^{3+}, etc. [30].

14. Calculate the expectation values of the kinetic energy, potential energy, and the total energy of the H_2^+ molecule as a function of the internuclear distance R and the effective nuclear charge Z [31].

REFERENCES

1. B. Carnahan et al., "Applied Numerical Methods," Wiley, New York, 1969, pp 71-73.

2. B. Carnahan et al., ibid., p. 73-90.

3. C. Froberg, "Introduction to Numerical Analysis," Addison-Wesley, Reading, Mass., 1965, pp. 175-179.

4. A. H. Stroud and D. Secrest, "Gaussian Quadrature Formulas," Prentice-Hall, Englewood Cliffs, N. J., 1966.

5. C. Froberg, op. cit., pp. 181-190.

6. B. Carnahan et al., op. cit., pp. 100-127.

7. B. Carnahan et al., ibid., pp. 100-101.

8. A. H. Stroud and D. Secrest, op. cit., pp. 99-151.

9. B. Carnahan et al., op. cit., p. 104.

10. A. H. Stroud and D. Secrest, ibid., pp. 253-274.

11. R. C. Weast (ed.), "Handbook of Chemistry and Physics," 45th Ed., Chemical
    Rubber Co., Cleveland, 1964, p. A-140.

12. B. Carnahan et al., op. cit., p. 115.

13. A. H. Stroud and D. Secrest, op. cit., pp. 217-251.

14. B. Carnahan et al., op. cit., p. 116.

15. A. Ralston and H. S. Wilf (eds.), "Mathematical Methods for Digital Computers,"
    Wiley, New York, Vol. 1, Chapter 23, and Vol. 2, Chapter 6, and references
    therein.

16. G. E. Forsythe, M. A. Malcom, and C. B. Moler, "Computer Methods for Mathe-
    matical Computations," Prentice-Hall, Englewood Cliffs, N. J., 1977, pp.
    240-246.

17. C. Froberg, op. cit., pp. 199-200.

18. B. Carnahan et al., op. cit., p. 134.

19. R. C. Wext, op. cit., pp. A-142,143.

20. B. Carnahan et al., op. cit., p. 113, and references therein.

21. B. Carnahan et al., ibid., pp. 92-99.

22. E. Kregszig, "Advanced Engineering Mathematics," Wiley, New York, 1967,
    Chapter 8.

23. J. W. Cooper, "The Minicomputer in the Laboratory," Wiley, New York, 1977,
    Chapter 18.

24. W. J. Moore, "Physical Chemistry," 3rd Ed., Prentice-Hall, Englewood Cliffs,
    N. J., 1963, p. 206; a similar problem is given in the fourth edition, p. 115.

25. J. B. Dence, "Mathematical Techniques in Chemistry," Wiley, New York, 1975,
    p. 127.

26. T. R. Dickson, "The Computer and Chemistry," Freeman, San Francisco, 1968,
    pp. 133-134.

27. B. Carnahan et al., op. cit., p. 206.

28. J. O. Hirschfelder and C. N. Weygandt, Integrals required for the energy of H_3 and H_3^+, <u>J. Chem. Phys</u>., <u>6</u>, 806 (1938).

29. M. E. Blakemore and G. A. Hyslop, A numerical integration scheme for iterative calculations on atomic systems, <u>Int. J. Quant. Chem</u>., <u>11</u>, 325 (1977).

30. D. S. Alderdice and R. S. Watts, Calculation of atomic energy levels, <u>J. Chem. Educ</u>., <u>47</u>, 123 (1970).

31. A. G. Robiette, The variation theorem applied to H_2^+, <u>J. Chem. Educ</u>., <u>52</u>, 95 (1975).

CHAPTER 7

DIFFERENTIAL EQUATIONS

Time- and space-dependent physical phenomena are usually represented by differential equations. For example, the kinetic system

$$A + B \underset{k_2}{\overset{k_1}{\rightleftharpoons}} C + D$$

$$B + C \xrightarrow{k_3} D + E$$

is described by the following set of linear, first-order differential equations,

$$\frac{d[A]}{dt} = k_2[C][D] - k_1[A][B]$$

$$\frac{d[B]}{dt} = k_2[C][D] - k_1[A][B] - k_3[B][C]$$

$$\frac{d[C]}{dt} = k_1[A][B] - k_2[C][D] - k_3[B][C]$$

$$\frac{d[D]}{dt} = k_1[A][B] - k_2[C][D] + k_3[B][C]$$

and

$$\frac{d[E]}{dt} = k_3[B][C]$$

There is evidence [1] that the reaction

$$Tl^+ + 2Co^{3+} \longrightarrow Tl^{3+} + 2Co^{2+}$$

proceeds by this mechanism, where

$$A = Tl^+ \qquad B = Co^{3+}$$

$$C = Tl^{2+} \qquad D = Co^{2+}$$

$$E = Tl^{3+}$$

The solution to the set of five-coupled differential equations depends on the initial values,

$$[t_0; A(t_0), B(t_0), C(t_0), D(t_0), E(t_0)]$$

and the values of the rate constants, k_1, k_2, and k_3. If the rate constants span several orders of magnitude, the system is referred to as <u>stiff</u>. The methods outlined in this chapter approximate the solution to systems of differential equations provided they are not too stiff. The reader is referred to a more advanced text [2] for a discussion of numerical methods for stiff systems.

Three methods are considered in this chapter, the Euler method, the Runge-Kutta method, and the predictor-corrector method. These techniques are used to approximate the solution to initial value problems,

$$\frac{dy}{dx} = F(x,y) \qquad (x_0, y_0)$$

Here x_0 and y_0 are the initial values. Then the Runge-Kutta method is extended to solve systems of linear, first-order differential equations like the kinetics system above. Next the application of numerical methods to solving higher order differential equations is introduced, for example, the one-dimensional Schroedinger equation,

$$\frac{d^2 \Psi}{dx_2} + V(x)\Psi = E\Psi$$

Finally, the Monte-Carlo method is used to approximate the solution to a system of simultaneous equations.

7.1 EULER'S METHOD

Euler's method [3] for approximating the solution to a first-order differential equation is the forward difference approximation,

$$\frac{dy}{dx} = F(x,y) \approx \frac{\Delta y}{\Delta x} = \frac{y_{i+1} - y_i}{x_{i+1} - x_i}$$

Here y_{i+1} is the solution at x_{i+1}. Let $h = x_{i+1} - x_i$, then

$$\frac{y_{i+1} - y_i}{x_{i+1} - x_i} = \frac{y_{i+1} - y_i}{h} = F(x_i, y_i)$$

or

$$y_{i+1} \approx y_i + hF(x_i, y_i)$$

Given initial values (x_0, y_0),

$$y_1 = y_0 + hF(x_0, y_0)$$
$$y_2 = y_1 + hF(x_1, y_1)$$
$$\dots \dots \dots \dots \dots \dots$$
$$y_n = y_{n-1} + hF(x_{n-1}, y_{n-1})$$

Here

$$x_i = x_0 + ih \qquad i = 1,2,3,\dots,n$$

Problem 7.1. Use Euler's method to approximate the solution to the following
             initial value problem,

$$\frac{dy}{dx} = x + y \qquad (x_0 = y_0 = 0)$$

Find $y(x)$ for $x = 0.1, 0.2, 0.3, \ldots, 1.0$.

 Solution. Arbitrarily choosing $h = 0.1$,

$$y_1 = y(0.1) \approx y_0 + hF(x_0, y_0)$$

$$= 0 + 0.1F(0,0)$$

$$= 0$$

$$y_2 \approx y_1 + hF(x_1, y_1)$$

$$= 0 + 0.1F(0.1, 0)$$

$$= 0 + 0.1(0.1 + 0)$$

$$= 0.01$$

$$y_3 \approx y_2 + hF(x_2, y_2)$$

$$= 0.01 + 0.1F(0.2, 0.01)$$

$$= 0.01 + 0.1(0.2 + 0.01)$$

$$= 0.031$$

The analytical solution to this differential equation is

$$y = e^x - x - 1$$

The exact solution and numerical solution using Euler's method
with $h = 0.1$ are tabulated below.

x	y(Euler)	y(exact)
0	0	0
0.1	0	0.0052
0.2	0.01	0.0214
0.3	0.0310	0.0499
0.4	0.0641	0.0918
0.5	0.1105	0.1487
0.6	0.1716	0.2221
0.7	0.2487	0.3138
0.8	0.3436	0.4255
0.9	0.4579	0.5596
1.0	0.5937	0.7183

The method has clearly failed to accurately provide the solution.
The remedy is to reduce the step size h and try again. Choosing

$$h = 0.05,$$

$$y(0.05) \approx y_0 + hF(x_0, y_0)$$

$$= 0 + 0.05F(0,0)$$

$$= 0$$

$$y(0.1) \approx y(0.05) + 0.05F(0.05,0)$$

$$= 0 + 0.0025$$

$$= 0.0025$$

The value 0.0025 is still not a very good approximation to the analytical solution (0.0052), however, it is considerably closer than the value obtained with $h = 0.1$ (0). Now try $h = 0.025$,

$$y(0.025) \approx 0 + 0.025(0 + 0) = 0$$

$$y(0.05) \approx 0 + 0.025(0.025 + 0) = 0.000625$$

$$y(0.075) \approx 0.000625 + 0.025(0.05 + 0.00625) = 0.00189$$

$$y(0.1) \approx 0.00189 + 0.025(0.075 + 0.00189) = 0.00381$$

The method is improving as the step size decreases. The results of several binary step-size reductions is shown below:

h	Y(0.1)
0.1	0.0
0.05	0.0025
0.025	0.00381
0.0125	0.00449
0.00625	0.00483
0.003125	0.00500
0.0015625	0.00508
0.00078125	0.00513
0.000390625	0.00515
0.0001953125	0.00516

Analytically, $y(0.1) = 0.005171$

Problem 7.2. Write a program to approximate the solution to the initial value problem,

$$\frac{dy}{dx} = F(x,y) \qquad (x_0, y_0)$$

using an iterative Euler method.

Solution. See the following program.

7.1.1. EULER

Introduction

EULER approximates the solution to a first-order linear differential equation
with initial values using an iterative Euler method. The input to the program
includes the following: XO and YO, the initial values x_0 and y_0; XF, the final
value to be assigned to x; HO, the initial step size; HOUT, the output step size;
and TOL, the stability criterion. EULER evaluates the solution over the range

   $XO \leq X \leq XF$

and tabulates the solution for

   X = XO + i*HOUT   i = 1,2,3,...,NOUT

where

   NOUT = (XF-XO)/HOUT

The function F(x,y) is coded in the function subprogram, FUNK(X,Y).

   Method

EULER approximates the solution in successive steps of HO. The first
approximation is

   Y(XO+HO) = YO + HO*FUNK(X,Y)

The stability of the first approximation is tested by a binary reduction of the
step size,

   H = HO/2

Then two steps (XO+H and XO+2*H) are taken to obtain the second approximation. The
test for stability uses the following definition:

$$REL = \left| \frac{Y(H) - Y(HO)}{Y(H)} \right|$$

If REL $\leq$ TOL, then the solution is said to be stable. If this test fails, H is
reduced by successive factors of two until either a stable solution is found or the
procedure is terminated. ITMAX iterations are allowed before the program is stopped.
If the solution is stable, EULER assigns new "initial values."

   XO = XO + HO

   YO = YO + HO

and repeats the procedure. The results are printed when XO + HO is equal to one
of the values, XO + i*HOUT, i = 1, 2, 3, ..., NOUT. The value of H required to
obtain a stable solution is also printed.

   EULER consists of a main program and two subprograms,

   SUBROUTINE EULER(XO,YO,Y,H,K)

and

    FUNCTION FUNK(X,Y)

The subroutine EULER contains the code to take K steps starting with XO and YO, and
terminating with X = XO + KH. The value of K is 2^i, i = 0, 1, 2, 3, ..., ITMAX.
ITMAX has been arbitrarily set to 20, so the minimum value assigned to H is $XO/2^{20}$
or HO/1,048,570.

    ## Sample Execution
ENTER XO,YO,XF,HO,HOUT,TOL
> 0   0   1   0.1   0.1   0.0001

HO = 0.100E+00   HOUT = 0.100E+00   TOL = 0.100E-03

H	X	Y
0.000E+00	0.000E+00	0.00000E+00
0.610E-05	0.100E+00	0.51706E-02
0.244E-04	0.200	0.21401E-01
0.488E-04	0.300	0.49853E-01
0.977E-04	0.400	0.91811E-01
0.977E-04	0.500	0.14870
0.195E-03	0.600	0.22208
0.195E-03	0.700	0.31369
0.195E-03	0.800	0.42544
0.391E-03	0.900	0.55945
0.391E-03	1.00	0.71806

    ## Listing

```
C
C
C                     EULER
C
C    EULER APPROXIMATES THE SOLUTION TO A FIRST ORDER
C    DIFFERENTIAL EQUATION USING AN ITERATIVE
C    EULER ALGORITHM
C
C
C  GLOSSARY:
C
C         XO   -  INITIAL VALUE OF X
C         XF   -  FINAL VALUE OF X
C         YO   -  INITIAL VALUE OF Y
C         HO   -  MAX IMUM STEP SIZE
C       HOUT   -  OUTPUT STEP SIZE
C        TOL   -  STABILITY CRITERION
C      ITMAX   -  MAXIMUM NUMBER OF BINARY REDUCTIONS (20)
C       NOUT   -  NUMBER OF OUTPUT VALUES
C       NINT   -  NUMBER OF STEPS OF SIZE HO BETWEEN OUTPUT VALUES
C
C  THE DIFFERENTIAL EQUATION IS CODED IN THE FUNCTION
C  SUBPROGRAM  FUNK(X,Y).
C
C  THE EULER ALGORITHM IS CODED IN THE SUBROUTINE
C  SUBPROGRAM EULER
```

```
C
C     AUTHOR:  K. J. JOHNSON
C
C
      WRITE(6,11)
      READ(5,22)XO,YO,XF,HO,HOUT,TOL
      H=0.
      WRITE(6,33)HO,HOUT,TOL,H,XO,YO
C
C   SET CONSTANTS
C
      NOUT=(XF-XO)/HOUT
      NINT=HOUT/HO
      ITMAX=20
C
C   OUTERMOST LOOP (100) CONTROLS OUTPUT STEP SIZE  (HOUT)
C
      DO 100 IOUT=1,NOUT
C
C   MIDDLE LOOP (10) CONTROLS INTERNAL STEP SIZE  (H)
C
      DO 10 I=1,NINT
      K=1
C
C   CALL EULER TO GET FIRST APPROXIMATION
C
      CALL EULER(XO,YO,YOLD,HO,K)
C
C  INNER LOOP (20) TESTS THE STABILITY OF THE INITIAL APPROXIMATION
C
      DO 20 IHLF=1,ITMAX
      K=2*K
      H=HO/K
      CALL EULER(XO,YO,Y,H,K)
C
C   WATCH FOR SPECIAL CASE, YOLD=0.0 LEGITIMATELY
C
      IF(YOLD.EQ.0.0)YOLD=1.
      IF(ABS((Y-YOLD)/YOLD).LE.TOL)GO TO 50
      YOLD=Y
   20 CONTINUE
C
C   ERROR CONDITION - UNSTABLE IN ITMAX ITERATIONS
C
      WRITE(6,99)
      STOP
   50 XO=XO+HO
      YO=Y
   10 CONTINUE
      WRITE(6,44)H,XO,YO
  100 CONTINUE
      STOP
C
C   FORMATS
C
   11 FORMAT(//'  ENTER XO,YO,XF,HO,HOUT,TOL'/)
   22 FORMAT(6E)
   33 FORMAT(//'   HO = ',G10.3,5X,'HOUT = ',G10.3,
```

```
      1 5X,'TOL = ',G10.3///9X,'H',14X,'X',11X,'Y'//2G15.3,G15.5)
   44 FORMAT(2G15.3,G15.5)
   99 FORMAT(//'  SORRY.  UNABLE TO FIND A STABLE SOLUTION'/)
      END
C
C       EULER
C
      SUBROUTINE EULER(XO,YO,Y,H,K)
      Y=YO
      X=XO
      DO 10 I=1,K
      Y=Y+H*FUNK(X,Y)
   10 X=X+H
      RETURN
      END
C
C     FUNK
C
      FUNCTION FUNK(X,Y)
      FUNK=X+Y
      RETURN
      END
```

Discussion

EULER including the subprograms requires approximately 265 words of core. This sample execution took approximately 1 sec. The only change required to approximate the solution to another differential equation is to change the statement

```
      FUNK = X + Y
```

in the subprogram FUNK.

There is an optimization feature of note in this program. EULER takes one step (HO) at a time, iteratively reducing the step size until a stable solution is found. An alternative approach is to evaluate the solution over the entire range of X, test for stability, and if the stability criterion is not satisfied, recompute the solution over the entire range of X with a smaller step size. It is considerably more efficient to take one step at a time.

The EULER method is the simplest numerical method for solving differential equations. However, for accurate work small step sizes are required. This means relatively long computation times, and possibly, incorrect results due to accumulated round-off errors. The following methods are usually preferred.

7.2 THE RUNGE-KUTTA METHOD

The Euler method requires only one function evaluation per step,

$$y_{i+1} \approx y_i + hF(x_i, y_i)$$

The Runge-Kutta algorithms [4-6] involve more than one function evaluation per step. The general expression for the Runge-Kutta algorithms is

$$y_{i+1} \approx y_i + Q(x,y,h)$$

Here $Q(x,y,h)$ is an approximation to the integral

$$\int_{x_i}^{x_{i+1}} F(x,y) \, dx$$

The second-, third-, and fourth-order Runge-Kutta algorithms are:

Second order:

$$y_{i+1} \approx y_i + hF(x_i + \frac{h}{2}, y'_{i+1})$$

where

$$y'_{i+1} = y_i + \frac{h}{2} F(x_i, y_i)$$

Third order:

$$y_{i+1} \approx y_i + \frac{1}{6}(k_1 + 4k_2 + k_3)$$

where

$$k_1 = hF(x_i, y_i)$$
$$k_2 = hF(x_i + \frac{h}{2}, y_i + \frac{k_1}{2})$$
$$k_3 = hF(x_i + h, y_i + 2k_2 - k_1)$$

Fourth order:

$$y_{i+1} \approx y_i + \frac{1}{6}(k_1 + 2k_2 + 2k_3 + k_4)$$

where

$$k_1 = hF(x_i, y_i)$$
$$k_2 = hF(x_i + \frac{h}{2}, y_i + \frac{k_1}{2})$$
$$k_3 = hF(x_i + \frac{h}{2}, y_i + \frac{k_2}{2})$$
$$k_4 = hF(x_i + h, y_i + k_3)$$

The second-, third-, and fourth-order algorithms require two, three, and four function evaluations per step, respectively.

Problem 7.3.   Use the second-, third-, and fourth-order Runge-Kutta algorithms to solve the differential equation of Problem 7.1,

$$\frac{dy}{dx} = x + y \qquad (x_0 = y_0 = 0)$$

Use $h = 0.1$ and find the solution at $x = 0.1$.

<u>Solution.</u>   Second order:

$$y_1' = y_0 + \frac{h}{2}F(x_0,y_0)$$

$$= 0 + \frac{0.1}{2}(0 + 0)$$

$$= 0$$

$$y_1 \approx y_0 + hF(x_0 + \frac{h}{2},y_1')$$

$$= 0 + 0.1(0.05 + 0)$$

$$= 0.005$$

Third order:

$$k_1 = hF(x_0,y_0)$$

$$= 0.1(0 + 0)$$

$$= 0$$

$$k_2 = hF(x_0 + \frac{h}{2},\ y_0 + \frac{k_1}{2})$$

$$= 0.1(0.05 + 0)$$

$$= 0.005$$

$$k_3 = hF(x_0 + h,\ y_0 + 2k_2 - k_1)$$

$$= 0.1[0.1 + 2(0.005)]$$

$$= 0.011$$

$$y_1 \approx y_0 + \frac{1}{6}(k_1 + 4k_2 + k_3)$$

$$= 0.0 + \frac{1}{6}[0 + 4(0.005) + 0.011]$$

$$= 0.005167$$

Fourth order:

$$k_1 = hF(x_0,y_0)$$

$$= 0.01(0 + 0)$$

$$= 0.0$$

$$k_2 = hF(x_0 + \frac{h}{2},\ y_0 + \frac{k_1}{2})$$

$$= 0.1(0.05 + 0)$$

$$= 0.005$$

$$k_3 = hF(x_0 + \frac{h}{2},\ y_0 + \frac{k_2}{2})$$

$$= 0.1(0.05 + 0.0025)$$

$$= 0.00525$$

$$k_4 = hF(x_0 + h, y_0 + k_3)$$
$$= 0.1(0.1 + 0.00525)$$
$$= 0.010525$$

$$y_1 \quad y_0 + \frac{1}{6}(k_1 + 2k_2 + 2k_3 + k_4)$$
$$\equiv 0 + \frac{0 + 2(0.005) + 2(0.00525) + 0.010525}{6}$$
$$= 0.005171$$

Analytically,

$$y(0.1) = 0.0051709$$

Note the substantially increased accuracy of the fourth-order Runge-Kutta method compared to Euler's method. The former evaluated $y(0.1)$ correctly to four significant figures with $h = 0.1$, whereas Euler's method required $h = 0.1/2^8$ (with 2^8 associated function evaluations) to attain four-significant-figure accuracy.

Problem 7.4. Write an iterative fourth-order Runge-Kutta routine.

Solution. See the following program.

7.2.1. RUNKUT

Introduction

RUNKUT approximates the solution to first-order, linear differential equations with initial values using an iterative fourth-order Runge-Kutta routine. The input specifications are exactly the same as those in EULER (Section 7.1.1).

Method

RUNKUT approximates the solution in successive steps of HO. The first approximation to Y(XO + HO) is

    Y(XO + HO) = YO + [T1 + 2*T2 + 2*T3 + T4]/6

where

    T1 = HO*FUNK(XO,YO)
    T2 = HO*FUNK(XO+HO/2,YO+T1/2)
    T3 = HO*FUNK(XO+HO/2,YO+T2/2)
    T4 = HO*FUNK(XO+H,YO+T3)

Here FUNK is the function subprogram in which the function is coded. The stability of this approximation is tested by successive binary step-size reductions following the same strategy as EULER. RUNKUT consists of a main program and two subprograms, SUBROUTINE RUNKUT(XO,YO,Y,H,K) and FUNCTION FUNK(X,Y).

Sample Execution

```
ENTER XO,YO,XF,HO,HOUT,TOL
> 0   0   1   0.1   0.1   0.0001
```

```
     HO =  0.100E+00      HOUT =  0.100E+00      TOL =        0.100E-03

          H                  X                  Y

     0.000E+00          0.000E+00          0.00000E+00
     0.500E-01          0.100E+00          0.51709E-02
     0.500E-01          0.200             0.21403E-01
     0.500E-01          0.300             0.49859E-01
     0.500E-01          0.400             0.91825E-01
     0.500E-01          0.500             0.14872
     0.500E-01          0.600             0.22212
     0.500E-01          0.700             0.31375
     0.500E-01          0.800             0.42554
     0.500E-01          0.900             0.55960
     0.500E-01          1.00              0.71828
```

Listing

```
C
C                    RUNKUT
C
C     RUNKUT APPROXIMATES THE SOLUTION TO A FIRST ORDER
C     DIFFERENTIAL EQUATION USING AN ITERATIVE
C     FOURTH-ORDER RUNGE-KUTTA ALGORITHM
C
C  INPUT:  XO  -  INITIAL VALUE OF X
C          XF  -  FINAL VALUE OF X
C          HO  -  MAXIMUM STEP SIZE
C          YO  -  INITIAL VALUE OF Y
C        HOUT  -  OUTPUT STEP SIZE
C         TOL  -  STABILITY CRITERION
C
C  THE DIFFERENTIAL EQUATION IS CODED IN THE FUNCTION
C  SUBPROGRAM  FUNK(X,Y).
C
C  THE FOURTH-ORDER RUNGE-KUTTA ROUTINE IS CODED
C  IN THE SUBROUTINE RUNKUT
C
C   AUTHOR:  K. J. JOHNSON
C
C
      WRITE(6,11)
      READ(5,22)XO,YO,XF,HO,HOUT,TOL
      H=0.
      WRITE(6,33)HO,HOUT,TOL,H,XO,YO
C
C  SET CONSTANTS
C
      NOUT=(XF-XO)/HOUT
      NINT=HOUT/HO
      ITMAX=20
C
```

```
C     OUTERMOST LOOP (100) CONTROLS OUTPUT STEP SIZE   (HOUT)
C
      DO 100 IOUT=1,NOUT
C
C     MIDDLE LOOP (10) CONTROLS INTERNAL STEP SIZE   (H)
C
      DO 10 I=1,NINT
      K=1
C
C     CALL RUNKUT TO GET FIRST APPROXIMATION
C
      CALL RUNKUT(XO,YO,YOLD,HO,K)
C
C  INNER LOOP (20) TESTS THE STABILITY OF THE INITIAL APPROXIMATION
C
      DO 20 IHLF=1,ITMAX
      K=2*K
      H=HO/K
      CALL RUNKUT(XO,YO,Y,H,K)
C
C    WATCH FOR SPECIAL CASE, YOLD=0.0 LEGITIMATELY
C
      IF(YOLD.EQ.0.0)YOLD=1.
      IF(ABS((Y-YOLD)/YOLD).LE.TOL)GO TO 50
      YOLD=Y
   20 CONTINUE
C
C    ERROR CONDITION - UNSTABLE IN ITMAX ITERATIONS
C
      WRITE(6,99)
      STOP
   50 XO=XO+HO
      YO=Y
   10 CONTINUE
      WRITE(6,44)H,XO,YO
  100 CONTINUE
      STOP
C
C   FORMATS
C
   11 FORMAT(//'  ENTER XO,YO,XF,HO,HOUT,TOL')
   22 FORMAT(6E)
   33 FORMAT(//'     HO = ',G10.3,5X,'HOUT = ',G10.3,
     1 5X,'TOL = ',G15.3///9X,'H',14X,'X',11X,'Y'//2G15.3,G15.5)
   44 FORMAT(2G15.3,G15.5)
   99 FORMAT(//'  SORRY.  UNABLE TO FIND A STABLE SOLUTION'/)
      END
C
C     RUNKUT
C
      SUBROUTINE RUNKUT(XO,YO,Y,H,K)
      Y=YO
      X=XO
      H2=H/2.
      DO 10 I=1,K
      T1=H*FUNK(X,Y)
      T2=H*FUNK(X+H2,Y+T1/2.)
      T3=H*FUNK(X+H2,Y+T2/2.)
```

```
      T4=H*FUNK(X+H,Y+T3)
      Y=Y+(T1+2.*T2+2.*T3+T4)/6.
   10 X=X+H
      RETURN
      END
C
C     FUNK
C
      FUNCTION FUNK(X,Y)
      FUNK=X+Y
      RETURN
      END
```

Discussion

RUNKUT including the subprograms requires 325 words of core. This sample execution took approximately 0.2 sec. The stability criterion in this sample execution is such that only the default single binary step-size reduction (to 0.05) was required. The rate of convergence to a stable solution is indicated by the following table.

h	y(0.1)
0.1	0.0051708333
0.05	0.0051709125
0.025	0.0051709177
0.0125	0.0051709181
0.00625	0.0051709181

Analytically,

$y(0.1) = 0.005170918076$.

RUNKUT and Euler are identical except for the subroutines, the associated CALL statements, and some comment statements. Both use the optimization feature of taking one step at a time (see the Discussion Section of EULER, Section 7.1.1). The greater efficiency of the fourth-order routine in this example is manifest. However, RUNKUT requires four function evaluations per step, compared to Euler's one function evaluation. For some applications the simpler EULER method may be preferable.

7.3. THE PREDICTOR-CORRECTOR METHOD

The predictor-corrector method [7] is a widely used method for approximating differential equations. The following algorithm gives results that are comparable in accuracy to the fourth-order Runge-Kutta method of Section 7.2.

Consider the following differential equation with two initial values,

$$\frac{dy}{dx} = F(x,y) \qquad (x_0,y_0) \ \ \text{and} \ \ (x_1,y_1)$$

One of the predictor-corrector algorithms is the following:
Let

$$y_1' = F(x_1,y_1)$$

then the predictor equation is

$$P_2 = y_0 + 2hy_1'$$

Now let

$$p_2' = F(x_2, p_2)$$

then the corrector equation is

$$c_2 = y_1 + \frac{h}{2}(p_2' + y_1')$$

The error in the corrector equation is

$$E_c = \frac{p_2 - c_2}{5}$$

and the solution $y_2 = y(x_0 + 2h)$ is

$$y_2 = c_2 + E_c$$

The predictor-corrector methods are not self-starting, they require two initial values, (x_0, y_0) and (x_1, y_1). Typically a Runge-Kutta routine is used to provide the second initial value. However, the predictor-corrector algorithm requires only two function evaluations per step, $F(x_1, y_1)$ and $F(x_2, y_2)$. Consequently, once started, if the solution is stable, then the predictor-corrector method is faster than a fourth-order Runge-Kutta routine.

Problem 7.5. Use the predictor-corrector algorithm given above to solve the differential equation of Problem 7.1 given the following initial values:
values:

$$\frac{dy}{dx} = x + y \qquad (0,0) \text{ and } (0.1, 0.00517)$$

Solution. $x_0 = 0$ $y_0 = 0$
$\qquad\quad x_1 = 0.1$ $y_1 = 0.00517$
$\qquad\quad x_2 = 0.2$ $x_2 = ?$

With h = 0.1,

$$y_1' = F(x_1, y_1)$$
$$= 0.1 + 0.00517$$
$$= 0.10517$$

$$P_2 = y_0 + 2hy_1'$$
$$= 0 + 2(0.1)(0.10517)$$
$$= 0.021034$$

$$p_2' = F(x_2, y_1')$$
$$= 0.2 + 0.021034$$
$$= 0.221034$$

$$c_2 = y_1 + \frac{h}{2}(p_2' + y_1')$$

$$= 0.00517 + \frac{0.1}{2}(0.221034 + 0.10517)$$

$$= 0.02148$$

$$E_c = \frac{p_2 - c_2}{5}$$

$$= \frac{0.021034 - 0.02148}{5}$$

$$= -0.000089$$

$$y_2 = c_2 + E_c$$

$$= 0.02148 - 0.000089$$

$$= 0.02139$$

Analytically,

$$y(0.2) = 0.021403$$

Problem 7.6. Write a noniterative predictor-corrector routine using this algorithm. Incorporate sections of RUNKUT to start the procedure. Tabulate the error term as part of the output.

Solution. See the following program.

7.3.1. PRECOR

Introduction

PRECOR approximates the solution to a first-order, linear differential equation with initial values,

$$\frac{dx}{dy} = F(x,y) \qquad (x_0, y_0)$$

using the predictor-corrector algorithm discussed above. The fourth-order Runge-Kutta routine RUNKUT (Section 7.2.1) is used to start the predictor-corrector method. The input includes the initial values x_0 and y_0; the final value to be assigned to the independent variable, XF; the output step size, HOUT; and the internal step size, H. PRECOR prints the input for verification purposes and tabulates the solution including the error term.

Method

The procedure outlined above is followed. The number of values printed in the table is

$$NOUT = (XF-XO)/HOUT$$

The number of steps that must be taken before each output value is reached is

    NINT = HOUT/H

The total number of steps taken is

    N = NOUT*NINT

For example, if HOUT = 0.1, h = 0.05, XO = 0.0, and XF = 1.0, then

    NOUT = (1-0)/0.1 = 10
    NINT = 0.1/0.05 = 2

and

    N = 10*2 = 20

    The program is not iterative. The compromise between machine time and numerical stability is made by the user with his choice of internal step size. The error term provides a monitor of the errors accumulating in the marching method.

    Sample Execution


```
ENTER XO,YO,XF,HOUT AND H
NOTE:  HOUT MUST BE AN INTEGRAL MULTIPLE OF H
> 0   0   1   0.1   0.05

   H =  0.500E-01        HOUT =  0.100E+00

          X                    Y                   ERROR

      0.000E+00          0.00000E+00
      0.100              0.51703E-02            -1.08E-05
      0.200              0.21401E-01            -1.18E-05
      0.300              0.49855E-01            -1.31E-05
      0.400              0.91819E-01            -1.44E-05
      0.500              0.14871                -1.60E-05
      0.600              0.22211                -1.76E-05
      0.700              0.31374                -1.95E-05
      0.800              0.42552                -2.15E-05
      0.900              0.55958                -2.38E-05
       1.00              0.71826                -2.63E-05
```


    Listing

```
C
C                       PRECOR
C
C     PRECOR APPROXIMATES THE SOLUTION TO A FIRST ORDER
C     DIFFERENTIAL EQUATION USING A PREDICTOR-CORRECTOR
C     ALOGORITHM
C
C   INPUT:   XO   -   INITIAL VALUE OF X
C            YO   -   INITIAL VALUE OF Y
C            XF   -   FINAL VALUE OF X
C            HOUT -   OUTPUT STEP SIZE
```

```
C            H   -   STEP SIZE
C
C
C    THE PROCEDURE IS STARTED BY A FOURTH-ORDER RUNGE-KUTTA
C    ROUTINE CODED IN THE SUBPROGRAM RUNKUT
C
C    THE PREDICTOR-CORRECTOR ROUTINE IS CODED IN THE SUBPROGRAM PRECOR
C
C    THE DIFFERENTIAL EQUATION IS CODED IN THE FUNCTION FUNK(X,Y)
C
C      AUTHOR:   K. J. JOHNSON
C
       WRITE(6,11)
       READ(5,22)XO,YO,XF,HOUT,H
       IF(AMOD(HOUT,H).NE.0.)STOP
       WRITE(6,33)H,HOUT,XO,YO
C
C    SET CONSTANTS
C
       NOUT=(XF-XO)/HOUT
       NINT=HOUT/H
       N=NOUT*NINT
C
C    USE RUNKUT TO GET Y1   ( Y1=Y(XO+H) )
C
       CALL RUNKUT(XO,YO,Y1,H)
       X=XO+H
       IF(NINT.EQ.1)WRITE(6,44)X,Y1
C
C      PREDICTOR-CORRECTOR METHOD
C
       DO 100 I=2,N
       CALL PRECOR(XO,YO,Y1,Y2,H,ERR)
       X=X+H
       IF(MOD(I,NINT).EQ.0)WRITE(6,44)X,Y2,ERR
       XO=XO+H
       YO=Y1
       Y1=Y2
  100 CONTINUE
       STOP
   11 FORMAT(/'  ENTER XO,YO,XF,HOUT AND H',/,
      1' NOTE:  HOUT MUST BE AN INTEGRAL MULTIPLE OF H')
   22 FORMAT(5E)
   33 FORMAT(//'    H = ',G10.3,5X,'  HOUT = ',G10.3,
      1  ///10X,'X',15X,'Y',17X,'ERROR'//G15.3,G18.5)
   44 FORMAT(G15.3,G18.5,1PE18.2)
       END
C
C            RUNKUT
C
C    FOURTH-ORDER RUNGE-KUTTA ROUTINE
C
       SUBROUTINE RUNKUT(X,Y,Y1,H)
       H2=H/2.
       T1=H*FUNK(X,Y)
       T2=H*FUNK(X+H2,Y+T1/2.)
       T3=H*FUNK(X+H2,Y+T2/2.)
       T4=H*FUNK(X+H,Y+T3)
       Y1=Y+(T1+2.*T2+2.*T3+T4)/6.
       RETURN
       END
```

```
C
C            PRECOR
C
C    PREDICTOR-CORRECTOR ROUTINE
C
C    PRECOR APPROXIMATES Y2 GIVEN YO AND Y1
C
      SUBROUTINE PRECOR(XO,YO,Y1,Y2,H,ERR)
      X1=XO+H
      X2=X1+H
      FUNK1=FUNK(X1,Y1)
      P2=YO + 2.*H*FUNK1
      C2=Y1 + H/2.*(FUNK1 + FUNK(X2,P2))
      ERR=0.2*(P2-C2)
      Y2=C2+ERR
      RETURN
      END
C
C    FUNK
C
      FUNCTION FUNK(X,Y)
      FUNK=X+Y
      RETURN
      END
```

Discussion

PRECOR requires 365 words of core. This sample execution took approximately
0.2 sec. The following output shows the improvement in the results when the
internal step size is reduced by a factor of 10, to 0.005.

H = 0.500E-02   HOUT = 0.100E+00

X	Y	ERROR
0.000E+00	0.00000E+00	
0.100E+00	0.51709E-02	-1.14E-08
0.200	0.21403E-01	-1.26E-08
0.300	0.49859E-01	-1.40E-08
0.400	0.91825E-01	-1.55E-08
0.500	0.14872	-1.68E-08
0.600	0.22212	-1.90E-08
0.700	0.31375	-2.09E-08
0.800	0.42554	-2.24E-08
0.900	0.55960	-2.68E-08
1.00	0.71828	-2.68E-08

7.4. SYSTEMS OF DIFFERENTIAL EQUATIONS

The Euler, Runge-Kutta, and predictor-corrector methods discussed in Section 7.3 can be extended to solve systems of simultaneous, first-order, linear differential equations with initial values. For example,

$$\frac{dy_1}{dx} = F_1(x; \; y_1, y_2, y_3, \ldots, y_n)$$

$$\frac{dy_2}{dx} = F_2(x; \; y_1, y_2, y_3, \ldots, y_n)$$

$$\ldots\ldots\ldots\ldots\ldots\ldots\ldots\ldots\ldots$$

$$\frac{dy_n}{dx} = F_n(x; \; y_1, y_2, y_3, \ldots, y_n)$$

with initial values $(x_0; \; y_{10}, y_{20}, y_{30}, \ldots, y_{n0})$. The solution to this system using Euler's method is considered first.

Let

$$\frac{dy_i}{dx} = F_i(x, y_j) \qquad i, j = 1, 2, 3, \ldots, n$$

Then

$$\frac{dy_i}{dx} \approx \frac{\Delta y_i}{\Delta x} = \frac{y_i(x_0 + h) - y_{i0}}{h} = F_i(x_0, y_{j0})$$

so

$$y_i(x_0 + h) \approx y_{i0} + hF_i(x_0, y_{j0})$$

Problem 7.7. Consider the following kinetic system:

$$A \; \underset{k_2}{\overset{k_1}{\rightleftharpoons}} \; B$$

Let $A_0 = 1.0$, $B_0 = 0.0$, $k_1 = 10$, and $k_2 = 5$. Calculate $A(0.1)$ and $B(0.1)$ to three significant figures using Euler's method and compare the results with the analytical solution [8]:

$$A(t) = A_0 - m(1 - \exp[-(k_1 + k_2)t])$$

$$B(t) = B_0 + m(1 - \exp[-(k_1 + k_2)t])$$

$$= A_0 + B_0 - A(t)$$

where

$$m = \frac{k_1 A_0 - k_2 B_0}{k_1 + k_2}$$

Solution. The differential equations that define this system are

$$\frac{dA}{dt} = k_2B - k_1A = 5B - 10A$$

$$\frac{dB}{dt} = -\frac{dA}{dt} = 10A - 5B$$

Using Euler's method with h = 0.1,

$$\begin{aligned}
A(0.1) &= A_0 + h(5B_0 - 10A_0)\\
&= 1.0 + 0.1[5(0) - 10(1)]\\
&= 1.0 - 1.0\\
&= 0
\end{aligned}$$

$$\begin{aligned}
B(0.1) &= B_0 + 0.1(10A_0 - 5B_0)\\
&= 0 + 0.1(10 - 0)\\
&= 1.0
\end{aligned}$$

Next try h = 0.05,

$$\begin{aligned}
(0.05) &= A_0 + 0.05(5B_0 - 10A_0)\\
&= 1.0 + 0.05[5(0) - 10(1)]\\
&= 0.05
\end{aligned}$$

$$\begin{aligned}
B(0.05) &= B_0 + 0.05(10A_0 - 5B_0)\\
&= 0 + 0.05(10 - 0)\\
&= 0.05
\end{aligned}$$

$$\begin{aligned}
A(0.1) &= A(0.05) + 0.05[5B(0.05) - 10A(0.05)]\\
&= 0.05 + 0.05[5(0.05) - 10(0.05)]\\
&= 0.375
\end{aligned}$$

$$\begin{aligned}
B(0.1) &= B(0.05) + 0.05[10A(0.05) - 5B(0.05)]\\
&= 0.05 + 0.05[10(0.05) - 5(0.05)]\\
&= 0.625
\end{aligned}$$

Continuing to reduce the step size gives the following results:

h	A(0.1)	B(0.1)
0.1	0	1.0
0.05	0.375	0.625
0.025	0.435	0.565
0.0125	0.460	0.540
0.00625	0.471	0.529
0.003125	0.477	0.523

The analytical solution is

$$m = \frac{k_1A_0 - k_2B_0}{k_1 + k_2} = \frac{10(1) - 5(0)}{10 + 5}$$

$$= \frac{10}{15} = 0.6667$$

$$A(0.1) = A_0 - m[1 - \exp((k_1 + k_2)t)]$$
$$= 1.0 - 0.6667[1 - \exp(-15 \times 0.1)]$$
$$= 1.0 - 0.6667(0.7769)$$
$$= 0.482$$

$$B(0.1) = A_0 + B_0 - A(0.1)$$
$$= 1.0 - 0.482 = 0.518$$

The fourth-order Runge-Kutta method is extended to solve a system of differential equations as follows [9]:

$$y_i(x_0 + h) \approx y_{i0} + Q(x,y,h)$$
$$\approx y_{i0} + \frac{k_{i1} + 2k_{i2} + 2k_{i3} + k_{i4}}{6} \qquad i = 1,2,3,\ldots,n$$

where

$$k_{i1} = hF_i(x_0, y_{i0})$$
$$k_{i2} = hF_i(x_0 + \frac{h}{2}, y_{i0} + \frac{k_{i1}}{2})$$

$$k_{i3} = hF_i(x_0 + \frac{h}{2}, y_{i0} + \frac{k_{i2}}{2})$$
$$k_{i4} = hF_i(x_0 + h, y_{i0} + k_{i3}) \qquad i = 1,2,3,\ldots,n$$

These equations are applied in parallel, one step at a time, as demonstrated in the following problem.

Problem 7.8. Solve Problem 7.7 using the fourth-order Runge-Kutta algorithm.

Solution. With h = 0.1,

$$A(0.1) = A_0 + \frac{1}{6}(k_{11} + 2k_{12} + 2k_{13} + k_{14})$$

and

$$B(0.1) = B_0 + \frac{1}{6}(k_{21} + 2k_{22} + 2k_{23} + k_{24})$$

where

$$k_{11} = hF_1(t_0, A_0, B_0)$$
$$= 0.1 \, F_1(0,1,0)$$
$$= 0.1[5(0) - 10(1)]$$
$$= -1.0$$

$$k_{21} = hF_2(t_0, A_0, B_0)$$
$$= 0.1 \, F_2(0,1,0)$$
$$= 0.1[10(1) - 5(0)]$$
$$= 1.0$$

$$k_{12} = hF_1[t_0 + \frac{h}{2}, A_0 + \frac{k_{11}}{2}, B_0 + \frac{k_{21}}{2}]$$
$$= 0.1\,F_1\,(0.05, 0.5, 0.5)$$
$$= 0.1[5(0.5) - 10(0.5)]$$
$$= -0.25$$

$$k_{22} = hF_2(t_0 + \frac{h}{2}, A_0 + \frac{k_{11}}{2}, B_0 + \frac{k_{21}}{2})$$
$$= 0.1F_2(0.05, 0.5, 0.5)$$
$$= 0.25$$

$$k_{13} = hF_1(t_0 + \frac{h}{2}, A_0 + \frac{k_{12}}{2}, B_0 + \frac{k_{22}}{2})$$
$$= 0.1\,F_1\,(0.05, 0.875, 0.125)$$
$$= -0.8125$$

$$k_{23} = hF_2(t_0 + \frac{h}{2}, A_0 + \frac{k_{12}}{2}, B_0 + \frac{k_{22}}{2})$$
$$= 0.1\,F_2\,(0.05, 0.875, 0.125)$$
$$= 0.8125$$

$$k_{14} = hF_1(t_0 + h, A_0 + k_{13}, B_0 + k_{23})$$
$$= 0.1\,F_1(0.1, 0.1875, 0.8125)$$
$$= 2.1875$$

$$k_{24} = hF_2(t_0, A_0 + k_{13}, B_0 + k_{23})$$
$$= 0.1\,F_2\,(0.1, 0.1875, 0.8125)$$
$$= -0.21875$$

Then

$$A(0.1) = A_0 + \frac{1}{6}(k_{11} + 2k_{21} + 2k_{31} + k_{41})$$
$$= 1.0 + \frac{1}{6}(-1 - 0.5 - 1.625 + .21875)$$
$$= 1.0 - 0.484 = 0.516$$

and

$$B(0.1) = B_0 + \frac{1}{6}(k_{12} + 2k_{22} + 2k_{32} + k_{42})$$
$$= 0 + \frac{1}{6}(1 + 0.5 + 1.625 - 0.21875)$$
$$= 0.484$$

Successive step-size reductions give the following results:

h	A(0.1)	B(0.1)
0.1	0.516	0.484
0.05	0.483	0.517
0.025	0.482	0.518

Analytically, $A(0.1) = 0.482$ and $B(0.1) = 0.518$.

Problem 7.9. Write a program to solve systems of simultaneous differential equations
             using the iterative Runge-Kutta method. Demonstrate the algorithm
             using the kinetic scheme introduced in Section 7.1.

     Solution. See the following program.

7.4.1. RUNGE

Introduction

     RUNGE uses the fourth-order Runge-Kutta method to solve a system of linear,
first-order differential equations with initial values. The input includes the
number of differential equations (NDIM); the initial values (XO,YO(I), I = 1,2,3,...,
NDIM); the range of the independent variable (XF); the output step size (HOUT); the
internal step size (H); and a stability criterion (TOL). The program prints the
input data for verification purposes and then tabulates the solution starting with
the initial values and incrementing the independent variable by the specified output
step size until the solution at XF is found.

     Method

     Let n be the number of differential equations in the system. Then the problem
to be solved is

$$\frac{dy_1}{dx} = F_1(x;\ y_1,y_2,\ldots,y_n) = DERY(1)$$

$$\frac{dy_2}{dx} = F_2(x;\ y_1,y_2,\ldots,y_n) = DERY(2)$$
....

$$\frac{dy_n}{dx} = F_n(x;\ y_1,y_2,\ldots,y_n) = DERY(n)$$

The initial values are x_0, $y_1(x_0)$, $y_2(x_0)$, ..., $y_n(x_0)$. RUNGE prints the solution
to this system over the range

     XO $\leq$ X $\leq$ XF

in steps of HOUT. If H is the specified internal step size, and TOL is the relative
stability criterion, then the following procedure is used to integrate the system
of differential equations:

1. Given the initial values, the fourth-order method is used to approximate
    the solution at X = XO + H.
2. This approximation is tested by reducing the step size by a factor of 2.
    Two steps are now made, and the result is compared to the previous solution
    as follows:

$$\sum_{i=1}^{n} \left| \frac{Y_i(HO/2) - Y_i(HO)}{Y_i(HO/2)} \right|$$

Here $Y_i(HO)$ and $Y_i(HO/2)$ refer to the solution using step size HO and HO/2, respectively. If this sum is less than the specified stability criterion TOL, then a stable solution has been found.

3. If the value of X is equal to one of the output values,

      XO + jHOUT   j = 1,2,3,...

then the program prints the solution. If not, new initial values are defined,

      XO = XO + HO

      $Y_{i0} = Y_i(XO + HO)$

and the procedure continues from step 1.

4. If the stability test fails, then a second binary step size reduction is made, four integration steps are taken, and the solution is compared to that obtained using the previous value of H. The successive binary step size reduction continues for ITMAX iterations. If a stable solution is not found, the program terminates.

Sample Execution

```
                        1.0
            A  +  B   <=====>   C  +  D
                       0.25

                        0.5
            B  +  C    ----->   D  +  E

   A(0) = B(0) = 1.0     C(0) = D(0) = E(0) = 0
ENTER NDIM, XO, XF, HOUT, H, AND TOL
NOTE: HOUT MUST BE AN INTEGRAL MULTIPLE OF H.
> 5   0   20   1   1   .0001

 ENTER THE INITIAL VALUES: (Y0(I),I=1,NDIM)
> 1   1

    NDIM =  5                 HOUT =      1.00

    H =     1.00                 TOL =   0.100E-03
```

X	H	Y(I),	I = 1,2,3, ... ,NDIM		
0.00	0.00E+00	1.00	1.00	0.000E+00	0.000E+00	0.000E+00
1.00	0.63E-01	0.534	0.454	0.387	0.546	0.793E-01
2.00	0.13	0.432	0.280	0.416	0.720	0.152
3.00	0.13	0.409	0.208	0.391	0.792	0.200
4.00	0.25	0.408	0.172	0.357	0.828	0.236
5.00	0.25	0.413	0.150	0.324	0.850	0.263
6.00	0.25	0.420	0.135	0.295	0.865	0.285
7.00	0.50	0.427	0.124	0.270	0.876	0.303
8.00	0.50	0.433	0.114	0.249	0.886	0.319
9.00	0.50	0.438	0.106	0.230	0.894	0.332
10.00	0.50	0.443	0.994E-01	0.214	0.901	0.343
11.00	0.50	0.447	0.934E-01	0.200	0.907	0.353
12.00	0.50	0.450	0.881E-01	0.188	0.912	0.362
13.00	0.50	0.453	0.833E-01	0.177	0.917	0.370
14.00	0.50	0.456	0.790E-01	0.167	0.921	0.377
15.00	0.50	0.458	0.751E-01	0.159	0.925	0.383
16.00	0.50	0.460	0.716E-01	0.151	0.928	0.389
17.00	0.50	0.462	0.684E-01	0.144	0.932	0.394
18.00	0.50	0.464	0.655E-01	0.137	0.935	0.399
19.00	0.50	0.466	0.628E-01	0.131	0.937	0.403
20.00	0.50	0.467	0.603E-01	0.126	0.940	0.407

Listing

```
C
C
C                             RUNGE
C
C     THIS PROGRAM SOLVES SYSTEMS OF FIRST ORDER DIFFERENTIAL EQUATIONS
C     WITH  INITIAL VALUES USING AN ITERATIVE FOURTH-ORDER
C     RUNGE-KUTTA ALGORITHM
C
C     PROGRAM ARRANGEMENT:
C
C          MAIN PROGRAM - I/O, CONTROL OF ITERATIVE PROCEDURE
C          SUBR. FCT    - DIFFERENTIAL EQUATIONS
C          SUBR. RK4    - FOURTH-ORDER RUNGE-KUTTA ROUTINE
C
C
C     GLOSSARY:
C
C     NDIM        -  NUMBER OF DIFFERENTIAL EQUATIONS
C     XO          -  INITIAL VALUE OF INDEP. VARIABLE
C     XF          -  FINAL VALUE OF INDEP. VARIABLE
C     HO          -  INITIAL STEP SIZE
C     HOUT        -  OUTPUT STEP SIZE
C     TOL         -  STABILITY CRITERION
C     ITMAX       -  MAXIMUM NUMBER OF BINARY STEP-SIZE REDUCTIONS
C     SUM         -  SUM OF RELATIVE CHANGES IN SOLUTION VECTOR
C     YO(10)      -  INITIAL VALUES
C     YHOLD(10)   -  TEMPORARY STORAGE
C     Y(10)       -  SOLUTION TO THE SYSTEM
C     DERY(10)    -  DIFFERENTIAL EQUATIONS (FCT)
C
C
C        AUTHOR:  K. J. JOHNSON
C
C
      DIMENSION Y(10),YHOLD(10),YO(10)
```

```
C
C     INPUT
C
      WRITE(6,11)
      READ(5,22) NDIM,XO,XF,HOUT,HO,TOL
      WRITE(6,33)
      READ(5,44)(YO(I),I=1,NDIM)
C
C     PRINT INPUT FOR VERIFICATION
C
      WRITE(6,55)NDIM,HOUT,HO,TOL
      H=0.
      WRITE(6,66)XO,H,(YO(I),I=1,NDIM)
C
C   SET CONSTANTS
C
      NOUT=(XF-XO)/HOUT
      NINT=HOUT/HO
      ITMAX=10
C
C   OUTERMOST LOOP (100) CONTROLS OUTPUT STEP SIZE (HOUT)
C
      DO 100 IOUT=1,NOUT
C
C   MIDDLE LOOP (10) CONTROLS INTERNAL STEP SIZE (H)
C
      DO 10 I=1,NINT
      K=1
C
C   CALL RK4 TO GET FIRST APPROXIMATION
C
      CALL RK4(XO,YO,YHOLD,HO,NDIM,K)
C
C   INNER LOOP (20) TESTS THE STABILITY OF THE FIRST APPROXIMATION
C
      DO 20 IHLF=1,ITMAX
      K=2*K
      H=HO/K
      CALL RK4(XO,YO,Y,H,NDIM,K)
      SUM=0.
      DO 40 IJK=1,NDIM
   40 SUM=SUM+ABS((Y(IJK)-YHOLD(IJK))/Y(IJK))
      IF (SUM.LT.TOL)GO TO 50
      DO 45 KJJ=1,NDIM
   45 YHOLD(KJJ)=Y(KJJ)
   20 CONTINUE
C
C ERROR CONDITION - UNSTABLE IN ITMAX STEPS
C
      WRITE(6,99)
      STOP
   50 XO=XO+HO
      DO 60 KJJ=1,NDIM
   60 YO(KJJ)=Y(KJJ)
   10 CONTINUE
      WRITE(6,66)XO,H,(Y(KJJ),KJJ=1,NDIM)
  100   CONTINUE
C
C     FORMATS
```

```
C
   11 FORMAT(/' ENTER NDIM, XO, XF, HOUT, H, AND TOL  '/' NOTE:',
     1' HOUT MUST BE AN INTEGRAL MULTIPLE OF H.'/)
   22 FORMAT( I,5F)
   33 FORMAT(/'  ENTER THE INITIAL VALUES: (YO(I),I=1,NDIM)'/)
   44 FORMAT(10F)
   55 FORMAT(/5X,'NDIM = ',I2,15X,'HOUT = ',G12.3//5X'H =',
     1G12.3,10X,'TOL =',G12.3,
     2////4X,'X',6X,'H',10X,'Y(I),   I = 1,2,3, ... ,NDIM'//)
   66 FORMAT(F6.2,G11.2,5G11.3,/10X,5G11.3)
   99 FORMAT(//' FAILED TO FIND STABLE SOLN. IN ITMAX TRIES'/)
      END
C
C                    RK4
C
C
C     FOURTH-ORDER RUNGE-KUTTA ROUTINE
C
      SUBROUTINE RK4(XO,YO,Y,H,NDIM,K)
      DIMENSION Y(10),YO(10),YY(10),DERY(10),RK(10)
      X=XO
      DO 2 I=1,NDIM
    2 Y(I)=YO(I)
      DO 100 JJ=1,K
      CALL FCT(X,Y,DERY)
      DO 10 I=1,NDIM
      YY(I)=Y(I)+H*DERY(I)/2.
   10 RK(I)=DERY(I)
      X=X+H/2.
      CALL FCT(X,YY,DERY)
      DO 20 I=1,NDIM
      YY(I)=Y(I)+H*DERY(I)/2.
   20 RK(I)=RK(I)+2.*DERY(I)
      CALL FCT(X,YY,DERY)
      DO 30 I=1,NDIM
      YY(I)=Y(I)+H*DERY(I)
   30 RK(I)=RK(I)+2.*DERY(I)
      X=X+H/2
      CALL FCT(X,YY,DERY)
      DO 40 I=1,NDIM
      RK(I)=H*(RK(I)+DERY(I))
   40 Y(I)=Y(I)+RK(I)/6.
  100 CONTINUE
      RETURN
      END
C
C         FCT
C
C
C     THE DIFFERENTIAL EQUATIONS ARE
C     CODED IN DERY(NDIM)
C
      SUBROUTINE FCT(X,Y,DERY)
      DIMENSION DERY(10),Y(10)
      REAL K(3)
      DATA K/1.,0.25,0.5/
      A=Y(1)
      B=Y(2)
      C=Y(3)
      D=Y(4)
```

```
      E=Y(5)
      RATE1=K(1)*A*B
      RATE2=K(2)*C*D
      RATE3=K(3)*C*B
      DERY(1)=RATE2-RATE1
      DERY(2)=RATE2-RATE1-RATE3
      DERY(3)=RATE1-RATE2-RATE3
      DERY(4)=RATE1-RATE2+RATE3
      DERY(5)=RATE3
      RETURN
      END
```

Discussion

RUNGE including the two subroutines requires approximately 600 words of core. This sample execution took approximately 0.3 sec.

The iterative procedure during the first step was monitored using the following statement:

```
40  SUM=SUM+ABS((Y(IJK)-YHOLD(IJK))/Y(IJK)
    WRITE(6,666)IHLF,K,H,SUM
666 FORMAT(2I5,2G15.5)
```

The following results were obtained.

IHLF	K	H	SUM
1	2	0.5	0.63
2	4	0.25	0.0095
3	8	0.125	0.0038
4	16	0.0625	0.00019

The accuracy of the numerical solution to the chemical kinetics system can be verified by mass balance,

$$A_0 + B_0 + C_0 + D_0 + E_0 = A(t) + B(t) + C(t) + D(t) + E(t)$$

For example, at $t = 20$,

$$1.0 + 1.0 = 0.467 + 0.0603 + 0.126 + 0.940 + 0.407$$
$$= 2.0$$

The output corresponding to this sample execution of RUNGE is plotted in Figure 7.1.

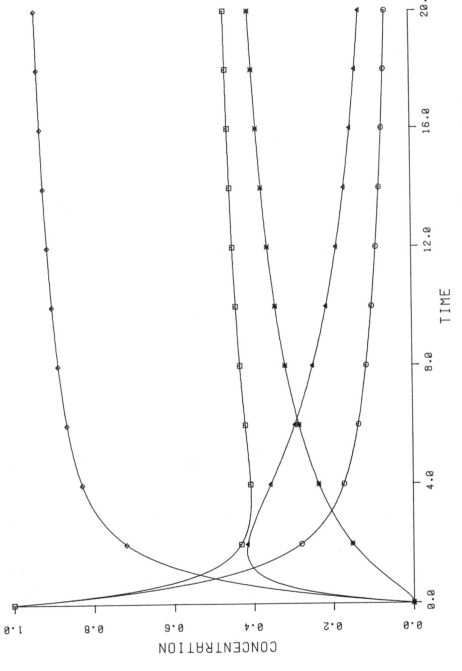

Figure 7.1 Solution to Problem 7.9: A = "□"; B = "⊙"; C = "△"; D = "◇"; and E = "*".

Problem 7.10. Write a program to calculate trajectories for the following exchange
          reaction:

$$H_a + H_b{-}H_c \longrightarrow H_a{-}H_b + H_c$$

          Assume the reaction occurs in one dimension.

     Solution. [10, 11]. See the following program.

7.4.2. EXCHG

Introduction

     EXCHG calculates trajectories for the hydrogen atom-hydrogen molecule exchange
reaction. There are four input parameters:

     EBC: The initial energy of the $H_b{-}H_c$ molecule. It is assumed that this
               energy is all potential energy, that is, that $H_b{-}H_c$ is at a classical
               turning point.

     JWHICH: 0 or 1 depending on whether the $H_a{-}H_b$ molecule is at the minimum or
               maximum separation.

     VA: Initial velocity of the atom H_a.

     YO(2):   The initial value for the distance between H_a and the center of mass
               of $H_b{-}H_c$.

The program prints the input data for verification purposes and generates a table
containing the time, the internuclear separations $H_a{-}H_b$ and $H_b{-}H_c$, two momentum
values (see below), the potential energy, and the total energy.

     Method
     The classical solution to this problem involves Hamilton's generalized equations
of motion,

$$H(Q_i, P_i) = T(Q_i, P_i) + V(Q_i, P_i)$$

where H, T, and V represent total, kinetic, and potential energy, respectively; Q_i
is a generalized coordinate; and P_i is the associated generalized momentum. The
time dependence of Q_i and P_i are given by

$$\frac{dP_i}{dt} = -\frac{\partial H}{\partial Q_i} \qquad \frac{dQ_i}{dt} = \frac{\partial H}{\partial P_i}$$

The choice of coordinate system that minimizes the number of differential equations
is as follows:

$$Q_1 = X_c - X_b = R_{bc}$$

$$Q_2 = \frac{1}{2}(X_b + X_c) - X_a$$

$$Q_3 = \frac{1}{3}(X_a + X_b + X_c)$$

Here X_j is the distance of the jth nucleus from the origin, Q_2 is the distance between H_a and the center of mass of H_b-H_c, and Q_3 is the center of mass of the triatomic system. This coordinate system is illustrated in Figure 7.2. The inter-nuclear distance R_{ab} is

$$R_{ab} = X_b - X_a = Q_2 - \frac{Q_1}{2}$$

The classical Hamiltonian is

$$H(Q_i, P_i) = \frac{1}{2m}(2P_1^2 + \frac{3}{2}P_2^2 + \frac{1}{3}P_3^2) + V(Q_1, Q_2, Q_3)$$

where m is the mass of the hydrogen atom. The potential function is given by Sato's function [12],

$$V = \frac{A_0}{4 + 4S}\left\{3(A_1 + A_3 + A_5) - 2(A_2 + A_4 + A_6) - \frac{1}{2}\sqrt{c_1^2 + c_2^2 + c_3^2}\right\}$$

where

A_0 = an adjustable parameter (here 0.175)
S = the value of the overlap integral (here 0.25)
$A_1 = \exp(-2.008(R_{ab} - R_0))$
$A_2 = \exp(-1.004(R_{ab} - R_0))$
$A_3 = \exp(-2.008(R_{bc} - R_0))$
$A_4 = \exp(-1.004(R_{bc} - R_0))$
$A_5 = \exp(-2.008(R_{ac} - R_0))$
$A_6 = \exp(-1.004(R_{ac} - R_0))$
$C_1 = A_1 - 6A_2$
$C_2 = A_3 - 6A_4$
$C_3 = A_5 - 6A_6$

The system can be described by the following four differential equations,

$$\frac{dQ_1}{dt} = \frac{dy_1}{dt} = \frac{2P_1}{m} \qquad \frac{dQ_2}{dt} = \frac{dy_2}{dt} = \frac{3P_2}{2m}$$

$$\frac{dP_1}{dt} = \frac{dy_3}{dt} = \frac{\partial V}{\partial Q_1} \qquad \frac{dP_2}{dt} = \frac{dy_4}{dt} = \frac{\partial V}{\partial Q_2}$$

The initial values are assigned as follows. YO(1) is the internuclear separation in the H_a-H_b molecule. Assuming all energy is initially potential energy, the classical turning points can be calculated from the Morse potential function [13],

$$V(r) = EBC = D_e[1 - \exp(-a(R - R_0))]^2$$

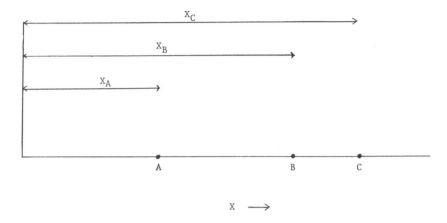

<u>Figure 7.2.</u> Coordinate system for the linear H_3 system.

Here D_e = 4.747 eV and a = 1.004 atomic units, so

$$EBC = 4.747[1 - \exp(-1.004(R - 1.402))]^2$$

$$1 - \exp(-1.004(R - 1.402)) = \sqrt{\frac{EBC}{4.747}} = \pm S'$$

$$\exp(-1.004(R - 1.402)) = 1 \pm S'$$

Then

$$R = 1.402 - \frac{\ln(1 \pm S')}{1.004}$$

For example, if EBC = 1 eV,

$$S' = \sqrt{\frac{1}{4.747}} = 0.459$$

The minimum classical turning point is

$$R = 1.402 - \frac{\ln(1 + 0.459)}{1.004}$$

$$= 1.03 \text{ atomic units}$$

and the maximum classical turning point is

$$R = 1.402 - \frac{\ln(1 - 0.459)}{1.004}$$

$$= 2.01 \text{ atomic units}$$

YO(2) is the distance between H_a and the center of the $H_b\text{–}H_c$ bond. This value
is one of the input parameters. The suggested range for this distance is between
4 and 9 atomic units (approximately 7 to 17 Å). YO(3) is the momentum associated
with Q_1, which is R_{bc}. Since the energy of the $H_a\text{–}H_b$ molecule is assumed to be all
potential energy, this momentum is zero. Finally, YO(4) is the momentum associated
with Q_2,

$$YO(4) = P_2$$
$$= M_2 V_2$$
$$= -\frac{2(1837)}{3} VR \frac{2.42 \times 10^{-17}}{0.529 \times 10^{-8}}$$
$$= -5.60 \times 10^{-6} VR$$

Here P_2, M_2, and V_2 are the momentum, mass, and velocity associated with the coordinate Q_2. VR is the relative velocity of H_a, and is an input parameter. The suggested range for VR is between 0.5×10^6 and 2.0×10^6 cm/sec. The value $2(1837)/3$ is the reduced mass of the triatomic system, and the two constants are conversion factors for time and distance from seconds and centimeters into atomic units.

Sample Execution

```
ENTER THE INITIAL ENERGY OF THE HB-HC MOLECULE IN EV.
(SUGGESTED RANGE:  0.25 - 1.5 EV)
>1


ENTER 0 OR 1 DEPENDING ON WHETHER HB-HC IS AT A
MIMIMUM OR MAXIMUM CLASSICAL TURNING POINT.
>1


ENTER THE INITIAL VELOCITY OF THE HA ATOM IN CM/SEC.
(SUGGESTED RANGE:  0.5E6 - 2E6 CM/SEC )
>1.E6


ENTER THE INITIAL VALUE FOR THE DISTANCE BETWEEN THE HA ATOM
AND THE MIDPOINT OF THE HB-HC MOLECULE IN ATOMIC UNITS.
(SUGGESTED RANGE:  4 - 9 ATOMIC UNITS)
>6


THE INITIAL ENERGY OF THE MOLECULE IS     1.00

HB-HC IS AT THE MAXIMUM CLASSICAL TURNING POINT.

THE INITIAL VELOCITY OF THE HYDROGEN ATOM IS     1.00E+06

THE DISTANCE BETWEEN THE ATOM AND MOLECULE IS     6.00
```

TIME	RBC	RAB	P1	P2	V	ETOT
0.0	2.01	4.99	0.000E+00	-5.60	-2.98	-2.63
50.0	1.92	4.81	-3.48	-5.54	-3.15	-2.63
100.0	1.64	4.73	-6.53	-5.47	-3.59	-2.63
150.0	1.25	4.70	-6.93	-5.39	-3.66	-2.63
200.0	1.02	4.60	-0.270	-5.29	-2.94	-2.63
250.0	1.23	4.28	6.77	-5.17	-3.60	-2.63
300.0	1.62	3.88	6.65	-5.00	-3.56	-2.63
350.0	1.91	3.53	3.68	-4.81	-3.09	-2.63
400.0	2.02	3.29	0.323	-4.62	-2.87	-2.63
450.0	1.94	3.14	-2.96	-4.43	-2.98	-2.63

500.0	1.70	3.08	−5.98	−4.14	−3.35	−2.63
550.0	1.32	3.11	−7.12	−3.67	−3.53	−2.63
600.0	1.04	3.11	−1.97	−3.08	−2.79	−2.63
650.0	1.17	2.94	5.87	−2.39	−3.20	−2.63
700.0	1.54	2.67	6.97	−1.63	−3.38	−2.63
750.0	1.87	2.45	4.88	−1.13	−3.00	−2.63
800.0	2.08	2.30	3.03	−1.27	−2.78	−2.63
850.0	2.22	2.16	2.19	−2.16	−2.75	−2.63
900.0	2.34	1.98	2.41	−3.72	−2.87	−2.63
950.0	2.50	1.71	3.61	−5.68	−3.18	−2.63
1000.0	2.73	1.34	4.69	−6.34	−3.40	−2.63
1050.0	2.96	1.05	2.91	−1.11	−2.77	−2.63
1100.0	3.01	1.16	−0.488	7.15	−3.20	−2.63
1150.0	2.97	1.53	−0.685	8.77	−3.49	−2.63
1200.0	2.97	1.84	0.844	6.55	−3.12	−2.63
1250.0	3.06	2.00	2.58	3.54	−2.87	−2.63
1300.0	3.25	1.99	4.34	0.376	−2.91	−2.63
1350.0	3.54	1.80	6.13	−2.79	−3.27	−2.63
1400.0	3.91	1.45	7.39	−4.87	−3.70	−2.63
1450.0	4.30	1.09	6.08	−1.90	−3.22	−2.63
1500.0	4.51	1.08	1.83	6.84	−3.20	−2.63
1550.0	4.55	1.43	0.334	10.1	−3.75	−2.63
1600.0	4.60	1.79	1.46	7.99	−3.37	−2.63
1650.0	4.72	1.99	3.19	4.67	−3.02	−2.63
1700.0	4.94	1.99	4.99	1.20	−3.01	−2.63
1750.0	5.26	1.81	6.71	−2.16	−3.35	−2.63
1800.0	5.67	1.47	7.90	−4.47	−3.78	−2.63
1850.0	6.08	1.10	6.69	−1.99	−3.34	−2.63
1900.0	6.33	1.07	2.40	6.63	−3.20	−2.63
1950.0	6.39	1.41	0.610	10.2	−3.80	−2.63
2000.0	6.45	1.77	1.59	8.32	−3.44	−2.63

```
    TIME                     INTERNUCLEAR DISTANCES

             0.000                      3.22                    6.45
               :                         :                       :
               :+----+----+----+----+----+----+----+----+----+----+
        0.0 :                      +                  -
       50.0 :                      +                  -   ,
      100.0 :                   +                      -
      150.0 :               +                      -
      200.0 :            +                         -
      250.0 :              +                   -
      300.0 :                 +               -
      350.0 :                +             -
      400.0 :                 +         -
      450.0 :                +        -
      500.0 :              +         -
      550.0 :            +           -
      600.0 :          +             -
      650.0 :           +         -
      700.0 :            +       -
      750.0 :              +   -
      800.0 :             + -
      850.0 :               -
      900.0 :            -  +
      950.0 :          -     +
     1000.0 :         -         +
     1050.0 :         -           +
```

```
      1100.0 :                -                      +
      1150.0 :                   -                   +
      1200.0 :                  -                    +
      1250.0 :                      -                  +
      1300.0 :                    -                   +
      1350.0 :                    -                 +
      1400.0 :                 -                   +
      1450.0 :            -                      +
      1500.0 :            -                      +
      1550.0 :             -                     +
      1600.0 :                 -                     +
      1650.0 :                 -                      +
      1700.0 :                -                        +
      1750.0 :                -                         +
      1800.0 :             -                             +
      1850.0 :           -                                +
      1900.0 :          -                                 +
      1950.0 :           -                                  +
      2000.0 :                  -                           +
```

Listing (See Section 2.1 for PLOT and Section 7.4.2 for RK4)

```
C
C
C                         EXCHG
C
C     THIS PROGRAM CALCULATES TRAJECTORIES FOR THE HYDROGEN
C     ATOM - HYDROGEN MOLECULE EXCHANGE REACTION
C
C         HA     +     HA - HB   ------>    HA - HB    +    HC
C
C
C      THE POTENTIAL ENERGY IS CALCULATED USING SATO'S FUNCTION
C
C      THE MORSE POTENTIAL FUNCTION FOR THE HYDROGEN MOLECULE IS
C
C         V(R) = DE*[ 1 - EXP( -A*(R-RO) ) ]**2
C
C     WHERE    DE = 4.747 EV    A = 1.004    AND RO = 1.402 ATOMIC UNITS
C
C        1 ATOMIC UNIT OF ENERGY IS 27.21 EV
C        1 ATOMIC UNIT OF LENGTH IS 0.529 ANGSTROMS
C        1 ATOMIC UNIT OF TIME IS 2.42E-17 SEC
C
C        GLOSSARY  (SEE RUNGE IN SECTION 7.4.1):
C
C     TIME(40)     -   TIME VECTOR FOR PLOT
C     RIJ(40,2)    -   INTERNUCLEAR DISTANCES FOR PLOT
C     MINMAX(2)    -   "MIN" OR "MAX"-IMUM CLASSICAL TURNING POINT
C     A(6),BB(3)   -   TERMS IN THE SATO POTENTIAL ENERGY FUNCTION
C     C(3),S       -      "     "     "     "       "        "
C     C1           -   CONSTANTS IN THE MORSE POTENTIAL FUNCTION
C     C3           -   PARAMETER IN THE SATO FUNCTION
C     SK           -   OVERLAP INTERGRAL, PARAMETER IN THE SATO FUNCTION
C     RO           -   EQUILIBRIUM INTERNUCLEAR SEPARATION IN HA-HB
C     EBC          -   INITIAL POTENTIAL ENERGY OF  HA-HB
C     JWHICH       -   MINIMUM OR MAXIMUM CLASSICAL TURNING POINT
C     VA           -   INITIAL VELOCITY OF HA
C     YO(2)        -   INITIAL VALUE FOR THE DISTANCE BETWEEN HA AND HB-HC
C     B            -   INTERNUCLEAR DISTANCE RAB
C     V            -   POTENTIAL ENERGY
C     ETOT         -   TOTAL ENERGY
C
```

```
C
C     PROGRAM STRUCTURE:
C
C           MAIN    -    I/O AND CONTROL
C           CALCV   -    SUBROUTINE TO CALCULATE THE POTENTIAL ENERGY
C           RK4     -    SUBROUTINE FOR RUNGE-KUTTA METHOD (SEC. 7.4.1)
C           FCT     -    DIFFERENTIAL EQUATIONS
C           PLOT    -    PLOT ROUTINE (SECTION 2.1)
C
C     AUTHORS:  S. LEVITT, M. HUGHES, F. LIN, AND K.J. JOHNSON
C
      DIMENSION Y(4),YHOLD(4),YO(4),TIME(41),RIJ(41,2),MINMAX(2)
      DATA NDIM,XO,XF,HO,HOUT,TOL,ITMAX/4,0.,2000.,50.,50.,.005,10/
      DATA MINMAX/'MIN','MAX'/
      COMMON/VTERMS/A(6),BB(3),C(3),S
      COMMON C1,C2,C3,RO,SK
C
C     SET PARAMETERS
C
      C1=1.004
      C2=2.008
      C3=.175
      RO=1.402
      SK=0.25
C
C     INPUT
C
      WRITE(6,11)
      READ(5,22)EBC
      WRITE(6,33)
      READ(5,44)JWHICH
      WRITE(6,55)
      READ(5,22)VR
      WRITE(6,66)
      READ(5,22)YO(2)
      WRITE(6,77)EBC,MINMAX(JWHICH+1),VR,YO(2)
C
C     CALCULATE THE REST OF THE INITIAL VALUES
C
      S=SQRT(EBC/4.747)
      IF(JWHICH.EQ.0)S=-S
      YO(1)=RO-(ALOG(1.0-S))/C1
      YO(3)=0.
      YO(4)=-5.60E-6*VR
      CALL CALCV(YO,B,V,ETOT)
      TIME(1)=XO
      RIJ(1,1)=YO(1)
      RIJ(1,2)=B
      V=27.21*V
      ETOT=27.21*ETOT
      WRITE(6,88)X0,YO(1),B,YO(3),YO(4),V,ETOT
C
C     OUTERMOST LOOP (100) CONTROLS OUTPUT STEP SIZE (HOUT)
C
      DO 100 IOUT=1,40
C
C     MIDDLE LOOP (10) CONTROLS INTERNAL STEP SIZE (H)
C
      DO 10 I=1,NINT
      K=1
```

```
C
C    CALL RK4 TO GET FIRST APPROXIMATION
C
      CALL RK4(XO,YO,YHOLD,HO,NDIM,K)
C
C    INNER LOOP (20) TESTS THE STABILITY OF THE FIRST APPROXIMATION
C
      DO 20 IHLF=1,ITMAX
      K=2*K
      H=HO/K
      CALL RK4(XO,YO,Y,H,NDIM,K)
      SUM=0.
      DO 40 IJK=1,NDIM
   40 SUM=SUM+ABS((Y(IJK)-YHOLD(IJK))/Y(IJK))
      IF (SUM.LT.TOL)GO TO 50
      DO 45 KJJ=1,NDIM
   45 YHOLD(KJJ)=Y(KJJ)
   20 CONTINUE
C
C  ERROR CONDITION - UNSTABLE IN ITMAX STEPS
C
      WRITE(6,99)
      STOP
   50 XO=XO+HO
      DO 60 KJJ=1,NDIM
   60 YO(KJJ)=Y(KJJ)
   10 CONTINUE
C
C    PRINT RESULTS
C
      CALL CALCV(Y,B,V,ETOT)
      V=27.21*V
      ETOT=27.21*ETOT
      WRITE(6,88)XO,Y(1),B,Y(3),Y(4),V,ETOT
      IP1=IOUT+1
      TIME(IP1)=XO
      RIJ(IP1,1)=Y(1)
      RIJ(IP1,2)=B
  100   CONTINUE
C
C    PLOT RESULTS
C
      CALL PLOT(TIME,RIJ,IP1,2)
      STOP
C
C    FORMATS
C
   11 FORMAT(//'  ENTER THE INITIAL ENERGY OF THE HB-HC MOLECULE',
     1 ' IN EV.'/'  (SUGGESTED RANGE:  0.25 - 1.5 EV)')
   22 FORMAT(E)
   33 FORMAT(//'  ENTER 0 OR 1 DEPENDING ON WHETHER HB-HC IS',
     1' AT A '/'  MIMIMUM OR MAXIMUM CLASSICAL TURNING POINT.')
   44 FORMAT(I)
   55 FORMAT(//'  ENTER THE INITIAL VELOCITY OF THE HA ATOM',
     1' IN CM/SEC.'/'  (SUGGESTED RANGE:  0.5E6 - 2E6 CM/SEC )')
   66 FORMAT(//'  ENTER THE INITIAL VALUE FOR THE DISTANCE',
     1' BETWEEN THE HA ATOM'/,'  AND THE MIDPOINT OF THE ',
     2'HB-HC MOLECULE IN ATOMIC UNITS.',
     3/'  (SUGGESTED RANGE:  4 - 9 ATOMIC UNITS)')
```

```
   77 FORMAT(//'   THE INITIAL ENERGY OF THE MOLECULE IS',G12.3,
      1 //'   HB-HC IS AT THE ',A3,'IMUM CLASSICAL TURNING POINT.',
      2//'   THE INITIAL VELOCITY OF THE HYDROGEN ATOM IS',1PE12.2,//
      3 '   THE DISTANCE BETWEEN THE ATOM AND MOLECULE IS',  G12.3,
      4   ///4X,'TIME',7X,'RBC',7X,'RAB',7X,'P1',
      5 8X,'P2',8X,'V',8X,'ETOT'/)
   88 FORMAT(F8.1,4X,6G10.3)
   99 FORMAT(//'   FAILED TO FIND STABLE SOLN. IN ITMAX TRIES'/)
      END
C
C          FCT
C
C
C      THE DIFFERENTIAL EQUATIONS ARE
C      CODED IN DERY(NDIM)
C
      SUBROUTINE FCT(X,Y,DERY)
      DIMENSION DERY(4),Y(4)
      COMMON C1,C2,C3,RO,SK
      COMMON/VTERMS/A(6),BB(3),C(3),S
      Q1=Y(1)
      Q2=Y(2)
      P1=Y(3)
      P2=Y(4)
      CALL CALCV(Y,B,V,ETOT)
      R=C1*C3/(4.+4.*SK)
      F1=R*(-6.*A(1)+2.*A(2)+(A(1)-3.*A(2))*(C(1)-C(3))/S)
      F2=R*(-6.*A(3)+2.*A(4)+(A(3)-3.*A(4))*(C(2)-C(1))/S)
      F3=R*(-6.*A(5)+2.*A(6)+(A(5)-3.*A(6))*(C(3)-C(2))/S)
      DERY(1)=2.*P1/1837.
      DERY(2)=3.*P2/3674.
      DERY(3)=.5*(F1-F3)-F2
      DERY(4)=-F1-F3
      RETURN
      END
C
C      CALCV
C
C   V(HA,HB,HC)   IS CALCULATED USING SATO'S FUNCTION
C
C
      SUBROUTINE CALCV(Y,B,V,ETOT)
      DIMENSION Y(4)
      COMMON/VTERMS/A(6),BB(3),C(3),S
      COMMON C1,C2,C3,RO,SK
      B=ABS(Y(2)-.5*Y(1))
      Q1=Y(1)
      Q2=Y(2)
      P1=Y(3)
      P2=Y(4)
      A(1)=EXP(-C2*((Q2-.5*Q1)-RO))
      A(2)=EXP(-C1*((Q2-.5*Q1)-RO))
      A(3)=EXP(-C2*(Q1-RO))
      A(4)=EXP(-C1*(Q1-RO))
      A(5)=EXP(-C2*((Q2+.5*Q1)-RO))
      A(6)=EXP(-C1*((Q2+.5*Q1)-RO))
      BB(1)=A(1)-6.*A(2)
      BB(2)=A(3)-6.*A(4)
      BB(3)=A(5)-6.*A(6)
      C(1)=BB(1)-BB(2)
```

```
C(2)=BB(2)-BB(3)
C(3)=BB(3)-BB(1)
S=SQRT(0.5*(C(1)**2+C(2)**2+C(3)**2))
V=(C3/(4.+4.*SK))*(3.*(A(1)+A(3)+A(5))-2.*(A(2)+A(4)+A(6))-S)
EK1=Y(3)**2/1837.
EK2=Y(4)**2/2448.
ETOT=V+EK1+EK2
RETURN
END
```

Discussion

EXCHG, including the subroutines, requires approximately 1350 words of core. This sample execution took approximately 1 sec. Two sample executions, one showing an exchange and the other showing no exchange, are plotted in Figure 7.3

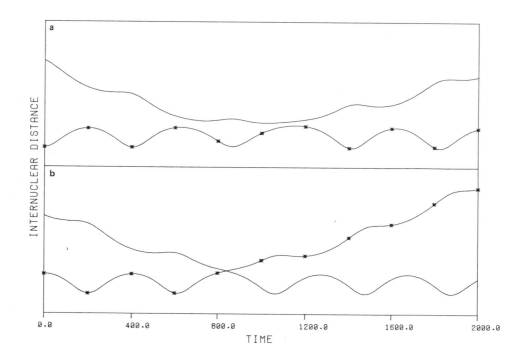

Figure 7.3 (a) Same set of parameters as sample execution. (b) Same set of parameters as in (a) except the molecule is initially at the minimum classical separation.

7.5. HIGHER ORDER DIFFERENTIAL EQUATIONS

   The numerical methods discussed in the preceding section can also be applied to higher order differential equations with initial values. For example, consider the following second-order differential equation,

$$A \frac{d^2y}{dx^2} + B \frac{dy}{dx} + C y + D = 0$$

Here A, B, C, and D are constants or functions of x, and d^2y/dx^2 is the second derivative of y with respect to x. The equation is solved by the following substitution, Let

$$z = \frac{dy}{dx}$$

then

$$\frac{dz}{dx} = \frac{d^2y}{dx^2}$$

Given initial values x_0, y_0, and z_0, the second-order equation can be expressed as a system of two first-order equations

$$\frac{dy}{dx} = z \qquad\qquad (x_0, y_0, z_0)$$

$$\frac{dz}{dx} = \frac{d^2y}{dx^2} = \frac{-(D + Cy + Bz)}{A} \qquad (x_0, y_0, z_0)$$

Problem 7.11. Solve the differential equation

$$y'' + F(x)y' + G(x)y + H(x) = 0$$

where

$$F(x) = \frac{1}{x} \qquad G(x) = \frac{1}{x^2}$$

and

$$H(x) = -(9x + \frac{x^3 + 1}{x^2})$$

The initial values are $x_0 = 1$, $y_0 = 2$, and $(\frac{dy}{dx})_0 = y_0' = 3$.

Solution. The analytical solution is $y = x^3 + 1$. The numerical solution is obtained with the following modification to the subroutine FCT in RUNGE (Section 7.4.1). Let $dy/dx = z$, then

$$\frac{dz}{dx} = \frac{d^2y}{dx^2} = y''$$
$$= -[F(x)z + G(x)y + H(x)]$$
$$= -\frac{z}{x} + \frac{y}{x^2} - 9x - \frac{x^3 + 1}{x^2}$$

The corresponding statements in FCT are

```
SUBROUTINE FCT(X,Y,DERY)
DIMENSION DERY(10),Y(10)
DERY(1) = Y(2)
DERY(2)=-(Y(2)/X + Y(1)/(X*X) - 9.*X - (X**3+1.)/(X*X) )
RETURN
END
```

The output from RUNGE is shown below.

ENTER NDIM, XO, XF, HOUT, H, AND TOL
NOTE: HOUT MUST BE AN INTEGRAL MULTIPLE OF H.
>   2 1 10 1 1   .0001

 ENTER THE INITIAL VALUES: YO(I),I=1,NDIM)
> 2   3

     NDIM = 2   HOUT =   1.00

     H =   1.00   TOL =   0.100E-03

     X   H   Y(I),   I = 1,2,3, ... ,NDIM

    1.00   0.00E+00   2.00   3.00
    2.00   0.13   9.00   12.0
    3.00   0.25   28.0   27.0
    4.00   0.25   65.0   48.0
    5.00   0.50   126.   75.0
    6.00   0.50   217.   108.
    7.00   0.50   344.   147.
    8.00   0.50   513.   192.
    9.00   0.50   730.   243.
   10.00   0.50   0.100E+04   300.

Note that Y(1) is $x^3 + 1$ and that Y(2) is $3x^2$, which is the first derivative of y
with respect to x.

Problem 7.12.   Write a program to solve the Schrödinger equation for a vibrating
                diatomic molecule assuming the harmonic oscillator approximation to
                the potential function.

     Solution.[14, 15]. See the following program.

7.5.1. SCHRO

Introduction

     SCHRO solves the following second-order differential equation

$$- \frac{\hbar^2}{2\mu} \frac{d^2\psi(x)}{dx^2} + V(x)\,\psi(x) = E\psi(x)$$

Here $\hbar$ is Planck's constant divided by 2π, μ is the reduced mass of the diatomic molecule, ψ is the wavefunction for one of the allowed vibrational eigenfunctions, E is E is the energy corresponding to that state, and V(x) is approximated by the harmonic oscillator potential,

$$V(x) = \frac{k}{2} x^2$$

where k is the Hooke's law restoring force. The reduced mass is

$$\mu = \frac{M_a M_b}{M_a + M_b}$$

where M_a and M_b are the masses of the atoms. This differential equation has an analytical solution [12, 13]. The energy levels are given by

$$E(v) = (v + 0.5)\, \hbar \sqrt{\frac{k}{\mu}} \qquad v = 0,1,2,3,\ldots$$

The eigenfunctions are given by the modified Hermite polynomials. SCHRO is designed to numerically determine eigenfunctions corresponding to a given eigenvalue. The input to the program includes the reduced mass of the diatomic molecule in atomic mass units, the force constant in dynes per centimeter. The vibrational quantum number, v, and the energy in electron volts. The program prints these data for verification purposes, the numerical solution to the differential equation, and a plot of the results.

Method

Rearranging the Schrodinger equation,

$$\frac{d^2 \psi(x)}{dx^2} = \frac{2\mu}{\hbar^2} \left\{ \frac{k}{2} x^2 - E \right\} \psi(x)$$

Let $A = 2\mu E/\hbar^2$ and $B = \sqrt{\mu k}/\hbar$. Then

$$\frac{d^2 \psi(x)}{dx^2} = (B^2 x^2 - A)\, \psi(x)$$

Now, introducing the dimensionless variable [15]

$$z = \sqrt{B}\, x$$

the equation becomes

$$\frac{d^2 \psi(z)}{dz^2} = \left(z^2 - \frac{A}{B} \right) \psi(z)$$

The units used here are μ in atomic mass units; k in dynes/cm; E in electron volts; and distance (x) in centimeters. In these units,

$$\frac{A}{B} = \frac{4\pi E\sqrt{\mu/k}}{h}$$

$$= \frac{4(3.14159)E(eV)(1.602 \times 10^{-12})}{6.626 \times 10^{-27}}\left\{\frac{\mu(amu)(1.66 \times 10^{-24})}{k}\right\}^{1/2}$$

$$= 3915.12\ E(eV)\left\{\frac{\mu(amu)}{k}\right\}^{1/2}$$

and

$$x = \frac{z}{\sqrt{B}} = \frac{z\sqrt{h}}{[\mu(amu)k]^{1/4}}$$

$$= \frac{z[6.626 \times 10^{-27}/2(3.14159)]^{1/2}}{[\mu(amu)(1.66053 \times 10^{-24})k]^{1/4}}$$

$$= \frac{2.861 \times 10^{-8}\ z}{[\mu(amu)k]^{1/4}}$$

For example, for CO,

$$\mu = \frac{12.0(16.0)}{12.0 + 16.0} = 6.8607$$

$$k = 1.87 \times 10^{6}\ dynes/cm$$

So

$$\frac{A}{B} = 3915.12E(eV)\left\{\frac{6.8607}{1.87 \times 10^{6}}\right\}^{1/2}$$

$$= 7.499\ E(eV)$$

and

$$x = \frac{2.861 \times 10^{-8}\ z}{[6.8607(1.87 \times 10^{6})]^{1/4}} = 4.780 \times 10^{-10}\ z$$

The range of x is determined by the classical turning points,

$$\frac{k}{2}\ x_t^2 = (v + \tfrac{1}{2})\ \hbar\sqrt{\frac{k}{\mu}}$$

or

$$x_t = \pm\frac{4.046 \times 10^{-8}\ \sqrt{v + 1/2}}{[k\ \mu(amu)]^{1/4}}$$

For example, for CO in the ground vibrational state, $v = 0$, the classical turning points are $R \pm x_t$ or $R \pm 4.78 \times 10^{-10}$ cm, where R is the equilibrium internuclear distance in CO.

SCHRO reads μ, k, v, E, and the initial values, $\psi(0)$, and $(d\psi/dx)_0$. The program then computes the constants A and B, the classical turning point x_t, and integrates the system of equations from $x = 0$ to $x = 2.5x_t$.

Sample Execution

ENTER THE REDUCED MASS (AMU), AND
THE FORCE CONSTANT (DYNE/CM)
> 6.8607 1.87E6

ENTER THE VIBRATIONAL QUANTUM NUMBER,AND
THE ENERGY (EV)
> 2   0.66676

ENTER THE INITIAL VALUES: U(0) AND DU/DX (0)
> -1

REDUCED MASS =   6.8607   FORCE CONSTANT = 0.18700E+07

VIBR. QUANTUM   = 2   ENERGY = 0.66676

X (CM)	U(X)	DU(X)/DX
0.00E+00	-1.00	0.000E+00
1.34E-10	-0.811	1.30
2.67E-10	-0.321	2.09
4.01E-10	0.286	2.12
5.34E-10	0.803	1.50
6.68E-10	1.09	0.576
8.02E-10	1.13	-0.257
9.35E-10	0.982	-0.766
1.07E-09	0.739	-0.918
1.20E-09	0.493	-0.814
1.34E-09	0.294	-0.597
1.47E-09	0.159	-0.379
1.60E-09	7.749E-02	-0.212
1.74E-09	3.438E-02	-0.106
1.87E-09	1.373E-02	-4.832E-02
2.00E-09	4.498E-03	-2.166E-02
2.14E-09	-1.343E-04	-1.418E-02
2.27E-09	-5.085E-03	-2.521E-02
2.41E-09	-1.814E-02	-8.082E-02
2.54E-09	-6.501E-02	-0.306
2.67E-09	-0.253	-1.27


```
              X (CM)                              U(X)

                 -1.00                    0.667E-01                    1.13
                    :                         :                          :
                    :+----+----+----+----+----+----+----+----+----+----+
        0.00E+00 :+                          -
        1.34E-10 :      +                    -
        2.67E-10 :               +           -
        4.01E-10 :                           -       +
        5.34E-10 :                           -            +
        6.68E-10 :                           -              +
        8.02E-10 :                           -              +
        9.35E-10 :                           -           +
        1.07E-09 :                           -        +
```

```
      1.20E-09 :                                    -                  +
      1.34E-09 :                                    -           +
      1.47E-09 :                                    -      +
      1.60E-09 :                                   - +
      1.74E-09 :                                   -+
      1.87E-09 :                                   -+
      2.00E-09 :                                   -+
      2.14E-09 :                                   -
      2.27E-09 :                                   -
      2.41E-09 :                                   -
      2.54E-09 :                                 +-
      2.67E-09 :                          +       -
```

Listing

```
C
C
C                        SCHRO
C
C  THIS PROGRAM APPROXIMATES THE SOLUTION TO THE ONE-DIMENSIONAL
C  SCHROEDINGER EQUATION FOR A VIBRATING DIATOMIC MOLECULE
C
C        (-H**2/4*PI**2*MU)*D2U/DX2 + (1/2)K*X**2*U = E*U
C
C   INTRODUCING A DIMENSIONLESS INDEPENDENT VARIABLE,
C
C        D2U/DZ2 = (Z**2 - C1)*U
C
C   WHERE   Z = SQRT(B)*X
C           B = SQRT(MU*K)/H/(2*PI)
C          C1 = A/B
C           A = 2*MU*E/(H/2*PI)**2
C
C SO
C           C1 = (2/H/2*PI)*E*SQRT(MU/K)
C              = 3915.12*E*SQRT(MU/K)
C
C      X = Z/SQRT(B)   (CM)
C        = Z*C2 = 2.861E-8/(MU*K)**(1/4)
C
C   UNITS:  E IN EV,  X IN CM    (SCALED BY 10**8 FOR OUTPUT)
C           MU IN AMU     AND      K IN DYNES/CM
C
C
C
C   GLOSSARY (SEE RUNGE IN SECTION 7.4.1):
C
C   U(21)           - WAVE FUNCTION TO BE PLOTTED
C   XPLT(21)        - X IN ANGSTROMS FOR PLOT
C   AMU             - REDUCED MASS IN AMU
C   AK              - FORCE CONSTANT IN DYNES/CM
C   V               - VIBRATIONAL QUANTUM NUMBER
C   E               - EIGENVALUE IN EV
C   C1,C2           - SEE ABOVE
C   XT              - CLASSICAL TURNING POINT (CM)
C   ZF              - MAXIMUM VALUE OF Z
C
C
C   PROGRAM ARRANGEMENT:
C
C        MAIN PROGRAM - I/O, CONTROL OF ITERATIVE PROCEDURE
```

```
C                SUBR. FCT    - DIFFERENTIAL EQUATIONS
C                SUBR. RK4    - FOURTH-ORDER RUNGE-KUTTA ROUTINE
C                SUBR. PLOT   - PLOT ROUTINE (SECTION 2.1)
C
C
C
C        AUTHOR:  K. J. JOHNSON
C
C
      DIMENSION Y(2),YHOLD(2),YO(2),U(21,2),XPLT(21)
      DATA NDIM,ZO,TOL/2,0.,0.001/
      COMMON C1
C
C    INPUT
C
      WRITE(6,11)
      READ(5,22)AMU,AK
      WRITE(6,23)
      READ(5,22)V,E
      WRITE(6,33)
      READ(5,22)(YO(I),I=1,NDIM)
      IV=V
      WRITE(6,44)AMU,AK,IV,E
      C1=3915.12*E*SQRT(AMU/AK)
      ROOT=(AMU*AK)**(0.25)
      C2=2.861E-8/ROOT
      XT=4.046E-8*SQRT(V+0.5)/ROOT
      ZF=2.5*XT/C2
      HOUT=ZF/20
      HO=HOUT
C
C    PRINT INPUT FOR VERIFICATION
C
      WRITE(6,55)
      H=0.
      WRITE(6,66)ZO,(YO(I),I=1,NDIM)
      XPLT(1)=0.
      U(1,1)=YO(1)
      U(1,2)=0.
C
C    SET CONSTANTS
C
      NOUT=(ZF-ZO)/HOUT
      NINT=HOUT/HO
      ITMAX=10
C
C  OUTERMOST LOOP (100) CONTROLS OUTPUT STEP SIZE (HOUT)
C
      DO 100 IOUT=1,NOUT
C
C  MIDDLE LOOP (10) CONTROLS INTERNAL STEP SIZE (H)
C
      DO 10 I=1,NINT
      K=1
C
C  CALL RK4 TO GET FIRST APPROXIMATION
C
      CALL RK4(ZO,YO,YHOLD,HO,NDIM,K)
C
C  INNER LOOP (20) TESTS THE STABILITY OF THE FIRST APPROXIMATION
```

```
      DO 20 IHLF=1,ITMAX
      K=2*K
      H=HO/K
      CALL RK4(ZO,YO,Y,H,NDIM,K)
      SUM=0.
      DO 40 IJK=1,NDIM
   40 SUM=SUM+ABS((Y(IJK)-YHOLD(IJK))/Y(IJK))
      IF (SUM.LT.TOL)GO TO 50
      DO 45 KJJ=1,NDIM
   45 YHOLD(KJJ)=Y(KJJ)
   20 CONTINUE
C
C  ERROR CONDITION - UNSTABLE IN ITMAX STEPS
C
      WRITE(6,99)
      STOP
   50 ZO=ZO+HO
      DO 60 KJJ=1,NDIM
   60 YO(KJJ)=Y(KJJ)
   10 CONTINUE
C
C    PRINT RESULTS - CONVERT FROM DIMENSIONLESS Z TO X IN CM
C
      X=C2*ZO
      WRITE(6,66)X,(Y(KJJ),KJJ=1,NDIM)
      JJJ=IOUT+1
      XPLT(JJJ)=X
      U(JJJ,1)=Y(1)
      U(JJJ,2)=0.
  100 CONTINUE
      CALL PLOT(XPLT,U,21,2)
C
C    FORMATS
C
   11 FORMAT(/'  ENTER THE REDUCED MASS (AMU), AND',/
     1/'  THE FORCE CONSTANT (DYNE/CM)')
   22 FORMAT(2E)
   23  FORMAT(/'  ENTER THE VIBRATIONAL QUANTUM NUMBER,',
     1'AND'/'  THE ENERGY (EV)')
   33 FORMAT(/'  ENTER THE INITIAL VALUES:  U(0) AND  DU/DX (0)')
   44 FORMAT(//'  REDUCED MASS = ',G12.5,10X,'FORCE',
     1' CONSTANT = ',G12.5//'  VIBR. QUANTUM  =',I3,
     217X,'ENERGY = ',G12.5//)
   55 FORMAT(//4X,'X (CM)',10X,'U(X)',13X, 'DU(X)/DX'/)
   66 FORMAT(1PE10.2,2G18.3)
   99 FORMAT(//'  FAILED TO FIND STABLE SOLN. IN ITMAX TRIES'/)
      END
C
C        FCT
C
C
C
C    THE DIFFERENTIAL EQUATIONS ARE
C    CODED IN DERY(NDIM)
C
      SUBROUTINE FCT(X,Y,DERY)
      DIMENSION DERY(2),Y(2)
      COMMON C1
      DERY(1)=Y(2)
      DERY(2)=(X**2-C1)*Y(1)
      RETURN
      END
```

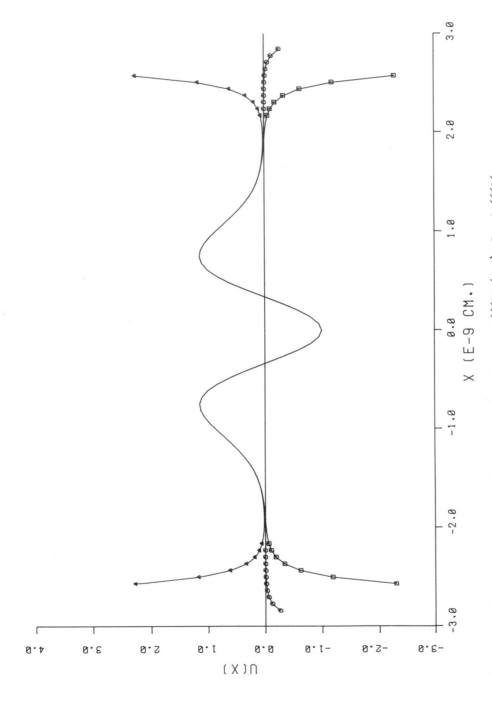

Figure 7.4 Wavefunctions for v = 2 corresponding to E = 0.66675 (top), E = 0.66676 (middle), and E = 0.66677 (bottom).

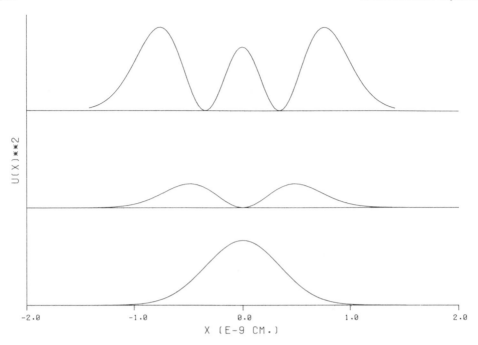

Figure 7.5 Squares of the wavefunctions of CO with v = 0 (bottom), v = 1 (middle), and v = 2 (top).

Discussion

SCHRO, including the subroutines RK4 and FCT, requires approximately 1000 words of core. This sample execution took approximately 0.3 sec of CPU time.

The sensitivity of the wavefunction to the eigenvalue is illustrated in Figure 7.4. The squares of the first three eigenfunctions for CO are plotted in Figure 7.5.

7.6. THE MONTE-CARLO METHOD

The Monte-Carlo method for approximating an integral was discussed in Section 6.5. This technique can also be used to solve differential equations. The Monte-Carlo technique is applied to chemical kinetics [16-18] by interpreting the rate equations as probabilities. For example, consider the first-order decay process,

$$A \xrightarrow{k} B \qquad \frac{dA}{dt} = -kA$$

The solution is

$$A_t = A_0 e^{-kt}$$

where A_0 is the initial concentration of A. The Monte-Carlo method utilizes a constant which symbolizes the volume of the reaction vessel. A random number generator is used to obtain a large number of uniformly distributed random numbers in the range 0 to 1. The rate constant k and the initial concentration A_0 are scaled to values between 0 and 1. An <u>attempt at reaction</u> can be defined in terms of the comparison.

$\qquad$ RAN(X) : kA

The reaction will occur if the random number is less than or equal to the value kA. In that case the concentration of A is reduced and that of B is increased as follows:

$$A = A - \frac{1}{V} \qquad B = B + \frac{1}{V}$$

where V is the volume parameter. As the concentration of A decreases the probability of reaction decreases. If the volume parameter and the corresponding number of attempts at reaction are large enough, then the results are reliable as shown in Figure 7.6.

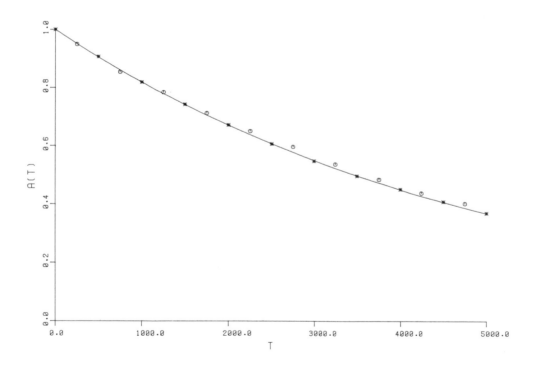

Figure 7.6 Monte-Carlo method for a first-order reaction (see text).

Let N be the number of attempts at reaction, and V be the volume parameter. In
Figure 7.6, the asterisks were obtained using V = N = 5000, the circles using
V = N = 500, and the solid line using the exponential function,

$$A_t = e^{-0.0002t} \qquad (0 \le t \le 5000)$$

The data plotted in Figure 7.6 were generated by the following program.

```
C
C
C      SIMULATION OF     A  ----->  B
C
       REAL K
       DATA K,A,I/2*1.,0/
       WRITE(6,11)
       READ(5,22)N,V
       WRITE(6,33)N,V
       V=1./V
       NINT=N/20
       WRITE(6,44)I,A
       DO 20 JJ=1,20
       DO 10 I=1,NINT
       IF(RAN(XXX).GT.A)GO TO 10
       A=A-V
       IF(A.LE.0.)STOP
   10  CONTINUE
       ITIME=JJ*NINT
       WRITE(6,44)ITIME,A
   20  CONTINUE
       STOP
   11  FORMAT(//'   ENTER N AND V')
   22  FORMAT(I,E)
   33  FORMAT(//'    N = ',I6, 5X,'V = ',F10.2,
      1///5X,'T',8X,'A(T)'/)
   44  FORMAT(I7,G15.3)
       END
```

A sample execution of this program is given next.

```
 ENTER N AND V
>5000 5000

  N =    5000     V =     5000.00

     T         A(T)

     0        1.00
   250        0.952
   500        0.907
   750        0.862
  1000        0.819
  1250        0.782
  1500        0.742
  1750        0.705
  2000        0.671
  2250        0.640
  2500        0.606
  2750        0.574
  3000        0.547
```

3250	0.520
3500	0.496
3750	0.473
4000	0.451
4250	0.429
4500	0.408
4750	0.388
5000	0.369

Problem 7.13. Write a program to solve the kinetic system solved by RUNGE in
            Section 7.4.1 using the Monte-Carlo method.

    Solution. See the following program.

7.6.1 MONTE

Introduction

    MONTE uses the Monte-Carlo method to approximate the solution to the kinetic
system

$$A + B \underset{k_2}{\overset{k_1}{\rightleftharpoons}} C + D$$

$$B + C \xrightarrow{k_3} D + E$$

The input includes the initial concentrations of the five species $CO(I)$, $I = 1, 2,$
..., 5); the rate constants $K(I)$, $I = 1, 2, 3$; the number of attempts at reaction
N, and the volume parameter V. The program prints the input values for verification
purposes, a table showing the relative concentrations as a function of number of
attempts at reaction, and a plot of the results.

    Method
    There are three rate equations,

$$Rate_1 = k_1 AB$$

$$Rate_2 = k_2 CD$$

$$Rate_3 = k_3 BC$$

At each attempt at reaction one of the three rate expressions is compared to a
random number. If the normalized rate expression exceeds the random number, the
reaction proceeds and reactant and product concentrations are modified by the
appropriate volume term. Since the reactions are coupled, it is necessary to test
for reaction in random order.

Sample Execution

ENTER THE INITIAL CONCENTRATIONS OF A,B,C,D, AND E
> 1 1

ENTER THE RATE CONSTANTS K1,K2, AND K3
> 1   .25   .5

ENTER THE NUMBER OF ATTEMPTS AT REACTION AND THE EFFECTIVE VOLUME
>250000 25000

```
   A =    1.00          B =     1.00          C =    0.000E+00

   D =    0.000E+00     E =    0.000E+00

  K1 =    1.00         K2 =     0.250        K3 =    0.500

  NUMBER OF RX. ATTEMPTS =    250000           VOLUME =    25000.
```

RELATIVE CONCENTRATIONS

T	A	B	C	D	E
	.00	1.00	0.000E+00	0.000E+00	0.000E+00
12500	0.856	0.850	0.139	0.150	0.548E-02
25000	0.751	0.733	0.231	0.267	0.179E-01
37500	0.673	0.640	0.294	0.360	0.332E-01
50000	0.614	0.565	0.337	0.435	0.488E-01
62500	0.569	0.505	0.366	0.495	0.642E-01
75000	0.532	0.454	0.389	0.546	0.783E-01
87500	0.504	0.411	0.403	0.589	0.929E-01
100000	0.483	0.377	0.411	0.623	0.106
112500	0.465	0.348	0.417	0.652	0.118
125000	0.451	0.321	0.419	0.679	0.130
137500	0.440	0.299	0.419	0.701	0.141
150000	0.432	0.281	0.416	0.719	0.151
162500	0.426	0.266	0.414	0.734	0.160
175000	0.422	0.254	0.410	0.746	0.168
187500	0.418	0.242	0.407	0.758	0.175
200000	0.416	0.231	0.400	0.769	0.184
212500	0.413	0.222	0.395	0.778	0.191
225000	0.411	0.212	0.390	0.788	0.199
237500	0.410	0.204	0.384	0.796	0.206
250000	0.408	0.196	0.380	0.804	0.212

```
        TIME                  RELATIVE CONCENTRATION

          0.0                      0.50                    1.00
          :                         :                       :
          :+----+----+----+---_+----+----+----+----+----+----+
   0.0 :E                          \                        B
12500.0 :E       D                                     B
25000.0 : E             CD                      BA
37500.0 :   E               C  D            B A
50000.0 :   E                 C    D     B  A
62500.0 :     E               C      D  A
75000.0 :       E             C   B    D
87500.0 :         E           CB   A     D
```

```
100000.0 :       E            B C  A        D
112500.0 :       E          B   C A          D
125000.0 :       E          B   C A        D
137500.0 :        E        B    CA          D
150000.0 :         E      B     CA            D
162500.0 :         E    B      C              D
175000.0 :         E    B      C              D
187500.0 :          E  B       CA           D
200000.0 :          E  B       CA           D
212500.0 :            EB       CA          D
225000.0 :            EB       C A          D
237500.0 :            E        C A           D
250000.0 :            BE       CA             D
```

Listing

```
C                          MONTE
C
C
C     MONTE-CARLO SIMULATION OF THE KINETIC SCHEME
C
C
C                         K1
C          A  +  B     <=======>   C  +  D
C                         K2
C
C
C                         K3
C          B  +  C     ------->   D  +  E
C
C
C
C    GLOSSARY:
C
C
C     CONC(21,5)  -  RELATIVE CONCENTRATIONS FOR PLOT
C     T(21)       -  NUMBER OF ATTEMPTS AT REACTION FOR PLOT
C     CO(5)       -  RELATIVE CONCENTRATIONS; CO(1)=A, CO(2)=B,
C                    CO(3)=C, CO(4)=D, AND CO(5)=E
C     K(3)        -  RATE CONSTANTS K1,K2 AND K3
C     CMAX        -  MAXIMUM OF THE FIVE INITIAL CONCENTRATIONS
C     KMAX        -  MAXIMUM OF THE THREE RATE CONSTANTS
C     N           -  NUMBER OF ATTEMPTS AT REACTION
C     V           -  VOLUME OF REACTION VESSEL
C
C
C    BOTH   N AND   V   SHOULD EXCEED 1000 FOR REASONABLE STATISTICS.
C
C    IF AN ATTEMPT AT REACTION IS SUCCESSFUL, SUBROUTINE RX IS
C    CALLED TO MODIFY THE REACTANT AND PRODUCT CONCENTRATIONS.
C
C     AUTHOR:  K. J. JOHNSON
C
C
C
      DIMENSION CONC(21,5),T(21),CO(5)
      REAL K(3),KMAX
      EQUIVALENCE (A,CO(1)),(B,CO(2)),(C,CO(3)),
     1 (D,CO(4)),(E,CO(5))
C    INITIALIZE THE RANDOM NUMBER GENERATOR
C
      CALL TIME(X,Y)
      CALL SETRAN(Y)
C
C    INPUT
C
```

```
       WRITE(6,11)
       READ(5,22) (CO(J),J=1,5)
       WRITE(6,33)
       READ(5,22) (K(J),J=1,3)
       WRITE(6,44)
       READ(5,55)N,V
       WRITE(6,66)(CO(J),J=1,5),(K(J),J=1,3),N,V
C
C    SET CONSTANTS
C
       IT=0
       T(1)=0.
       NINT=N/20
       V=1./V
C
C      NORMALIZE CONCENTRATIONS AND RATE CONSTANTS
C
       CMAX=AMAX1( A,B,C,D,E )
       KMAX=AMAX1( K(1), K(2), K(3) )
       DO 10 J=1,5
       CO(J)=CO(J)/CMAX
   10 CONC(1,J)=CO(J)
       DO 20 I=1,3
   20 K(I)=K(I)/KMAX
       WRITE(6,77) IT,(CONC(1,J),J=1,5)
C
C    OUTER LOOP (100) CONTROLS OUTPUT
C
       DO 100 I=2,21
       DO 90 J=1,NINT
C
C    REACTION 1, 2, OR 3?
C
       RNX=RAN(X)
       IF(RNX .LT. 0.3333) GO TO 30
       IF(RNX .LT. 0.6667) GO TO 40
C
C    REACTION 1:    A  +  B ----- K1 ------>  C  +  D
C
       IF( RAN(X) .LE. K(1)*A*B ) CALL RX(A,B,C,D,V)
       GO TO 90
C
C     REACTION 2:    C  +  D ---- K2 ----->  A  +  B
C
   30 IF( RAN(X) .LE. K(2)*C*D ) CALL RX(C,D,A,B,V)
       GO TO 90
C
C     REACTION 3:    B +  C ---- K3 ----->  D  +  E
C
   40 IF(RAN(X) .LE. K(3)*B*C ) CALL RX(B,C,D,E,V)
   90 CONTINUE
       DO 95 J=1,5
   95 CONC(I,J)=CO(J)
       IT=(I-1)*NINT
       T(I)=IT
  100 WRITE(6,77)IT,(CONC(I,J),J=1,5)
C
C     PLOT RESULTS
C
```

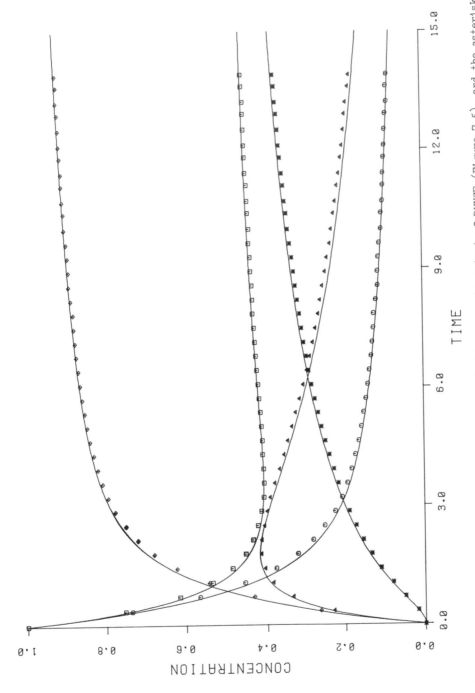

Figure 7.7 Comparison of MONTE and RUNGE. The smooth lines are the output of RUNGE (Figure 7.5), and the asterisks were obtained from MONTE using N = 1,000,000 and V = 25,000. The unit of time corresponds to 70,000 attempts at reaction

```
      CALL PLOT(T,CONC,21,5)
      STOP
11 FORMAT(' ENTER THE INITIAL CONCENTRATIONS OF A,B,C,D, AND E')
22 FORMAT(5E)
33 FORMAT(//' ENTER THE RATE CONSTANTS K1,K2, AND K3')
44 FORMAT(//' ENTER THE NUMBER OF ATTEMPTS AT REACTION AND THE',
   1 ' EFFECTIVE VOLUME')
55 FORMAT(I,E)
66 FORMAT(//'   A = ',G12.3,6X,'B = ',G12.3,6X,'C = ',
   1 G12.3//'   D = ',G12.3,6X,'E = ',G12.3//,
   2 '   K1 = ',G12.3,5X,'K2 = ',G12.3,5X,'K3 = ',G12.3,
   3 //'  NUMBER OF RX. ATTEMPTS = ',I8,12X,'VOLUME = ',G12.5,
   4 ////26X,'RELATIVE CONCENTRATIONS'//4X,'T',5X,
   5 ' A',12X,'B',13X,'C',13X,'D',13X,'E'/)
77 FORMAT(I5,5G13.3)
      END
C
C
C          SUBROUTINE   RX
C
C     R1  +  R2  ------->  P1 + P2
C
      SUBROUTINE RX(R1,R2,P1,P2,V)
      R1=R1-V
      R2=R2-V
      P1=P1+V
      P2=P2+V
      RETURN
      END
```

Discussion

MONTE, including the two subroutines, requires approximately 900 words of core. This sample execution took approximately 14 sec. These results are compared to the output from RUNGE in Figure 7.7.

Monte-Carlo techniques have been used to simulate the liquid-state electrode phenomena, including kinetics and diffusion processes, and to approximate the solution to quantum mechanical problems [19-33].

7.7. ADDITIONAL PROBLEMS

1. Approximate the solution to the following differential equations over the range $x_1(h)x_f$, where $x_1 = x_0 + h$, h = step size, and x_f = the final value of x.

(a) $y' = x - y^2$ x = 0.2(0.2)1   y(0) = 1

(b) $y' = \frac{1}{x+y}$ x = 0.5(0.5)2   y(0) = 1

(c) $y' = \sin x + \cos y$ x = 3(0.5)4   y(2.5) = 0

(d) $y'' = xy$ $x = 0.5(0.5)1$ $y(0) = 0$ $y'(0) = 1$

(e) $y'' = \dfrac{x^2 - y^2}{1 + (y')^2}$ $x = 0.5(0.5)1.5$ $y(0) = 1$ $y'(0) = 0$

(f) $y' = xz + 1$ $x = 0.3(0.3)0.9$ $y(0) = 0$ $z(0) = 1$

     $z' = -xz$

Solutions [34].

  (a) $y(0.2) = 0.8512$, $y(0.4) = 0.7798$, $y(0.6) = 0.7620$, $y(0.8) = 0.7834$, $y(1) = 0.8334$

  (b) $y(0.5) = 1.3571$, $y(1.0) = 1.5837$, $y(1.5) = 1.7555$, $y(2.0) = 1.8956$

  (c) $y(3.0) = 0.649$, $y(3.5) = 0.935$, $y(4.0) = 0.941$

  (d) $y(0.5) = 0.50521$, $y(1) = 1.08508$

  (e) $y(0.5) = 0.8891$, $y(1.0) = 0.6674$, $y(1.5) = 0.5794$

  (f) $y(0.3) = 0.3448$, $z(0.3) = 0.9900$, $y(0.6) = 0.7739$, $z(0.6) = 0.9121$, $y(0.9) = 1.2551$, $z(0.9) = 0.6808$

2. The error function is usually defined as the integral

$$\mathrm{erf}(x) = \frac{2}{\sqrt{\pi}} \int_0^x e^{-t^2}\, dt$$

It can also be defined [35] as the solution to the differential equation

$$y' = \frac{2}{\sqrt{\pi}} e^{-x^2}\, dx$$

Calculate $\mathrm{erf}(x)$ as a function of x and compare the results with tabulated values.

3. Simulate a kinetic scheme of interest by integrating the system of differential equations given initial concentrations or partial pressures and values for the rate constants. Compare the numerical calculations with the analytical results where possible [36-38].

4. Simulate the pyrolysis of ethane in a tubular reactor [39],

$$C_2H_6 \longrightarrow C_2H_4 + H_2$$

The parameters include the cross-sectional area of the tube, the concentration of ethane, the heat capacity, the rate constant, the length measured from the reactor inlet, the inlet feed rate, the temperature, the reactor volume, and the total pressure.

5. Simulate the production of ethylene oxide,

$$8C_2H_4 + 9O_2 \longrightarrow 6C_2H_4O + 4CO_2 + 4H_2O$$

using the model described in Ref. 40 and references therein.

6. Simulate the flash photolysis of H_2O vapor in the presence of a small amount of oxygen gas in an atmosphere of helium or argon [41].

7. Write a program to simulate the spectrophotometric study of the rate of the reaction [42],

$$Cr(H_2O)_5Br^{2+} + U^{3+} \longrightarrow Cr^{2+} + U^{4+} + Br^-$$

8. Simulate the formation of ozone and photochemical smog in the atmosphere [43-46].

9. Write a program to provide a student user with all the data required to determine the kinetic parameters of the system [47],

$$A + B \underset{k_2}{\overset{k_1}{\rightleftharpoons}} C + D$$

10. Simulate the production of I_2 formed by the reaction of HI with free radicals which are produced at a constant rate [48].

11. A simple mathematical model to describe the ecosystem consisting of rabbits that have an infinite food supply and foxes that prey on the rabbits is the following [49]:

$$\frac{dR}{dt} = 2R - aRF \qquad (R(0) = R_0)$$

$$\frac{dF}{dt} = -F + aRF \qquad (F(0) = F_0)$$

Here t is time, R is the number of rabbits, and F is the number of foxes. Solve this system for various values of a, R_0, and F_0. The number of rabbits can be prevented from growing indefinitely by changing the first equation to the following:

$$\frac{dR}{dt} = 2(1 - \frac{R}{R_{max}})R - aRF$$

REFERENCES

1. K. G. Ashurst and W. C. E. Higginson, Kinetics of the reaction between cobalt (III) and thallium (I) in aqueous perchloric acid, J. Chem. Soc., 1956, 343.

2. G. W. Gear, "Numerical Initial Value Problems in Ordinary Differential Equations," Prentice-Hall, Englewood Cliffs, N.J., 1971.

3. B. Carnahan, H. Luther, and J. Wilkes, "Applied Numerical Methods," Wiley, New York, 1969, pp. 343-360.

4. B. Carnahan et al., ibid., pp. 361-380.

5. C. Froberg, "Introduction to Numerical Analysis," Addison-Wesley, Reading, Mass., 1965, pp. 244-248.

6. G. Forsythe, M. Malcolm, and C. Moler, "Computer Methods for Mathematical Computations," Prentice-Hall, Englewood Cliffs, N. J., 1977, pp. 121-147.

7. B. Carnahan et al., op. cit., pp. 384-404.

8. Z. G. Szabo, in "Comprehensive Chemical Kinetics" (C. H. Bamford and C. F. H. Tipper, eds.), Vol. 2, Elsevier, New York, 1969.

9. B. Carnahan et al., op. cit., pp. 365-380.

10. G. L. Hemphill and J. M. White, Trajectory calculations in chemical kinetics, J. Chem. Educ., $\underline{49}$, 121 (1972).

11. R. D. Levine and R. B. Bernstein, "Molecular Reaction Dynamics," Oxford University Press, London, 1974.

12. S. Sato, On a new method of drawing the potential energy surface, J. Chem. Phys., $\underline{23}$, 592 (1955).

13. J. C. Davis, Jr., "Advanced Physical Chemistry," Ronald, New York, 1965, pp. 105-110, 344 ff.

14. F. D. Tabbutt, Analog and Hybrid Computations, in "Computer-Assisted Instruction in Chemistry -- Part A: General Approach" (J. Mattson et al., eds.), Dekker, New York, 1974, p. 90ff.

15. S. R. LaPaglia, "Introduction to Quantum Chemistry," Harper and Row, New York, 1971, p. 203ff.

16. D. A. Dixon and R. H. Shafer, Computer simulations of kinetics by the Monte-Carlo technique, J. Chem. Educ., $\underline{50}$, 648 (1973).

17. A. F. Para and E. Lazzarini, Some simple classroom experiments on the Monte-Carlo method, J. Chem. Educ., $\underline{51}$, 336 (1974).

18. L. J. Soltzberg and F. G. Weber, Caution in using Monte-Carlo kinetics modeling, J. Chem. Educ., $\underline{51}$, 576 (1974).

19. J. Kushick and B. J. Berne, Molecular dynamics methods: continuous potentials, Mod. Theor. Chem., $\underline{6}$, 41 (1977).

20. M. Sisido and K. Shimada, Computer simulation of intramolecular electron transfer in $\alpha N(CH_2)_n \alpha N^- \cdot$ system, J. Amer. Chem. Soc., $\underline{99}$, 7785 (1977).

21. M. Gottlieb, Application of computer simulation techniques to macromolecular theories, Comput. Chem., 1, 155 (1977).

22. W. B. Streett and K. E. Gubbins, Liquids of linear molecules: computer simulation and theory, Ann. Rev. Phys. Chem., 28, 373 (1977).

23. P. S. Y. Cheung and J. G. Powles, The properties of liquid nitrogen. V. Computer simulation with quadruple interaction, Mol. Phys., 32, 1383 (1976).

24. M. A. Winnik, Computer simulation of intramolecular hydrogen abstraction in the photochemistry of p-benzophenonecarboxylate esters, J. Amer. Chem. Soc., 96, 4843 (1974).

25. M. E. Starzak, Computer examination of some basic probability models, J. Chem. Educ., 51, 717 (1974).

26. J. R. Lubbers et al., Digital simulation of a rotating photoelectrode, Anal. Chem., 46, 865 (1974).

27. P. Empedocles, Fundamental theory of gases, liquids, and solids by computer simulation, J. Chem. Educ., 51, 593 (1974).

28. D. D. Fitts, "Statistical Thermodynamics of Liquids", in MTP International Review of Science, Physical Chemistry, Series One, Vol. 1, Theoretical Chemistry (W. B. Brown, ed.), Butterworth, London, 1972.

29. S. W. Feldberg, Ditigal simulation of electrical surface boundary phenomena: Multiple electron transfer and absorption, in "Computers in Chemistry and Instrumentation" (J. S. Matson et al., ed.), Vol. 2, Dekker, New York, 1972, pp. 185-215, and other articles in this volume.

30. K. B. Prater and A. J. Bard, Rotating ring-disk electrodes. I. Fundamentals of the digital simulation approach. Disk and ring transients and collection efficiencies, J. Electrochem. Soc., 117, 207 (1970).

31. V. S. Steckline, Computer demonstrated Lenz-Ising model, J. Chem. Educ., 47, 128 (1970).

32. S. Feldberg, Digital simulation: A general method for solving electrochemical diffusion-kinetics problems, in "Electroanalytical Chemistry" (A. J. Bard, ed.), Vol. 3, Dekker, New York, 1969, pp. 199-295.

33. M. N. Rosenbluth and A. Rosenbluth, Monte-Carlo calculation of the average extension of molecular chains, J. Chem. Phys., 23, 356 (1955).

34. C. Froberg, op. cit., p. 260.

35. G. Forsythe et al., op. cit., p. 148.

36. D. Edelson, Symposium on reaction mechanisms, models, and computers, J. Phys. Chem., 81, 2309 (1977).

37. C. H. Bamford and C. Tipper (eds.), "Comprehensive Chemical Kinetics," Vol. 2, Elsevier, New York, 1969.

38. A. A. Frost and R. G. Pearson, "Kinetics and Mechanism," 2nd ed., Wiley, New York, 1961.

39. B. Carnahan et al., op. cit., p. 353ff.

40. B. Carnahan et al., ibid., p. 423.

41. S. M. Koop and P. J. Ogren, HO_2 kinetics in simple systems, J. Chem. Educ., 53, 128 (1976).

42. W. C. Child, A computer simulation of a kinetics experiment, J. Chem. Educ., 50, 290 (1973).

43. A. C. Baldwin et al., Photochemical smog. Rate parameters and computer simulations, J. Phys. Chem., 81, 2483 (1977).

44. B. J. Hubert, Computer modeling of photochemical smog formation, J. Chem. Educ., 51, 644 (1974).

45. R. F. Gould (ed.), "Photochemical Smog and Ozone Reactions," Advances in Chemistry Series, No. 113, American Chemical Society, Washington, D. C., 1972.

46. J. P. Chesick, Effects of water and nitrogen dioxides on stratospheric ozone shield, J. Chem. Educ., 49, 722 (1972).

47. J. C. Merrill et al., A computer-simulated experiment in complex order kinetic expressions, J. Chem. Educ., 52, 166 (1975).

48. R. H. Schuler, Calculator-plotter routines for numerical integration of kinetic expressions, J. Chem. Educ., 52, 166 (1975).

49. G. Forsythe et al., op. cit., p. 353ff.

CHAPTER 8

EIGENVALUES AND EIGENVECTORS

The n x n matrix A is said to be Hermitian if all elements are real and if A is symmetric, that is,

$$A = A^{\dagger}$$

where "$\dagger$" denotes complex conjugate transpose. For a given Hermitian matrix, A, there exist n real constants,

$$\lambda_1, \lambda_2, \lambda_3, \ldots, \lambda_n$$

called _eigenvalues_, and n associated real vectors,

$$X_1, X_2, X_3, \ldots, X_n$$

called _eigenvectors_, such that

$$AX_i = \lambda_i X_i \qquad \lambda i = 1, 2, 3, \ldots, n$$

Two numerical methods for determining eigenvalues and eigenvectors of Hermitian matrices are considered here, evaluation by determinants, and the Jacobi method. The latter method is then applied to three chemical problems: determination of the three principal moments of inertia of a nonlinear molecule; solution to the three-spin proton NMR problem; and Hückel molecular orbital calculations.

8.1. EVALUATION BY DETERMINANTS

Given $AX = \lambda X$, then

$$(A - \lambda I)X = 0 \qquad\qquad (8.1)$$

Equation (8.1) represents a homogeneous system of simultaneous equations for which a nontrivial solution ($X \neq 0$) exists only if

$$|A - \lambda I| = 0$$

The eigenvalues of A are determined by expanding this determinant in terms of λ and finding the roots of the resulting characteristic equation. The eigenvectors are then determined by substitution into Eq. (8.1).

Problem 8.1. Find the eigenvalues and eigenvectors of the following matrix:

$$A = \begin{bmatrix} 1 & 2 \\ 2 & -1 \end{bmatrix}$$

Solution.

$$|A - \lambda I| = \begin{vmatrix} 1 - \lambda & 2 \\ 2 & -1 - \lambda \end{vmatrix} = 0$$

The characteristic equation is

$$(1 - \lambda)(-1 - \lambda) - 4 = 0$$

$$\lambda^2 - 5 = 0$$

$$\lambda = \pm\sqrt{5}$$

Arbitrarily assign

$$\lambda_1 = +\sqrt{5} \qquad \text{and} \qquad \lambda_2 = -\sqrt{5}$$

Then X_1 is determined as follows:

$$AX_1 = \lambda_1 X_1$$

Let the elements of X_1 be x_{11} and x_{21}. Then

$$\begin{bmatrix} 1 & 2 \\ 2 & -1 \end{bmatrix} \begin{pmatrix} x_{11} \\ x_{21} \end{pmatrix} = \sqrt{5} \begin{pmatrix} x_{11} \\ x_{21} \end{pmatrix}$$

or

$$x_{11} + 2x_{21} = \sqrt{5}x_{11}$$

$$2x_{11} - x_{21} = \sqrt{5}x_{21}$$

From the second of these equations,

$$x_{11} = \frac{1 + \sqrt{5}}{2} x_{21}$$

Then, in terms of x_{21},

$$X_1 = \begin{pmatrix} \dfrac{1 + \sqrt{5}}{2} x_{21} \\ x_{21} \end{pmatrix}$$

Arbitrarily assigning $x_{21} = 1$,

$$X_1 = \begin{pmatrix} \dfrac{1 + \sqrt{5}}{2} \\ 1 \end{pmatrix}$$

Similarly,

$$X_2 = \begin{pmatrix} \dfrac{1 - \sqrt{5}}{2} \\ 1 \end{pmatrix}$$

These eigenvalues and eigenvectors are readily verified by substitution into the equation

$$AX_i = \lambda_i X_i$$

For example, for i = 1

$$AX_1 = \lambda_i X_i$$

$$\begin{bmatrix} 1 & 2 \\ 2 & -1 \end{bmatrix} \begin{pmatrix} \dfrac{1 + \sqrt{5}}{2} \\ 1 \end{pmatrix} = \sqrt{5} \begin{pmatrix} \dfrac{1 + \sqrt{5}}{2} \\ 1 \end{pmatrix}$$

$$\begin{pmatrix} \dfrac{5(1 + \sqrt{5})}{2} \\ \sqrt{5} \end{pmatrix} = \begin{pmatrix} \dfrac{5(1 + \sqrt{5})}{2} \\ \sqrt{5} \end{pmatrix}$$

Note that the eigenvectors are not unique. The eigenvector, cX, where c is any real constant, is also an eigenvector of A. The method of determinants is convenient for systems of low order. The JACOBI method avoids the problems associated with finding roots of characteristic equations of high order.

8.2 THE JACOBI METHOD

A matrix, D, is diagonal if

$$D_{ij} = 0 \qquad i \neq j$$

The eigenvalues of a diagonal matrix are the diagonal elements,

$$\lambda_i = D_{ij}$$

For example,

$$D = \begin{bmatrix} 1.23 & 0 & 0 \\ 0 & 4.56 & 0 \\ 0 & 0 & 7.89 \end{bmatrix}$$

The characteristic equation for D is

$$(1.23 - \lambda)(4.56 - \lambda)(7.89 - \lambda) = 0$$

The roots of this cubic are $\lambda_1 = 1.23$, $\lambda_2 = 4.56$, and $\lambda_3 = 7.89$.

The Jacobi method [1, 2] is an iterative procedure for diagonalizing an n x n Hermitian matrix. The method uses a series of orthogonal transformations. An n x n matrix, M, is orthogonal if its transpose is also its inverse,

$$M^t = M^{-1}$$

For example, the matrix

$$M = \begin{bmatrix} 1 & 0 & 0 \\ 0 & \cos\theta & \sin\theta \\ 0 & -\sin\theta & \cos\theta \end{bmatrix}$$

is orthogonal because

$$M^{t}M = I$$

$$\begin{bmatrix} 1 & 0 & 0 \\ 0 & \cos\theta & -\sin\theta \\ 0 & \sin\theta & \cos\theta \end{bmatrix} \begin{bmatrix} 1 & 0 & 0 \\ 0 & \cos\theta & \sin\theta \\ 0 & -\sin\theta & \cos\theta \end{bmatrix} = \begin{bmatrix} 1 & 0 & 0 \\ 0 & 1 & 0 \\ 0 & 0 & 1 \end{bmatrix}$$

An orthogonal transformation of a matrix A is the sequence of operations

$$M^{t}AM = B$$

M is chosen so that one off-diagonal element of A is annihilated by the orthogonal transformation. Then a second matrix M is chosen so that another off-diagonal element of A is annihilated. This procedure is continued until the resulting transformed matrix is diagonal,

$$M_{r}^{t} \cdots M_{3}^{t}M_{2}^{t}M_{1}^{t}AM_{1}M_{2}M_{3} \cdots M_{r} = D$$

or

$$S^{t}AS = D$$

Problem 8.2 Annihilate the largest off-diagonal element in the following matrix:

$$A = \begin{bmatrix} 1 & 2 & -1 & 3 \\ 2 & -3 & 4 & 6 \\ -1 & 4 & 3 & 0 \\ 3 & 6 & 0 & 13 \end{bmatrix}$$

Solution. The largest off-diagonal element is $A_{24} = A_{42} = 6$. To annihilate this element choose the orthogonal matrix

$$M = \begin{bmatrix} 1 & 0 & 0 & 0 \\ 0 & \cos\theta & 0 & \sin\theta \\ 0 & 0 & 1 & 0 \\ 0 & -\sin\theta & 0 & \cos\theta \end{bmatrix}$$

Let p and q be the row and column indices of the elements to be annihilated. Let B be the result of the orthogonal transformation:

$$B = M^{t}AM$$

The elements of B are as follows:

$$B_{ik} = A_{ik} \qquad i,k \neq p,q \tag{8.2a}$$

$$B_{pk} = A_{pk} \cos\theta - A_{qk} \sin\theta \tag{8.2b}$$

$$B_{qk} = A_{pk} \sin \theta + A_{qk} \cos \theta \tag{8.2c}$$

$$B_{ip} = A_{ip} \cos \theta - A_{iq} \sin \theta \tag{8.2d}$$

$$B_{iq} = A_{ip} \sin \theta + A_{iq} \cos \theta \tag{8.2e}$$

$$B_{pp} = A_{pp} \cos^2 \theta + A_{qq} \sin^2 \theta - 2A_{pq} \sin \theta \cos \theta \tag{8.2f}$$

$$B_{qq} = A_{pp} \sin^2 \theta + A_{qq} \cos^2 \theta + 2A_{pq} \sin \theta \cos \theta \tag{8.2g}$$

$$B_{pq} = A_{pp} \sin^2 \theta + A_{qq} \cos^2 \theta + 2A_{pq} \sin \theta \cos \theta \tag{8.2h}$$

The angle θ is chosen so that $B_{pq} = B_{qp} = 0$.

$$B_{pq} = 0 = (A_{pp} - A_{qq}) \sin \theta \cos \theta + A_{pq} (\cos^2 \theta - \sin^2 \theta) \tag{8.3}$$

Using

$$\sin 2\theta = 2 \sin \theta \cos \theta$$

and

$$\cos 2\theta = \cos^2 \theta - \sin^2 \theta$$

Equation (8.3) becomes

$$0 = \frac{1}{2} (A_{pp} - A_{qq}) \sin 2\theta + A_{pq} \cos 2\theta$$

or

$$\tan 2\theta = \frac{-2 A_{pq}}{A_{pp} - A_{qq}}$$

and finally,

$$\theta = \frac{1}{2} \tan^{-1} \left\{ \frac{-2 A_{pq}}{A_{pp} - A_{qq}} \right\}$$

Now that the angle θ is defined, $\sin \theta$ and $\cos \theta$ can be calculated, and the orthogonal transformation can be performed. However, there are two problems with this result. The use of the trigonometric functions may introduce truncation errors. Also, special provision must be made for the special case,

$$A_{pp} = A_{qq}$$

These difficulties can be avoided with the following trigonometric substitutions. Let

$$\lambda = -A_{pq} \qquad \mu = \frac{A_{pp} - A_{qq}}{2}$$

and

$$\omega = \text{sgn}(\mu) \frac{\lambda}{\sqrt{\lambda^2 + \mu^2}}$$

Here

$$\mu = \begin{cases} +1 & \text{if } \mu \geq 0 \\ -1 & \text{if } \mu < 0 \end{cases}$$

Then

$$\sin \theta = \frac{\omega}{[2(1 + \sqrt{1 - \omega^2})]^{1/2}}$$

and

$$\cos \theta = \sqrt{1 - \sin^2 \theta}$$

These algebraic equations avoid the errors associated with the trigonometric functions, and if $A_{pp} = A_{qq}$, no division by zero is attempted.

Returning now to the solution to the problem,

$$p = 2 \qquad q = 4$$

$$\lambda = -A_{24} = -6 \qquad \mu = \frac{A_{22} - A_{44}}{2} = -8$$

and

$$\text{sgn}(\mu) = -1$$

Then

$$\omega = (-1) \cdot \frac{-6}{\sqrt{36 + 64}} = 0.6$$

$$\sin \theta = \frac{0.6}{[2(1 + \sqrt{1 - 0.36})]^{1/2}} = 0.316$$

and

$$\cos \theta = \sqrt{1 - (0.316)^2} = 0.949$$

The transformation

$$M^t AM = B$$

can be evaluated either by matrix multiplication or using Eq. (8.2). The result is

$$B = \begin{bmatrix} 1 & 0.95 & -1 & 3.48 \\ 0.95 & -5 & 3.80 & 0 \\ -1 & 3.80 & 3 & 1.26 \\ 3.48 & 0 & 1.26 & 15 \end{bmatrix}$$

The largest off-diagonal element in B is now B_{23}. A second orthogonal transformation designed to annihilate this element gives

$$C = \begin{bmatrix} 1 & 1.25 & -0.58 & 3.48 \\ 1.25 & -6.51 & 0 & -0.47 \\ -0.58 & 0 & 4.51 & 1.17 \\ 3.48 & -0.47 & 1.17 & 15 \end{bmatrix}$$

Note that only the elements in row p and column q are affected by a given transformation. Note also that the element C_{24} is no longer zero. However the Jacobi

method does converge [1]. After 16 transformations the following approximately
diagonal matrix results

$$
\begin{bmatrix}
15.91 & -1.1 \times 10^{-9} & 0.0 & -1.1 \times 10^{-14} \\
-1.1 \times 10^{-9} & 4.584 & -1.7 \times 10^{-13} & 3.8 \times 10^{-22} \\
0 & -1.7 \times 10^{-13} & 0.2775 & -1.0 \times 10^{-23} \\
-1.1 \times 10^{-14} & 3.8 \times 10^{-22} & -1.0 \times 10^{-23} & -6.770
\end{bmatrix}
$$

The eigenvalues are

$$\lambda_1 = 15.91$$
$$\lambda_2 = 4.584$$
$$\lambda_3 = 0.2775$$
$$\lambda_4 = -6.770$$

The matrix product

$$S = M_1 M_2 M_3 \cdots M_r$$

is also orthogonal, and the ith column of S is the ith eigenvector of the Hermitian
matrix [1]. In this example,

$$
S = \begin{bmatrix}
0.2224 & -0.2061 & 0.9354 & 0.1821 \\
0.3313 & 0.3112 & 0.1603 & -0.8762 \\
0.08543 & 0.9158 & 0.1080 & 0.3774 \\
0.9129 & -0.1484 & -0.2962 & 0.2383
\end{bmatrix}
$$

For example, the eigenvector corresponding to λ_1 is

$$
X_1 = \begin{pmatrix}
0.2224 \\
0.3313 \\
0.8543 \\
0.9129
\end{pmatrix}
$$

The accuracy of this set of eigenvalues is readily checked by evaluating

$$AX_i = \lambda_i X_i$$

For example, for i = 1,

$$
AX_1 = \begin{pmatrix}
3.54 \\
5.27 \\
1.36 \\
14.5
\end{pmatrix}
$$

and

$$
\lambda_1 X_1 = 15.91 \begin{pmatrix}
0.2224 \\
0.3313 \\
0.08543 \\
0.9130
\end{pmatrix} = \begin{pmatrix}
3.54 \\
5.27 \\
1.36 \\
14.5
\end{pmatrix}
$$

The accuracy can also be checked by regenerating the original matrix

$$S^t AS = D$$

so

$$SDS^t = A$$

Problem 8.3. Write a program to evaluate the eigenvalues and eigenvectors of a
            Hermitian matrix. Check the results by comparing AX_i and $\lambda_i X_i$.

Solution. See the following program.

8.2.1 JACOBI

Introduction

        JACOBI computes the eigenvalues and eigenvectors of an n x n Hermitian (real
and symmetric) matrix using the Jacobi method. The input consists of the order of
the matrix, n, and the n x n elements, entered one row at a time. JACOBI prints
the matrix for verification purposes, tabulates the eigenvalues and eigenvectors,
and then verifies them by calculating AX_i, $\lambda_i X_i$, and the difference vector AX_i -
$\lambda_i X_i$, where A is the original Hermitian matrix, X_i is the ith eigenvector, λ_i is
the corresponding eigenvalue, and i = 1, 2, 3, ..., n.

    Method

        The method is that outlined above. JACOBI consists of a main program and the
subroutine HDIAG. The calling sequence is

        CALL HDIAG(A,U,N,IGEN)

Here U is the N x N matrix containing the eigenvectors stored columnwise (S), and
IGEN is a switch set to 0 if only eigenvalues are wanted. Any other value assigned
to IGEN will generate both eigenvalues and eigenvectors. HDIAG uses only the
elements on and above the main diagonal of A. The eigenvalues are returned as the
diagonal elements of A.

    Sample Execution

 ENTER THE ORDER OF THE MATRIX
 (ZERO ENDS THE PROGRAM)
>4

 ENTER THE MATRIX - ONE ROW PER LINE
>10 9 7 5
>9 10 8 6
>7 8 10 7
>5 6 7 5

THE INPUT MATRIX IS:

10.00	9.000	7.000	5.000
9.000	10.00	8.000	6.000
7.000	8.000	10.00	7.000
5.000	6.000	7.000	5.000

THE EIGENVALUES ARE:

| 30.29 | 3.858 | 0.8431 | 0.1015E-01 |

THE CORRESPONDING EIGENVECTORS ARE (COLUMNWISE):

0.5209	-0.6254	0.5676	0.1237
0.5520	-0.2716	-0.7603	-0.2086
0.5286	0.6149	0.3017	-0.5016
0.3803	0.3963	-0.9331E-01	0.8304

I	A * X	LAMBDA * X	DIFF
1	15.8	15.8	0.834E-06
	16.7	16.7	0.954E-06
	16.0	16.0	0.477E-06
	11.5	11.5	0.358E-06
2	-2.41	-2.41	-0.596E-07
	-1.05	-1.05	0.149E-07
	2.37	2.37	0.179E-06
	1.53	1.53	0.447E-07
3	0.479	0.479	-0.745E-08
	-0.641	-0.641	0.373E-07
	0.254	0.254	0.484E-07
	-0.787E-01	-0.787E-01	0.186E-07
4	0.126E-02	0.126E-02	0.431E-07
	-0.212E-02	-0.212E-02	0.186E-07
	-0.509E-02	-0.509E-02	-0.693E-07
	0.843E-02	0.843E-02	0.509E-07

ENTER THE ORDER OF THE MATRIX
(ZERO ENDS THE PROGRAM)
>0

Listing

```
C
C                         JACOBI
C
C     JACOBI DIAGONALIZES HERMITIAN (REAL & SYMMETRIC) MATRICES
C     BY THE JACOBI METHOD.
C
C     THE ORTHOGONAL TRANSFORMATIONS ARE DONE IN THE SUBROUTINE
C     HDIAG (SEE BELOW).
C
C   GLOSSARY:
```

```
C
C      A(10,10)    -   HERMITIAN MATRIX
C      U(10,10)    -   ARRAY CONTAINING EIGENVECTORS STORED COLUMNWISE
C      X(10)       -   WORK ARRAY
C      D(10)       -   EIGENVALUES
C      N           -   ORDER OF A
C
C      HDIAG DOES NOT USE THE ELEMENTS TO THE LEFT OF THE MAIN DIAGONAL
C
C      HDIAG RETURNS THE EIGENVALUES ON THE DIAGONAL OF A
C
C
C        JACOBI ALSO CHECKS THE EIGENVALUES AND EIGENVECTORS
C        BY COMPARING  A*X(I) AND LAMBDA(I)*X(I)
C
C    AUTHOR:  K. J. JOHNSON
C
C
       DIMENSION A(10,10),U(10,10),X(10),D(10)
C
C      INPUT
C
    1  WRITE(6,3)
       READ(5,5) N
       IF(N.LE.0) STOP
       IF(N.GT.10)GO TO 100
    7  WRITE(6,8)
       DO 9 I=1,N
    9  READ(5,10) (A(I,J),J=1,N)
C
C      PRINT THE INPUT MATRIX
C
       WRITE(6,11)
       DO 13 I=1,N
   13  WRITE(6,24) (A(I,J),J=1,N)
C
C      SAVE THE DIAGONAL
C
       DO 20 I=1,N
   20  D(I)=A(I,I)
C
C      CALL HDIAG TO DIAGONALIZE THE MATRIX A
C
       IGEN=0
       CALL HDIAG(A,N,IGEN,U)
C
C      PRINT THE EIGENVALUES AND EIGENVECTORS
C
       WRITE(6,21) (A(I,I),I=1,N)
       WRITE(6,22)
       DO 23 I=1,N
   23  WRITE(6,24) (U(I,J),J=1,N)
C
C      CHECK        A      *X      = LAMBDA *X
C          OR
C                   A(I,*)*U(*,I) = A(I,I)*U(*,I)
C
       DO 50 NR=1,N
       DO 40 I =1,N
```

```
      SUM=0.0
      DO 30 J =1,N
      IF(NR-J) 25,26,28
   25 SUM=SUM+A(J,NR)*U(J,I)
      GO TO 30
   26 SUM=SUM+D(J)*U(J,I)
      GO TO 30
   28 SUM=SUM+A(NR,J)*U(J,I)
   30 CONTINUE
   40 X(I)=SUM
      D(NR)=A(NR,NR)
      DO 45 J=1,N
   45 A(NR,J)=X(J)
   50 CONTINUE
C
      DO 70 I=1,N
      DO 70 J=1,N
   70 U(J,I)=D(I)*U(J,I)
C
C     PRINT THE RESULTS
C
      WRITE(6,72)
      DO 90 I=1,N
      D1 = A(1,I)-U(1,I)
      WRITE(6,75) I,A(1,I),U(1,I),D1
      DO 90 J=2,N
      D1 = A(J,I)-U(J,I)
   90 WRITE(6,83) A(J,I),U(J,I),D1
      GO TO 1
  100 WRITE(6,101) N
      GO TO 1
    3 FORMAT(//,' ENTER THE ORDER OF THE MATRIX '/'  (ZERO ENDS TH
     .E PROGRAM)'/)
    5 FORMAT(I)
    8 FORMAT(/,'  ENTER THE MATRIX - ONE ROW PER LINE'/)
   10 FORMAT(10F)
   11 FORMAT(//,'  THE INPUT MATRIX IS:',//)
   21 FORMAT(///,' THE EIGENVALUES ARE:',//,(10G13.4))
   22 FORMAT(/' THE CORRESPONDING EIGENVECTORS ARE (COLUMNWISE):'/)
   24 FORMAT(10G13.4)
   72 FORMAT(///,5X,'I',9X,'A * X',8X,'LAMBDA * X',8X,'DIFF')
   75 FORMAT(/,I6,2X,3G15.3)
   83 FORMAT(8X,3G15.3)
  101 FORMAT(//,' THIS PROGRAM CANNOT HANDLE A MATRIX OF ORDER',I3,/,
     .' STARTING OVER')
      END
C
C
C                     HDIAG
C
C
C     HDIAG DIAGONALIZES HERMITIAN (REAL & SYMMETRIC) MATRICES
C     BY THE JACOBI METHOD.  THE CALLING SEQUENCE IS:
C
C         CALL HDIAG (H,N,IGEN,U)
C
C         WHERE    H    =  THE HERMITIAN MATRIX
C                  N    =  THE ORDER OF H
C                  IGEN =  0 FOR BOTH EIGENVALUES AND EIGENVECTORS,
C                          1 FOR ONLY EIGENVALUES
C             AND  U    =  EIGENVECTORS
```

```
C
C        THE EIGENVALUES ARE THE DIAGONAL ELEMENTS OF H
C        HDIAG OPERATES ONLY ON THE ELEMENTS OF H THAT ARE TO THE
C        RIGHT OF THE MAIN DIAGONAL. THUS, ONLY A TRIANGULAR
C        SECTION MUST BE STORED IN THE ARRAY H.
C
C     AUTHORS:   F. CORBATO AND M. MERWIN (SEE K. WIBERG, "COMPUTER
C                PROGRAMS FOR CHEMISTS", BENJAMIN, 1965, PP.48-50).
C
C                THE ORIGINAL CODE HAS BEEN MODIFIED BY K. J. JOHNSON
C
C
C
       SUBROUTINE  HDIAG (H,N,IGEN,U)
       DIMENSION H(10,10), U(10,10), X(10), IQ(10)
C
C     INITIALIZE U
C
       IF(IGEN.NE.0)GO TO 15
       DO 10 I=1,N
       DO 10 J=1,N
       U(I,J)=0.
       IF(I.EQ.J) U(I,J)=1.
  10   CONTINUE
  15   NR = 0
       IF(N.LE.1)RETURN
C
C     SCAN FOR LARGEST OFF-DIAGONAL ELEMENT IN EACH ROW
C     X(I) CONTAINS LARGEST ELEMENT IN ITH ROW
C     IQ(I) HOLDS SECOND SUBSCRIPT DEFINING POSITION OF ELEMENT
C
       NMI1=N-1
       DO 30 I=1,NMI1
       X(I) = 0.
       IPL1=I+1
       DO 30 J=IPL1,N
       IF(X(I).GT.ABS( H(I,J))) GO TO 30
       X(I)=ABS(H(I,J))
       IQ(I)=J
  30   CONTINUE
C
C     SET INDICATOR FOR SHUT-OFF:  RAP=2**-27,  NR=NO.OF ROTATIONS
C
       RAP = 7.45058060E-9
       HDTEST = 1.7E38
C
C     FIND THE MAXIMUM OF X(I)'S FOR PIVOT ELEMENT
C     TEST FOR END OF PROBLEM
C
  40   DO 70  I=1,NMI1
       IF(I.LE.1)GO TO 60
       IF(XMAX.GE.X(I))GO TO 70
  60 XMAX=X(I)
       IPIV=I
       JPIV=IQ(I)
  70   CONTINUE
C
C   IS MAX. X(I) EQUAL TO ZERO? IF LESS THAN HDTEST, REVISE HDTEST
```

```
C
      IF(XMAX.LE.0.)RETURN
      IF(HDTEST.LE.0.)GO TO 90
      IF(XMAX.GT.HDTEST)GO TO 148
 90   HDIMIN = ABS ( H(1,1) )
      DO 110  I=2,N
      IF( HDIMIN .LE. ABS(H(I,I)) ) GO TO 110
      HDIMIN=ABS (H(I,I))
110   CONTINUE
      HDTEST = HDIMIN*RAP
C
C     RETURN IF MAX.H(I,J) LESS THAN (2**-27)*ABS (H(K,K)-MIN)
C
      IF(HDTEST.GE.XMAX)RETURN
148   NR=NR+1
C
C     COMPUTE TANGENT, SINE AND COSINE, H(I,I), H(J,J)
C
150   TANG=SIGN (2.0, (H(IPIV,IPIV)-H(JPIV,JPIV)))*H(IPIV,JPIV)/
    . ( ABS( H(IPIV,IPIV)-H(JPIV,JPIV) )
    . + SQRT( (H(IPIV,IPIV)-H(JPIV,JPIV) )**2
    . + 4.0*H(IPIV,JPIV)**2))
      COSINE=1.0/SQRT(1.0+TANG**2)
      SINE=TANG*COSINE
      HII=H(IPIV,IPIV)
      H(IPIV,IPIV)=COSINE**2*(HII+TANG*(2.*H(IPIV,JPIV)+TANG*
    .H(JPIV,JPIV)))
      H(JPIV,JPIV)=COSINE**2*(H(JPIV,JPIV)-TANG*(2.*H(IPIV,JPIV)-TANG*
    .HII))
      H(IPIV,JPIV)=0.
C
C     PSEUDO RANK THE EIGENVALUES
C     ADJUST SINE AND COS FOR COMPUTATION OF H(IK) AND U(IJ)
C
      IF ( H(IPIV,IPIV) .GE. H(JPIV,JPIV)) GO TO 153
      HTEMP = H(IPIV,IPIV)
      H(IPIV,IPIV) = H(JPIV,JPIV)
      H(JPIV,JPIV) = HTEMP
C
C     RECOMPUTE SINE AND COSINE
C
      HTEMP = SIGN (1.0, -SINE) * COSINE
      COSINE = ABS(SINE)
      SINE = HTEMP
153   CONTINUE
C
C     INSPECT THE IQS BETWEEN I+1 AND N-1 TO DETERMINE
C     WHETHER A NEW MAXIMUM VALUE SHOULD BE COMPUTED SINCE
C     THE PRESENT MAXIMUM IS IN THE I OR J ROW.
C
      DO 350 I=1,NMI1
      IF(I-IPIV) 210,350,200
200   IF (I.EQ.JPIV) GO TO 350
210   IF(IQ(I).EQ.IPIV) GO TO 240
230   IF(IQ(I).NE.JPIV) GO TO 350
240   K=IQ(I)
      HTEMP=H(I,K)
      H(I,K)=0.
      IPL1=I+1
      X(I) = 0.0
```

```
C       SEARCH IN DEPLETED ROW FOR NEW MAXIMUM
C
        DO 320 J=IPL1,N
        IF ( X(I) .GT. ABS( H(I,J) )) GO TO 320
        X(I) = ABS(H(I,J))
        IQ(I)=J
 320    CONTINUE
        H(I,K)=HTEMP
 350    CONTINUE
        X(IPIV) = 0.0
        X(JPIV) = 0.0
C
C       CHANGE THE ORDER ELEMENTS OF H
C
        DO 530 I=1,N
        IF (I-IPIV) 370,530,420
 370    HTEMP = H(I,IPIV)
        H(I,IPIV) = COSINE*HTEMP + SINE*H(I,JPIV)
        IF ( X(I) .GE. ABS( H(I,IPIV) )) GO TO 390
        X(I) = ABS( H(I,IPIV) )
        IQ(I) = IPIV
 390    H(I,JPIV) = -SINE*HTEMP + COSINE*H(I,JPIV)
        IF ( X(I) .GE. ABS( H(I,JPIV) )) GO TO 530
 400    X(I) = ABS( H(I,JPIV) )
        IQ(I) = JPIV
        GO TO 530
 420    IF(I-JPIV) 430,530,480
 430    HTEMP = H(IPIV,I)
        H(IPIV,I) = COSINE*HTEMP + SINE*H(I,JPIV)
        IF ( X(IPIV) .GE. ABS( H(IPIV,I) )) GO TO 450
        X(IPIV) = ABS( H(IPIV,I) )
        IQ(IPIV) = I
 450    H(I,JPIV) = -SINE*HTEMP + COSINE*H(I,JPIV)
        IF ( X(I) - ABS( H(I,JPIV) )) 400,530,530
 480    HTEMP = H(IPIV,I)
        H(IPIV,I) = COSINE*HTEMP + SINE*H(JPIV,I)
        IF ( X(IPIV) .GE. ABS( H(IPIV,I) )) GO TO 500
        X(IPIV) = ABS( H(IPIV,I) )
        IQ(IPIV) = I
 500    H(JPIV,I) = -SINE*HTEMP + COSINE*H(JPIV,I)
        IF ( X(JPIV) .GE. ABS( H(JPIV,I) )) GO TO 530
        X(JPIV) = ABS( H(JPIV,I) )
        IQ(JPIV) = I
 530    CONTINUE
C
C       TEST FOR COMPUTATION OF EIGENVECTORS
C
        IF(IGEN.NE.0)GO TO 40
        DO 550 I=1,N
        HTEMP = U(I,IPIV)
        U(I,IPIV) = COSINE*HTEMP + SINE*U(I,JPIV)
 550    U(I,JPIV) = -SINE*HTEMP + COSINE*U(I,JPIV)
        GO TO 40
1000    RETURN
        END
```

Discussion

The main program and the subroutine HDIAG require 684 and 615 words of memory, respectively. This sample execution required approximately 0.3 sec.

Problem 8.4. Write a program to calculate the entropy of a molecule in the gas phase using statistical mechanics [3-7].

Solution. The following program uses the Jacobi method to diagonalize the 3 x 3 Hermitian matrix, the eigenvalues of which are the three principal moments of inertia of a nonlinear molecule. The moments of inertia are needed to calculate the rotational partition coefficient, which in turn is used to calculate the rotational entropy of the molecule.

8.2.2 ENTROPY

Introduction

ENTROPY calculates the translational, rotational, and vibrational components of the total entropy of a molecule in the gas phase. The input includes the following: the number of atoms in the molecule; the linearity of the molecule (linear or nonlinear): the temperature; the pressure (atm); the symmetry factor; the multiplicity of the ground electronic state; the fundamental vibration frequencies; the mass of each atom (atomic mass units); and the x, y, and z coordinates of each atom relative to an arbitrary origin.

Method

A number of assumptions have been made. Ideal gas behavior is assumed in the partition functions in that the energy levels involved in the system are for the individual molecules and are not dependent on molecular interactions. The assumption that the gaseous molecule is in its ground electronic state allows the calculation of the entropy from the ground translational, vibrational, and rotational quantum energy levels with zero contribution from the electronic quantum energy levels. No internal rotation or other degrees of freedom are included. The effects of nuclear spin and anharmonicity are neglected. The translational, rotational, and vibrational motions are assumed to be mutually independent.

The total partition function is given by

$$Q_{tot} = Q_{trans} \, Q_{rot} \, Q_{vib}$$

Classical expressions for the translational and rotational energy system are assumed, since the separation of energy levels is small. The quantum mechanical expression for the vibrational energy level system is used, since there the separation of the energy levels is relatively large.

A glossary of the symbols used in the partition functions and the three contributions to the total entropy is given next.

Symbol	Description
g	Multiplicity of the ground electronic state
k	Boltzmann's constant
R	Ideal gas constant
T	Absolute temperature
P	Pressure (atm)
e	$2.71828\cdots$
h	Planck's constant
n	Number of rotational degrees of freedom (2 for linear molecules, 3 for nonlinear molecules)
m	Mass of the molecule (M/A, where M is the molecular weight)
A	Avogadro's number
v	Volume of the molecule
N	Number of atoms in the molecule
ν_i	ith fundamental vibration frequency
I_a, I_b, I_c	Principal moments of inertia
	Number of indistinguishable positions into which the molecule can be turned by simple rotations ($C_1, C_i, C_s = 1$; $C_n = n$; $D_n = 2n$; $C_{\infty V} = 1$; $D_{\infty V} = 2$; T, $T_d = 12$; $O_h = 24$; $S_6 = 3$)
Q	Partition coefficient

The entropy is given by the following expression:

$$S = R[\ln Q + T(\frac{d\ln Q}{dT})_V]$$

1. <u>Translational component</u>.

$$Q_{trans} = \frac{veg(2\pi mkT)^{3/2}}{h^3}$$

Substituting M/A for m, kT/P for v, and using

$$P(atm) = 1.0132 \times 10^6 \; P \; (dyne/cm^2)$$

$$S_{trans} = R \ln \left\{ \frac{g(2\pi)^{3/2} k^{5/2} M^{3/2}}{A^{3/2} h^3 P(1.0132 \times 10^6)} \right\}$$

$$+ R \left\{ \frac{5}{2} \ln T + T \frac{5}{2T} \right\}$$

$$S_{trans} = R \ln g + 1.5 R \ln M - R \ln P + 2.5 R \ln T - 2.315$$

2. <u>Vibrational component</u>.

$$Q_{vib} = \prod_{i=1}^{3N-3-n} [1 - \exp \frac{-h\nu_i}{kT}]$$

$$S_{vib} = R [- \sum_{i=1}^{3N-3-n} \ln(1 - \exp \frac{-h\nu_i}{kT})$$

$$+ \sum_{i=1}^{3N-3-n} \frac{h\nu_i}{kT} \left[\exp \frac{h\nu_i}{kT} - 1 \right]^{-1}$$

3. <u>Rotational component</u>. For linear molecules,

$$Q_{rot} = \frac{1}{\pi \sigma} \frac{8\pi^3 I k T}{h^2}$$

and

$$S_{rot} = R \ln \frac{IT}{\sigma} + R \ln \frac{8\pi^2 k}{h^2} + R$$

$$= R \ln \frac{IT}{\sigma} + 177.672$$

Here I is the moment of inertia of the linear molecule.

For nonlinear molecules,

$$Q_{rot} = \frac{1}{\pi \sigma} \left[\frac{8\pi^3 kT (I_A I_B I_C)^{1/3}}{h^2} \right]^{3/2}$$

and

$$S_{rot} = R \left\{ \frac{1}{2} \ln I_A I_B I_C - \ln \sigma + \frac{3}{2} \ln T \right\} + 267.645$$

The three components of the moment of inertia, I_A, I_B, and I_C, are calculated from an arbitrary set of coordinates using the rigid molecule method of Hirschfelder [8]. For nonlinear molecules the moments of inertia are the eigenvalues of the following matrix.

$$\begin{bmatrix} A - I_A & -D & -E \\ -D & B - I_B & -F \\ -E & -F & C - I_C \end{bmatrix}$$

where

$$A = \sum_{i=1}^{N} m_i (y_i^2 + z_i^2) - \frac{1}{M} \left\{ \sum_{i=1}^{N} m_i y_i \right\}^2 - \frac{1}{M} \left\{ \sum_{i=1}^{N} m_i z_i \right\}^2$$

$$B = \sum_{i=1}^{N} m_i(x_i^2 + z_i^2) - \frac{1}{M}\left\{\sum_{i=1}^{N} m_i y_i\right\}^2 - \frac{1}{M}\left\{\sum_{i=1}^{N} m_i z_i\right\}^2$$

$$C = \sum_{i=1}^{N} m_i(x_i^2 + y_i^2) - \frac{1}{M}\left\{\sum_{i=1}^{N} m_i x_i\right\}^2 - \frac{1}{M}\left\{\sum_{i=1}^{N} m_i y_i\right\}^2$$

$$D = \sum_{i=1}^{N} m_i\left(x_i y_i - \frac{1}{M}\left\{\sum_{i=1}^{N} m_i x_i\right\}\left\{\sum_{i=1}^{N} m_i z_i\right\}\right.$$

$$E = \sum_{i=1}^{N} m_i x_i z_i - \frac{1}{M}\left\{\sum_{i=1}^{N} m_i x_i\right\}\left\{\sum_{i=1}^{N} m_i z_i\right\}$$

$$F = \sum_{i=1}^{N} m_i y_i z_i - \frac{1}{M}\left\{\sum_{i=1}^{N} m_i y_i\right\}\left\{\sum_{i=1}^{N} m_i z_i\right\}$$

$$M = \sum_{i=1}^{N} m_i$$

Here m_i is the mass of atom i and the coordinates x_i, y_i, and z_i are the coordinates of the ith atom relative to the arbitrary origin of the molecule. If the molecule contains more than two atoms, the subroutine HDIAG is called to calculate I_A, I_B, and I_C by diagonalization of this Hermitian matrix.

    4. <u>Total entropy</u>.

$$S_{tot} = S_{trans} + S_{vib} + S_{rot}$$

<u>Sample Execution</u> ($CHCl_3$)

```
 ENTER THE NUMBER OF ATOMS IN THE MOLECULE
>5

 TYPE 1 IF THE MOLECULE IS LINEAR, 2 OTHERWISE
>2

 ENTER T (DEGREES KELVIN) AND P (ATM.)
>298 1

 WHAT IS THE SYMMETRY FACTOR FOR THE MOLECULE?
>3

 WHAT IS THE MULTIPLICITY OF THE GROUND ELECTRONIC STATE?
>1


 OK, THE NONLINEAR MOLECULE HAS  5 ATOMS

 NOW ENTER THE 3N-6 =  9  FUNDAMENTAL FREQUENCIES
  (CM-1, SIX PER LINE)
>3019 1216 1216 757 757 668
>368 261 261
```

FINALLY, ENTER THE ATOMIC WEIGHTS (GRAMS) AND THE CARTESIAN
COORDINTES (ANGSTROMS) FOR EACH ATOM

```
>1.008 0 0 1.093
>12.012
>35.453 1.467 .847 -.512
>35.453 0 -1.694 -0.512
>35.453 -1.467 .847 -.512
```

        T =   298.0   DEGREES KELVIN

        P =   1.000   ATM

        THE SYMMETRY FACTOR IS   3

        THE MULTIPLICITY OF THE GROUND ELECTRONIC STATE IS   1

        THE   9 FUNDAMENTALS ARE:

3019.	1216.	1216.	757.0	757.0	668.0
368.0	261.0	261.0			

MASS	X	Y	Z
1.0080	0.0000	0.0000	1.0930
12.0120	0.0000	0.0000	0.0000
35.4530	1.4670	0.8470	-0.5120
35.4530	0.0000	-1.6940	-0.5120
35.4530	-1.4670	0.8470	-0.5120

THE THREE MOMENTS OF INERTIA ARE (G-CM2):

        IXX =   26.209   E-39
        IYY =   26.207   E-39
        IZZ =   50.676   E-39

AND THE ENTROPIES ARE (ENTROPY UNITS):

        TRANSLATIONAL =   40.243
        ROTATIONAL   =   25.156
        VIBRATIONAL   =   5.3092
        TOTAL   =   70.709

Listing

```
C
C   THIS PROGRAM CALCULATES ENTROPIES OF GASEOUS MOLECULES
C          USING SPECTROSCOPIC DATA
C
C
C
C   INPUT SPECIFICATIONS
C
C       N= NUMBER OF ATOMS
C       L= 1 FOR LINEAR MOLECULE, 2 FOR NONLINEAR MOLECULE
C       T= TEMPERATURE IN DEGREES KELVIN
C       P=PRESSURE IN ATM
```

```
C          ISYM=SYMMETRY NUMBER OF MOLECULE
C          MASS=MOLECULAR WEIGHT OF MOLECULE
C          X,Y,Z=CARTESIAN COORDINATES OF ATOMS
C          G=MULTIPLICITY OF GROUND STATE
C
C
C   ASSUMPTIONS
C
C          1 IDEAL GAS LAW
C          2 TRANSLATION, ROTATION, VIBRATION MUTUALLY INDEPENDENT
C          3 MOLECULE IN GROUND ELECTRONIC STATE
C          4 NO INTERNAL ROTATIONS, OR BARRIERS TO ROTATION
C          5 TRANSLATION AND ROTATION OBEY CLASSICAL LAWS
C          6 VIBRATION IS CHARACTERIZED BY HARMONIC MOTION
C
C
C   METHOD
C
C      PARTITION FUNCTIONS QTRANS, QROT, QVIB ARE EVALUATED
C      STRANS,SROT,SVIB CAN THEN BE CALCULATED
C      IA,IB,IC ARE EVALUATED BY DIAGONALIZATION OF THE REAL
C      SYMMETRIC MATRIX
C      CF.   MOELWYN-HUGHES, PHYSICAL CHEMISTRY,1961,PERGAMON PRESS
C      CF.   COLTHUP,N.B., ET AL., INTRODUCTION TO INFRARED AND
C         RAMAN SPECTROSCOPY,1964,ACADEMIC PRESS
C
C
C   UNITS
C
C      FUNDAMENTAL FREQUENCIES (HERTZ)
C      MOMENTS OF INERTIA (IXX,IYY,IZZ)  (G-CM**2)
C      MASS   (ATOMIC MASS UNITS)
C      PRESSURE  (ATMOSPHERES)
C      ENTROPY (STRANS,SROT,SVIB,STOT)    (ENTROPY UNITS)
C      TEMPERATURE    ( DEGREES KELVIN )
C      CARTESIAN COORDINATES (X,Y,Z)    (ANGSTROMS)
C
C
       IMPLICIT REAL (I-N)
       DIMENSION X(25),Y(25),Z(25),M(25),NU(70),H(3,3),U(3,3)
       INTEGER I,ISYM,N,G,L,NNU
C
C   INPUT
C
       WRITE(6,11)
       READ (5,22) N
       IF(N.EQ. 0) STOP
       WRITE(6,33)
       READ(5,22)L
       GO TO (4,5) L
    4  NNU=N*3-5
       GO TO  6
    5  NNU=N*3-6
    6  WRITE(6,44)
       READ(5,55) T,PA
       WRITE(6,66)
       READ (5,22) ISYM
       SYM=ISYM
       WRITE(6,77)
       READ(5,22) G
```

```
          GG=G
          GO TO (7,8) L
    7     WRITE(6,88) N,NNU
          GO TO 15
    8     WRITE(6,99)N,NNU
   15     READ(5,55) (NU(I),I=1,NNU)
          WRITE(6,111)
          READ(5,56)(M(I),X(I),Y(I),Z(I),I=1,N)
C
C    WRITE ALL VALUES FOR VERIFICATION PURPOSES
C
          WRITE(6,122)T,PA,ISYM,G,NNU,(NU(I),I=1,NNU)
          WRITE(6,124)(M(I),X(I),Y(I),Z(I),I=1,N)
C
C    CALCULATE THE TRANSLATIONAL ENTROPY
C
          WT=0.0E0
          MASS=0.0E0
          DO 10   I=1,N
          MASS=MASS +M(I)
          M(I)=M(I)/6.02257E0
          WT=WT+M(I)
   10     CONTINUE
          RC=1.9872E0
          STRANS=((ALOG(MASS)*RC*1.5E0)+(ALOG(T)*RC*2.5E0))
         .-(ALOG(PA)*RC)-2.315E0 +RC*ALOG(GG)
C
C     CALCULATE THE VIBRATIONAL ENTROPY
C
          VIBSUM=0.0E0
          CONST=1.43868E0/T
          DO 20 I=1,NNU
          X1=CONST*NU(I)
   20     VIBSUM=VIBSUM +X1/(EXP(X1)-1.E0)-ALOG(1.E0-EXP(-X1))
          SVIB=VIBSUM*RC
C
C     CALCULATE THE ROTATIONAL ENTROPY
C
          BX=0.0E0
          BY=0.0E0
          BZ=0.0E0
          CX=0.0E0
          CY=0.0E0
          CZ=0.0E0
          DXZ=0.0E0
          DXY=0.0E0
          DYZ=0.0E0
          DO 30 I=1,N
          BX=BX+(Y(I)*Y(I)+Z(I)*Z(I))*M(I)
          BY=BY+(X(I)*X(I)+Z(I)*Z(I))*M(I)
          BZ=BZ+(X(I)*X(I)+Y(I)*Y(I))*M(I)
          CX=CX+M(I)*X(I)
          CY=CY+M(I)*Y(I)
          CZ=CZ+M(I)*Z(I)
          DXY=DXY+M(I)*X(I)*Y(I)
          DXZ=DXZ+M(I)*X(I)*Z(I)
   30     DYZ=DYZ+M(I)*Y(I)*Z(I)
          H(1,1)=BX-(CY*CY+CZ*CZ)/WT
          H(2,2)=BY-(CX*CX+CZ*CZ)/WT
          H(3,3)=BZ-(CX*CX+CY*CY)/WT
```

```
         H(1,2)=DXY-(CX*CY)/WT
         H(1,3)=DXZ-(CX*CZ)/WT
         H(2,3)=DYZ-(CY*CZ)/WT
         CALL HDIAG(H,3,1,U)
         INERT1=H(1,1)
         INERT2=H(2,2)
         INERT3=H(3,3)
         IF (L.EQ.2) GO TO 50
C
C     LINEAR CASE
C
         WRITE(6,133) INERT1,INERT2,INERT3
         IF(INERT1.LE.1.0E-5) INERT1=INERT2
         SR1=177.672E0+(ALOG(T))*RC-ALOG(SYM)*RC
         SR3=RC*ALOG(INERT1)+RC*(-89.80082E0)
         SROT=SR1+SR3
         GO TO 100
C
C     NONLINEAR CASE
C
  50     WRITE(6,133) INERT1,INERT2,INERT3
         SR4=267.645 +RC*3./2.*ALOG(T)-RC*ALOG(SYM)
         SR5=INERT1*INERT2*INERT3
         SR6=.5E0*RC*(ALOG(SR5))+.5E0*RC*(-269.4024E0)
         SROT=SR4+SR6
 100     STOT=SVIB+SROT+STRANS
C
C       OUTPUT
C
         WRITE(6,144) STRANS,SROT,SVIB,STOT
         STOP
C
C     FORMATS
C
  11     FORMAT(///'  ENTER THE NUMBER OF ATOMS IN THE MOLECULE')
  22     FORMAT(I)
  33     FORMAT(/'  TYPE 1 IF THE MOLECULE IS LINEAR, 2 OTHERWISE')
  44     FORMAT(/'  ENTER T (DEGREES KELVIN) AND P (ATM.)')
  55     FORMAT(6F)
  56     FORMAT(4F)
  66     FORMAT(/'  WHAT IS THE SYMMETRY FACTOR FOR THE MOLECULE?')
  77     FORMAT(/'  WHAT IS THE MULTIPLICITY OF THE GROUND',
        . ' ELECTRONIC STATE?')
  88     FORMAT(//'  OK, THE LINEAR ',
        .'MOLECULE HAS ',I2, ' ATOMS',//'  NOW ENTER THE',
        .' 3N-5  ',I2,' FUNDAMENTAL FREQUENCIES',/'   (CM-1 ',
        . ' SIX PER LINE ) ')
  99     FORMAT(//'  OK, THE NONLINEAR MOLECULE HAS ',I2,
        .' ATOMS',//'  NOW ENTER THE 3N-6 = ',I2,
        2'  FUNDAMENTAL FREQUENCIES (CM-1, SIX PER LINE)')
 111     FORMAT(/'  FINALLY, ENTER THE ATOMIC',
        . ' WEIGHTS (GRAMS) AND THE CARTESIAN',/
        . '  COORDINTES (ANGSTROMS) FOR EACH ATOM'/)
 122     FORMAT(/,10X,'T =',G15.4,' DEGREES KELVIN   ',//,10X,'P =',
        . G15.4,'  ATM',/
        ./,10X,'THE SYMMETRY FACTOR IS ',I2,/
        ./,10X,'THE MULTIPLICITY OF THE GROUND ELECTRONIC STATE IS ',I2,
        .//,10X,'THE ',I2,' FUNDAMENTALS ARE:',//
        .13(/6G12.4))
```

```
124   FORMAT(//9X, 'MASS',11X,'  X ',11X,'  Y ',11X,'  Z '//
     .25(4F15.4/))
133   FORMAT(///'  THE THREE MOMENTS OF INERTIA ARE (G-CM2):'
     .,//,15X,'IXX =',G15.5,' E-39',/,15X,'IYY =',G15.5,' E-39',
     . /,15X,'IZZ =',G15.5,' E-39')
144   FORMAT(//'  AND THE ENTROPIES ARE (ENTROPY UNITS):',
     .//,15X,'TRANSLATIONAL =',G15.5,/15X,'ROTATIONAL    =',
     .G15.5,/,15X,'VIBRATIONAL   =',G15.5,/,15X,
     .'TOTAL          =',G15.5,/)
      END
```

Discussion

ENTROPY including HDIAG with dimensions reduced to 3 x 3 requires approximately 1800 words of core. This sample execution took approximately 0.2 sec.

8.3 SIMULATION OF NMR SPECTRA

The procedures for calculating NMR spectra have been presented in a number of books [9-14]. The program documented in Section 2.8 contains algorithms that simulate proton NMR spectra for which, to a first approximation, the spin systems have a closed-form solution. The following program simulates the proton NMR spectrum arising from three magnetically inequivalent protons, referred to as the ABC spin system. For example, the molecule,

$$CH_2CHCN$$

contains three protons which give rise to an ABC splitting pattern [15].

8.3.1 ABC

Introduction

The input to ABC consists of the six parameters that define the ABC spin system. These are the three chemical shifts ν_A, ν_B, and ν_C and three coupling constants J_{ab}, J_{ac}, and J_{bc}. The program prints the input data for verification purposes, a table containing the frequencies and intensities of the allowed transitions, and, on option, a teletype plot of the spectrum.

Method

Let I_z represent the z-component of the proton nuclear magnetic moment, I_z can assume one of two values, denoted α and β,

$$I_z = +\frac{1}{2} \qquad (\alpha)$$

$$I_z = -\frac{1}{2} \qquad (\beta)$$

These two nuclear spin states can be considered in terms of the alignment of the nuclear spin, either parallel to (state α) or antiparallel to (state β) the external applied magnetic field. The ABC case has three proton spins. Let F_z be the total spin

$$F_z = \sum_{j=1}^{3} I_{zj}$$

Then eight combinations of these three proton nuclear spin alignments are possible, as indicated in Table 8.1. State 1 is the state in which all three nuclear spins are aligned with the applied magnetic field,

$$\phi_1 = \alpha\alpha\alpha$$

Here ϕ represents the nuclear spin wavefunction. The transition frequencies and intensities are derived from the eigenvalues and eigenvectors of the following nuclear spin Hamiltonian equation,

$$\mathcal{K}\psi = E\psi$$

Here $\mathcal{K}$ is an 8 x 8 Hermitian matrix, the spin Hamiltonian; ψ is the wavefunction that is a linear combination of the eight possible spin states; and E is the energy eigenvalue corresponding to one of the eight states.

The diagonal elements of $\mathcal{K}$ are given by the following equation:

$$\mathcal{K}_{mm} = \sum_{i=1}^{3} \nu_i I_{zi} + \sum_{i=1}^{3} \sum_{j=i+1}^{3} J_{ij} I_{zi} I_{zj}$$

Here m = 1, 2, 3, ..., 8. This equation generates the following set of diagonal elements:

$$\mathcal{K}_{11} = \frac{1}{2}(\nu_1 + \nu_2 + \nu_3) + \frac{1}{4}(J_{12} + J_{13} + J_{23})$$

$$\mathcal{K}_{22} = \frac{1}{2}(\nu_1 + \nu_2 - \nu_3) + \frac{1}{4}(J_{12} - J_{13} - J_{23})$$

$$\mathcal{K}_{33} = \frac{1}{2}(\nu_1 - \nu_2 + \nu_3) + \frac{1}{4}(-J_{12} + J_{13} - J_{23})$$

Table 8.1 ABC Nuclear Spin States

State k	Spin function ϕ_k	Total spin F_z
1	$\alpha\alpha\alpha$	+3/2
2	$\alpha\alpha\beta$	+1/2
3	$\alpha\beta\alpha$	+1/2
4	$\beta\alpha\alpha$	+1/2
5	$\alpha\beta\beta$	-1/2
6	$\beta\alpha\beta$	-1/2
7	$\beta\beta\alpha$	-1/2
8	$\beta\beta\beta$	-3/2

$$\mathcal{K}_{44} = \frac{1}{2}(-\nu_1 + \nu_2 + \nu_3) + \frac{1}{4}(-J_{12} - J_{13} + J_{23})$$

$$\mathcal{K}_{55} = \frac{1}{2}(\nu_1 - \nu_2 - \nu_3) + \frac{1}{4}(-J_{12} - J_{13} + J_{23})$$

$$\mathcal{K}_{66} = \frac{1}{2}(-\nu_1 + \nu_2 - \nu_3) + \frac{1}{4}(-J_{12} + J_{13} - J_{23})$$

$$\mathcal{K}_{77} = \frac{1}{2}(-\nu_1 - \nu_2 + \nu_3) + \frac{1}{4}(J_{12} - J_{13} - J_{23})$$

$$\mathcal{K}_{88} = \frac{1}{2}(-\nu_1 - \nu_2 - \nu_3) + \frac{1}{4}(J_{12} + J_{13} + J_{23})$$

The off-diagonal elements of $\mathcal{K}$ are given by the equation

$$\mathcal{K}_{mn} = Q_{ij}J_{ij} \qquad m,n = 1,2,3,\ldots,8$$

Here Q_{ij} is 1/2 if the spin functions for states i and j differ only in the interchange of spins i and j. Otherwise, Q_{ij} is zero. Consider, for example, $\mathcal{K}_{12}$. States 1 and 2, $\alpha\alpha\alpha$ and $\alpha\alpha\beta$, differ only with respect to the spin of the third proton. Therefore $\mathcal{K}_{12} = 0$. On the other hand, states 2 and 4, $\alpha\alpha\beta$ and $\beta\alpha\alpha$ have spins 1 and 3 interchanged. Therefore $\mathcal{K}_{23} = J_{13}/2$. The complete spin Hamiltonian matrix is given below.

$$
\begin{bmatrix}
\mathcal{K}_{11} & 0 & 0 & 0 & 0 & 0 & 0 & 0 \\
0 & \mathcal{K}_{22} & J_{23}/2 & J_{13}/2 & 0 & 0 & 0 & 0 \\
0 & J_{23}/2 & \mathcal{K}_{33} & J_{12}/2 & 0 & 0 & 0 & 0 \\
0 & J_{13}/2 & J_{12}/2 & \mathcal{K}_{44} & 0 & 0 & 0 & 0 \\
0 & 0 & 0 & 0 & \mathcal{K}_{55} & J_{12}/2 & J_{13}/2 & 0 \\
0 & 0 & 0 & 0 & J_{12}/2 & \mathcal{K}_{66} & J_{23}/2 & 0 \\
0 & 0 & 0 & 0 & J_{13}/2 & J_{23}/2 & \mathcal{K}_{77} & 0 \\
0 & 0 & 0 & 0 & 0 & 0 & 0 & \mathcal{K}_{88}
\end{bmatrix}
$$

For example, consider the ABC spectrum with

$$\nu_1 = 15 \text{ Hz} \qquad \nu_2 = 25 \qquad \nu_3 = 50$$

$$J_{12} = 7.5 \text{ Hz} \qquad J_{13} = 2.0 \qquad J_{23} = 4.5$$

The spin Hamiltonian is

$$
\begin{bmatrix}
48.5 & 0 & 0 & 0 & 0 & 0 & 0 & 0 \\
0 & -4.75 & 2.25 & 1.00 & 0 & 0 & 0 & 0 \\
0 & 2.25 & 17.5 & 3.75 & 0 & 0 & 0 & 0 \\
0 & 1.00 & 3.75 & 28.75 & 0 & 0 & 0 & 0 \\
0 & 0 & 0 & 0 & -31.25 & 3.75 & 1.00 & 0 \\
0 & 0 & 0 & 0 & 3.75 & -22.5 & 2.25 & 0 \\
0 & 0 & 0 & 0 & 1.00 & 2.25 & 5.25 & 0 \\
0 & 0 & 0 & 0 & 0 & 0 & 0 & -41.5
\end{bmatrix}
$$

HDIAG is used to find the eigenvalues and eigenvectors of this matrix. They are

State	1	2	3	4
Eigenvalue	48.50	29.96	16.53	-4.987
Eigenvector	1	0	0	0
	0	0.0466	0.0865	0.9952
	0	0.296	0.950	-0.0964
	0	0.954	-0.299	-0.0188
	0	0	0	0
	0	0	0	0
	0	0	0	0
	0	0	0	0

State	5	6	7	8
Eigenvalue	5.478	-21.34	-32.64	-41.50
Eigenvector	0	0	0	0
	0	0	0	0
	0	0	0	0
	0	0	0	0
	0.0358	0.344	0.938	0
	0.0849	0.934	-0.346	0
	0.996	-0.0920	-0.0042	0
	0	0	0	0

The selection rule for nuclear spin transitions is

$$\Delta F_z = \pm 1$$

There are, accordingly, 15 allowed transitions for the ABC case. For example, the transition,

$$\phi_1 \longleftarrow \phi_2$$

is allowed because the change in total spin is $\pm 1 \, (3/2 - 1/2)$. The frequency of this transition is

$$\Delta \nu = \nu_1 - \nu_2 = 48.50 - 29.96 = 18.54 \text{ Hz}$$

The transition energy in ergs is

$$E = h \, \Delta \nu$$

where h is Planck's constant. The 15 allowed transition frequencies are

ϕ_i	ϕ_j	Transition frequency	ϕ_i	ϕ_j	Transition frequency
2	1	18.54	7	3	49.16
3	1	31.97	5	4	-10.46
4	1	53.49	6	4	16.35
5	2	24.48	7	4	27.65
6	2	51.30	8	5	46.98
7	2	62.60	8	6	20.16
5	3	11.05	8	7	8.86
6	3	37.87			

Transition $5 \leftarrow 4$ is the only transition involving the interchange of all three spins and occurs with an anomalous frequency.

The corresponding transition frequencies are calculated using the following expression:

$$I_{lm} = \left\{ \sum_{i=1}^{8} \sum_{j=1}^{8} c_{il} X_{ij} c_{jm} \right\}^{2}$$

Here c_{il} and c_{jm} are the coefficients of the l and m eigenvectors and X_{ij} has the value 1.0 if ϕ_i and ϕ_j differ in the spin coordinates of one spin, 0 otherwise. For example, consider the intensity of the transition between nuclear spin states 2 and 5. The eigenvectors are

$$c_2 = \begin{pmatrix} 0 \\ 0.0466 \\ 0.296 \\ 0.0954 \\ 0 \\ 0 \\ 0 \\ 0 \end{pmatrix} \quad \text{and} \quad c_5 = \begin{pmatrix} 0 \\ 0 \\ 0 \\ 0 \\ 0.0358 \\ 0.0849 \\ 0.996 \\ 0 \end{pmatrix}$$

The intensity of this transition is

$$I_{52} = \left\{ \sum_{i=1}^{8} \sum_{j=1}^{8} c_{i5} X_{ij} c_{j2} \right\}^{2}$$

Since $c_{i5} = 0$ for i = 1, 2, 3, 4, and 8, there are no more than nine nonvanishing terms in this expression.

$$I_{52} = [c_{55}(X_{52}c_{22} + X_{53}c_{32} + X_{54}c_{42})$$
$$+ c_{65}(X_{62}c_{22} + X_{63}c_{32} + X_{64}c_{42})$$
$$+ c_{75}(X_{72}c_{22} + X_{73}c_{32} + X_{74}c_{42})]^{2}$$

Three of these terms vanish because the spin functions differ by more than one spin orientation.

$$X_{54} = X_{63} = X_{72} = 0$$

Then,

$$I_{52} = [0.0358(0.0466 + 0.296) + 0.0849(0.0406 + 0.954)$$
$$+ 0.996(0.296 + 0.954)]^{2} = 1.80$$

The intensities corresponding to the 15 transition frequencies in this example are tabulated below.

Line	Frequency	Intensity	Line	Frequency	Intensity
1	18.54	1.68	9	49.16	1.09
2	31.87	0.55	10	-10.46	0.00
3	53.49	0.77	11	16.35	1.52
4	24.48	1.80	12	27.65	0.26
5	51.30	0.88	13	46.98	1.25
6	62.60	0.00	14	20.16	1.41
7	11.05	0.45	15	8.86	0.35
8	37.87	0.01			

ABC also calculates, on option, a line spectrum consisting of 14 overlapping Lorentzian bands. (The anomalous transition is ignored.) The line shape function is

$$I(\nu) = \frac{I_0 \omega}{\omega^2 + (\nu - \nu_0)^2}$$

Here $I(\nu)$ is the intensity of a Lorentzian band at frequency ν, I_0 is the intensity at the transition frequency $(\nu = \nu_0)$, and ω iw the half-width at half-maximum intensity. It is assumed here that ω is the same for each of the 14 bands in the spectrum. The calculated spectrum is the sum of these overlapping bands,

$$I_{tot}(\nu) = \sum_{j=1}^{14} I_j(\nu)$$

Sample Execution (CH_2CHCN)

THE ABC CASE

ENTER THE CHEMICAL SHIFTS OF A, B, AND C (IN HZ.)
>342 364.6 372.2

ENTER THE COUPLING CONSTANTS JAB, JAC, JBC
>11.75 17.9 0.91

CHEMICAL SHIFTS: A = 342.00   B = 364.60   C = 372.20

COUPLING CONSTANTS: JAB = 11.75   JAC = 17.90   AND   JBC = 0.91

LINE	FREQUENCY	INTENSITY	LINE	FREQUENCY	INTENSITY
1	324.13	0.33	9	365.42	2.12
2	328.73	0.06	10	367.94	0.67
3	335.46	0.55	11	370.02	0.91
4	340.84	0.53	12	372.54	0.26
5	347.58	0.31	13	382.14	0.41
6	352.17	2.22	14	384.65	0.52
7	358.69	0.54	15	402.50	0.01
8	361.20	2.53			

```
C
C                           ABC
C
C    THIS PROGRAM SIMULATES ABC NMR SPECTRA
C
C    THE 8X8 SPIN HAMILTONIAN IS DEFINED IN TERMS
C    OF THE THREE CHEMICAL SHIFTS AND THREE COUPLING CONSTANTS.
C
C       INPUT:   SHIFTA,SHIFTB,SHIFTC
C         AND    JAB,JAC,JBC
C
C    THE SPIN HAMILTONIAN IS DIAGONALIZED BY THE
C    SUBROUTINE HDIAG
C
C    AUTHORS:   D. L. DOERFLER AND K. J. JOHNSON
C
C
         REAL JAB, JAC, JBC, MAXI, LINE(60), INTEN(15), LOW, INCRE
         DIMENSION H(8,8), U(8,8), FREQ(15)
         DATA LINE,INTEN,FREQ,NPOINT,W,STAR/90*0.,60,0.5,'*'/
         DATA H/64*0./
         W2=W**2
C
C    INPUT
C
         WRITE(6,15)
         READ(5,20) SHIFTA, SHIFTB, SHIFTC
         IF(SHIFTA+SHIFTB+SHIFTC.EQ.0.0) STOP
         WRITE(6,25)
         READ(5,20) JAB,JAC,JBC
         WRITE(6,30) SHIFTA,SHIFTB,SHIFTC,JAB,JAC,JBC
C
C    CALCULATE THE DIAGONAL ELEMENTS OF H
C
         H(1,1) = (+SHIFTA+SHIFTB+SHIFTC)/2.0 + (+JAB+JBC+JAC)/4.0
         H(2,2) = (+SHIFTA+SHIFTB-SHIFTC)/2.0 + (+JAB-JBC-JAC)/4.0
         H(3,3) = (+SHIFTA-SHIFTB+SHIFTC)/2.0 + (-JAB-JBC+JAC)/4.0
         H(4,4) = (-SHIFTA+SHIFTB+SHIFTC)/2.0 + (-JAB+JBC-JAC)/4.0
         H(5,5) = (+SHIFTA-SHIFTB-SHIFTC)/2.0 + (-JAB+JBC-JAC)/4.0
         H(6,6) = (-SHIFTA+SHIFTB-SHIFTC)/2.0 + (-JAB-JBC+JAC)/4.0
         H(7,7) = (-SHIFTA-SHIFTB+SHIFTC)/2.0 + (+JAB-JBC-JAC)/4.0
         H(8,8) = (-SHIFTA-SHIFTB-SHIFTC)/2.0 + (+JAB+JBC+JAC)/4.0
C
C    CALCULATE THE OFF DIAGONAL ELEMENTS OF H
C
         H(2,3) = JBC/2.0
         H(2,4) = JAC/2.0
         H(3,4) = JAB/2.0
         H(5,6) = H(3,4)
         H(5,7) = H(2,4)
         H(6,7) = H(2,3)
C
C    CALL MATRIX DIAGONALIZATION SUBROUTINE
C
         CALL HDIAG(H,8,0,U)
C
C    CALCULATE THE FREQUENCY AND INTENSITY FOR THE 15 TRANSITIONS
C
         DO 90 I=1,15
         IF(I.LE.3)GO TO 50
         IF(I.LE.12)GO TO 60
         J=I-8
         FREQ(I)=H(J,J)-H(8,8)
         INTEN(I)=((U(5,J)+U(6,J)+U(7,J))*U(8,8))**2
         GO TO 90
   50    J=I+1
         FREQ(I)=H(1,1)-H(J,J)
         INTEN(I)=((U(2,J)+U(3,J)+U(4,J))*U(1,1))**2
         GO TO 90
   60    J=(MOD(I-1,3))+2
         K=((I-1)/3)+4
         FREQ(I)=H(J,J)-H(K,K)
         INTEN(I)=(U(2,J)*U(5,K)+U(2,J)*U(6,K)+U(3,J)*U(5,K)+U(3,J)*U(7,K)
        .+U(4,J)*U(6,K)+U(4,J)*U(7,K))**2
   90    CONTINUE
C
C    ORDER THE LINES IN INCREASING FREQUENCY
C
         DO 110 I=1,14
         K=I
         J=K+1
  100    IF(FREQ(K).LE.FREQ(J))GO TO 110
         TEMP = FREQ(J)
C        EXCHANGE FREQUENCIES
         FREQ(J) = FREQ(K)
         FREQ(K) = TEMP
C        EXCHANGE INTENSITIES
         TEMP = INTEN(J)
         INTEN(J) = INTEN(K)
         INTEN(K) = TEMP
         K=K-1
         IF(K.LE.0)GO TO 110
         J=K+1
         GO TO 100
```

```
   110 CONTINUE
C
C      WRITE OUT HEADING AND RESULTS
C
       WRITE(6,111) (L, FREQ(L), INTEN(L), L=1,15)
C
C    SEE IF A TELETYPE PLOT IS DESIRED
C
       WRITE(6,112)
       READ(5,113)IPLT
       IF(IPLT.NE.1)STOP
C
C      CONSTRUCT THE GRAPH IN MEMORY
C      IGNORE THE FIRST OR LAST TRANSITION
C      IF THE INTENSITY IS LESS THAN 0.01
C
       LOW=FREQ(1)
       HI=FREQ(15)
       IF(INTEN(1).LT.0.01) LOW=FREQ(2)
       IF(INTEN(15).LT.0.01)HI=FREQ(14)
       INCRE=1.2*(HI-LOW)/FLOAT(NPOINT)
       POINT=LOW-2.*INCRE
       DO 120 I=1,NPOINT
       POINT=POINT+INCRE
       DO 119 J=1,15
   119 LINE(I)=LINE(I)+INTEN(J)*W/
      1( W2  + (POINT - FREQ(J) ) **2 )
   120 CONTINUE
C
C      FIND THE MAXIMUM INTENSITY
C
       MAXI=0.0
       DO 145 I=1,NPOINT
   145 IF(LINE(I).GT.MAXI) MAXI=LINE(I)
       SCALE=MAXI/50.0
C
C    PRINT THE TELETYPE PLOT
C
       WRITE(6,115)
       POINT=LOW-2.*INCRE
       DO 150 I=1,NPOINT
       POINT=POINT+INCRE
       LEN=LINE(I)/SCALE + 1.5
       RELINT=2.*LINE(I)/SCALE
   150 WRITE(6,122)POINT,RELINT,(STAR,L=1,LEN)
C
    15 FORMAT(//' THE ABC CASE',//,' ENTER THE CHEMICAL SHIFTS OF A, B,
      . AND C (IN HZ.)')
    20 FORMAT( 3F )
    25 FORMAT(//' ENTER THE COUPLING CONSTANTS JAB, JAC, JBC ')
    30 FORMAT(//'  CHEMICAL SHIFTS:',7X,'A =', F7.2,5X,'B =',
      .F7.2,5X,'C =',F7.2//' COUPLING CONSTANTS:  JAB =',F7.2,
      .3X, 'JAC =',F7.2, 3X, 'AND  JBC =',F7.2)
   111 FORMAT(//,5X,' LINE', 4X, 'FREQUENCY', 4X, 'INTENSITY', //,
      . 15 (5X,I4, 2F12.2,/))
   112 FORMAT(//' ENTER 1 TO OBTAIN A TELETYPE PLOT')
   113 FORMAT(I)
   115 FORMAT(//' FREQUENCY',2X,'RELATIVE INTENSITY'/)
   122 FORMAT(2F8.2,2X,51A1)
       END
```

Discussion

ABC and HDIAG (modified to dimension 8 x 8) require approximately 1600 words of core. This sample execution took approximately 0.5 sec.

The simulated CH_2CHCN spectrum is plotted in Figure 8.1 using a half-width parameter in the Lorentzian line shape function that gives a reasonably close match to the experimental 60 MHz spectrum [15].

8.4   HUCKEL MOLECULAR ORBITAL THEORY

A relatively simple quantum mechanical theory has been developed [16-20] to approximate the molecular parameters of planar, conjugated hydrocarbons. Only π electrons are considered. Each carbon atom in the molecule contributes to the π bonding via $2p_z$ atomic orbitals. Molecular orbitals are constructed by taking linear combinations of these atomic orbitals (the LCAO method),

$$\psi = \sum_{i=1}^{n} c_i \phi_i$$

Here ψ is one of the molecular orbitals, n is the number of carbon atoms in the conjugated system, and c_i is a constant denoting the contribution of the $2p_z$

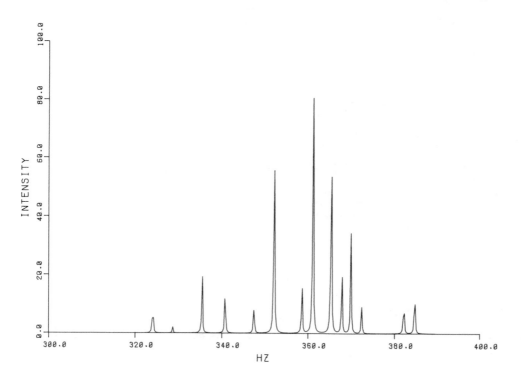

Figure 8.1 Simulated NMR spectrum for CH_2CHCN.

function, ϕ_i, to that molecular orbital. For example, for ethylene, C_2H_4, the 2π electrons occupy the molecular orbital,

$$\psi_1 = c_1\phi_1 + c_2\phi_2$$

The coefficients of the molecular orbital and the corresponding relative energy values are determined by solving the eigenvalue equation,

$$\mathcal{K}\psi = E\psi$$

$\mathcal{K}$ is the appropriate Hamiltonian operator for the π-electron system, expressed in matrix form, and E is the energy eigenvalue corresponding to a particular energy level. The variation method [17] is used to express this eigenvalue equation in terms of the secular determinant,

$$\begin{vmatrix} H_{11} - ES_{11} & H_{12} - ES_{12} & H_{13} - ES_{13} & \cdots & H_{1n} - ES_{1n} \\ H_{21} - ES_{21} & H_{22} - ES_{22} & H_{23} - ES_{23} & \cdots & H_{2n} - ES_{2n} \\ \cdots & \cdots & \cdots & \cdots & \cdots \\ H_{n1} - ES_{n1} & H_{n2} - ES_{n2} & H_{n3} - ES_{n3} & \cdots & H_{nn} - ES_{nn} \end{vmatrix}$$

Here H_{ii} is the coulomb integral,

$$H_{ii} = \int \psi_i \mathcal{K}\psi_i \, d\tau$$

H_{ij} is the resonance integral,

$$H_{ij} = \int \psi_i \mathcal{K}\psi_j \, d\tau$$

and S_{ij} is the exchange integral,

$$S_{ij} = \int \psi_i \psi_j \, d\tau$$

The approximations made in Huckel molecular orbital theory are

$$H_{ii} = \alpha$$

$$H_{ij} = \beta \qquad \text{if atoms i and j are bonded,}$$

$$H_{ij} = 0 \qquad \text{otherwise}$$

and

$$S_{ij} = 1 \qquad \text{if i = j}$$

$$S_{ij} = 0 \qquad \text{otherwise}$$

Here α and β are constants. For example, consider the molecule 3-methylene-1,4-pentadiene, C_6H_8,

$$
\begin{array}{c}
C_4 \\
\| \\
C_3 \\
C_2 \quad C_5 \\
\|2 \quad \|5 \\
C_1 \quad C_6
\end{array}
$$

The secular determinant is

$$
\begin{vmatrix}
\alpha - E & \beta & 0 & 0 & 0 & 0 \\
\beta & \alpha - E & \beta & 0 & 0 & 0 \\
0 & \beta & \alpha - E & \beta & \beta & 0 \\
0 & 0 & \beta & \alpha - E & 0 & 0 \\
0 & 0 & \beta & 0 & \alpha - E & \beta \\
0 & 0 & 0 & 0 & \beta & \alpha - E
\end{vmatrix}
$$

with the substitution $x = (\alpha - E)/\beta$,

$$
\begin{vmatrix}
x & 1 & 0 & 0 & 0 & 0 \\
1 & x & 1 & 0 & 0 & 0 \\
0 & 1 & x & 1 & 1 & 0 \\
0 & 0 & 1 & x & 0 & 0 \\
0 & 0 & 1 & 0 & x & 1 \\
0 & 0 & 0 & 0 & 1 & x
\end{vmatrix}
$$

The roots of the sixth-order characteristic equation are the eigenvalues of the following matrix:

$$
A = \begin{bmatrix}
0 & 1 & 0 & 0 & 0 & 0 \\
1 & 0 & 1 & 0 & 0 & 0 \\
0 & 1 & 0 & 1 & 1 & 0 \\
0 & 0 & 1 & 0 & 0 & 0 \\
0 & 0 & 1 & 0 & 0 & 1 \\
0 & 0 & 0 & 0 & 1 & 0
\end{bmatrix}
$$

This Hermitian matrix is the same as the secular determinant, except that the diagonal elements are zero instead of x. The six eigenvalues of A, determined by HDIAG (Section 8.2.1), and the corresponding molecular energy levels in terms of α and β are given below.

Eigenvalue (x)	Energy
1.932	$E_1 = \alpha + 1.932\beta$
1.000	$E_2 = \alpha + 1.000\beta$
0.518	$E_3 = \alpha + 0.518\beta$
-0.518	$E_4 = \alpha - 0.518\beta$
-1.000	$E_5 = \alpha - 1.000\beta$
-1.932	$E_6 = \alpha - 1.932\beta$

Since α and β are both negative quantities, E_1 lies lowest in energy. The neutral molecule contains six π electrons. The molecular orbital energy level diagram for the π electrons is shown in Figure 8.2. The total π- electron energy is

$$E_\pi = 2E_1 + 2E_2 + 2E_3 = 6\alpha + 6.899\beta$$

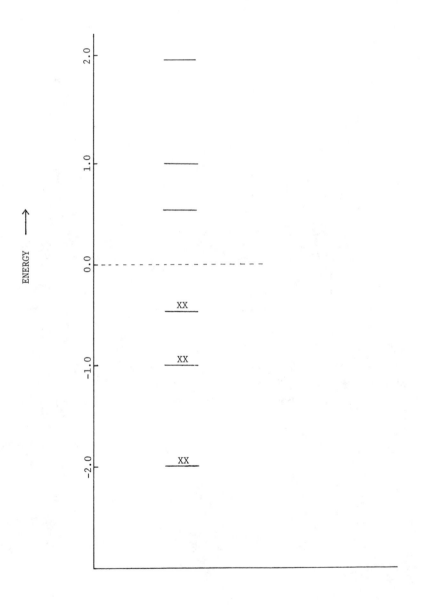

Figure 8.2   Energy Level Diagram for C_6H_8.

The total π-electron energy for ethylene, $H_2C\!=\!\!=\!\!CH_2$, is $2\alpha + 2\beta$. Therefore, 3-methylene-1,4-pentadiene is more stable than three ethylene molecules by

$$6\alpha + 6.899\beta - 3(2\alpha + 2\beta)$$

or 0.899β. This is the delocalization energy and can be correlated with the heat of formation of the molecule.

The coefficients of the molecular orbitals, c_i, are the eigenvectors of the matrix. These eigenvectors are

ψ_1	ψ_2	ψ_3	ψ_4	ψ_5	ψ_6
0.230	-0.500	0.444	0.444	-0.500	0.230
0.444	-0.500	0.230	-0.230	0.500	-0.444
0.628	0.000	-0.325	-0.325	0.000	0.628
0.325	0.000	-0.628	0.628	-0.000	-0.325
0.444	0.500	0.230	-0.230	-0.500	-0.444
0.230	0.500	0.444	0.444	0.500	0.230

For example, the third molecular orbital, the highest occupied orbital in the molecule, is

$$\psi_3 = 0.44\,\phi_1 + 0.23\,\phi_2 - 0.32\,\phi_3 - 0.63\,\phi_4 + 0.23\,\phi_5 + 0.44\,\phi_6$$

Note that there are two nodes in this molecular orbital, one between atoms 2 and 3, and the other between atoms 4 and 5.

The π-electron bond order P_{rs} is given by the equation

$$P_{rs} = \sum_{i=1}^{n'} n_i c_{ir} c_{is}$$

Here n' is the number of occupied molecular orbitals, n_i is the number of electrons in the ith molecular orbital (0, 1, or 2), and c_{ir} and c_{is} are the ith components of ψ_r and ψ_s. For example, the bond order between atoms 1 and 2 is

$$P_{12} = \sum_{i=1}^{3} n_i c_{i1} c_{i2}$$

$$= n_1 c_{11} c_{12} + n_2 c_{21} c_{22} + n_3 c_{31} c_{32}$$

$$= 2(0.230)(0.44) + 2(-0.500)(-0.500) + 2(0.444)(0.230)$$

$$= 0.908$$

Similarly,

Bond	1-2	2-3	3-4	4-5	5-6
π bond order	0.908	0.408	0.816	0.408	0.908

The π bonding in the molecule can be represented as follows:

The π-electron density at each carbon, q_i, is calculated using the formula

$$q_r = \sum_{i=1}^{n'} n_i c_{ri}^2$$

For example,

$$q_1 = \sum_{i=1}^{3} n_i c_{i1}^2$$

$$= n_1 c_{11}^2 + n_2 c_{21}^2 + n_3 c_{31}^2$$

$$= 2(0.23)^2 + 2(-0.5)^2 + 2(0.44)^2$$

$$= 1.0$$

All atomic charges are unity for this molecule with six π electrons. If an additional electron is added to form the anion, then the π-electron densities are as indicated in the following figure:

```
      C4   1.39
     ‖4
     C3,  1.11
   ⟋⟍3⟍
 C2     C5  1.05
 ‖2     ‖5
 C1     C6  1.20
```

The free valence F_r of each carbon atom is calculated using the equation

$$F_r = \sqrt{3} - N_r$$

where N_r is the sum of all π bond orders for atom r. For example, for atom 3 of the neutral molecule,

$$F_r = \sqrt{3} - N_r$$
$$= 1.73 - (0.408 + 0.816 + 0.408) = 0.098$$

The six free valencies are

```
      C4   0.916
     ‖4
     C3,  0.098
   ⟋⟍3⟍
 C2     C5  0.416
 ‖2     ‖5
 C1     C6  0.824
```

The following program performs these Huckel molecular orbital calculations.

8.4.1 HMO

Introduction

HMO uses the techniques described above to find the π molecular orbital energy levels, the LCAO wavefunctions, the bond orders, electron densities, and free valencies for planar, unsaturated hydrocarbons. The input consists of the n x n

Hermitian matrix corresponding to the secular determinant, and the number of π electrons. The program allows the user to change the number of π electrons to observe the bonding parameters of radicals and ionic species. The program prints the matrix for verification purposes, the eigenvalues (π molecular orbital levels in terms of α and β), the associated eigenvectors (LCAO-MOs), the total π energy, the bond order for each carbon-carbon bond in the molecule, and the π-electron density and the free valence of each carbon atom in the molecule.

Method

   The Hermitian matrix is diagonalized by the subroutine HDIAG. The molecular parameters are calculated from the eigenvalues and eigenvectors as described above.

Sample Execution

ENTER THE ORDER OF THE MATRIX   (ZERO ENDS THE PROGRAM)
>6

ENTER THE MATRIX BY ROWS
>0 1
>1 0 1
>0 1 0 1 1
>0 0 1 0
>0 0 1 0 0 1
>0 0 0 0 1

THE INPUT MATRIX OF ORDER   6 IS:

```
  0.00    1.00    0.00    0.00    0.00    0.00
  1.00    0.00    1.00    0.00    0.00    0.00
  0.00    1.00    0.00    1.00    1.00    0.00
  0.00    0.00    1.00    0.00    0.00    0.00
  0.00    0.00    1.00    0.00    0.00    1.00
  0.00    0.00    0.00    0.00    1.00    0.00
```

I	EIGENVALUE(I)		EIGENVECTOR(I)		
1	1.932	0.230	0.444	0.628	0.325
		0.444	0.230		
2	1.000	-0.500	-0.500	0.000	0.000
		0.500	0.500		
3	0.518	0.444	0.230	-0.325	-0.628
		0.230	0.444		
4	-0.518	0.444	-0.230	-0.325	0.628
		-0.230	0.444		
5	-1.000	-0.500	0.500	0.000	-0.000
		-0.500	0.500		
6	-1.932	0.230	-0.444	0.628	-0.325
		-0.444	0.230		

ENTER THE NUMBER OF PI ELECTRONS   (ZERO STARTS OVER)
>6

THE PI ENERGY IS   6 ALPHA   6.899 BETA.

BOND   A   -   B   BOND ORDER FOR 6 PI ELECTRONS

        1   -   2   0.908
        2   -   3   0.408
        3   -   4   0.816
        3   -   5   0.408
        5   -   6   0.908

ATOM   PI ELECTRON DENSITY   FREE VALENCE

    1   1.000   0.824
    2   1.000   0.416
    3   1.000   0.099
    4   1.000   0.916
    5   1.000   0.416
    6   1.000   0.824

ENTER THE NUMBER OF PI ELECTRONS   (ZERO STARTS OVER)
>7

THE PI ENERGY IS   7 ALPHA   6.381 BETA.

BOND   A   -   B   BOND ORDER FOR 7 PI ELECTRONS

        1   -   2   0.806
        2   -   3   0.483
        3   -   4   0.612
        3   -   5   0.483
        5   -   6   0.806

ATOM   PI ELECTRON DENSITY   FREE VALENCE

    1   1.197   0.926
    2   1.053   0.443
    3   1.106   0.154
    4   1.394   1.120
    5   1.053   0.443
    6   1.197   0.926

ENTER THE NUMBER OF PI ELECTRONS   (ZERO STARTS OVER)
>0

ENTER THE ORDER OF THE MATRIX   (ZERO ENDS THE PROGRAM)
>0

    Listing

C
C   HMO
C
C   THIS PROGRAM SOLVES SIMPLE HUCKEL MOLECULAR ORBITAL PROBLEMS
C
C
C   INPUT: SECULAR DETERMINANT AND NUMBER OF PI ELECTRONS
C
C

```
C      OUTPUT:  EIGENVALUES AND EIGENVECTORS, TOTAL PI ENERGY,
C               PI BOND ORDERS, ELECTRON DENSITIES AND FREE VALENCE
C
C
C      AUTHORS:  D. L. DOERFLER AND K. J. JOHNSON
C
C
       DIMENSION H(25,25), U(25,25), D(25)
C
       WRITE(6,10)
C
       READ(5,15) N
       NN = N-1
C
       IF(N.LE.0)STOP
       WRITE(6,20)
C
       DO 25 I=1,N
    25 READ(5,30) (H(I,J),J=1,N)
C
       WRITE(6,35) N
C
       DO 40 I=1,N
    40 WRITE(6,45) (H(I,J),J=1,N)
C
C      CALL SUBROUTINE HDIAG
C
       CALL HDIAG(H,N,0,U)
C
       WRITE(6,49)
C
C      ORDER EIGENVALUES IN DECREASING ORDER    (INCREASING ENERGY)
C
       DO 56 I=1,NN
       K=I
       J=K+1
    50 IF( H(K,K) .GE. H(J,J) ) GO TO 56
C      EXCHANGE THE EIGENVALUES
       TEMP = H(J,J)
       H(J,J) = H(K,K)
       H(K,K) = TEMP
C      EXCHANGE THE EIGENVECTORS
       DO 52 L=1,N
       TEMP = U(L,J)
       U(L,J) = U(L,K)
    52 U(L,K) = TEMP
    53 K=K-1
       IF( K.LE.0 )GO TO 56
       J=K+1
       GO TO 50
    56 CONTINUE
C
       DO 65 I=1,N
       WRITE(6,57)I,H(I,I),(U(J,I),J=1,N)
    65 CONTINUE
C      READ IN THE NUMBER OF ELECTRONS
C
    77 WRITE(6,80)
C
       READ(5,82) J
C
```

```
C      CHECK THE NUMBER OF ELECTRONS FOR TOO FEW OR TOO MANY
C      IF J=0 START OVER
C
       IF(J.LE.0) GO TO 1
       IF( J.GT.2*N) GO TO 200
C
C      CALCULATE THE PI ENERGY
C
       PI = 0.0
       DO 87 I=1,J
       K = (I+1)/2
    87 PI = PI + H(K,K)
C
C      WRITE OUT THE PI ENERGY
C
       WRITE(6,88) J, PI
C
C      WRITE OUT TITLE
C
       WRITE(6,89) J
C
C   THIS SECTION FINDS DEGENERATE ENERGY LEVELS
C                 I =  FIRST MO OF THE DEGENERATE SET
C                 L =  THE NUMBER OF DEGENERATE MO'S
C                 K =  THE LAST MO OF THE DEGENERATE SET
C                 II=  THE NUMBER OF FILLED MO'S (NON-DEGENERATE)
C
C
       JJ=(J+1)/2
       DO 95 I=1,JJ
       L=1
       K=I+1
C
C      DO NOT COMPARE THE LAST MO WITH THE NEXT HIGHEST
C
       IF( K.GT.N )GO TO 95
C
C      ARE THE TWO MO'S DEGENERATE
C
    90 IF(ABS(H(I,I)-H(K,K))-.0001) 91,91,92
    91 L=L+1
       K=K+1
       IF(K-N) 90,90,93
    92 IF(L-1) 95,95,93
C
C      ARE THESE DEGENERATE LEVELS FILLED?
C
    93 II=I-1
       K=K-1
       IF(2*(II+L)-J) 95,96,97
    95 CONTINUE
C
C      COME HERE IF THERE ARE NO DEGENERACIES
C      ADJUST FOR PARTIALLY FILLED MO'S
C
    96 II=J/2
       LL=1
       K=JJ
       GO TO 102
C
```

```
C
C     COME HERE FOR PARTIALLY FILLED DEGENERATE MO'S
C
   97 LL=J-2*II
C
C     NOW THERE ARE L DEGENERATE LEVELS CONTAINING LL ELECTRONS
C
C     ADJUSTMENT FACTOR FOR PARTIALLY FILLED DEGENERATE MO'S
C
  102 FACTOR=FLOAT(LL)/FLOAT(L)
      DO 145 I=1,N
      DO 143 L=I,N
C
C   IS THIS AN ELECTRON DENSITY CALCULATION?  127 ==> YES,  105 ==> NO
C
  105 IF(H(L,I)) 109,143,109
  109 H(L,I)=0.0
C
C     CALCULATION LOOP FOR BOND ORDERS
C
      DO 125 M=1,K
      IF(M.GT.II)GO TO 115
      H(L,I)=H(L,I)+U(L,M)*U(I,M)*2.0
      GO TO 125
  115 H(L,I)=H(L,I)+U(L,M)*U(I,M)*FACTOR
  125 CONTINUE
C
C     OUTPUT BOND ORDERS
C
      WRITE(6,126) I,L,H(L,I)
      GO TO 143
C
C     CALCULATION LOOP FOR ELECTRON DENSITIES
C
  127 D(I)=0.0
      DO 138 M=1,K
      IF(M.GT.II)GO TO 133
      D(I)=D(I)+U(L,M)*U(I,M)*2.0
      GO TO 138
  133 D(I)=D(I)+U(L,M)*U(I,M)*FACTOR
  138 CONTINUE
  143 CONTINUE
  145 CONTINUE
C
C     WRITE OUT TITLE
C
      WRITE(6,173)
C
C     CALCULATE THE FREE VALENCE FOR EACH ATOM
C     FREE VALENCE = SQRT(3) - TOTAL BONDING
C
      DO 181 I=1,N
      FR = 1.73205081
      DO 180 L=1,N
```

```
C
C        DO NOT COUNT THE BONDING OF AN ATOM WITH ITSELF
C        NOTE:   BOND ORDER A-B = BOND ORDER B-A
C
         IF(I-L) 175,180,177
     175 FR=FR-H(L,I)
         GO TO 180
     177 FR=FR-H(I,L)
     180 CONTINUE
C
C        OUTPUT ELECTRON DENSITIES AND FREE VALENCE
C
     181 WRITE(6,183) I,D(I),FR
C
C        ASK FOR A NEW NUMBER OF ELECTRONS
C
         GO TO 77
C
     200 WRITE(6,201) J,N
C
C        ASK AGAIN FOR THE NUMBER OF ELECTRONS.
C
         GO TO 77
C
C
      10 FORMAT(//,' ENTER THE ORDER OF THE MATRIX    (ZERO ENDS THE
        . PROGRAM)')
      15 FORMAT( I )
      20 FORMAT(/,' ENTER THE MATRIX BY ROWS ')
      30 FORMAT( 25F )
      35 FORMAT(//,' THE INPUT MATRIX OF ORDER',I3,' IS:',//)
      45 FORMAT( 10F7.2 )
      49 FORMAT(//2X,'I       EIGENVALUE(I)',18X,'EIGENVECTOR(I)')
      57 FORMAT(/I3,F12.3,7(T19,4F12.3/))
      80 FORMAT(//,' ENTER THE NUMBER OF PI ELECTRONS    (ZERO STARTS
        . OVER)')
      82 FORMAT( 2I )
      88 FORMAT(//,' THE PI ENERGY IS', I3, ' ALPHA', F9.3, ' BETA.')
      89 FORMAT(//' BOND  A  -  B',5X,'BOND ORDER FOR',I3,' PI ELECTRONS'/)
     126 FORMAT(I8,'  -  ',I2,F13.3)
     173 FORMAT(///,' ATOM  PI ELECTRON DENSITY',5X,'FREE VALENCE',/)
     183 FORMAT(I4,F13.3,F23.3)
     201 FORMAT( I4, ' ELECTRONS IS TOO MANY TO BE PLACED IN',I3,' MO''S.')
         END
```

Discussion

HMO and HDIAG (modified to dimension 25 x 25) require approximately 2700 words of core. This sample execution took approximately 0.3 sec.

8.5 ADDITIONAL PROBLEMS

1. Determine the eigenvalues and eigenvectors of the following symmetric matrices:

(a) $\begin{bmatrix} 10 & 9 & 7 & 5 \\ 9 & 10 & 8 & 6 \\ 7 & 8 & 10 & 7 \\ 5 & 6 & 7 & 5 \end{bmatrix}$
   (b) $\begin{bmatrix} 6 & 4 & 4 & 4 \\ 4 & 6 & 1 & 4 \\ 4 & 1 & 6 & 4 \\ 1 & 4 & 4 & 6 \end{bmatrix}$

(c) The 8 x 8 matrix with $A_{ij} = 2I_{ij}$, except

$$A_{i+1,i} = A_{i,i+1} = -1.0 \qquad i = 1,2,3,\ldots,8$$

Solutions.   (a) Eigenvalues: 30.289, 0.843, 3.858, 0.0101

Eigenvectors	0.521	0.568	-0.625	-0.124
stored	0.552	-0.760	-0.272	0.209
columnwise:	0.529	0.320	0.615	0.502
	0.380	-0.093	0.396	-0.830

(b) Eigenvalues: 15.0, -1.0, 5.0, 5.0

Eigenvectors	0.500	0.500	-0.267	0.655
stored	0.500	-0.500	-0.655	-0.267
columnwise:	0.500	-0.500	0.655	0.267
	0.500	0.500	0.267	-0.655

(c) The eigenvalues are

3.879	3.532	3.000	2.347	1.653	1.000	0.468	0.121

The corresponding eigenvectors are (columnwise)

0.161	-0.303	-0.408	-0.464	-0.464	-0.408	-0.303	0.161
-0.303	0.464	0.408	0.161	-0.161	-0.408	-0.464	0.303
0.408	-0.408	0.000	0.408	0.408	0.000	-0.408	0.408
-0.464	0.161	-0.408	-0.303	0.303	0.408	-0.161	0.464
0.464	0.161	0.408	-0.303	-0.303	0.408	0.161	0.464
-0.408	-0.408	0.000	0.408	-0.408	0.000	0.408	0.408
0.303	0.464	-0.408	0.161	0.161	-0.408	0.464	0.303
-0.161	-0.303	0.408	-0.464	0.464	-0.408	0.303	0.161

2. Write a program to diagonalize symmetric matrices using the Givens-Householder method [21].

REFERENCES

1. J. Greenstadt, The determination of the characteristic roots of a matrix by the Jacobi method, in "Mathematical Methods for Digital Computers" (A. Ralston and H. Wilf, eds.), Vol. 1, Wiley, New York, 1967, Chapter 7.

2. B. Carnahan, H. Luther, and J. Wilkes, "Applied Numerical Methods," Wiley, New York, 1969, pp. 250-261.

3. E. A. Moelwyn-Hughes, "Physical Chemistry," Pergamon Press, New York, 1961, p. 343ff.

4. N. B. Colthup et al., "Introduction to Infrared and Raman Spectroscopy,"
    Academic Press, New York, 1964.

5. G. M. Barrow, "Physical Chemistry," McGraw-Hill, New York, 1961, pp. 86, 251,
    355ff.

6. W. J. Moore, "Physical Chemistry," 4th ed., Prentice-Hall, Englewood Cliffs,
    N. J., pp. 185ff.

7. J. C. Davis, Jr., "Advanced Physical Chemistry," Ronald Press, New York, 1965,
    p. 130ff.

8. J. O. Hirshfelder, J. Chem. Phys., 8, 431 (1940).

9. K. B. Wiberg, "Computer Programming for Chemistry," Benjamin, New York, 1965,
    p. 189ff.

10. K. B. Wiberg, "Physical Organic Chemistry," Wiley, New York, 1964, p. 483ff.

11. J. A. Pople, W. G. Schneider, and H. Bernstein, "High Resolution Nuclear
    Magnetic Resonance," McGraw-Hill, New York, 1959, pp. 130, 132.

12. J. W. Emsley, J. Feeney, and L. H. Sutcliffe, "High Resolution Nuclear Magnetic
    Resonance," McGraw-Hill, New York, 1959, pp. 130, 132.

13. K. B. Wiberg and B. J. Nist, "Interpretation of NMR Spectra," Benjamin, New
    York, 1962.

14. D. F. Detar (ed.), "Computer Programs for Chemistry," Vol. 1, Benjamin, New
    York, 1968, Chapters 2-4.

15. C. L. Wilkins and C. E. Klopfenstein, "Simulation of NMR Spectra," J. Chem.
    Educ., 43, 10 (1966).

16. K. B. Wiberg, "Computer Programming for Chemists," op. cit., p. 215 ff.

17. K. B. Wiberg, "Physical Organic Chemistry," op.cit., pp. 42ff, 64ff.

18. H. H. Greenwood, "Computing Methods in Quantum Organic Chemistry," Wiley, New
    York, 1972.

19. A. Streitwieser, "Molecular Orbital Theory for Organic Chemists," Wiley, New
    York, 1961.

20. C. A. Coulson and A. Streitwieser, "Dictionary of π Electron Calculations,"
    Pergamon Press, New York, 1965.

21. J. Ortega, The Givens-Householder method for symmetric matrices, in "Mathematical
    Methods for Digital Computers" (A. Ralston and H. Wilf, eds.), Vol. 2, Wiley,
    New York, 1966, p. 94ff.

CHAPTER 9

MISCELLANEOUS TOPICS

The emphasis of this book is on the applications of computing techniques and numerical analysis to chemistry. It seems appropriate, however, to mention some additional examples of the impact of computing technology on chemistry. This chapter presents a brief overview of the following topics: graphics, computer-assisted instruction (CAI), computer-assisted test construction (CATC), computer-managed instruction (CMI), computers in the laboratory, artificial intelligence, and information retrieval.

9.1 GRAPHICS

This area may be arbitrarily divided into three categories: graphics using terminals and line printers; graphics using incremental plotters; and interactive graphics. Many of the programs in this book used the "teletype-graphics" routine, PLOT (Section 2.1). The majority of the remaining figures were plotted using an incremental plotter, the Calcomp plotter. Interactive graphics refers to those systems which support both the software and the hardware to allow the user to specify how his results are to be plotted in an interactive mode. The interested reader is referred to a number of books and articles for applications of graphics in chemistry [1-18]. The following program was used to prepare the Calcomp plots in this book.

9.1.1 CPLOT

Introduction

CPLOT performs the following functions required to generate a Calcomp plot:

1. Read the specifications of the plot size in centimeters.
2. Read the data specifying the plot title, the labels used for the ordinate and abscissa, and the size of the numbers used on the axes.
3. Read the number of units into which the axes are to be divided, the values at the origin, and the increment for both abscissa and ordinate.
4. Calculate the dimensions in centimeters of the graph.
5. Read the data for each curve from a disk file.

Here the plot refers to the entire area, including the curve(s), the axes, and all labels. The curve refers to the area bounded by the ordinate and abscissa.

Method

CPLOT prompts the user for the following input data. All dimensions are in centimeters. The values used in the sample execution (see below) are included in parentheses.

XSIZE,YSIZE   Size of the plot (26,19)

NSETS   Number of curves (2)

SNUNIT,YNUNIT   Number of units into which the axes are to be divided i.e.,
                     number of tic marks, not counting the origin (5,5)

XNOSIZ,XNO1,XINC   Size (height) of the digits used to index the abscissa; abscissa
                     value at the origin; and spacing between tic marks on the
                     abscissa (0.25,0,20.0)

YNOSIZ,YNO1,YINC   Size of the digits used to index the ordinate; value of the
                     ordinate at the origin, and increment between tic marks on
                     the ordinate (0.25,0,0.2)

ITITLE,TTLSIZ   Number and size of characters in the plot title (17,0.5)

TITLE   The plot title (TWO SITE EXCHANGE)

IXTTL,XTLSIZ   Number and size of characters in the abscissa label (1,0.3)

XTITLE   The abscissa label (V)

IYTITL,YTLSIZ   Number and size of characters in the ordinate label (4,0.3)

YTITLE   The ordinate label (I(V))

IAXES   1 draws axes through the logical origin, 0 does not (0)

NPTS   Number of (x,y) coordinates for the first curve (100)

IPOS,SYSIZ,KSYM   If IPOS = -2, then a connecting straight line is drawn between
                     symbols; if IPOS = -1, then the symbols are not connected by
                     straight lines; SYMSIZ is the size of the plotting character;
                     and KSYM is an integer representing one of several Calcomp
                     plotting symbols (-2,1,0.01).

X(I),Y(I)   The NPTS coordinates (see below)

NPTS   Number of points for the second curve (50)

IPOS,SYMSIZ,KSYM   Same as above (-1,1,0.25)

CPLOT uses the following Calcomp subroutines:

1. GRAPH(X,Y): X and Y are the dimensions of the plot in centimeters.

2. ORIGIN(X,Y): X and Y are the coordinates of the origin of the plot.

3. SYMBOL(X,Y,SIZE,STRING,DEG,N): X and Y are the coordinates of the lower
    left-hand corner of the first character in STRING, SIZE is the height,
    STRING is the alphameric vector to be printed, DEG is the orientation in
    degrees relative to the abscissa, and N is the number of characters in
    STRING.

4. SYMBOL(X,Y,SIZE,KSYM,DEG,IPOS): This is an alternate calling sequence to
    SYMBOL. Here X, Y, SIZE, DEG, and IPOS have been previously defined.
    KSYM is an index representing one of the following 15 centered symbols:

    0   1   2   3   4   5   6   7   8   9   10   11   12   13   14

5. PENUP: Raise the pen tip.

6. PENDN: Lower the pen tip.

7. PLOT(X,Y,IPEN): X and Y are the coordinates of the position to which the
    pen is to be moved, and IPEN is 3,2,1, or 0. If IPEN is 3, the pen tip
    is up when the pen moves; if IPEN is 2, the pen tip is down when the pen
    moves; if IPEN is 1 or 0, there is no change in the current position of
    the pen tip.

8. NUMBER(X,Y,SIZE,FPN,DEG,N): X and Y are the coordinates of the lower left-
    hand corner of the first digit in the number, SIZE is the height of the
    number, FPN is the floating point number, DEG is the orientation in degrees
    relative to the abscissa, and N is the number of significant digits after
    the decimal point.

9. ENDPAG: Closes the disk file that has been generated.

The first section of CPLOT is input (see Listing). Then GRAPH and ORIGIN are
called. Next the size of the graph is calculated. Then the coordinates of the
lower left-hand corner of the first character in the abscissa label are calculated
(TX,TY), and the label is drawn by SYMBOL. Then the pen is raised and loop 10 is
executed. This loop draws the abscissa, makes the tic marks, and prints the abscissa
values at the tic marks. Next the 20 loop draws the ordinate, makes the tic marks,
and prints the ordinate values at the tic marks. Then the ordinate and plot labels
are drawn. The axes are drawn through the logical origin of the graph if IAXES = 1,
otherwise control is branched to statement 38. The outer loop (50) controls the
number of curves to be drawn. The inner loop (40) reads the coordinates of each
point from the disk file, FOR04.DAT, and draws the curves point by point. After the
last curve has been drawn the plot disk file is closed by ENDPAG.

For the purposes of this documentation, the two-site NMR exchange system [11] is plotted. The function is

$$I(\nu) = \frac{C\tau(\nu_a - \nu_b)^2}{[(\nu_a + \nu_b)/2 - \nu]^2 + 4\pi^2\tau^2(\nu - \nu_a)^2(\nu - \nu_b)^2}$$

Here C is a normalizing constant, τ is the lifetime of the exchanging proton, ν is the observed frequency, ν_a and ν_b are the observed frequencies of protons a and b in the absence of exchange. The program used to generate the data is given next.

```
C
C      PROGRAM TO GENERATE DATA FOR CPLOT
C
C      THE FIRST 100 POINTS ARE (V,Y)   OR   (V, I(V) )
C      THE NEXT 50 POINTS ARE (JJ,Z)    OR   (V, I(V) )
C      THE LATTER 50 POINTS ARE GENERATED BY SUPERIMPOSING A
C      UNIFORMLY DISTRIBUTED +/- 10% RANDOM ERROR ON THE
C      CORRESPONDING POINTS FROM THE SET OF 100
C
       DIMENSION Z(50)
       VA=33.3
       VB=66.7
       T=3./( 2*3.14159*(VB-VA) )
       C1=3.8*T*(VB-BA)**2
       C2=(VA+VB)/2.
       C3=4.*3.14159**2*T*T
       DO 100 I=1,100
       V=I
       Y=C1/( (C2-V)**2 + C3*(VA-V)**2*(VB-V)**2 )
       IF(MOD(I,2).EQ.0) Z(I/2)=Y*(0.9+0.2*RAN(XXX))
  100 WRITE(4,11)V,Y
       DO 200 I=1,50
       JJ=I*2
  200 WRITE(4,11)JJ,Z(I)
       STOP
   11 FORMAT(2G12.3)
       END
```

Some of the output from this program follows.

```
    1.00      0.624E-02
    2.00      0.683E-02
    3.00      0.749E-02
    4.00      0.824E-02
    5.00      0.908E-02
```

(rest of output omitted)

```
   96.0      0.824E-02
   97.0      0.749E-02
   98.0      0.683E-02
   99.0      0.624E-02
  100.       0.571E-02
           2  0.641E-02
           4  0.862E-02
           6  0.103E-01
           8  0.119E-01
```

```
              10   0.142E-01
   (rest of output omitted)
              92   0.121E-01
              94   0.104E-01
              96   0.791E-02
              98   0.654E-02
             100   0.554E-02
```

Sample Execution

ENTER 1 TO GET A DESCRIPTION OF INPUT, OTHERWISE 0
>1
             1

INPUT SPECIFICATIONS:

```
XSIZE,YSIZE     - LENGTH OF X AND Y AXES IN CENTIMETERS
NSETS           - NO. OF SEPARATE CURVES ON THE SAME PLOT
SYMSIZ          - SIZE IN CENTIMETERS OF PLOTTING SYMBOL
XNUNIT,YNUNIT   - NO. OF UNIT LENGTHS ON AXIS
XNOSIZ,YNOSIZ   - SIZE OF DIGITS USED TO NUMBER X AND Y AXES
XNO1,YNO1       - X AND Y VALUES AT THE PLOT ORIGIN
XINC,YINC       - NUMERICAL INCREMENT OF ONE UNIT ON AXIS
ITITLE,IXTITL,- NO. OF LETTERS RESPECTIVELY IN THE 3 LABELS
   IYTITL
TTLSIZ,XTLSIZ,- SIZE OF LETTERS IN 3 LABELS
   YTLSIZ
TITLE,XTITLE, - VECTOR CONTAINING THE LABEL, UP TO 50 CHARS.
   YTITLE
 IAXES          -  1: DRAWS AXES THROUGH (0,0); 0: NO AXES
 NPTS           - NO. OF DATAPOINTS (X,Y) IN CURVE
 IPOS           - PLOT OPTION:-2 CONNECTS SYMBOLS WITH LINE
                            -1 DOES NOT CONNECT POINTS
 X,Y            - CO-ORDINATES OF POINTS; READ FROM FOR04.DAT
```


ENTER XSIZE AND YSIZE. XSIZE=0 ENDS. (2E).
>26,19

   26.000   19.000

ENTER NSETS. (I)
>2

             2

ENTER XNUNIT AND YNUNIT. (2E)
>5,5

   5.0000   5.0000

ENTER XNOSIZ,XNO1,AND XINC. (3E)
>0.25,0,20

  0.25000   0.00000E+00   20.000

```
ENTER YNOSIZ,YNO1, AND YINC.    (3E)
>0.25,0,0.2


 0.25000        0.00000E+00   0.20000

ENTER ITITLE AND TTLSIZ FOR PLOT TITLE    (I,E)
 (IF 0 LETTERS, NO TITLE WILL BE WRITTEN.)
>17,0.5

           17   0.50000

ENTER TITLE OF PLOT   (10A5)
>TWO SITE EXCHANGE

      TWO SITE EXCHANGE

ENTER IXTITL AND XTLSIZ FOR ABSCISSA LABEL   (I,E)
 (IF 0 LETTERS, X AND Y WILL NOT BE LABELLED.)
>1,0.3

            1   0.30000

ENTER ABSCISSA LABEL.    (10A5)
>V

      V

ENTER IYTITL AND YTLSIZ FOR ORDINATE LABEL.   (I,E)
>4,0.3

            4   0.30000

ENTER ORDINATE LABEL.    (10A5)
>I(V)

      I(V)

ENTER IAXES. 1 DRAWS AXES, 0 DOES NOT.   (I)
>0

            0

CURVE NO.  1.   ENTER NPTS    (I)
>100

         100

ENTER -2 TO CONNECT POINTS, -1 OTHERWISE.
ALSO ENTER KSYM AND SYMSIZ    (2I,E)
>-2,1,0.01

          -2            1  0.10000E-01

(X(I),Y(I),I=1,NPTS) NOW BEING READ FROM FILE FOR04.DAT.      (2E)


CURVE NO.  2.   ENTER NPTS    (I)
>50
```

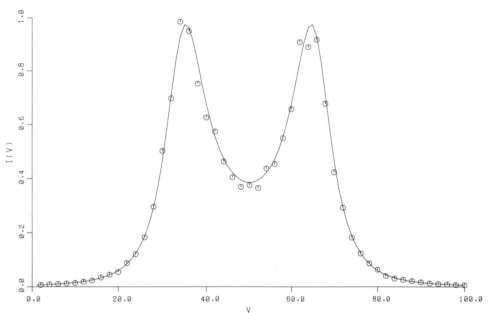

Figure 9.1   Sample execution of CPLOT.

```
              50

ENTER -2 TO CONNECT POINTS, -1 OTHERWISE.
ALSO ENTER KSYM AND SYMSIZ    (2I,E)
>-1,1,0.25

              -1            1  0.25000

(X(I),Y(I),I=1,NPTS) NOW BEING READ FROM FILE FOR04.DAT.    (2E)
```

The output from the Calcomp plotter is shown in Figure 9.1.

Listing

```
C                              CPLOT
C
C     PROGRAM TO PLOT ANY FUNCTION OR ANY ARRAY OF FUNCTIONS.  THE
C     PROGRAM PLOTS UP TO 99 CURVES ON ONE GRAPH OR ANY NUMBER OF
C     SEPARATE GRAPHS.
C
C
C     GLOSSARY:
C
C        XSIZE,YSIZE    - TOTAL SIZE OF THE PLOT
C        NSETS          - NUMBER OF SEPARATE CURVES TO APPEAR ON GRAPH
C        KSYM           - CALCOMP PLOTTER SYMBOL TO BE USED ON THE
C                         CURRENT CURVE.
```

```
C          SYMSIZ          - SIZE IN CENTIMETERS OF SYMBOL TO BE PLOTTED
C          XNUNIT,YNUNIT - NUMBER OF UNIT DISTANCES ON THE RESPECTIVE
C                            AXES (MAY CONTAIN A FRACTIONAL PART)
C          XUNIT,YUNIT   - LENGTH OF ON UNIT ON X OR Y AXIS
C          XINC,YINC     - NUMERICAL INCREMENT BETWEEN TWO UNIT MARKS ON
C                            X OR Y AXIS
C          XNO1,YNO1     - FIRST NUMBER USED IN NUMBERING X OR Y AXIS
C          XORIG,YORIG   - CALCULATED POSITION OF THE NEW ABSOLUTE ORIGIN
C                            OF THE GRAPH
C          SIZEX,SIZEY   - THE LENGTH OF THE X AND Y AXES ON THE GRAPH
C          X,Y           - CO-ORDINATES OF SYMBOLS TO BE PLOTTED; READ
C                            FROM FOR04.DAT (2E)
C          ITITLE,IXTITL,- NUMBER OF LETTERS IN PLOT TITLE, ABSCISSA
C             IYTITL        LABEL AND ORDINATE LABEL, RESPECTIVELY
C          TTLSIZ,XTLSIZ,- SIZE OF LETTERS IN EACH OF THE THREE LABELS
C             YTLSIZ
C          TITLE,XTITLE, - VECTOR CONTAINING THE THREE LABELS
C             YTITLE
C          TX,TY         - USED THREE TIMES; THE STARTING LOCATION OF THE
C                            THREE LABELS
C          IGRAFX,IGRAFY - NUMBER OF UNIT MARKS ALONG EACH AXIS
C          GRAFX,GRAFY   - CURRENT POSITION ALONG BORDER TO PLACE A UNIT
C                            MARK
C          XNUM,YNUM     - CURRENT VALUE USED TO NUMBER A UNIT MARK
C          IAXES         - FLAG: IAXES=1, DRAW AXES THROUGH (0,0)
C                                IAXES=0, NO AXES DRAWN
C          XFACTR,YFACTR - SCALING FACTORS TO ADJUST DATA TO ACTUAL SIZE
C                            OF GRAPH
C          IPOS          - FLAG TO SEE IF SYMBOLS TO BE CONNECTED
C
C
C          AUTHORS:  D.E. HAWKINS, JR. AND K.J. JOHNSON
C
       INTEGER TITLE(10),XTITLE(10),YTITLE(10)
       WRITE(6,11)
       READ(5,22)L
       WRITE(6,66)L
       IF(L.EQ.1)WRITE(6,33)
C
C      INPUT DATA TO DEFINE PAGE SIZE, NO. OF CURVES, SIZE OF DIGITS,
C      AND NUMBERING SEQUENCES
C
       WRITE(6,44)
       READ(5,22)XSIZE,YSIZE
       IF(XSIZE.LE.0)STOP
       WRITE(6,66)XSIZE,YSIZE
       WRITE(6,77)
       READ(5,22)NSETS
       WRITE(6,66)NSETS
       WRITE(6,99)
       READ(5,22)XNUNIT,YNUNIT
       WRITE(6,66)XNUNIT,YNUNIT
       WRITE(6,111)
       READ(5,22)XNOSIZ,XNO1,XINC
       WRITE(6,66)XNOSIZ,XNO1,XINC
       WRITE(6,122)
       READ(5,22)YNOSIZ,YNO1,YINC
       WRITE(6,66)YNOSIZ,YNO1,YINC
       PAUSE
C
```

```
C      CALL GRAPH TO DEFINE SIZE OF CALCOMP PAGE
C
       CALL GRAPH(XSIZE+5.,YSIZE+5.)
       WRITE(6,133)
C
C      INPUT LENGTH OF PLOT TITLE AND SIZE OF LETTERS; IF 0 LETTERS, NO
C      TITLE
C
       READ(5,22)ITITLE,TTLSIZ
       WRITE(6,66)ITITLE,TTLSIZ
       IF(ITITLE.EQ.0)GO TO 4
       J=(ITITLE-1)/5+1
       WRITE(6,144)
C
C      INPUT TITLE OF PLOT
C
       READ(5,155)(TITLE(I),I=1,J)
       WRITE(6,156)(TITLE(I),I=1,J)
     4 WRITE(6,166)
C
C      GET THE SAME INFORMATION FOR THE ABSCISSA AND ORDINATE LABELS
C
       READ(5,22)IXTITL,XTLSIZ
       WRITE(6,66)IXTITL,XTLSIZ
       IF(IXTITL.EQ.0)GO TO 6
       J=(IXTITL-1)/5+1
       WRITE(6,177)
       READ(5,155)(XTITLE(I),I=1,J)
       WRITE(6,156)(XTITLE(I),I=1,J)
       WRITE(6,188)
       READ(5,22)IYTITL,YTLSIZ
       WRITE(6,66)IYTITL,YTLSIZ
       J=(IYTITL-1)/5+1
       WRITE(6,199)
       READ(5,155)(YTITLE(I),I=1,J)
       WRITE(6,156)(YTITLE(I),I=1,J)
       PAUSE
C
C      CALCULATE THE ABSOLUTE ORIGIN OF THE GRAPH
C
     6 XORIG=1.27+2*(YTLSIZ+YNOSIZ)
       YORIG=1.27+2*(XTLSIZ+XNOSIZ)
       CALL ORIGIN(XORIG+2.5,YORIG+2.5)
C
C      CALCULATE LENGTH OF AXES OF GRAPH
C
       SIZEX=XSIZE-XORIG-1.25
       SIZEY=YSIZE-YORIG-1.5-2.5*TTLSIZ
       IF(IXTITL.EQ.0)GO TO 8
C
C      CALCULATE THE POSITION OF THE ABSCISSA LABEL AND WRITE LABEL
C
       TX=(SIZEX-IXTITL*XTLSIZ)/2
       TY=1.016-YORIG
       CALL SYMBOL(TX,TY,XTLSIZ,XTITLE,0.,IXTITL)
C
C      SET UP INFORMATION NEEDED TO DRAW ABSCISSA OF GRAPH
C
     8 IGRAFX=XNUNIT+1
```

```
          GRAFX=0
          XUNIT=SIZEX/XNUNIT
          XNUM=XNO1
          Y=-0.25-2*XNOSIZ
          CALL PENUP
C
C     DRAW ABSCISSA, TIC MARKS, AND NUMBER THEM
C
          DO 10 I=1,IGRAFX
          CALL PLOT(GRAFX,0.,1)
          CALL NUMBER(GRAFX-XNOSIZ,Y,XNOSIZ,XNUM,0.,1)
          CALL PLOT(GRAFX,-.25,3)
          CALL PLOT(GRAFX,0.,2)
          XNUM=XNUM+XINC
       10 GRAFX=GRAFX+XUNIT
C
C     SET UP INFORMATION NEEDED TO DRAW ORDINATE
C
          IGRAFY=YNUNIT+1
          YUNIT=SIZEY/YNUNIT
          GRAFY=YUNIT*IGRAFY
          X=-0.25-2*YNOSIZ
          YNUM=YNO1+IGRAFY*YINC
          CALL PENUP
C
C     DRAW AND NUMBER ORDINATE
C
          DO 20 I=1,IGRAFY
          GRAFY=GRAFY-YUNIT
          YNUM=YNUM-YINC
          CALL PLOT(0.,GRAFY,1)
          CALL NUMBER(X,GRAFY-YNOSIZ,YNOSIZ,YNUM,90.,1)
          CALL PLOT(-0.25,GRAFY,3)
       20 CALL PLOT(0.,GRAFY,2)
C
C     CALCULATE POSITION OF ORDINATE LABEL AND DRAW
C
          IF(IYTITL.EQ.0)GO TO 32
          TX=1.016-XORIG
          TY=(SIZEY-YTLSIZ*IYTITL)/2
          CALL SYMBOL(TX,TY,YTLSIZ,YTITLE,90.,IYTITL)
C
C     CALCULATE POSITION OF PLOT TITLE AND DRAW
C
       32 IF(ITITLE.EQ.0)GO TO 34
          TX=(SIZEX-ITITLE*TTLSIZ)/2
          TY=SIZEY+2.*TTLSIZ
          CALL SYMBOL(TX,TY,TTLSIZ,TITLE,0.,ITITLE)
C
C     SET UP SCALING FACTORS TO FIT DATA TO ACTUAL SIZE OF GRAPH
C
       34 XFACTR=XUNIT/XINC
          YFACTR=YUNIT/YINC
          XNO1=XNO1*XFACTR
          YNO1=YNO1*YFACTR
C
C     ARE LOGICAL AXES DESIRED?
C
          WRITE(6,211)
          READ(5,22)IAXES
```

```
      WRITE(6,66)IAXES
      IF(IAXES.EQ.0)GO TO 38
      IF(XNO1.GE.0.OR.(SIZEX+XNO1)/XFACTR.LE.0)GO TO 36
C
C     DRAW ORDINATE FROM TOP DOWN.(THE LAST ENTRY WAS PROBABLY THE
C     TITLE OR THE ORDINATE LABEL)
C
C
      CALL PLOT(-XNO1,SIZEY,3)
      CALL PLOT(-XNO1,0.,2)
   36 IF(YNO1.GE.0.OR.(SIZEY+YNO1)/YFACTR.LE.0)GO TO 38
C
C     DRAW ABSCISSA FROM RIGHT TO LEFT.(THE NEXT ENTRY, THE CURVE(S),
C     PROBABLY START AT THE LEFT)
C
      CALL PLOT(SIZEX,-YNO1,3)
      CALL PLOT(0.,-YNO1,2)
C
C     OUTER LOOP CONTROLS NO.OF CURVES TO APPEAR ON PLOT
C
   38 DO 50 M=1,NSETS
      WRITE(6,222)M
C
C     READ NO.OF POINTS IN CURVE
C
      READ(5,22)NPTS
      WRITE(6,66)NPTS
      WRITE(6,233)
C
C   ARE SYMBOLS TO BE CONNECTED?  READ SYMBOL, AND SYMBOL SIZE.
C
C
      READ(5,22)IPOS,KSYM,SYMSIZ
      IF(SYMSIZ.EQ.0)SYMSIZ=SYMOLD
      SYMOLD=SYMSIZ
      WRITE(6,66)IPOS,KSYM,SYMSIZ
      PAUSE
      WRITE(6,244)
      READ(4,22)X,Y
      CALL SYMBOL(X*XFACTR-XNO1,Y*YFACTR-YNO1,SYMSIZ,KSYM,0.,-1)
C
C     INNER LOOP READS COORDINATES FROM UNIT 4, AND TRANSFORMS
C     THE COORDINATE SYSTEM TO THAT OF GRAPH
C
      IF(NPTS.LE.1)GO TO 50
      DO 40 I=2,NPTS
      READ(4,22)X,Y
   40 CALL SYMBOL(X*XFACTR-XNO1,Y*YFACTR-YNO1,SYMSIZ,KSYM,0.,IPOS)
   50 CONTINUE
C
C     CLOSE THE PLOTTER DISK FILE AND TERMINATE
C
      CALL ENDPAG
      STOP
   11 FORMAT(/' ENTER A 1 TO GET A DESCRIPTION OF INPUT. OTHERWISE 0')
   22 FORMAT(3G)
   33 FORMAT(/' INPUT SPECIFICATIONS:'/
     1/' XSIZE,YSIZE    - LENGTH OF X AND Y AXES IN CENTIMETERS'
     2/' NSETS          - NO. OF SEPARATE CURVES ON THE SAME PLOT'
     3/' SYMSIZ         - SIZE IN CENTIMETERS OF PLOTTING SYMBOL'
```

```
     4/' XNUNIT,YNUNIT - NO. OF UNIT LENGTHS ON AXIS'
     5/' XNOSIZ,YNOSIZ - SIZE OF DIGITS USED TO NUMBER X AND Y AXES'
     6/' XNO1,YNO1    - X AND Y VALUES AT THE PLOT ORIGIN'
     7/' XINC,YINC    - NUMERICAL INCREMENT OF ONE UNIT ON AXIS'
     8/' ITITLE,IXTITL, - NO. OF LETTERS RESPECTIVELY IN THE 3 LABELS'
     9/'    IYTITL'
     1/' TTLSIZ,XTLSIZ,- SIZE OF LETTERS IN 3 LABELS'
     2/'    YTLSIZ'
     3/' TITLE,XTITLE, - VECTOR CONTAINING THE LABEL, UP TO 50 CHARS.'
     4/'    YTITLE'
     5/'  IAXES        -  1: DRAWS AXES THROUGH (0,0); 0: NO AXES'
     6/'  NPTS         - NO. OF DATAPOINTS (X,Y) IN CURVE'
     7/'  IPOS         - PLOT OPTION:-2 CONNECTS SYMBOLS WITH LINE'
     8/'                       -1 DOES NOT CONNECT POINTS'
     9/',X,Y          - CO-ORDINATES OF POINTS; READ FROM '
     1'FOR04.DAT'/)
  44 FORMAT(/' ENTER XSIZE AND YSIZE.  XSIZE=0 ENDS.  (2E).')
  66 FORMAT(3G13.5)
  77 FORMAT(/' ENTER NSETS.  (I)')
  99 FORMAT(/' ENTER XNUNIT AND YNUNIT.  (2E)')
 111 FORMAT(/' ENTER XNOSIZ,XNO1,AND XINC.   (3E)')
 122 FORMAT(/' ENTER YNOSIZ,YNO1, AND YINC.   (3E)')
 133 FORMAT(/' ENTER ITITLE AND TTLSIZ FOR PLOT TITLE    (I,E)'/
     1'  (IF 0 LETTERS, NO TITLE WILL BE WRITTEN.)')
 144 FORMAT(/' ENTER TITLE OF PLOT  (10A5)')
 155 FORMAT(10A5)
 156 FORMAT(5X,10A5)
 166 FORMAT(/' ENTER IXTITL AND XTLSIZ FOR ABSCISSA LABEL  (I,E)'/
     1'  (IF 0 LETTERS, X AND Y WILL NOT BE LABELLED.)')
 177 FORMAT (/' ENTER ABSCISSA LABEL.   (10A5)')
 188 FORMAT(/' ENTER IYTITL AND YTLSIZ FOR ORDINATE LABEL.  (I,E)')
 199 FORMAT(/' ENTER ORDINATE LABEL.   (10A5)')
 211 FORMAT(/' ENTER IAXES. 1 DRAWS AXES, 0 DOES NOT.  (I)')
 222 FORMAT(/' CURVE NO. ',I2,'.  ENTER NPTS   (I)')
 233 FORMAT(/' ENTER -2 TO CONNECT POINTS, -1 OTHERWISE.',
     1/'  ALSO ENTER SYM AND SYMSIZ   (2I,E)')
 244 FORMAT(/'  (X(I),Y(I),I=1,NPTS) NOW BEING READ FROM FILE FOR04.',
     1'DAT.   (2E)')
     END
```

Discussion

CPLOT requires approximately 1500 words of core. This sample execution took
approximately 0.3 sec. This includes the time required to write the disk file
that is subsequently read by the Calcomp software.

9.2 COMPUTER-ASSISTED INSTRUCTION (CAI)

A number of books and general articles on the applications of computers to chemistry
in general and chemical education in particular have been published [19-141]. This
set of references is ordered as follows: general references [19-41]; interactive
CAI [42-70]; course augmentation by computers [52, 71-122]; and computer-simulated
experiments [123-141]. For the purposes of this discussion, CAI means the application of computing technology to simulate and augment conventional modes of instruction. Two types of CAI will be mentioned:

   1.   Interactive, skill-acquisition CAI

   2.   Simulation and data reduction programs, including simulated experiments.

9.2.1   Interactive CAI

CAI usually connotes the attempt to simulate human tutors using computer programs
that are interactive, highly branched, and emphasize the acquisition of certain
chemical skills and concepts by repetitive tutorial drill. For example, an
instructor might write a computer program to drill students on solubility concepts
and calculations. Suppose the following question is desired:

> What is the solubility in moles per liter of PbI_2 at $25^{\circ}C$ in a 0.25 $\underline{M}$
> NaI solution? The solubility product for PbI_2 at 25° is 8.3×10^{-9}.
> (Ignore complexation and hydrolysis equilibria.)

The program would contain lists of slightly soluble salts, the corresponding
solubility products, and soluble salts containing common ions. The program would
generate numbers using random number generators over the appropriate ranges. For
example, if the desired range of concentrations of the common ion is

$$0.10 \leq \text{concentration} \leq 0.95$$

if only two significant figures were desired, and the concentration is to be a
multiple of 0.05, then the following statement would be appropriate:

   CONC = 0.05*(2. + AINT(18.*RAN(X)))

The program could branch to sets of statements contingent upon the following student
responses:

   1.   Print a positively reinforcing message in response to the correct answer.

   2.   Print a tutorial message in response to "HELP."

   3.   Branch to a harder problem in response to "SKIP."

   4.   Print appropriate responses to the following anticipated incorrect
        responses:
        a. $K_{sp}/CONC$
        b. $(K_{sp}/4)^{1/3}$
        c. $K_{sp}/2CONC$

   5.   Print a detailed solution if the student fails to respond correctly in
        three attempts.

   It soon becomes evident that writing highly branched interactive CAI programs
is an extremely labor-intensive undertaking. The task is considerably simplified
if a CAI author language is available, for example, TUTOR, CATALYST/PIL, or COURSE-
WRITER. However, such general purpose languages as Fortran, BASIC, and APL have
also been used.

Table 9.1 contains a description of the 46 CAI lessons written by the author and his collaborators in the CATALYST/PIL languages (AQEQ.1-7 means there are seven lessons on aqueous equilibrium, AQEQ.1, AQEQ.2, etc.).

Table 9.1 Library of CAI Chemistry Lessons at the University of Pittsburgh (Fall, 1977)

Lesson	Description
AQEQ.1-7	Aqueous equilibrium concepts and problems including properties of acids and bases, weak and strong electrolytes, dissociation equilibria, hydrolysis equilibria pH, buffers, polyprotic acid equilibria solubility calculations, and complex ion formation
BEER.1	Beer's law calculations
BOHR.1	Bohr atom concepts and calculations
BONDING.1-2	Periodic trends, quantum numbers, and magnetic properties
CONC.1-3	Concentration units (molarity and molality), neutralization reactions, and solution stoichiometry problems
COLLIG.1	Freezing point depression, boiling point elevation, and molecular weight determination
ELECT.1-3	Electrochemistry problems including mole-faraday, amp-coulomb, and current-mass conversions; electrolysis calculations; equivalent-electrical work conversions; cell potentials; and equilibrium constant calculations
EXPERT.1	Simulation of a stoichiometry experiment to determine the empirical formula of a compound of copper and sulfur
EXPERT.2	Simulation of an experiment to determine Avogadro's number
GAS.1-3	Gas law calculations (Boyle, Charles, Gay-Lussac, Graham, Dalton, and the combined gas law)
GASEQ.1	Gas-phase equilibrium problems
KINETI.1-2	Determination of rate law and activation energy; simulation of the kinetic system $$A \underset{k_2}{\overset{k_1}{\rightleftharpoons}} B$$
LECHAT.1	Drill of LeChatelier's principle
NAMES.1	Drill of nomenclature of inorganic ions and acids
NUKES.1	Energy changes in nuclear reactions
PH.1	Drill of pH - H^+ and pOH - OH^- calculations
REDOX.1	Balancing oxidation-reduction half-reactions in acidic and basic solutions

Table 9.1 (continued)

Lesson	Description
STOICH.1-6	Molecular weight determinations, gram-mole and molecule-mole conversions, empirical formulas, percent composition, atomic weight of an element from its isotopic composition, balancing equations, and weight-weight stoichiometry
THERMO.1-2	Enthalpy, entropy, and free energy changes in chemical reactions
TITRAT.1-3	Simulation of three acid-base titrations: (1) strong acid-strong base, (2) weak acid-strong base, and (3) weak base-strong acid
UNITS.1-3	Unit manipulations, temperature conversions, density, weight percentage, and specific heat problems

9.2.2 Simulation and Data Reduction Programs

This book contains a number of programs that have been used to enrich the undergraduate chemistry curriculum at the University of Pittsburgh and elsewhere. The simulation programs allow students to solve chemistry problems they might not otherwise encounter because of the mathematical complexity. The data reduction programs, for example, TITR, LINEAR, and NONLIN, allow students to more thoroughly reduce experimental data then would be feasible without computer assistance.

One use of computers not emphasized in this book is the simulation of experiments. Consider,for example, the determination of the percent by weight of acetic acid in a sample of commercial vinegar. A computer program could be written which generates the following "experimental data":

.EX ACETIC

This program simulates the determination of the percentage of percentage of acetic acid in a sample of vinegar.

Enter your identification number.
123456789

I.   Standardization of NaOH Solution

Three samples of KHP were weighed and dissolved using 50.00 ml of H_2O. The following titration data were obtained.

Titration	Weight(g)	Volume NaOH (ml)
1	1.7757	23.18
2	2.1866	28.55
3	2.4228	31.63

II.   Titration of Vinegar

Three 10.00-ml aliquots of vinegar were diluted to 100.00 ml
with H_2O. The following titration data were obtained.

Titration	Volume NaOH (ml)
1	24.87
2	24.54
3	24.96

Now 123456789, calculate the average value of the
percentage by weight of acetic acid in the vinegar.

Also, calculate the standard deviation of this average.

The program ACEANS will check your results.

There are four variables to be specified in this simulated experiment; the
percentage acetic acid in the vinegar sample; the concentration of the titrant
(NaOH); the weights of the primary standard, potassium acid phthalate (KHP); and
the "experimental error" that is superimposed on the calculated volume of NaOH.
The program ACEANS prompts the user for his results and accepts the values if the
relative error in the weight percentage and in the standard deviation of the
average is within a specified tolerance. A program of this sort could be used as
an homework assignment or, if this experiment is performed in the course, as a
pre-lab assignment. However, a more exciting application of such programs is to
make available to students, through the computer, chemical concepts that they
would not otherwise encounter in the laboratory because of expense, safety, or
time constraints.
    Table 9.2 contains descriptions of computer-simulated experiments developed
by the author and his collaborators.

Table 9.2 Library of Simulated Experiments for General Chemistry (Fall, 1977)

Program	Description
AB	Simulation of the reversible kinetic system

$$A \underset{k_2}{\overset{k_1}{\rightleftharpoons}} B$$

Program	Description
	The student calculates E_{a1} and the enthalpy change for the reaction.
ACETIC	Simulates the determination of the weight percentage acetic acid in vinegar by NaOH titration. The student calculates the average value and the standard deviation.
CHROM	Simulates the chromatographic determination of an unknown ternary mixture.

Table 9.2 (continued)

Program	Description
EMF	Simulates the determination of the standard enthalpy and entropy changes for the solubility reaction, $$MX \rightleftharpoons M^+ + X^-$$ by measuring the potential of the indicator electrode, $M^+|M$ as a function of temperature.
H3A	Simulates the determination of pK_1, pK_2, and pK_3 from titration data.
JOB	Simulates Job's method for determining the stoichiometry and formation constant of a coordination complex.
KA	Simulates the spectrophotometric determination of the dissociation constant of an indicator.
KC	Simulates the equilibrium system $$A + B \rightleftharpoons C$$ The student determines the equilibrium constant from spectrophotometric data.
KEQ	Simulates the equilibrium system $$aA + bB \rightleftharpoons cC + dD$$ The student determines the stoichiometry and the equilibrium constant.
KF	Simulates the determination of the overall formation constant of the amine complex $$M^{2+} + iNH_3 \rightleftharpoons M(NH_3)_i^{2+}$$ by potentiometric titration of $M(NO_3)_2$ and NH_4NO_3 using NaOH.
KINETI	Calculates concentration-time profiles for the reaction $$A + B + C \longrightarrow Products$$ The student determines the rate law and the activation energy of the reaction.
MIXTUR	Simulates the spectrophotometric determination of an unknown mixture containing three species in solution.
NEUTRO	Simulates the analysis of an unknown binary mixture by neutron activation analysis.
PHOS	Simulates the analysis of a mixture containing two of the four species H_3PO_4, NaH_2PO_4, Na_2HPO_4 and Na_3PO_4 by acid-base titration.
POTENT	Simulates the potentiometric titration of an ore sample containing from using MnO_4^-.
QUAL	Simulates a subset of the inorganic qualitative analysis scheme. The student identifies up to 4 of 13 cations from groups 1, 2, and 3.
SALTS	Simulates the analysis of an unknown mixture consisting of NaCl, NaBr, NaI, and $NaNO_3$ by potentiometric titration with standard $AgNO_3$.

Table 9.2 (continued)

Program	Description
SECOND	Simulates the determination of the second-order rate constant for the reaction $$A \ + \ B \ \longrightarrow Products$$ The absorption spectra of A and B are calculated and plotted as a function of time.
STEEL	Simulates the spectrophotometric determination of Mn in a steel sample.
THERMO	Simulates the thermodynamic study of the dissociation reaction $$HA \rightleftharpoons H^+ \ + \ A^- \qquad K_a$$ The pH of the solution as a function of temperature is tabulated. The student calculates the enthalpy and entropy changes for the reaction.

These programs provide access by students to chemical concepts and calculations that can enrich both the lecture and laboratory components of the general chemistry course.

9.3 COMPUTER-ASSISTED TEST CONSTRUCTION

Computer programs have been written to assist instructors in the generation of examination questions [52, 142-159]. The usual approach is to prepare, in machine-readable form, several thousand thoroughly tested questions. This item bank is categorized by topic, subtopic, difficulty level, item type, etc., and information retrieval techniques are used to search the data base and retrieve the desired number of questions. The obvious difficulty with this approach is the laborious undertaking of first collecting or writing and then preparing in machine-readable form several thousand questions. An alternative approach is to write computer programs to generate the questions. For example, an Fortran subroutine could be written to print the following item format:

> What volume of concentrated _____ solution must be diluted to prepare _____ liters of a solution that is _____ molar in _____? The concentrated _____ solution is _____ percent by weight and has density _____ g/ml.
>> a.
>> b.
>> c.
>> d.
>> e.

The subroutine would select a compound from a list, provide the corresponding weight percentage and solution density, calculate the volume and concentration of the desired final solution using random numbers distributed over the appropriate ranges, calculate the correct answer and four distractors, and print the question. For example,

> What volume of a concentrated NaOH solution must be diluted to prepare 3.50 liters of a solution that is 0.100 molar in NaOH? The concentrated NaOH solution is 50.0 percent by weight and has density 1.525 g/ml.

>   a. 9.18 ml
>   b. 18.4 ml
>   c. 21.4 ml
>   d. 42.7 ml
>   e. 50.0 ml

The author and his collaborators have written approximately 350 Fortran subroutines designed to generate multiple-choice general chemistry examination questions. Most of the subroutines use this format, i.e., they print five numeric options. Two other formats are also used. In the following example the answers are given in set-up format.

> A hypothetical element has the following isotopic composition:

Isotope	Mass	Percent Abundance
76	75.9217	27.6
78	77.9212	72.4

> What is the atomic weight of the element?
>   a. $27.6 \times 75.9217 + 72.4 \times 77.9212$
>   b. $(75.9217 + 77.9212)/2$
>   c. $0.276 \times 75.9217 + 0.724 \times 77.9212$
>   d. $0.276 \times 77.9212 + 0.724 \times 75.9217$
>   e. None of the above.

The subroutine that generates this question is coded to generate either three or four distractors so that "None of the above" is the correct answer 20% of the time. The third format is for nonnumeric question, for example,

> Consider the following half-reactions (reductions) in 1.0 $\underline{M}$ H^+:
>
> $Pb^{2+} \rightleftharpoons Pb$ $U^{4+} \rightleftharpoons U^{3+}$
>
> $Mn^{3+} \rightleftharpoons Mn^{2+}$ $H_3PO_3 \rightleftharpoons H_3PO_2$
>
> $Ag^+ \rightleftharpoons Ag$ $HNO_2 \rightleftharpoons N_2O$
>
> Given the following observations:
>   1. HNO_2 is a stronger oxidizing agent than H_3PO_3.
>   2. U^{3+} reduces both HNO_2 and H_3PO_3.
>   3. Neigher HNO_2 nor H_3PO_3 oxidizes Mn^{2+} to Mn^{3+}.
>   4. Ag^+ oxidizes H_3PO_2 but does not oxidize N_2O.
>   5. Ag^+ but not H_3PO_3 oxidizes Pb to Pb^{2+}.

Which of the following sequences is the correct order of decreasing strength of the metallic species as reducing agents?

a. $Mn^{2+} > Ag > Pb > U^{3+}$

b. $U^{3+} > Pb > Ag > Mn^{2+}$

c. $U^{3+} > Ag > Pb > Mn^{2+}$

d. $Mn^{2+} > Ag > U^{3+} > Pb$

e. None of the above.

The 350 items span 29 topics usually covered in general chemistry courses. The items are further categorized by difficulty as "easy," "of moderate difficulty," and "hard" questions. The coverage is indicated in Table 9.3.

A book containing sample executions of these subroutines is available [142]. A more detailed description of this and other CATC systems can be found in the literature [52, 142-159].

Table 9.3 University of Pittsburgh CATC System Coverage (Fall, 1977)

Topic	Easy	Moderate	Hard
1. Units and conversion factors	10	0	0
2. Atoms and molecules	14	10	4
3. Chemical reactions	2	4	5
4. Redox reactions	3	4	0
5. Faraday's laws	0	1	3
6. Colligative properties	0	8	0
7. Heats of reaction	4	5	1
8. Solution stoichiometry	4	8	6
9. Mixed stoichiometry	0	0	11
10. Laboratory stoichiometry	0	4	0
11. General gas laws	6	10	1
12. Dalton's and Graham's laws	2	4	4
13. Kinetic and molecular theory	0	0	6
14. Nomenclature	4	0	0
15. Solids	2	3	4
16. Atomic structure	1	5	5
17. Bonding	0	18	5
18. Strong acids and bases	12	8	1
19. Weak acids and bases	4	14	8
20. Buffers	2	4	2
21. Solubility	2	6	6
22. Thermodynamics	0	6	14
23. Gas-phase equilibria	2	3	2
24. Electrochemistry	0	5	7
25. Kinetics	4	1	4
26. Nuclear chemistry	3	9	0
27. Coordination chemistry	0	10	4
28. Organic chemistry	4	0	0
29. Descriptive chemistry and miscellaneous	1	0	1

9.4 COMPUTER-MANAGED INSTRUCTION (CMI)

Computer programs have been used to assist instructors with the clerical tasks associated with large enrollment courses [160-166]. A brief description of some of the CMI programs developed and by the author and his collaborators is given in Table 9.4.

The availability of such programs makes feasible such innovative teaching methods as self-paced courses using repetitive testing. They can also be used, as the following program demonstrates, to prepare current class standing for diagnostic and motivational purposes.

9.4.1 GRADER

Introduction

GRADER prepares a comprehensive record of student scores to assist an instructor keep class records, monitor student performance, and assign letter grades. The class record includes the students' names, identification number, and component scores.

Method

The following variables constitute the input to GRADER:

Input variable (format)	Description
SETNME (A10)	Class code (arbitrary)
NSTUD (I)	Number of students
NSCORS (I)	Number of component scores
NUNIT (I)	Input unit (disk file containing the student records)
(W(I),I=1,NSCORS) (7F)	Weighting factors
STUD(250,11)	Student records
STUD(I,J),J=1,4 (4A5)	Student's name
STUD(I,J),J=5,11 (7F)	Component scores

Table 9.4 University of Pittsburgh CMI Library (Fall, 1977)

Program	Description
CAIDMS	A set of programs to analyze information written to the disk file CAI.USE by students using the library of CAI programs (Section 9.2.1)
PROMPT	Program to prompt an instructor for the matrix of scores to be assigned in partial-credit scoring of multiple-choice exams
SCORER	Program to apply the partial-credit matrix to the student responses, compute scores, an item analysis, the average, and sort lists of scores
MASTER	Program to facilitate updating the master list of student scores
GRADER	Program to prepare a comprehensive record of student scores to monitor student performance and help assign letter grades (see Section 9.4.1)

The matrix STUD(NSTUD,NSCORS) is read from the disk file with unit designation NUNIT. A glossary of the remaining key variables is given next.

Variable	Description
TOTAL(250)	Total weighted score
AVG(J) J=1,NSCORS	Averages of component scores
AVG(NSCORS+1)	Total score average
ANO(J), J=1,NSCORS	Number of nonzero component scores
ANO(NSCORS+1)	Number of nonzero total scores
LIST(250)	Index to original entries in TOTAL
IP(50)	Histogram frequencies

The total weighted score is computed using the formula

$$TOTAL(I) = \sum_{J=1}^{NSCORS} W(J)*STUDN(I,J+4)$$

where I = 1, 2, 3, ..., NSTUD.

If any component score is zero, that entry in TOTAL is set to zero, and the number of students is decremented by one so that the averages are computed ignoring zero scores. After the set of total scores has been computed, the program prints a table containing each student's name, identification number, total weighted score, and the component scores. The median, average scores, and the standard deviations of the average scores are also printed.

Next the subroutine SORT is called. SORT is Algorithm 347 published in "Communications of the Association of Computing Machinists" [5]. SORT is an inverted binary sort routine with calling sequence

CALL SORT(A,JJ,B)

where A is the vector of real numbers to be sorted in descending order, JJ is the number of components of A and B, and B is the index to the original vector A. For example, if originally A(5) = 10.5 and A(15) = 22.7, and these were the only occurrences of 10.5 and 22.7 in A, and if after sorting, A(2) = 22.7 and A(17) = 10.5, then B(2) = 15 and B(17) = 5.

The input file for the sample execution which follows is listed below.

```
STUDENT   A          138963732      83  76  59  46  51
STUDENT   B          112344616      82  78   0  80  63
STUDENT   C          175838622      72  93  64  87  66
STUDENT   D          103942515      48  80  63  57  50
STUDENT   E          173125488      48  81  58  67  62
STUDENT   F          291144788      47  65  56  42  68
STUDENT   G          172323100      66  73  84  42  46
STUDENT   H          214851407      92  92  64  79  58
```

STUDENT	I	157669900	61	48	54	99	46
STUDENT	J	174724855	89	64	53	75	89
STUDENT	K	102675932	73	50	47	41	74
STUDENT	L	213206623	42	53	84	73	79
STUDENT	M	204493233	54	79	90	57	40
STUDENT	M	260408682	65	51	71	95	43
STUDENT	O	276893654	70	65	95	79	67
STUDENT	P	111329827	99	56	82	48	97
STUDENT	Q	272379866	66	0	94	81	86
STUDENT	R	255675392	41	98	67	81	46
STUDENT	S	257627372	65	91	49	86	55
STUDENT	T	235323796	47	91	91	88	50
STUDENT	U	103875682	40	44	87	77	74
STUDENT	V	227703327	48	58	55	84	94
STUDENT	W	236228520	69	97	47	50	95
STUDENT	X	188339996	51	76	60	67	86
STUDENT	Y	296432602	52	46	99	58	67
STUDENT	Z	115345870	86	50	48	44	63

Note that students B and Q have zero entries.

Sample Execution

ENTER CLASS CODE, NUMBER OF STUDENTS, NUMBER OF SCORES,
THE INPUT UNIT, AND THE WEIGHTING FACTORS (I10,3I,7F)
>SAMPLE   26   5   25   .2 .2 .2 .2 .2

READING NAMES, ID. NUMBERS, AND COMPONENT
SCORES FROM INPUT UNIT 25   (4A5,I,5F)

CLASS CODE: SAMPLE   NO. OF STUDENTS = 26   NO. OF SCORES = 5

THE INPUT UNIT = 25   AND THE WEIGHTING FACTORS ARE:

 0.200   0.200   0.200   0.200   0.200

NAME		ID.	TOTAL SCORE	COMPONENT SCORES				
STUDENT	A	138963732	63.00	83.0	76.0	59.0	46.0	51.0
STUDENT	B	112344616	0.00	82.0	78.0	0.0	80.0	63.0
STUDENT	C	175838622	76.40	72.0	93.0	64.0	87.0	66.0
STUDENT	D	103942515	59.60	48.0	80.0	63.0	57.0	50.0
STUDENT	E	173125488	63.20	48.0	81.0	58.0	67.0	62.0
STUDENT	F	291144788	55.60	47.0	65.0	56.0	42.0	68.0
STUDENT	G	172323100	62.20	66.0	73.0	84.0	42.0	46.0
STUDENT	H	214851407	82.60	82.0	92.0	64.0	79.0	96.0
STUDENT	I	157669900	61.60	61.0	48.0	54.0	99.0	46.0
STUDENT	J	174724855	74.00	89.0	64.0	53.0	75.0	89.0
STUDENT	K	102675932	57.00	73.0	50.0	47.0	41.0	74.0
STUDENT	L	213206623	66.20	42.0	53.0	84.0	73.0	79.0
STUDENT	M	204493233	64.00	54.0	79.0	90.0	57.0	40.0
STUDENT	M	260408682	65.00	65.0	51.0	71.0	95.0	43.0
STUDENT	O	276893654	75.20	70.0	65.0	95.0	79.0	67.0
STUDENT	P	111329827	76.40	99.0	56.0	82.0	48.0	97.0
STUDENT	Q	272379866	0.00	66.0	0.0	94.0	81.0	86.0

```
STUDENT   R     255675392     66.60       41.0  98.0  67.0  81.0  46.0
STUDENT   S     257627372     69.20       65.0  91.0  49.0  86.0  55.0
STUDENT   T     235323796     73.40       47.0  91.0  91.0  88.0  50.0
STUDENT   U     103875682     64.40       40.0  44.0  87.0  77.0  74.0
STUDENT   V     227703327     67.80       48.0  58.0  55.0  84.0  94.0
STUDENT   W     236228520     71.60       69.0  97.0  47.0  50.0  95.0
STUDENT   X     188339996     68.00       51.0  76.0  60.0  67.0  86.0
STUDENT   Y     296432602     64.40       52.0  46.0  99.0  58.0  67.0
STUDENT   Z     115345870     58.20       86.0  50.0  48.0  44.0  63.0
```

MEDIAN: 65.00

AVERAGES: 66.90   63.3 70.2 68.8 68.6 67.4

STANDARD DEVIATIONS: 6.8   16.6 17.6 17.2 18.0 18.3

TOTAL WEIGHTED SCORES IN DESCENDING ORDER

```
 1        STUDENT   H        214851407        82.60
 2        STUDENT   C        175838622        76.40
 3        STUDENT   P        111329827        76.40
 4        STUDENT   O        276893654        75.20
 5        STUDENT   J        174724855        74.00
 6        STUDENT   T        235323796        73.40
 7        STUDENT   W        236228520        71.60
 8        STUDENT   S        257627372        69.20
 9        STUDENT   X        188339996        68.00
10        STUDENT   V        227703327        67.80
11        STUDENT   R        255675392        66.60
12        STUDENT   L        213206623        66.20
13        STUDENT   M        260408682        65.00
14        STUDENT   U        103875682        64.40
15        STUDENT   Y        296432602        64.40
16        STUDENT   M        204493233        64.00
17        STUDENT   E        173125488        63.20
18        STUDENT   A        138963732        63.00
19        STUDENT   G        172323100        62.20
20        STUDENT   I        157669900        61.60
21        STUDENT   D        103942515        59.60
22        STUDENT   Z        115345870        58.20
23        STUDENT   K        102675932        57.00
24        STUDENT   F        291144788        55.60
25        STUDENT   Q        272379866         0.00
26        STUDENT   B        112344616         0.00
```


DISTRIBUTION OF TOTAL WEIGHTED SCORES

```
        >    82.6   0
82.1  TO    82.6   1     *
81.5  TO    82.1   0
81.0  TO    81.5   0
80.4  TO    81.0   0
79.9  TO    80.4   0
79.4  TO    79.9   0
```

```
78.8 TO   79.4   0
78.3 TO   78.8   0
77.7 TO   78.3   0
77.2 TO   77.7   0
76.7 TO   77.2   0
76.1 TO   76.7   2    **
75.6 TO   76.1   0
75.0 TO   75.6   1    *
74.5 TO   75.0   0
74.0 TO   74.5   1    *
73.4 TO   74.0   0
72.9 TO   73.4   1    *
72.3 TO   72.9   0
71.8 TO   72.3   0
71.3 TO   71.8   1    *
70.7 TO   71.3   0
70.2 TO   70.7   0
69.6 TO   70.2   0
69.1 TO   69.6   1    *
68.6 TO   69.1   0
68.0 TO   68.6   0
67.5 TO   68.0   2    **
66.9 TO   67.5   0
66.4 TO   66.9   1    *
65.9 TO   66.4   1    *
65.3 TO   65.9   0
64.8 TO   65.3   1    *
64.2 TO   64.8   2    **
63.7 TO   64.2   1    *
63.2 TO   63.7   1    *
62.6 TO   63.2   1    *
62.1 TO   62.6   1    *
61.5 TO   62.1   1    *
61.0 TO   61.5   0
60.5 TO   61.0   0
59.9 TO   60.5   0
59.4 TO   59.9   1    *
58.8 TO   59.4   0
58.3 TO   58.8   0
57.8 TO   58.3   1    *
57.2 TO   57.8   0
56.7 TO   57.2   1    *
56.1 TO   56.7   0
55.6 TO   56.1   0    *
      <   55.6   0
```

Listing

```
C
C
C                    GRADER
C
C      GRADER PREPARES A COMPREHENSIVE RECORD OF STUDENT SCORES
C
C      GLOSSARY:
C
C          SETNME    - NAME OF SECTION      (A10)
C          NSTUD     - NO. OF STUDENTS      (I)
C          NSCORS    - NO. COMPONENT SCORES (I)
```

```
C                    NUNIT        - INPUT DEVICE (I)
C                    W(I)         - WEIGHTING FACTORS        (7F)
C                      I=1,NSCORS
C                    STUDN(I,J) - STUDENT NAME               (4A5)
C                      J=1,4
C                    STUDN(I,J) - COMPONENT SCORES           (7F)
C                      J=5,11
C                    TOTAL(I)     - TOTAL SCORE
C                    AVG(J)       - AVERAGES OF COMPONENT SCORES
C                      J=1,7
C                     NF          - NUMBER OF SCORES + 1
C                    AVG(NF)      - AVERAGE OF TOTAL WEIGHTED SCORES
C                    ANO(J)       - NUMBER OF NON-ZERO COMPONENT SCORES
C                      J=1,7           (.LE. NSCORS)
C                    ANO(NF) - NUMBER OF NON-ZERO TOTAL SCORES
C                    SCORSQ  -   SQUARES OF SCORES
C                    STDEV   -   STANDARD DEVIATIONS OF AVERAGES
C
C      THE DIMENSIONS OF STUDN,TOTAL AND LIST
C      DEPEND ON THE NUMBER OF STUDENTS.
C
C      THE DIMENSIONS OF W,AVG,AND ANO DEPEND ON THE
C      NUMBER OF COMPONENT SCORES.
C
C       AUTHORS:  D. E. HAWKINS, JR. AND K. J. JOHNSON
C
       DIMENSION STUDN(250,11),W(7),AVG(8),ANO(8),TOTAL(250),
      1 ID(250),SCORSQ(8),STDEV(8),LIST(250),IP(51)
       DOUBLE PRECISION SETNME
       DATA STAR/'*'/
       DATA TOTAL,AVG,SCORSQ/266*0./
       DATA IP/51*0/
C
C      INPUT FROM TERMINAL
C
       WRITE(6,22)
       READ(5,33)SETNME,NSTUD,NSCORS,NUNIT,(W(I),I=1,NSCORS)
       IF(NSCORS.GT.7 .OR. NSTUD.GT.250)STOP
       IF(NSTUD.LE.0)STOP
       WRITE(6,66)NUNIT,NSCORS
       JK=NSCORS+4
       DO 10 M=1,NSTUD
       READ(NUNIT,77)(STUDN(M,JJ),JJ=1,4),ID(M),
      1 (STUDN(M,JJ),JJ=5,JK)
   10 CONTINUE
       WRITE(6,44)SETNME,NSTUD,NSCORS,NUNIT,(W(JJ),JJ=1,NSCORS)
       WRITE(6,45)
C
C   INITIALIZE VARIABLES
C
       NF=NSCORS+1
       DO 2 I=1,NF
    2 ANO(I)=NSTUD
       DO 4 I=1,NSTUD
    4  LIST(I)=I
C
C    COMPUTE WEIGHTED TOTAL SCORE
C
       DO 30  M=1,NSTUD
       IFLAG=0
```

```
      DO 20 I=1,NSCORS
      SCORE=STUDN(M,I+4)
      IF(SCORE.EQ.0.) GO TO 15
      AVG(I)=AVG(I)+SCORE
      SCORSQ(I)=SCORSQ(I)+SCORE**2
      TOTAL(M)=TOTAL(M)+SCORE*W(I)
      GO TO 20
   15 IFLAG=1
      ANO(I)=ANO(I)-1
   20 CONTINUE
      IF(IFLAG)28,28,25
   25 TOTAL(M)=0.
      ANO(NF)=ANO(NF)-1
C
C PRINT ALL INFORMATION, INCLUDING TOTAL WEIGHTED SCORE
C
   28 WRITE(6,99)(STUDN(M,JJ),JJ=1,4),ID(M),TOTAL(M),
     1(STUDN(M,JJ),JJ=5,JK)
      SCORSQ(NF)=SCORSQ(NF)+ TOTAL(M)**2
   30 AVG(NF)=AVG(NF)+TOTAL(M)
C
C CALL SORT TO ORDER ACCORDING TO DESCENDING TOTAL WEIGHTED SCORES
C
      CALL SORT(TOTAL,NSTUD,LIST)
C
C COMPUTE MEDIAN
C
      LMEDN=(NSTUD+1)/2
      XMEDN=TOTAL(LMEDN)
C
C COMPUTE AVERAGES AND STANDARD DEVIATIONS
C
      DO 40 I=1,NF
      AVG(I)=AVG(I)/ANO(I)
   40 STDEV(I)=SQRT((SCORSQ(I)-ANO(I)*AVG(I)**2)/(ANO(I)-1))
C
C PRINT MEDIAN, AVERAGES, AND STANDARD DEVIATIONS
C
      WRITE(6,111)XMEDN,AVG(NF),(AVG(JJ),JJ=1,NSCORS)
      WRITE(6,112)STDEV(NF),(STDEV(JJ),JJ=1,NSCORS)
      WRITE(6,113)
C
C FIND RANGE OF SCORES
C
      HI=TOTAL(1)
      NLO=ANO(NF)
      ALO=TOTAL(NLO)
      WINDOW=(HI-ALO)/50.
C
C   PRINT NAMES AND TOTAL SCORE IN DESCENDING ORDER
C   AND DEVELOP THE HISTOGRAM
C
      N=1
      BOTTOM=HI-WINDOW
      DO 80  M=1,NSTUD
      MM = LIST(M)
      WRITE(6,133)M,(STUDN(MM,JJ),JJ=1,4),ID(MM),TOTAL(M)
      IF(TOTAL(M).LE.0.)GO TO 80
   75 IF(TOTAL(M).GE.BOTTOM)GO TO 78
      BOTTOM=BOTTOM-WINDOW
```

```
      N=N+1
      GO TO 75
   78 IP(N)=IP(N)+1
   80 CONTINUE
C
C     PRINT THE HISTOGRAM   (NOTE - KM .LE. 100)
C
      WRITE(6,144)
      BOTTOM=HI
      KM=0
      WRITE(6,146)HI,KM
      DO 90 KL=1,51
      HI=BOTTOM
      BOTTOM=HI-WINDOW
      KM=IP(KL)
      IF(KM.GT.0)GO TO 85
      WRITE(6,155)BOTTOM,HI,KM
      GO TO 90
   85 IF(KM.GT.100)KM=100
      WRITE(6,166)BOTTOM,HI,IP(KL),(STAR,IK=1,KM)
   90 CONTINUE
      KM=0
      WRITE(6,148)BOTTOM,KM
      STOP
   22 FORMAT(//'  ENTER CLASS CODE, NUMBER OF STUDENTS, NUMBER OF',
     1 ' SCORES',',/'  THE INPUT UNIT, AND THE WEIGHTING FACTORS',
     2'  (I10,3I,7F)')
   33 FORMAT(A10,3I,7F)
   44 FORMAT('  CLASS CODE:',2X,A10,3X,'NO. OF STUDENTS =',I4,4X,
     1 'NO. OF SCORES =',I3//,'  THE INPUT UNIT = ',I3,8X,
     2 ' AND THE WEIGHTING FACTORS ARE:'//,7G12.3)
   45 FORMAT(///5X,' NAME ',14X,' ID. ',4X,'TOTAL',11X,
     1' COMPONENT SCORES'/35X,'SCORE'/)
   66 FORMAT(//'  READING NAMES, ID. NUMBERS, AND COMPONENT',
     1/'  SCORES FROM INPUT UNIT',I3,5X,'(4A5,I,',I1,'F)'/)
   77 FORMAT(4A5,I,7F)
   99 FORMAT(4A5,2X,I10,F8.2,4X,7F6.1)
  111 FORMAT(//'  MEDIAN:',21X,F10.2,//'  AVERAGES:',
     1 21X,F8.2,4X,7F6.1)
  112 FORMAT(/'  STANDARD DEVIATIONS:',12X,F5.1,5X,7F6.1)
  113 FORMAT(///'  TOTAL WEIGHTED SCORES',
     1' IN DESCENDING ORDER'//)
  133 FORMAT(1X,I3,5X,4A5,I12,F10.2)
  144 FORMAT(//3X,'DISTRIBUTION OF TOTAL WEIGHTED SCORES'//)
  146 FORMAT(9X,'>',F7.1,I3)
  148 FORMAT(9X,'<',F7.1,I3)
  155 FORMAT(F7.1,' TO',F7.1,I3)
  166 FORMAT(F7.1,' TO',F7.1,I3,3X,100A1)
      END
C
C           SORT
C
C
C     THIS IS AN INVERTED BINARY SORT ROUTINE
C
C     C.F. ALGORITHM 347,  C.A.C.M., 12(3), 185(1969)
C
C     CALLING SEQUENCE:  CALL SORT(A,JJ,B)
C
C        WHERE     A  -  VECTOR OF REAL NUMBERS TO BE SORTED
C                  JJ -  LENGTH OF A
C        AND       B  -  INTEGER VECTOR OF INDICES TO A
```

```
C                        B(I)=I,   I=1,2,.....,JJ
C
C     A IS RETURNED SORTED IN DESCENDING ORDER AND
C     B CONTAINS THE CORRESPONDING INDICES TO THE ORIGINAL A
C
C     E.G., IF ON ENTRY  A(5)=10.5 AND AFTER SORTING
C     A(30)=10.5, THEN B(30)=5
C
      SUBROUTINE SORT(A,JJ,B)
      DIMENSION A(250),IU(16),IL(16)
      INTEGER B(250)
      M=1
      I=1
      J=JJ
    5 IF(I.GE.J)GO TO 70
   10 K=I
      IJ=(J+I)/2
      T=A(IJ)
      IF(A(I).GE.T)GO TO 20
      A(IJ)=A(I)
      A(I)=T
      T=A(IJ)
      IT=B(IJ)
      B(IJ)=B(I)
      B(I)=IT
   20 L=J
      IF(A(J).LE.T)GO TO 40
      A(IJ)=A(J)
      A(J)=T
      T=A(IJ)
      IT=B(IJ)
      B(IJ)=B(J)
      B(J)=IT
      IF(A(I).GE.T)GO TO 40
      A(IJ)=A(I)
      A(I)=T
      T=A(IJ)
      IT=B(IJ)
      B(IJ)=B(I)
      B(I)=IT
      GO TO 40
   30 A(L)=A(K)
      A(K)=TT
      IT=B(L)
      B(L)=B(K)
      B(K)=IT
   40 L=L-1
      IF(A(L).LT.T)GO TO 40
      TT=A(L)
   50 K=K+1
      IF(A(K).GT.T)GO TO 50
      IF(K.LE.L)GO TO 30
      IF(L-I.NE.J-K)GO TO 60
      IL(M)=I
      IU(M)=L
      I=K
      M=M+1
      GO TO 80
   60 IL(M)=K
      IU(M)=J
```

```
        J=L
        M=M+1
        GO TO 80
   70   M=M-1
        IF(M.EQ.0)RETURN
        I=IL(M)
        J=IU(M)
   80   IF(J-I.GE.11)GO TO 10
        IF(I.EQ.1)GO TO 5
        GO TO 91
   90   I=I+1
   91   IF(I.EQ.J)GO TO 70
        T=A(I+1)
        IF(A(I).GE.T)GO TO 90
        K=I
        IT=B(I+1)
  100   A(K+1)=A(K)
        B(K+1)=B(K)
        K=K-1
        IF(T.GT.A(K))GO TO 100
        A(K+1)=T
        B(K+1)=IT
        GO TO 90
        END
```

Discussion

The storage requirements for GRADER and SORT with the dimensions indicated are approximately 4350 and 280 36-bit words, respectively. This sample execution took approximately 0.3 sec. With 250 students, each with seven components scores, the sample execution was approximately 1 sec.

9.5 COMPUTERS IN THE LABORATORY

Experimental chemists have utilized computing techniques to reduce data as long as computers have been available. For example, crystallographers have determined complex crystal structures because computers have been available to perform the iterative calculations that would be attempted with extreme reluctance without such computing power. Also, before the availability of minicomputers, the use of the computer-average transient (CAT) technique allowed experimentalists to simultaneously improve both the quantity and quality of data by taking advantage of rapid analog-to-digital conversion, data storage, and signal averaging. In those days, the high cost of computers usually demanded that data be transferred to magnetic tape, punched tape, or punched cards for subsequent processing by a large central computing facility.

With the introduction of mini- and microcomputers, scientists could afford to interface their instruments to computers and thereby utilize the advantages of high-speed communication between computer and apparatus. This allowed not only for optimization of data collection and reduction but also for the control of the experiment itself through feedback loops. These computers could also communicate with auxiliary storage devices to retrieve and store files.

The power of the mini-computers and microprocessors has increased dramatically in the last decade, and there is every expectation that additional advances in solid-state technology will be reflected in more powerful and affordable computer systems and components for laboratory equipment. The interested reader is referred to a variety of books and journal articles [167-205] that describe in detail various aspects of this extremely important component of modern experimental science.

9.6 ARTIFICIAL INTELLIGENCE AND INFORMATION RETRIEVAL

Computers are now being used to assist chemists interpret experimental data, design synthetic pathways, and keep up with the current literature [206-265]. Pattern recognition techniques are being applied to the analysis of complex chemical systems. For example, an oil sample from a spill can be analyzed by mass spectrometry. The resulting data can be interpreted by a "learning machine" that has been "trained" using a library of standard spectra to recognize patterns and classify the components of the mixture.

One approach to the problem of computer-assisted organic synthesis is the use of an interactive graphics device to draw the structure of the target molecule. Then the synthesis program applies a set of chemical rules to decide on a set of likely precursor molecules. The user decides which of these precursors seems most reasonable, and the program generates a set of likely precursors for this compound. The procedure continues until a reasonable synthetic pathway has been devised.

Many chemists now routinely receive computer output containing bibliographic citations to the current literature. This current awareness service involves a number of components: a machine readable data base, e.g., "CA Condensates"; a profile containing key words and authors, chemical formulas, and chemical substructures; and the software to match the profile against the data base and print the set of citations that match. Retrospective search systems are also available. Here is an excellent example of how the machine can be used to relieve the extremely time consuming task of a thorough literature search. One fringe benefit of machine searching is the "serendipity effect," in which a relevant article is cited from a journal that would not otherwise have been noticed by the scientist. For example, an inorganic chemist might see a unique application for a particular coordination compound by reading an article published in a medical journal.

REFERENCES

1. M. D. Glick et al., Interactive graphics for structural chemistry, Comp. Chem., 1, 75 (1977).

2. P. G. Dittmar et al., An algorithmic computer graphics program for generating chemical structure diagrams, J. Chem. Inf. Comput. Sci., 17, 186 (1977).

3. J. B. Barrett, Experiments with computer assisted graphical displays, Educ. Chem., 14, 42 (1977).

4. K. Knowlton and L. Cherry, ATOMS: A three-D opaque molecule system for color pictures of space-filling or ball-and-stick models, Comput. Chem., 1, 161 (1977).

5. N. Thalmann and J. Weber, A new computer program for generating three-dimensional plots of electronic densities and related contour levels, Chimia, 31, 361 (1977).

6. R. S. Macomber, A simple three-dimensional perspective plotting program, J. Chem. Educ., 53, 279 (1976).

7. J. W. Moore and W. G. Daview, Illustration of some consequences of the indistinguishability of electrons: use of computer-generated dot density diagrams, J. Chem. Educ., 53, 426 (1976).

8. J. F. Remark, Computerized three-dimensional illustrations of gas equations, J. Chem. Educ., 52, 61 (1975).

9. J. R. Van Wazer and I. Absar, "Electron Densities in Molecules and Molecular Orbitals," Academic Press, New York, 1975.

10. R. H. Schuler, Calculator-plotter routines for numerical integration of kinetic expressions, J. Chem. Educ., 52, 116 (1975).

11. G. Beech, "Fortran IV in Chemistry," Wiley, New York, 1975, Appendix A.

12. A. Streitwieser and P. H. Owens, "Orbital and Electron Density Diagrams," Macmillan, New York, 1973.

13. D. Y. Curtin, Stereo pair drawings of cyrstal structures prepared by a desk calculator-computer, J. Chem. Educ., 50, 775 (1973).

14. P. Polihroniadis et al., Perspective drawings of molecular structures by computers, Comput. Biomed. Res., 6, 509 (1973).

15. W. J. Sorgensen and L. Salem, "The Organic Chemists Book of Orbitals," Academic Press, New York, 1973.

16. E. J. Corey et al., Computer-assisted synthetic analysis. Communication of chemical structure by interactive computer graphics, J. Amer. Chem. Soc., 94, 421 (1972).

17. L. Soltzberg, Teletype graphics for chemical education, J. Chem. Educ., 49, 357 (1972).

18. C. S. Ewig, J. Thomas Gerig, and D. O. Harris, An interactive on-line computing system as an instructional aid, J. Chem. Educ., 47, 97 (1970).

General

19. J. S. Mattsom, H. B. Mark, Jr., and H. S. MacDonald, Jr. (eds.), "Computers in
     Chemistry and Chemical Instrumentation," Dekker, New York, Vol. 1, "Computer
     Fundamentals for Chemists," 1973; Vol. 2, "Electrochemistry: Calculations,
     Simulation, and Instrumentation," 1972; Vol. 3, "Spectroscopy and Kinetics,"
     1973; Vol. 4, "Computer-Assisted Instruction in Chemistry," Part A, "General
     Approach," and Part B, "Applications," 1974; Vol. 5, "Laboratory Systems and
     Spectroscopy," 1977; and Vol. 6, "Polymer Chemistry," 1977.

20. C. Klopfenstein and C. L. Wilkins (eds.), "Computers in Chemical and Biochemical
     Research," Academic Press, New York, Vol. 1, 1972, and Vol. 2, 1974.

21. D. F. DeTar (ed.), "Computer Programs for Chemistry," Benjamin, New York, Vol. 1,
     1968; Vol. 2, 1969; Vol. 3, 1969; and Vol. 4, 1972.

22. Proceedings on the International Conferences on Computers in Chemical Research
     and Education: 1. DeKalb, Ill., 1971; 2. Ljubljana, Yugoslavia, 1973;
     3. Caracas, Venezuala, 1976.

23. Proceedings of the Conferences on Computers in the Undergraduate Curricula:
     1. Iowa City, Iowa, 1970; 2. Hanover, N. H., 1971; 3. Atlanta, Ga., 1972;
     4. Claremont, Calif., 1973; 5. Pullman, Wash., 1974; 6. Fort Worth, Tex., 1975;
     7. Binghamton, N. Y., 1976; and 8. East Lansing, Mich., 1977.

24. P. Lykos, "The Computer's Role in Undergraduate Chemistry Curricula," A UNESCO
     Report, Paris, 1976.

25. Quantum Chemistry Program Exchange (QPCE), Aerospace Research Applications
     Center, Indiana University, Bloomington, Ind.

26. P. A. Cauchon, "Chemistry with a Computer," Educomp, Hartford, Conn., 1976.

27. L. Soltzberg et al., "BASIC and Chemistry," Houghton Mifflin, Boston, 1975.

28. R. Roskos, "Problem Solving in Physical Chemistry," West, New York, 1975.

29. C. Wilkins et al., "Introduction to Computer Programming for Chemists - BASIC
     Version," Allyn and Bacon, Boston, 1974.

30. K. B. Wiberg, Use of computers, in "Techniques of Chemistry" (E. S. Lewis, ed.),
     Vol. VI, Wiley, New York, 1974.

31. R. A. G. Carrington (ed.), "Computers for Spectroscopists," Wiley, New York,
     1974.

32. T. L. Isenhour and P. J. Jurs, "Introduction to Computer Programming for
     Chemists," Allyn and Bacon, Boston, 1972.

33. F. Van Zeggeren and S. H. Storey, "The Computation of Chemical Equilibrium,"
     Cambridge University Press, London, 1970.

34. A. C. Wahl, Chemistry by computer, _Sci. Amer._, _222_, 54 (1970).

35. F. D. Tabbut, Computers in chemical education, _Chem. Eng. News_, _48_, 44 (1970).

36. The February issue of the _Journal of Chemical Education_, Vol. 47, pp. 87-166, 1970, was devoted to articles relating to computer applications in chemical education.

37. T. R. Dickson, "The Computer and Chemistry," Freeman, San Francisco, 1968.

38. D. Drysen et al., "Computer Calculation of Ionic Equilibria and Titration Procedures," Wiley, New York, 1968.

39. J. M. Prausmitz et al., Computer Calculation of Multicomponent Vapor-Liquid Equilibria," Prentice-Hall, Englewood Cliffs, N. J., 1967.

40. K. B. Wiberg, "Computer Programming for Chemists," Benjamin, New York, 1965.

41. P. A. D. de Maine and R. D. Seawright, "Digital Computer Programs for Physical Chemistry," Macmillan, New York, Vol. 1, 1963; Vol. 2, 1965.

CAI

42. M. Bishop, Computer assisted instruction in qualitative analysis, _J. Chem. Educ._, _54_, 689 (1977).

43. R. Chaby and S. G. Smith, The use of computer-based chemistry lessons, _J. Chem. Educ._, _54_, 745 (1977).

44. H. D. Faram and M. A. Wartell, Single concept iterative minicomputer programs for use in general chemistry, _J. Chem. Educ._, _54_, 150 (1977).

45. S. G. Smith and R. Chaby, Computer games in chemistry, _J. Chem. Educ._, _54_, 688 (1977).

46. M. Feldman and M. Bishop, Computer-assisted instruction in organic synthesis, _J. Chem. Educ._, _53_, 91 (1976).

47. P. B. Ayscough, CAL-Boon or burden?, _Chem. Brit._, _12_, 348 (1976).

48. D. J. Macero and L. N. Davis, Computer-assisted instruction at Syracuse University: an overview, _Comput. Geosci._, _2_, 107 (1976).

49. S. G. Smith and B. A. Sherwood, Educational uses of the PLATO computer system, _Science_, _192_, 344 (1976).

50. H. W. Orf, Computer-assisted instruction in organic synthesis, _J. Chem. Educ._, _52_, 295 (1975).

51. G. L. Breneman, Interactive general chemistry problem practice on small computers, _J. Chem. Educ._, _52_, 295 (1975).

52. K. J. Johnson, "Curriculum Enrichment with Computers," in J. S. Mattson et al. (eds.), op. cit., Vol. 4, Part A, 1974, p. 9ff.

53. S. Smith and J. R. Chesquiere, Computer-Based Teaching of Organic Chemistry, in J. S. Mattson, et al. (eds.), op. cit., Vol. 4, Part B, 1974, p. 51ff.

54. T. R. Dehner and B. E. Norcross, "Laboratory and Classroom Use of an Interactive Terminal Language (APL)," op. cit., Part B, 1974, p. 3ff.

55. R. L. Ellis and T. A. Atkinson, "Practical Considerations of Computer-Assisted Instruction," in J. S. Matson et al. (eds.), op. cit., Vol. 4, Part A, 1974, p. 245ff.

56. S. G. Smith, J. R. Ghesquiere, and R. A. Avner, The use of computers in the teaching of chemistry, J. Chem. Educ., 51, 243 (1974).

57. H. A. Clark, J. C. Marshall, and T. L. Isenhour, An approach to computer-assisted drill in synthetic organic chemistry, J. Chem. Educ., 50, 645 (1973).

58. S. J. Castleberry, G. H. Culp, and J. J. Lagowski, The impact of computer-based instruction methods in general chemistry, J. Chem. Educ., 50, 469 (1973).

59. G. L. Breneman, MCAI: Minicomputer-aided instruction, J. Chem. Educ., 50, 473 (1973).

60. W. Geist and D. P. Ripota, Computer-assisted instruction in chemistry, Fort. Chem. Forsch., 1973, 169.

61. P. B. Ayschough, Computer-based learning in the teaching laboratory, Chem. Brit., 9, 61 (1973).

62. L. N. Davis and D. J. Macero, Computer-assisted instruction in a chemical instrumental course, J. Chem. Educ., 49, 759 (1972).

63. C. G. Venier and M. G. Reinecke, Armchair unknowns: a simple CAI qual organic simulation, J. Chem. Educ., 49, 541 (1972).

64. J. Eskinazi and D. J. Macero, An interactive program for teaching pH and Logarithms, J. Chem. Educ., 49, 571 (1972).

65. R. C. Grandey, The use of computers to aid instruction in beginning chemistry, J. Chem. Educ., 48, 791 (1971).

66. S. K. Lower, Audio-tutorial and CAI aids, J. Chem. Educ., 47, 143 (1970).

67. L. B. Rodewald, C. H. Culp, and J. J. Lagowski, The use of computers in organic chemistry instruction, J. Chem. Educ., 47, 134 (1970).

68. S. Castleberry and J. J. Lagowski, Individualized instruction using computer techniques, J. Chem. Educ., 47, 91 (1970).

69. J. J. Lagowski, Computer-assisted instruction in chemistry, in "Computer-Assisted Instruction, Testing, and Guidance" (W. H. Holtzman, ed.), Harper and Row, New York, 1970.

70. R. C. Atkinson and H. A. Wilson, Computers in chemical education, _Science_, _162_, 73 (1968).

COURSE AUGMENTATION

71. C. J. Jameson, Computer-enriched modules for introductory chemistry, _J. Chem. Educ._, _54_, 238 (1977).

72. G. E. Knudson and D. Nimrod, Exact equation for calculating titration curves for dibasic salts, _J. Chem. Educ._, _54_, 351 (1977).

73. S. D. Daubert and S. F. Sontum, Computer simulation of the determination of amino acid sequences in polypeptides, _J. Chem. Educ._, _54_, 35 (1977).

74. B. G. Williams et al., A computer method for the construction of Eh-pH diagrams, _J. Chem. Educ._, _54_, 107 (1977).

75. J. Hefter and R. Zuehlke, Computer simulation of acid-base behavior, _J. Chem. Educ._, _54_, 63 (1977).

76. W. Gayle Rhodes et al., Ever-more sophisticated lies about bonding: a mini-course in HMO theory utilizing small computers, _J. Chem. Educ._, _54_, 687 (1977).

77. J. D. Herron, Computer programs designed to balance inorganic chemistry equations, _J. Chem. Educ._, _54_, 704 (1977).

78. E. Sagstuen, Computer program for simulating single crystal electron spin resonance spectra, _J. Chem. Educ._, _54_, 153 (1977).

79. E. E. Hach, Jr., Computer program for the interpretation of second-order kinetic data, _J. Chem. Educ._, _54_, 386 (1977).

80. J. S. Miller and S. Z. Goldberg, Powder pattern program, _J. Chem. Educ._, _54_, 54 (1977).

81. G. Gilbert et al., Computers in teaching: now and tomorrow, _J. Chem. Educ._, _54_, 13 (1977).

82. F. L. Barker and R. J. Fredericks, Development of computer simulations for use in a high school chemistry course, _J. Chem. Educ._, _54_, 113 (1977).

83. J. Sikorski and J. G. Schaffhausen, NMR chemical shift correlations: a student oriented project, _J. Chem. Educ._, _53_, 761 (1976).

84. H. Kaufmann-Goetz and G. Kaufmann, Two examples of programmed learning in inorganic chemistry, _J. Chem. Educ._, _53_, 179 (1976).

85. J. R. Llinas and R. Freze, Computer estimation of thermodynamic properties of real gases, J. Chem. Educ., 53, 288 (1976).

86. C. Czerlinski and S. Sikorski, Computer-based modeling in the teaching of steady-state enzyme kinetics, J. Chem. Inf. Comput. Sci., 16, 30 (1976).

87. T. J. Moore, Computer program of the group multiplication table for the tetrahedral point group, J. Chem. Educ., 53, 44 (1976).

88. C. Leforestier and O. Kahn, A computer program to determine the molecular point group and the symmetry adapted orbitals, Comp. Chem., 1, 13 (1976).

89. S. M. Koop and P. J. Ogren, HO_2 kinetics in simple systems, J. Chem. Educ., 53, 128 (1976).

90. J. M. Anderson, Computer simulation of chemical kinetics, J. Chem. Educ., 53, 561 (1976).

91. P. R. Smith, A computer-assisted learning package in reactor kinetics, J. Inst. Nucl. Eng., 17, 147 (1976).

92. H. W. Orf, Computer-assisted instruction in organic synthesis, J. Chem. Educ., 52, 464 (1975).

93. M. Bader, Computers applied to physical chemistry instruction, in J. S. Mattson et al. (eds.), op. cit., Vol. 4, Part B, 1974, p. 83ff.

94. M. E. Starzak, Computer examination of some basic probability models, J. Chem. Educ., 51, 717 (1974).

95. P. E. Stevenson, Use of a Time-Shared Computer for Live Classroom Demonstration of Chemical Principles, in J. S. Mattson et al. (eds.), op. cit., Vol. 4, Part B, 1974, p. 131ff.

96. L. R. Sherman, Application of Canned Computer Programs to the Undergraduate Chemical Curriculum, in J. S. Mattson et al. (eds.), op. cit., Vol. 4, Part B, 1974, p. 173ff.

97. R. W. Collins, On-Line Classroom Computing in Chemistry Education Via Video Projection of Teletype Output, in J. S. Mattson et al.(eds.), op. cit., Vol. 4, Part A, 1974, p. 151ff.

98. D. O. Harris, An On-Line Graphics System as an Instructional Aid in Physical Chemistry, C. Klopfenstein and C. Wilkins (eds.), op. cit., Vol. 2, 1974, p. 77f.

99. B. J. Hubert, Computer modeling of photochemical smog formation, J. Chem. Educ., 51, 644 (1974).

100. R. H. Herber and Y. Hazong, Interactive digital computing in undergraduate physical chemistry, J. Chem. Educ., 51, 245 (1974).

101. P. Schettler, A computer problem in statistical thermodynamics, J. Chem. Educ., 51, 250 (1974).

102. J. L. Hogg, Computer programs for chemical kinetics: an annotated bibliography, J. Chem. Educ., 51, 109 (1974).

103. P. Empedocles, Fundamental theory of gases, liquids, and solids by computer simulation, J. Chem. Educ., 51, 593 (1974).

104. D. W. Beistel, A computer-oriented course in chemical spectroscopy, J. Chem. Educ., 50, 145 (1973).

105. C. Marzzacco and M. Waldman, Heat capacity calculations of diatomic molecules, J. Chem. Educ., 50, 444 (1973).

106. R. H. Galyan and V. A. Ryan, Computer-aided undergraduate radiochemistry instruction utilizing time sharing data terminals, J. Chem. Educ., 49, 591 (1972).

107. B. J. Duke, Use of the Huckel molecular orbital programs in chemistry, J. Chem. Educ., 49, 703 (1972).

108. G. Wise, Computer programs for undergraduate physical chemistry students, J. Chem. Educ., 49, 559 (1972).

109. T. J. MacDonald, B. J. Barker, and J. A. Caruso, Computer evaluation of titrations by Gran's method, J. Chem. Educ., 49, 200 (1972).

110. W. B. Eilsen, Interactive computer programs for equilibrium and kinetics, J. Chem. Educ., 48, 414 (1971).

111. R. M. Scott, Use of computers in teaching the theory of chromatography, J. Chromatogr., 60, 313 (1971).

112. D. L. Peterson and M. E. Fuller, Physical chemistry students discover the computer, J. Chem. Educ., 48, 314 (1971).

113. M. Bader, Computer programs in undergraduate chemistry, J. Chem. Educ., 48, 175 (1971).

114. L. J. Soltsberg, Qualitative computing in elementary chemical education, J. Chem. Educ., 48, 449 (1971).

115. K. J. Johnson, Numerical methods in chemistry, J. Chem. Educ., 47, 819 (1970).

116. R. E. Jensen et al., Determination of successive ionization constants, J. Chem. Educ., 47, 147 (1970).

117. W. E. Bennett, Computation of the number of isomers and their structures in coordination compounds, Inorg. Chem., 8, 1325 (1969).

118. H. B. Herman and D. E. Leyden, Use of computers in analytical chemistry courses, J. Chem. Educ., 45, 524 (1968).

119. D. F. De Tar, Simplified computer programs for teaching complex reaction mechanisms, J. Chem. Educ., 44, 191 (1967), and A computer program for making steady-state calculations, J. Chem. Educ., 44, 193 (1967).

120. R. C. Reiter and J. E. House, Jr., Teaching computer methods in undergraduate chemistry courses, J. Chem. Educ., 45, 465 (1968).

121. C. L. Wilkins and C. E. Klopfenstein, Simulation of NMR spectra, J. Chem. Educ., 43, 10 (1966).

122. O. T. Zajicek, A computer program for use in teaching analytical chemistry, J. Chem. Educ., 42, 622 (1965).

SIMULATED EXPERIMENTS

123. M. Yuan, A gas chromatographic simulation program, J. Chem. Educ., 54, 364 (1977).

124. A. I. Rosen, A computer program designed to balance inorganic chemical equations, J. Chem. Educ., 54, 707 (1977).

125. Ben-Zion, Simulated NMR spectrometry and shift reagents, J. Chem. Educ., 54, 669 (1977).

126. D. Rosenthal and D. Arnold, Simulation of experimental data, the design of experiments, and the analysis of results, J. Chem. Educ., 54, 323 (1977).

127. H. M. Bell, Computer-assisted analysis of infrared spectra, J. Chem. Educ., 53, 26 (1976).

128. M. Ben-Zion, KINDAT and kinetics: introducing undergraduates to the computer, J. Chem. Educ., 53, 436 (1976).

129. A. F. Para and E. Lazzarini, Some simple classroom experiments on the Monte-Carlo method, J. Chem. Educ., 51, 336 (1976).

130. J. C. Merill et al., A computer-simulated experiment in complex order kinetics, J. Chem. Educ., 52, 528 (1975).

131. J. J. Uebel and G. A. Heavener, Simulation of qual organic: an interactive computer program, J. Chem. Educ., 52, 136 (1975).

132. C. G. Venier and M. G. Reinecke, Computer-Simulating Unknowns, in J. S. Mattson et al. (eds.), op. cit., Vol. 4, Part B, 1974, p. 159ff.

133. W. C. Child, A computer simulation of a kinetics experiment, J. Chem. Educ., 50, 290 (1973).

134. A. C. Norris et al., A computer package for physical chemistry experiments, J. Chem. Educ., 50, 489 (1973).

135. G. H. Coleman, Computer simulation of countercurrent distribution experiments, J. Chem. Educ., 50, 825 (1973).

136. J. R. Garbarino and M. A. Wartell, The Rutherford scattering experiment: CAI in the laboratory, J. Chem. Educ., 50, 792 (1973).

137. O. Runquist et al., Programmable calculators-simulated experiments, J. Chem. Educ., 49, 265 (1972).

138. N. S. Craig et al., Computer experiments, J. Chem. Educ., 48, 310 (1971).

139. R. E. Jensen et al., Determination of successive ionization constants: a computer-assisted laboratory experiment, J. Chem. Educ., 47, 147 (1970).

140. R. H. Schendeman, Computer simulation of experimental data, J. Chem. Educ., 45, 665 (1968).

141. G. F. Pollnow and A. J. Hopfinger, A computer experiment in microwave spectroscopy, J. Chem. Educ., 45, 528 (1968).

TESTING AND GRADING

142. K. J. Johnson and L. M. Epstein, "General Chemistry Examination Questions," 3rd ed., Burgess, Minneapolis, Minn., 1977.

143. J. W. Moore et al., Repeatable testing, J. Chem. Educ., 54, 276 (1977).

144. M. L. Borke and C. A. Loch, Computer-generated examinations, J. Chem. Educ., 54, 112 (1977).

145. J. L. Deutsch, Computer-assisted grading of quantitative laboratory experiments, J. Chem. Educ., 53, 308 (1976).

146. L. A. Nutter and W. J. Nieckarz, Jr., A computer-graded laboratory practical examination for general chemistry, J. Chem. Educ., 52, 514 (1975).

147. G. Lippey (ed.), "Computer-Assisted Test Construction," Educational Testing, Englewood Cliffs, N. J., 1974.

148. J. W. Moore et al., Computer-Generated Repeatable Tests in Chemistry, J. S. Mattson et al., (eds.), op. cit., Vol. 4, Part B, 1974, p. 189ff.

149. N. D. Yaney, Computerized Homework Preparation and Grading, in J. S. Mattson et al. (eds.), op. cit., Vol. 4, Part B, 1974, p. 189ff.

150. R. C. Johnson, Interactive laboratory report grading at a computer terminal, J. Chem. Educ., 50, 223 (1973).

151. J. G. Macmillan and M. Epstein, A versatile fortran program for multi-section courses, J. Chem. Educ., 50, 459 (1973).

152. D. E. Jones and F. E. Lytle, Computer-aided grading of quantitative unknowns, J. Chem. Educ., 50, 285 (1973).

153. M. A. Wartell and J. A. Hurlbut, A Fortran IV program for grading quantitative analysis unknowns, J. Chem. Educ., 49, 508 (1972).

154. J. W. Connolly, Automated homework grading for large general chemistry classes, J. Chem. Educ., 49, 262 (1972).

155. N. D. Yaney, Computer system for individual homework, J. Chem. Educ., 48, 276 (1971).

156. C. B. Leonard, Jr., Computer-graded examinations: evaluation of a teacher, J. Chem. Educ., 47, 149 (1970).

157. J. M. Thorne et al., Computerized scheduling of teaching assistants and graders and graders, J. Chem. Educ., 47, 152 (1970).

158. J. L. Deutsch and E. W. Deutsch, Computer-graded qualitative analysis experiments, J. Chem. Educ., 46, 649 (1969).

159. C. C. Hinckly and J. J. Lagowski, A versatile computer-graded examination, J. Chem. Educ., 43, 575 (1966).

CMI

160. K. R. Williams and Z. C. Martinez, A computerized system for the submission of quantitative analysis laboratory results, J. Chem. Educ., 54, 94 (1977).

161. K. Hartman and A. M. Fischer, An informative computerized laboratory records maintenance system, J. Chem. Educ., 54, 507 (1977).

162. G. E. Dunkleberger and R. W. Smith, A computer managed approach to individualized instruction, J. Chem. Educ., 53, 649 (1976).

163. W. L. Felty, Juggling the chemistry locker padlocks with computer assistance, J. Chem. Educ., 53, 692 (1976).

164. B. Z. Shakhashiri, CHEM TIPS, J. Chem. Educ., 52, 588 (1975).

165. J. L. Deutsch et al., Computerized chemistry stockroom inventory system, J. Chem. Educ., 49, 180 (1972).

166. N. D. Yaney, Computer system for individual homework, J. Chem. Educ., 48, 276 (1971).

INTERFACING

167. J. W. Cooper, "The Minicomputer in the Laboratory," Wiley, New York, 1977.

168. J. W. Cooper, Computers in NMR. III. Algorithm for continuously variable knob-controlled CRT display expansion, Comput. Chem., 1, 121 (1977).

169. C. M. Ridder and D. W. Margerum, Simultaneous kinetic analysis of multicomponent mixtures, Anal. Chem., 49, 2090 (1977).

170. R. E. Dessey, Computer networking--a rational approach to laboratory automation, Anal. Chem., 49, 1100A (1977).

171. I. M. Campbell et al., Software package to collect and process radiogas chromatographic data, Anal. Chem., 49, 1726 (1977).

172. R. S. Schwall et al., On-line fast Fourier transform faradaic admittance measurements, Anal. Chem., 49, 1805 (1977).

173. J. S. Mattson, Design and applications of an on-line minicomputer system for dispersive infrared spectroscopy, Anal. Chem., 49, 470 (1977).

174. H. L. Felkel, Jr., and H. L. Pardue, Design and evaluation of a random access Vidicon-Eschelle spectrometer and application to multicomponent determination by atomic absorption spectroscopy, Anal. Chem., 49, 1112 (1977).

175. M. P. Miller et al., Spectroscopic system for the study of fluorescent lanthanide probe ions in solids, Anal. Chem., 49, 1474 (1977).

176. J. A. Perry, M. F. Bryant, and H. V. Malmstadt, Microprocessor-controlled, scanning dye laser for spectrometric analytical systems, Anal. Chem., 49, 1702 (1977).

177. F. D. Fassett et al., Quantitation of secondary ion mass spectrometric images by microphotodensitometry and digital image processing, Anal. Chem., 49, 2322 (1977).

178. P. F. Seelig and H. N. Blount, A time-share based simulator to teach the use of digital computers, J. Chem. Educ., 52, 469 (1976).

179. R. Swanson et al., Determination of the Nyquist frequency--a computer-interfacing experiment, J. Chem. Educ., 52, 530 (1976).

180. Q. V. Thomas, L. Kregger, and S. P. Perone, Computer-assisted optimization of anodic stripping voltametry, Anal. Chem., 48, 761 (1976).

181. R. W. Spillman and H. V. Malmstadt, Computer-controlled programmable monochromator system with automated wavelength calibration and background correction, Anal. Chem., 48, 303 (1976).

182. M. J. E. Hewlins, Computerized mass spectrometry, Chem. Brit., 12, 341 (1976).

183. L. F. Whiting et al., An inexpensive digital data acquisition system for the teaching laboratory, J. Chem. Educ., 53, 786 (1976).

184. P. S. Shoenfeld and J. R. DeVoe, Statistical and mathematical methods in analytical chemistry, Anal. Chem., 48, 403R (1976).

185. J. Finkel, "Computer-Aided Experimentation," Wiley, New York, 1975.

186. C. L. Wilkins et al., "Digital Electronics and Laboratory Computer Experiments," Plenum Press, New York, 1975.

187. D. D. Glover, Laboratory Minicomputer Data system, in C. Klopfenstein and C. Wilkins (eds.), op. cit., Vol. 2, 1974, p. 102ff. (This volume contains several other articles on interfacing.)

188. B. J. Bulkin, E. H. Cole, and A. Oguerola, Laboratory automation--a case history, J. Chem. Educ., 51, A273 (1974).

189. G. A. Korn, "Minicomputers for Engineers and Scientists," McGraw-Hill, New York, 1973.

190. R. E. Dessey and T. A. Titus, Computer interfacing, Anal. Chem., 45, 124A (1973).

191. R. R. Markman et al., A data acquisition experiment for instrumental analysis, J. Chem. Educ., 50, 295 (1973).

192. F. G. Pater and S. P. Perone, Computer-controlled colorimetry, J. Chem. Educ., 50, 428 (1973).

193. L. N. Davis, C. E. Coffey, and D. J. Macero, Computer-enhanced laboratory experience, J. Chem. Educ., 50, 711 (1973).

194. H. Nau, J. K. Kelley, and K. Beimann, Determination of the amino acid sequence for the C-terminal cyanogen bromide fragment of actin by computer-assisted gas chromatography and mass spectrometry, J. Amer. Chem. Soc., 95, 7162 (1973).

195. S. P. Perone and D. O. Jones, "Digital Computers in Scientific Instrumentation: Applications to Chemistry," McGraw-Hill, New York, 1973.

196. S. P. Perone, Training Chemists in Laboratory Computing, in C. E. Klopfenstein and C. L. Wilkins (eds.), op. cit., Vol. 1, 1972, p. 2ff. (This volume contains several other articles on interfacing.)

197. D. O. Jones et al., On-line digital computer applications to kinetic analysis, J. Chem. Educ., 49, 717 (1972).

198. W. Wayne Black, "An Introduction to On-Line Computers," Gordon and Breach, Science Publishers, New York, 1971.

199. L. E. Brady, Laboratory computer analysis of mass metastable ion signals,
      J. Chem. Educ., 48, 469 (1971).

200. A Weisberger and B. W. Rossiter (eds.), "Techniques of Chemistry," Vol. 1,
      "Physical Methods of Chemistry," Part IB, Automatic Recording and Control,
      Computers in Chemical Research, Wiley-Interscience, New York, 1971.

201. S. P. Perone, On-line digital computer applications in gas chromatography,
      J. Chem. Educ., 48, 438 (1971).

202. S. P. Perone, Computer applications in the chemistry laboratory--a survey,
      Anal. Chem., 43, 1288 (1971).

203. S. P. Perone and J. F. Eagleston, Introduction of digital computers into the
      undergraduate laboratory, J. Chem. Educ., 48, 317 (1971).

204. W. B. Wiberg, The small computer in the laboratory, J. Chem. Educ., 47, 113
      (1970).

205. C. H. Orr and J. A. Norris (eds.), "Progress in Analytical Chemistry," Vol. 4,
      "Computers in Analytical Chemistry," Plenum Press, New York, 1970.

PATTERN RECOGNITION AND ARTIFICIAL INTELLIGENCE

206. M. G. Hutchings et al., The steric courses of chemical reactions. 3. Computer
      generation of product distributions, steric courses, and permutational isomers,
      J. Amer. Chem. Soc., 99, 7126 (1977).

207. H. B. Woodruff et al., Computer-assisted interpretation of carbon-13 nuclear
      magnetic resonance spectra applied to structure elucidation of natural products,
      Anal. Chem., 49, 2075 (1977).

208. H. L. Surprenant and C. N. Reilley, Uniqueness of carbon-13 nuclear magnetic
      resonance spectra of acyclic saturated hydrocarbons, Anal. Chem., 49, 1134
      (1977).

209. S. R. Lowry et al., Comparison of various K-nearest neighbor voting schemes
      with the self-training interpretive and retrieval system for identifying
      molecular substructures with mass spectral data, Anal. Chem., 49, 1720 (1977).

210. J. S. Mattson et al., Classification of petroleum pollutants by linear
      discriminant function analysis of infrared spectral patterns, Anal. Chem., 49,
      500 (1977).

211. I. K. Mun, R. Venkataraghauan, and F. W. McLafferty, Computer Assignment of
      elemental compositions of mass spectral peaks from isotopic abundance, Anal.
      Chem., 49, 1723 (1977).

212. C. L. Wilkins and T. R. Brunner, Classification of binary carbon-13 nuclear
     magnetic resonance spectra, Anal. Chem., 49, 2136 (1977).

213. H. B. Woodruff and E. Morton, Computer-assisted interpretation of infrared
     spectra, Anal. Chim. Acta, 95, 13 (1977).

214. J. A. Gallucci et al., Pentacoordinated molecules. 24. Computer simulation
     of phosphorane structures, J. Amer. Chem. Soc., 99, 5461 (1977).

215. C. A. Shelley and M. E. Munk, Computer perception of topological symmetry,
     J. Chem. Inf. Comput. Sci., 17, 110 (1977).

216. R. E. Carhart and D. H. Smith, Applications of artificial intelligence for
     chemical inference. XX. Intelligent use of constraints in computer-assisted
     structure elucidation, Comp. and Chem., 1, 79 (1977).

217. W. T. Wipke and W. J. Howe (eds.), "Computer-Assisted Organic Synthesis,"
     ACS Symposium Series, Vol. 61, American Chemical Society, Washington, 1977.

218. D. E. Smith (ed.), "Computer-Assisted Structure Elucidation," ACS Symposium
     Series, Vol. 54, American Chemical Society, Washington, 1977.

219. Y. Miyashita et al., Computer-assisted analysis of infrared spectra of
     nitrogen-containing organic compounds, J. Chem. Inf. Comput. Sci., 17, 228
     (1977).

220. E. J. Corey et al., Computer-assisted synthetic analysis, J. Amer. Chem. Soc.,
     98, 189, 203, and 210 (1976).

221. H. Abe and P. C. Jurs, Automated chemical structure analysis of organic
     molecules with a molecular structure generator and pattern recognition
     techniques, Anal. Chem., 47, 1829 (1975).

222. G. S. Zander, A. J. Stuper, and P. C. Jurs, Nonparametric feature selection
     in pattern recognition applied to chemical problems, Anal. Chem., 47, 1085
     (1975).

223. D. H. Smith, Applications of artificial intelligence for chemical inference,
     Anal. Chem., 47, 1177 (1975).

224. R. E. Carhart et al., Applications of artificial intelligence for chemical
     inference. XVII. An approach to computer-assisted elucidation of molecular
     structure, J. Amer. Chem. Soc., 97, 5755 (1975).

225. H. B. Woodruff, S. R. Lowry, and T. L. Isenhour, Bayesian decision theory
     applied to multicategory classification of binary infrared spectra, Anal.
     Chem., 46, 2150 (1974).

226. K. C. Chu, Applications of artificial intelligence to chemistry--use of pattern
     recognition techniques and cluster analysis to determine the pharmacological
     activity of some organic compounds, Anal. Chem., 46, 1181 (1974).

227. C. K. Johnson and C. J. Collins, An algebraic model for the rearrangements of
      2-bicyclo[2.2.1]heptyl cations, J. Amer. Chem. Soc., 96, 2514, 2524 (1974).

228. W. T. Wipke and T. M. Dyott, Stereochemically unique naming algorithm, J. Amer.
      Chem. Soc., 96, 4834 (1974).

229. B. R. Kowalski, Pattern Recognition in Chemical Research, in C. Klopfenstein
      and C. Wilkins (eds.), op. cit., Vol. 2, 1974, p. 2ff.

230. J. Dugendji and I. Ugi, An algebraic model for constitutional chemistry as a
      basis for chemical computer programs, Fort. Chem. Forsch., 39, 19 (1973).

231. A. J. Thakker, The coming of the computer age to organic chemistry. Recent
      approaches to systematic synthesis analysis, in "Topics in Current Chemistry,"
      Number 39, Fort. Chem. Eorsch., 39, 3 (1973).

232. T. L. Isenhour and P. C. Jurs, Learning Machines, in J. S. Mattson, et al.
      (eds.), op. cit., Vol. 1, 1973, p. 285ff.

233. B. R. Kowalski and C. F. Bender, The Hadamard transform and spectral analysis
      by pattern recognition, Anal. Chem., 45, 2234 (1973).

234. K. Wok, R. Venkataraghaven, and F. W. McLafferty, Computer-aided interpretation
      of mass spectra. III. A self-training interpretive and retrieval system,
      J. Amer. Chem. Soc., 95, 4158 (1973).

235. J. Lederberg et al., Applications of artificial intelligence for chemical
      inference. I. The number of possible organic compounds. Acyclic structures
      containing C, H, O, and N, J. Amer. Chem. Soc., 91, 2973 (1969).

236. E. J. Corey and W. T. Wipke, Computer-assisted design of complex organic
      syntheses, Science, 166, 178 (1969).

INFORMATION RETRIEVAL

237. J. Zupan et al., A new retrieval system for infrared spectra, Comp. and Chem.,
      1, 71 (1977).

238. M. M. Cone et al., Molecular structure IR comparison program for the identifi-
      cation of maximal common substructures, J. Amer. Chem. Soc., 99, 7668 (1977).

239. B. W. Tattershall, Identification of components of mixtures by retrospective
      computer subtraction of gas-phase infrared spectra, Anal. Chem., 49, 772 (1977).

240. G. W. Adamson and D. Bawden, A substructural analysis method for structure-
      activity correlation in heterocyclic compounds using Wiswesser line notation,
      J. Chem. Inf. Comput. Sci., 17, 164 (1977).

241. H. Skolnik, The role of chemical information science in computer-assisted chemical research, J. Chem. Inf. Comput. Sci., 17, 234 (1977).

242. S. R. Heller et al., A computer-based chemical information system, Science, 195, 253 (1977).

243. F. H. Allen and W. G. Town, The automatic generation of keywords from chemical compound names: preparation of a permuted name index with KWIC layout, J. Chem. Inf. Comput. Sci., 17, 9 (1977).

244. K. W. Yim et al., Chemical characterization via fluorescence spectral files: data compression by Fourier transformation, Anal. Chem., 49, 2069 (1977).

245. J. Zupan, Combined retrieval system for infrared, mass, and carbon-13 nuclear magnetic resonance spectra, Anal. Chem., 49, 2141 (1977).

246. B. E. Blaisdell, Automatic computer construction, maintainence and use of specialized joint libraries of mass spectra and retention indices from gas chromatograph-mass spectrometry systems, Anal. Chem., 49, 180 (1977).

247. G. F. Luteri and J. M. Denham, A computer searchable data file for identifying organic compounds, J. Chem. Educ., 48, 670 (1976).

248. R. C. Fox, Computer searching of infrared spectra using peak location and intensity data, Anal. Chem., 48, 717 (1976).

249. G. M. Pesyna et al., Probability based matching system using a large collection of reference mass spectra, Anal. Chem., 48, 1362 (1976).

250. B. S. Archuleta et al., Computer-assisted literature searching, J. Chem. Educ., 53, 639 (1976).

251. H. W. Davis, "Computer Representation of the Stereochemistry of Organic Molecules," Birkhauser, Basel, 1976.

252. R. J. Munn, A computer-based information retrieval system, J. Chem. Educ., 52, 662 (1975).

253. W. T. Wipke et al., Computer Representation and Manipulation of Chemical Information, Wiley-Interscience, New York, 1974.

254. C. H. Davis and J. E. Rush, "Information Retrieval and Documentation in Chemistry," Greenwood Press, Westpoint, Conn., 1974.

255. M. E. Williams, Information Storage and Retrieval for Chemists: Computerized Systems, Sources, and Services, in J. S. Mattson et al. (eds.), op. cit., Vol. 4, Part A, 1974, p. 193ff.

256. D. C. Veal, Computer techniques for retrieval of information from the chemical literature, Fort. Chem. Forsch., 39, 65 (1973).

257. E. M. Arnett and A. Kent (eds.), "Computer-Based Chemical Information,"
      Dekker, New York, 1973.

258. P. F. Rusch and J. J. Lagowski, A Practical Computer-Based Information Retrieval
      System, in C. E. Klopenfenstein and C. L. Wilkins (eds.), op. cit., Vol. 1, 1972,
      p. 233ff.

259. S. R. Heller and R. J. Feldmann, An interactive NMR chemical shift search
      program, J. Chem. Educ., 49, 291 (1972).

260. S. R. Heller et al., An interactive mass spectral search system, J. Chem. Educ.,
      49, 725 (1972).

261. N. S. Grunstra and K. J. Johnson, Implementation and evaluation of two computer-
      ized information retrieval systems at the University of Pittsburgh, J. Chem.
      Doc., 10, 272 (1970).

262. E. S. Smith, "The Wiswesser Line-Formula Chemical Notation," McGraw-Hill, San
      Francisco, 1968.

263. L. C. Thomas, "A New Chemical Structure Code for Data Storage and Retrieval in
      Molecular Spectroscopy," Heyden, London, 1968.

264. L. S. Rattet and J. H. Goldstein, Computerizing scientific bibliographies,
      J. Chem. Educ., 45, 734 (1968).

265. W. J. Wiswesser, "A Line-Formula Chemical Notation," Thomas Y. Crowell, New
      York, 1954.

INDEX

499